Reactions at Supports and Connections for a Three-Dimensional Structure

Seventh Edition
VECTOR MECHANICS
FOR ENGINEERS

Statics

FERDINAND P. BEER
Lehigh University

E. RUSSELL JOHNSTON, JR.
University of Connecticut

ELLIOT R. EISENBERG
The Pennsylvania State University

With the collaboration of
George H. Staab
The Ohio State University

Boston   Burr Ridge, IL   Dubuque, IA   Madison, WI   New York   San Francisco   St. Louis
Bangkok   Bogotá   Caracas   Kuala Lumpur   Lisbon   London   Madrid   Mexico City
Milan   Montreal   New Delhi   Santiago   Seoul   Singapore   Sydney   Taipei   Toronto

Higher Education

VECTOR MECHANICS FOR ENGINEERS: STATICS, SEVENTH EDITION

Published by McGraw-Hill, a business unit of The McGraw-Hill Companies, Inc., 1221 Avenue of the Americas, New York, NY 10020. Copyright © 2004, 1996, 1988, 1984, 1977, 1972, 1962 by The McGraw-Hill Companies, Inc. All rights reserved. No part of this publication may be reproduced or distributed in any form or by any means, or stored in a database or retrieval system, without the prior written consent of The McGraw-Hill Companies, Inc., including, but not limited to, in any network or other electronic storage or transmission, or broadcast for distance learning.

Some ancillaries, including electronic and print components, may not be available to customers outside the United States.

This book is printed on acid-free paper.

5 6 7 8 9 0 VNH/VNH 0 9 8 7 6 5

ISBN 0–07–230493–6

Publisher: *Elizabeth A. Jones*
Associate sponsoring editor: *Debra D. Matteson*
Marketing manager: *Sarah Martin*
Project manager: *Jane Mohr*
Production supervisor: *Sherry L. Kane*
Senior media project manager: *Stacy A. Patch*
Senior media technology producer: *Phillip Meek*
Designer: *David W. Hash*

Lead photo research coordinator: *Carrie K. Burger*
Photo research: *Sabina Dowell*
Supplement producer: *Brenda A. Ernzen*
Compositor: *TechBooks*
Art house: *Fine Line Illustrations, Inc.*
Typeface: *10.5/12 New Caledonia*
Printer: *Von Hoffmann Corporation*

Cover Photograph: Courtesy Central Tunnel/Artery Project, Boston, MA. The Leonard P. Zakim Bunker Hill Bridge, shown illuminated at night, is at the north end of Boston. The Storrow Drive Connector Bridge is on the left. This shot is looking north and was taken on January 30, 2002. The bridge crosses the Charles River and connects Boston and Charlestown, Massachusetts. Conceived by Swiss bridge designer Christian Menn, the 270-foot inverted Y-shaped towers reflect the shape of the Bunker Hill Monument in Charlestown. The 1457-foot-long bridge is the widest cable-stayed bridge in the world, and the first in the United States to use an asymmetrical, hybrid design. Eight lanes of traffic pass through the legs of the twin towers, while an additional two lanes, supported by steel beams, are cantilevered from the east side of the bridge. The 745-foot-long, 183-foot-wide main span is supported by girders and two planes of cables; the back spans between the towers and the shores are supported by a single plane of cables.

The credits section for this book begins on page 603 and is considered an extension of the copyright page.

Library of Congress cataloged the main title as follows:

Vector mechanics for engineers : statics and dynamics / Ferdinand P. Beer . . . [et al.].—7th ed.
        p.   cm.
    Includes index.
    ISBN 0–07–230491–X (combined title)—ISBN 0–07–230492–8 (Dynamics)—ISBN 0–07–230493–6 (Statics).
    1. Mechanics, Applied. 2. Vector analysis. 3. Statics. 4. Dynamics. I. Beer, Ferdinand Pierre, 1915–.
    II. Beer, Ferdinand Pierre, 1915– Vector mechanics for engineers.

TA350.V34   2004
620.1′05—dc21   2003044560
                                                 CIP

www.mhhe.com

About the Authors

As one of the most successful author teams in engineering education, Ferd Beer and Russ Johnston are often asked how, with one author at Lehigh and the other at the University of Connecticut, they happened to write their books together and how they managed to keep collaborating on their many books with many successive revisions.

The answer to this question is simple. Russ Johnston's first teaching appointment was in the Department of Civil Engineering and Mechanics at Lehigh University. There he met Ferd Beer, who had joined that department two years earlier and was in charge of the courses in mechanics.

Ferd was delighted to discover that the young man who had been hired chiefly to teach graduate structural engineering courses was not only willing but eager to help him reorganize the mechanics courses. Both believed that these courses should be taught from a few basic principles and that the various concepts involved would be best understood and remembered by the students if they were presented to them in a graphic way. Together they wrote lecture notes in statics and dynamics, to which they later added problems they felt would appeal to future engineers, and soon they produced the manuscript of the first edition of *Mechanics for Engineers*.

The second edition of *Mechanics for Engineers* and the first edition of *Vector Mechanics for Engineers* found Russ Johnston at Worcester Polytechnic Institute and the next editions at the University of Connecticut. In the meantime, both Ferd and Russ assumed administrative responsibilities in their departments, and both were involved in research, consulting, and supervising graduate students— Ferd in the area of stochastic processes and random vibrations and Russ in the area of elastic stability and structural analysis design. However, their interest in improving the teaching of the basic mechanics courses had not subsided, and they both taught sections of these courses as they kept revising their texts and began writing the manuscript of the first edition of their *Mechanics of Materials* text.

Their collaboration has spanned many years and many successful revisions of all of their textbooks, and Ferd and Russ's contributions to engineering education have earned them a number of honors and awards. They were presented with the Western Electric Fund Award for excellence in the instruction of engineering students by their respective regional sections of the American Society for Engineering Education, and they both received the Distinguished Educator Award from the Mechanics Division of the same society. Starting in 2001, the New Mechanics Educator Award of the Mechanics Division has been named in honor of the Beer and Johnston author team.

Ferd and Russ are delighted to bring a new co-author and collaborator to the Beer and Johnston team for the seventh edition of *Vector Mechanics for Engineers: Statics.* Elliot Eisenberg worked as a collaborator on the sixth edition and is now a co-author for *Statics.* George Staab developed the interactive tutorial to accompany the sixth and seventh editions, and now he joins the seventh edition team as a collaborator on both the *Statics* and *Dynamics* texts.

Ferdinand P. Beer. Born in France and educated in France and Switzerland, Ferd holds an M.S. degree from the Sorbonne and an Sc.D. degree in theoretical mechanics from the University of Geneva. He came to the United States after serving in the French army during the early part of World War II and taught for four years at Williams College in the Williams-MIT joint arts and engineering program. Following his service at Williams College, Ferd joined the faculty of Lehigh University where he taught for thirty-seven years. He held several positions, including University Distinguished Professor and chairman of the Department of Mechanical Engineering and Mechanics, and in 1995 Ferd was awarded an honorary Doctor of Engineering degree by Lehigh University.

E. Russell Johnston, Jr. Born in Philadelphia, Russ holds a B.S. degree in civil engineering from the University of Delaware and an Sc.D. degree in the field of structural engineering from the Massachusetts Institute of Technology. He taught at Lehigh University and Worcester Polytechnic Institute before joining the faculty of the University of Connecticut where he held the position of chairman of the Department of Civil Engineering and taught for twenty-six years. In 1991 Russ received the Outstanding Civil Engineer Award from the Connecticut Section of the American Society of Civil Engineers.

Elliot R. Eisenberg. Elliot holds a B.S. degree in engineering and an M.E. degree, both from Cornell University. He has focused his scholarly activities on professional service and teaching, and he was recognized for this work in 1992 when the American Society of Mechanical Engineers awarded him the Ben C. Sparks Medal for his contributions to mechanical engineering and mechanical engineering technology education and for service to the American Society for Engineering Education. Elliot taught for thirty-two years, including twenty-nine years at Penn State where he was recognized with awards for both teaching and advising.

George H. Staab. George holds B.S., M.S., and Ph.D. degrees in aeronautical engineering from Purdue University. He worked at Sikorsky Aircraft for three years and is currently an Associate Professor in Mechanical Engineering at the Ohio State University where he has taught for twenty-two years. His scholarly activities are focused on teaching and service, and he has received two college-wide awards: the Charles E. MacQuigg Outstanding Teaching Award in 1998 and the Boyer Teaching Award in 1999. George serves as the campus representative for the American Society for Engineering Education and is the faculty advisor for the student chapter of the American Society of Mechanical Engineering and several student project teams.

Contents

Preface   xii

Acknowledgments   xviii

List of Symbols   xxi

1
INTRODUCTION
1

1.1   What Is Mechanics?   2
1.2   Fundamental Concepts and Principles   2
1.3   Systems of Units   5
1.4   Conversion from One System of Units to Another   10
1.5   Method of Problem Solution   11
1.6   Numerical Accuracy   13

2
STATICS OF PARTICLES
15

2.1   Introduction   16

Forces in a Plane   16
2.2   Force on a Particle. Resultant of Two Forces   16
2.3   Vectors   17
2.4   Addition of Vectors   18
2.5   Resultant of Several Concurrent Forces   20
2.6   Resolution of a Force into Components   21
2.7   Rectangular Components of a Force. Unit Vectors   27
2.8   Addition of Forces by Summing x and y Components   30
2.9   Equilibrium of a Particle   35
2.10   Newton's First Law of Motion   36
2.11   Problems Involving the Equilibrium of a Particle.
        Free-Body Diagrams   36

Forces in Space   45
2.12   Rectangular Components of a Force in Space   45
2.13   Force Defined by Its Magnitude and Two Points on Its
        Line of Action   48

2.14   Addition of Concurrent Forces in Space   49
2.15   Equilibrium of a Particle in Space   57

Review and Summary for Chapter 2   64
Review Problems   67
Computer Problems   69

3
RIGID BODIES: EQUIVALENT SYSTEMS OF FORCES
73

3.1   Introduction   74
3.2   External and Internal Forces   74
3.3   Principle of Transmissibility. Equivalent Forces   75
3.4   Vector Product of Two Vectors   77
3.5   Vector Products Expressed in Terms of
Rectangular Components   79
3.6   Moment of a Force about a Point   81
3.7   Varignon's Theorem   83
3.8   Rectangular Components of the Moment of a Force   83
3.9   Scalar Product of Two Vectors   93
3.10   Mixed Triple Product of Three Vectors   95
3.11   Moment of a Force about a Given Axis   97
3.12   Moment of a Couple   107
3.13   Equivalent Couples   108
3.14   Addition of Couples   110
3.15   Couples Can Be Represented by Vectors   110
3.16   Resolution of a Given Force Into a Force at *O*
and a Couple   111
3.17   Reduction of a System of Forces to One Force
and One Couple   122
3.18   Equivalent Systems of Forces   123
3.19   Equipollent Systems of Vectors   124
3.20   Further Reduction of a System of Forces   124
*3.21   Reduction of a System of Forces to a Wrench   127

Review and Summary for Chapter 3   146
Review Problems   151
Computer Problems   153

4
EQUILIBRIUM OF RIGID BODIES
157

4.1   Introduction   158
4.2   Free-Body Diagram   159

Equilibrium in Two Dimensions   160
4.3   Reactions at Supports and Connections
for a Two-Dimensional Structure   160
4.4   Equilibrium of a Rigid Body in Two Dimensions   162
4.5   Statically Indeterminate Reactions. Partial Constraints   164
4.6   Equilibrium of a Two-Force Body   183
4.7   Equilibrium of a Three-Force Body   184

Equilibrium in Three Dimensions   191
4.8   Equilibrium of a Rigid Body in Three Dimensions   191

4.9   Reactions at Supports and Connections for a
       Three-Dimensional Structure   191

Review and Summary for Chapter 4   211
Review Problems   213
Computer Problems   215

5

DISTRIBUTED FORCES: CENTROIDS AND CENTERS OF GRAVITY
219

5.1   Introduction   220

       Areas and Lines   220
5.2   Center of Gravity of a Two-Dimensional Body   220
5.3   Centroids of Areas and Lines   222
5.4   First Moments of Areas and Lines   223
5.5   Composite Plates and Wires   226
5.6   Determination of Centroids by Integration   236
5.7   Theorems of Pappus-Guldinus   238
*5.8   Distributed Loads on Beams   248
*5.9   Forces on Submerged Surfaces   249

       Volumes   259
5.10   Center of Gravity of a Three-Dimensional Body.
       Centroid of a Volume   259
5.11   Composite Bodies   262
5.12   Determination of Centroids of Volumes by Integration   262

Review and Summary for Chapter 5   274
Review Problems   278
Computer Problems   281

6

ANALYSIS OF STRUCTURES
284

6.1   Introduction   285

       Trusses   286
6.2   Definition of a Truss   286
6.3   Simple Trusses   288
6.4   Analysis of Trusses by the Method of Joints   289
*6.5   Joints under Special Loading Conditions   291
*6.6   Space Trusses   293
6.7   Analysis of Trusses by the Method of Sections   303
*6.8   Trusses Made of Several Simple Trusses   304

       Frames and Machines   315
6.9   Structures Containing Multiforce Members   315
6.10   Analysis of a Frame   315
6.11   Frames Which Cease to Be Rigid When Detached
       from Their Supports   316
6.12   Machines   331

Review and Summary for Chapter 6   343
Review Problems   346
Computer Problems   349

7
FORCES IN BEAMS AND CABLES
353

*7.1   Introduction   354
*7.2   Internal Forces in Members   354

Beams   361
*7.3   Various Types of Loading and Support   361
*7.4   Shear and Bending Moment in a Beam   362
*7.5   Shear and Bending-Moment Diagrams   364
*7.6   Relations among Load, Shear, and Bending Moment   372

Cables   383
*7.7   Cables with Concentrated Loads   383
*7.8   Cables with Distributed Loads   384
*7.9   Parabolic Cable   385
*7.10 Catenary   394

Review and Summary for Chapter 7   402
Review Problems   405
Computer Problems   408

8
FRICTION
411

8.1   Introduction   412
8.2   The Laws of Dry Friction. Coefficients of Friction   412
8.3   Angles of Friction   415
8.4   Problems Involving Dry Friction   416
8.5   Wedges   431
8.6   Square-Threaded Screws   431
*8.7   Journal Bearings. Axle Friction   440
*8.8   Thrust Bearings. Disk Friction   442
*8.9   Wheel Friction. Rolling Resistance   443
*8.10 Belt Friction   450

Review and Summary for Chapter 8   461
Review Problems   464
Computer Problems   467

9
DISTRIBUTED FORCES: MOMENTS OF INERTIA
471

9.1   Introduction   472

Moments of Inertia of Areas   473
9.2   Second Moment, or Moment of Inertia, of an Area   473
9.3   Determination of the Moment of Inertia of an Area
       by Integration   474
9.4   Polar Moment of Inertia   475
9.5   Radius of Gyration of an Area   476
9.6   Parallel-Axis Theorem   483
9.7   Moments of Inertia of Composite Areas   484
*9.8   Product of Inertia   497
*9.9   Principal Axes and Principal Moments of Inertia   498

*9.10   Mohr's Circle for Moments and Products of Inertia   506

Moments of Inertia of Masses   512

9.11   Moment of Inertia of a Mass   512
9.12   Parallel-Axis Theorem   514
9.13   Moments of Inertia of Thin Plates   515
9.14   Determination of the Moment of Inertia of a Three-Dimensional Body by Integration   516
9.15   Moments of Inertia of Composite Bodies   516
*9.16   Moment of Inertia of a Body with Respect to an Arbitrary Axis through O. Mass Products of Inertia   531
*9.17   Ellipsoid of Inertia. Principal Axes of Inertia   532
*9.18   Determination of the Principal Axes and Principal Moments of Inertia of a Body of Arbitrary Shape   534

Review and Summary for Chapter 9   545
Review Problems   551
Computer Problems   554

10
METHOD OF VIRTUAL WORK
557

*10.1   Introduction   558
*10.2   Work of a Force   558
*10.3   Principle of Virtual Work   561
*10.4   Applications of the Principle of Virtual Work   562
*10.5   Real Machines. Mechanical Efficiency   564
*10.6   Work of a Force during a Finite Displacement   578
*10.7   Potential Energy   580
*10.8   Potential Energy and Equilibrium   581
*10.9   Stability of Equilibrium   582

Review and Summary for Chapter 10   592
Review Problems   595
Computer Problems   597

Appendix
FUNDAMENTALS OF ENGINEERING EXAMINATION
601

Photo Credits   603

Index   605

Answers to Problems   611

Preface

OBJECTIVES

The main objective of a first course in mechanics should be to develop in the engineering student the ability to analyze any problem in a simple and logical manner and to apply to its solution a few, well-understood, basic principles. It is hoped that this text, designed for the first course in statics offered in the sophomore year, and the volume that follows, *Vector Mechanics for Engineers: Dynamics,* will help the instructor achieve this goal.†

GENERAL APPROACH

Vector analysis is introduced early in the text and is used in the presentation and discussion of the fundamental principles of mechanics. Vector methods are also used to solve many problems, particularly three-dimensional problems where these techniques result in a simpler and more concise solution. The emphasis in this text, however, remains on the correct understanding of the principles of mechanics and on their application to the solution of engineering problems, and vector analysis is presented chiefly as a convenient tool.‡

Practical Applications Are Introduced Early. One of the characteristics of the approach used in these volumes is that mechanics of *particles* is clearly separated from the mechanics of *rigid bodies.* This approach makes it possible to consider simple practical applications at an early stage and to postpone the introduction of the more difficult concepts. For example:

- In *Statics*, statics of particles is treated first (Chap. 2); after the rules of addition and subtraction of vectors are introduced, the principle of equilibrium of a particle is immediately applied to practical situations involving only concurrent forces. The statics

---

†This text is available in a single volume, *Vector Mechanics for Engineers: Statics and Dynamics,* seventh edition.

‡In a parallel text, *Mechanics for Engineers: Statics,* fourth edition, the use of vector algebra is limited to the addition and subtraction of vectors.

of rigid bodies is considered in Chaps. 3 and 4. In Chap. 3, the vector and scalar products of two vectors are introduced and used to define the moment of a force about a point and about an axis. The presentation of these new concepts is followed by a thorough and rigorous discussion of equivalent systems of forces leading, in Chap. 4, to many practical applications involving the equilibrium of rigid bodies under general force systems.

- In *Dynamics*, the same division is observed. The basic concepts of force, mass, and acceleration, of work and energy, and of impulse and momentum are introduced and first applied to problems involving only particles. Thus, students can familiarize themselves with the three basic methods used in dynamics and learn their respective advantages before facing the difficulties associated with the motion of rigid bodies.

New Concepts Are Introduced in Simple Terms. Since this text is designed for the first course in statics, new concepts are presented in simple terms and every step is explained in detail. On the other hand, by discussing the broader aspects of the problems considered, and by stressing methods of general applicability, a definite maturity of approach is achieved. For example, the concepts of partial constraints and statical indeterminacy are introduced early and are used throughout.

Fundamental Principles Are Placed in the Context of Simple Applications. The fact that mechanics is essentially a *deductive* science based on a few fundamental principles is stressed. Derivations have been presented in their logical sequence and with all the rigor warranted at this level. However, the learning process being largely *inductive*, simple applications are considered first. For example:

- The statics of particles precedes the statics of rigid bodies, and problems involving internal forces are postponed until Chap. 6.
- In Chap. 4, equilibrium problems involving only coplanar forces are considered first and solved by ordinary algebra, while problems involving three-dimensional forces and requiring the full use of vector algebra are discussed in the second part of the chapter.

Free-Body Diagrams Are Used Both to Solve Equilibrium Problems and to Express the Equivalence of Force Systems. Free-body diagrams are introduced early, and their importance is emphasized throughout the text. They are used not only to solve equilibrium problems but also to express the equivalence of two systems of forces or, more generally, of two systems of vectors. The advantage of this approach becomes apparent in the study of the dynamics of rigid bodies, where it is used to solve three-dimensional as well as two-dimensional problems. By placing the emphasis on "free-body-diagram equations" rather than on the standard algebraic equations of motion, a more intuitive and more complete understanding of the fundamental principles of dynamics can be achieved. This approach, which was first introduced in 1962 in the first edition of *Vector Mechanics for*

Engineers, has now gained wide acceptance among mechanics teachers in this country. It is, therefore, used in preference to the method of dynamic equilibrium and to the equations of motion in the solution of all sample problems in this book.

A Four-Color Presentation Uses Color to Distinguish Vectors.
Color has been used, not only to enhance the quality of the illustrations, but also to help students distinguish among the various types of vectors they will encounter. While there was no intention to "color code" this text, the same color is used in any given chapter to represent vectors of the same type. Throughout *Statics*, for example, red is used exclusively to represent forces and couples, while position vectors are shown in blue and dimensions in black. This makes it easier for the students to identify the forces acting on a given particle or rigid body and to follow the discussion of sample problems and other examples given in the text.

A Careful Balance Between SI and U.S. Customary Units Is Consistently Maintained. Because of the current trend in the American government and industry to adopt the international system of units (SI metric units), the SI units most frequently used in mechanics are introduced in Chap. 1 and are used throughout the text. Approximately half of the sample problems and 60 percent of the homework problems are stated in these units, while the remainder are in U.S. customary units. The authors believe that this approach will best serve the need of students, who, as engineers, will have to be conversant with both systems of units.

It also should be recognized that using both SI and U.S. customary units entails more than the use of conversion factors. Since the SI system of units is an absolute system based on the units of time, length, and mass, whereas the U.S. customary system is a gravitational system based on the units of time, length, and force, different approaches are required for the solution of many problems. For example, when SI units are used, a body is generally specified by its mass expressed in kilograms; in most problems of statics it will be necessary to determine the weight of the body in newtons, and an additional calculation will be required for this purpose. On the other hand, when U.S. customary units are used, a body is specified by its weight in pounds and, in dynamics problems, an additional calculation will be required to determine its mass in slugs (or lb · s^2/ft). The authors, therefore, believe that problem assignments should include both systems of units.

The *Instructor's and Solutions Manual* provides six different lists of assignments so that an equal number of problems stated in SI units and in U.S. customary units can be selected. If so desired, two complete lists of assignments can also be selected with up to 75 percent of the problems stated in SI units.

Optional Sections Offer Advanced or Specialty Topics. A large number of optional sections have been included. These sections are indicated by asterisks and thus are easily distinguished from those which form the core of the basic statics course. They may be omitted without prejudice to the understanding of the rest of the text.

Among the topics covered in these additional sections are the reduction of a system of forces to a wrench, applications to hydrostatics, shear and bending-moment diagrams for beams, equilibrium of cables, products of inertia and Mohr's circle, the determination of the principal axes and moments of inertia of a body of arbitrary shape, and the method of virtual work. The sections on beams are especially useful when the course in statics is immediately followed by a course in mechanics of materials, while the sections on the inertia properties of three-dimensional bodies are primarily intended for students who will later study in dynamics the three-dimensional motion of rigid bodies.

The material presented in the text and most of the problems requires no previous mathematical knowledge beyond algebra, trigonometry, and elementary calculus; all the elements of vector algebra necessary to the understanding of the text are carefully presented in Chaps. 2 and 3. In general, a greater emphasis is placed on the correct understanding of the basic mathematical concepts involved than on the nimble manipulation of mathematical formulas. In this connection, it should be mentioned that the determination of the centroids of composite areas precedes the calculation of centroids by integration, thus making it possible to establish the concept of the moment of an area firmly before introducing the use of integration.

NEW TO THIS EDITION

While retaining the well-received approach and organization of previous editions, the *seventh edition* offers the following new features and improvements:

- Ninety percent of the homework problems for the seventh edition are new or revised. The emphasis on industry-related and discipline-specific questions of the new edition problems provides motivation for today's students.
- The computer problems have been revised to be used with popular computational software, and the number of problems has been increased. The computer problems, many of which are relevant to the design process, have been included in a special section at the end of each chapter. The problems focus on symbolic manipulation and plotting, as opposed to the programming-based computer problems in previous editions of the text.
- Numerous in-chapter photographs have been added to help students better visualize important concepts.
- Chapter outlines have been added to the introduction of each chapter to provide a preview of topics that will be covered in the chapter.
- A Fundamentals of Engineering Examination Appendix has been added for use when students prepare for the FE exam.

CHAPTER ORGANIZATION AND PEDAGOGICAL FEATURES

Chapter Introduction. Each chapter begins with an introductory section setting the purpose and goals of the chapter and describing in simple terms the material to be covered and its application to the solution of engineering problems. New chapter outlines provide students with a preview of chapter topics.

Chapter Lessons.   The body of the text is divided into units, each consisting of one or several theory sections, one or several sample problems, and a large number of problems to be assigned. Each unit corresponds to a well-defined topic and generally can be covered in one lesson. In a number of cases, however, the instructor will find it desirable to devote more than one lesson to a given topic.

Sample Problems.   The sample problems are set up in much the same form that students will use when solving the assigned problems. They thus serve the double purpose of amplifying the text and demonstrating the type of neat, orderly work that students should cultivate in their own solutions.

Solving Problems on Your Own.   A section entitled *Solving Problems on Your Own* is included for each lesson, between the sample problems and the problems to be assigned. The purpose of these sections is to help students organize in their own minds the preceding theory of the text and the solution methods of the sample problems, so that they can more successfully solve the homework problems. Also included in these sections are specific suggestions and strategies that will enable students to more efficiently attack any assigned problems.

Homework Problem Sets.   Most of the problems are of a practical nature and should appeal to engineering students. They are primarily designed, however, to illustrate the material presented in the text and to help students understand the principles of mechanics. The problems are grouped according to the portions of material they illustrate and are arranged in order of increasing difficulty. Problems requiring special attention are indicated by asterisks. Answers to 70 percent of the problems are given at the end of the book. Problems for which the answers are given are set in straight type in the text, while problems for which no answer is given are set in italic.

Chapter Review and Summary.   Each chapter ends with a review and summary of the material covered in that chapter. Marginal notes are used to help students organize their review work, and cross-references have been included to help them find the portions of material requiring their special attention.

Review Problems.   A set of review problems is included at the end of each chapter. These problems provide students further opportunity to apply the most important concepts introduced in the chapter.

Computer Problems.   Each chapter includes a set of problems designed to be solved with computational software. Many of these problems are relevant to the design process. In statics, for example, they may involve the analysis of a structure for various configurations and loadings of the structure or the determination of the equilibrium positions of a mechanism which may require an iterative method of solution. Developing the algorithm required to solve a given mechanics problem will benefit students in two different ways: (1) it will help them gain a better understanding of the mechanics principles involved; (2) it will provide them with an opportunity to apply their computer skills to the solution of a meaningful engineering problem.

SUPPLEMENTS

An extensive supplements package for both instructors and students is available with the text. Instructor resources include an instructor's solutions manual with complete solutions to all text problems; image sets with electronic files of all text art and photo images; PowerPoint lecture presentations for all text chapters; transparencies of additional solved problems; scripts in various computational software formats for all text computer problems; access to course management systems to accommodate your online course needs; and various other presentation and course organization resources.

Students have access to S.M.A.R.T. (Self-paced, Mechanics, Algorithmic, Review, and Tutorial), an online interactive tutorial with algorithmic quizzing which can also be used as a classroom presentation tool. Other student resources include FE Exam-style multiple-choice quizzes with feedback; a guide for using computational software packages in mechanics courses; and many more internet-based content and learning tools.

Please visit our *Vector Mechanics for Engineers: Statics and Dynamics* seventh edition Online Learning Center (OLC) at www.mhhe.com/beerjohnston7 for more information on the supplements available with this text.

ACKNOWLEDGMENTS

The authors wish to acknowledge the collaboration of George Staab to this seventh edition of *Vector Mechanics for Engineers* and thank him especially for his crucial role in making the extensive problem set revision possible.

A special thanks go to our colleagues who thoroughly checked the solutions and answers of all problems in this edition and then prepared the solutions for the accompanying *Instructor's and Solution Manual*: Richard H. Lance of Cornell University; Daniel W. Yannitell of Louisiana State University; Kenneth Oster of University of Missouri, Rolla; and Petru Petrina of Cornell University

The authors thank the many companies that provided photographs for this edition. We also wish to recognize the determined efforts and patience of our photo researcher Sabina Dowell.

The authors also thank the members of the staff at McGraw-Hill for their support and dedication during the preparation of this new edition. We particularly wish to acknowledge the contributions of Sponsoring Editor Debra Matteson and Project Manager Jane Mohr.

Finally, the authors gratefully acknowledge the many helpful comments and suggestions offered by users of the previous editions of *Vector Mechanics for Engineers*.

Ferdinand P. Beer
E. Russell Johnston, Jr.
Elliot R. Eisenberg

Acknowledgments

The McGraw-Hill engineering mechanics editorial and marketing team would like to join the authors in gratefully acknowledging the individuals listed in the authors' preface for their contributions to the seventh editions of the texts and solutions manuals. In addition, we would also like to thank the following reviewers, focus group attendees, and symposium attendees whose comments and suggestions over the past few years have not only helped us to develop the supplemental and media resources for the seventh editions of *Vector Mechanics for Engineers* but have also helped us to establish standards and goals for developing other texts, supplements, and new media resources for statics and dynamics courses.

REVIEWERS FROM THE UNITED STATES

Shaaban Abdallah
University of Cincinnati

Charles A. Barnes
University of Massachusetts, Lowell

Liang-Wu Cai
Kansas State University

Brian Coller
University of Illinois, Chicago

Stephen C. Cowin
City College of New York

Ali El-Zeiny
California State University, Fresno

Christopher M. Foley
Marquette University

John H. Forrester
University of Tennessee

Orlando Hankins
North Carolina State University

Yohannes Ketema
University of Minnesota

Sang-Soo Kim
Ohio University

Roger L. Ludin
California State University, San Luis Obispo

Mohammad Mahinfalah
North Dakota State University

James R. Matthews
University of New Mexico

Wilfrid Nixon
University of Iowa

Karim Nohra
University of South Florida

David B. Oglesby
University of Missouri, Rolla

G.E. Ramey
Auburn University

William F. Reiter, Jr.
Oregon State University

Ali A. Selim
South Dakota State University

William Seto
San Jose State University

Teong E. Tan
The University of Memphis

Francis M. Thomas
University of Kansas

T. Calvin Tszeng
Illinois Institute of Technology

Christine Valle
University of Maine

INTERNATIONAL REVIEWERS

Yap Fook Fah
Nanyang Technical University

Y.A. Khulief
*King Fahd University of
   Petroleum Minerals,
   Saudi Arabia*

Moktar Ngah
*Universiti Teknologi Malaysia,
   Malaysia*

Ted Stathopoulos
Concordia University

Wen-Fang Wu
*National Taiwan University,
   Taiwan*

Chou Chu Yang
*National Taiwan University,
   Taiwan*

STATICS AND DYNAMICS FOCUS GROUP ATTENDEES

Makola M. Abdullah
Florida A&M University

Riad Alakkad
University of Dayton

Kurt Anderson
Rensselaer Polytechnic Institute

George Buzyna
*Florida State University and
   Florida A&M University*

Ravindar Chona
Texas A&M University

Ted Conway
University of Central Florida

Julie Coonrod
University of New Mexico

Phillip Cornwell
Rose-Hulman Institute

K.L. Devries
University of Utah

DeRome O. Dunn
*North Carolina A&T State
   University*

Gil Emmert
University of Wisconsin

Brian Fabian
University of Washington

Roy James Hartfield, Jr.
Auburn University

Lih-Min Hsia
*California State University,
   Los Angeles*

S. Graham Kelly
The University of Akron

Sam Kiger
*University of Missouri,
   Columbia*

Jeff Kuo
*California State University,
   Fullerton*

Marc Mignolet
Arizona State University

Masami Nakagawa
Colorado School of Mines

Steven O'Hare
Oklahoma State University

Charles K. Roby
Clemson University

Scott Schiff
Clemson University

Geoff Schiflett
*University of Southern
    California*

Ganesh Thiagarajan
Louisiana State University

Francis Thomas
University of Kansas

Erik Thompson
Colorado State University

Josef S. Torok
Rochester Institute of Technology

Robert Weber
Marquette University

STATICS AND DYNAMICS SYMPOSIUM ATTENDEES

Subhash Anand
Clemson University

Stephen Bechtel
The Ohio State University

Sherrill Biggers
Clemson University

Pasquale Cinnella
Mississippi State University

Adel El-Safty
University of Central Florida

Walt Haisler
Texas A&M University

Kimberly M. Hill
*University of Illinois,
    Urbana–Champaign*

James Jones
Purdue University

Yohannes Ketema
University of Minnesota

Chuck Krousgrill
Purdue University

Mohammad Mahinfalah
North Dakota State University

Christine Masters
Pennsylvania State University

Dan Mendelsohn
The Ohio State University

Mark Orwat
United States Military Academy

Art Peterson
University of Alberta

Tad Pfeffer
*University of Colorado at
    Boulder*

Sadanandan Pradeep
Clarkson University

Robert Rankin
Arizona State University

Michael Shelton
*California State Polytechnic
    University*

Larry Silverberg
North Carolina State University

Bob Witt
*University of Wisconsin,
    Madison*

Wan-Lee Yin
Georgia Institute of Technology

Musharraf Zaman
University of Oklahoma

List of Symbols

a	Constant; radius; distance
$\mathbf{A}, \mathbf{B}, \mathbf{C}, \ldots$	Reactions at supports and connections
$A, B, C, \ldots$	Points
A	Area
b	Width; distance
c	Constant
C	Centroid
d	Distance
e	Base of natural logarithms
$\mathbf{F}$	Force; friction force
g	Acceleration of gravity
G	Center of gravity; constant of gravitation
h	Height; sag of cable
$\mathbf{i}, \mathbf{j}, \mathbf{k}$	Unit vectors along coordinate axes
$I, I_x, \ldots$	Moments of inertia
$\bar{I}$	Centroidal moment of inertia
$I_{xy}, \ldots$	Products of inertia
J	Polar moment of inertia
k	Spring constant
k_x, k_y, k_O	Radii of gyration
$\bar{k}$	Centroidal radius of gyration
l	Length
L	Length; span
m	Mass
$\mathbf{M}$	Couple; moment
$\mathbf{M}_O$	Moment about point O
$\mathbf{M}_O^R$	Moment resultant about point O
M	Magnitude of couple or moment; mass of earth
M_{OL}	Moment about axis OL
$\mathbf{N}$	Normal component of reaction
O	Origin of coordinates
p	Pressure
$\mathbf{P}$	Force; vector
$\mathbf{Q}$	Force; vector
$\mathbf{r}$	Position vector

r	Radius; distance; polar coordinate
$\mathbf{R}$	Resultant force; resultant vector; reaction
R	Radius of earth
$\mathbf{s}$	Position vector
s	Length of arc; length of cable
$\mathbf{S}$	Force; vector
t	Thickness
$\mathbf{T}$	Force
T	Tension
U	Work
$\mathbf{V}$	Vector product; shearing force
V	Volume; potential energy; shear
w	Load per unit length
$\mathbf{W}, W$	Weight; load
x, y, z	Rectangular coordinates; distances
$\bar{x}, \bar{y}, \bar{z}$	Rectangular coordinates of centroid or center of gravity
α, β, γ	Angles
γ	Specific weight
δ	Elongation
$\delta \mathbf{r}$	Virtual displacement
δU	Virtual work
$\boldsymbol{\lambda}$	Unit vector along a line
η	Efficiency
θ	Angular coordinate; angle; polar coordinate
μ	Coefficient of friction
ρ	Density
ϕ	Angle of friction; angle

VECTOR MECHANICS
FOR
ENGINEERS
Statics

C H A P T E R

Introduction

1

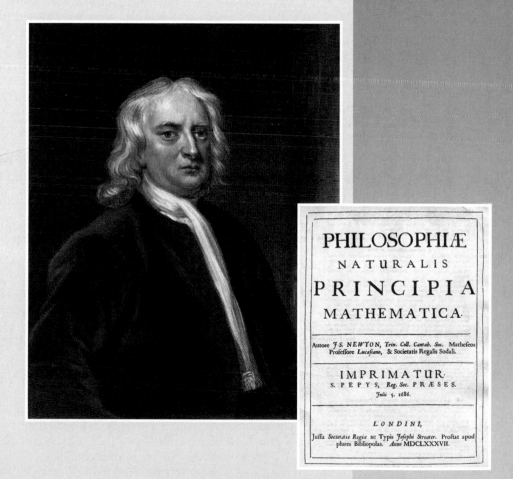

In the latter part of the seventeenth century, Sir Isaac Newton stated the fundamental principles of mechanics, which are the foundation of much of today's engineering. His *Principia,* published in 1687, summarized his work in terrestrial, celestial, and fluid mechanics.

INTRODUCTION

1.1   What Is Mechanics?
1.2   Fundamental Concepts and Principles
1.3   Systems of Units
1.4   Conversion from One System of Units to Another
1.5   Method of Problem Solution
1.6   Numerical Accuracy

1.1. WHAT IS MECHANICS?

Mechanics can be defined as that science which describes and predicts the conditions of rest or motion of bodies under the action of forces. It is divided into three parts: mechanics of *rigid bodies*, mechanics of *deformable bodies*, and mechanics of *fluids*.

The mechanics of rigid bodies is subdivided into *statics* and *dynamics*, the former dealing with bodies at rest, the latter with bodies in motion. In this part of the study of mechanics, bodies are assumed to be perfectly rigid. Actual structures and machines, however, are never absolutely rigid and deform under the loads to which they are subjected. But these deformations are usually small and do not appreciably affect the conditions of equilibrium or motion of the structure under consideration. They are important, though, as far as the resistance of the structure to failure is concerned and are studied in mechanics of materials, which is a part of the mechanics of deformable bodies. The third division of mechanics, the mechanics of fluids, is subdivided into the study of *incompressible fluids* and of *compressible fluids*. An important subdivision of the study of incompressible fluids is *hydraulics*, which deals with problems involving water.

Mechanics is a physical science, since it deals with the study of physical phenomena. However, some associate mechanics with mathematics, while many consider it as an engineering subject. Both these views are justified in part. Mechanics is the foundation of most engineering sciences and is an indispensable prerequisite to their study. However, it does not have the *empiricism* found in some engineering sciences, that is, it does not rely on experience or observation alone; by its rigor and the emphasis it places on deductive reasoning it resembles mathematics. But, again, it is not an *abstract* or even a *pure* science; mechanics is an *applied* science. The purpose of mechanics is to explain and predict physical phenomena and thus to lay the foundations for engineering applications.

1.2. FUNDAMENTAL CONCEPTS AND PRINCIPLES

Although the study of mechanics goes back to the time of Aristotle (384–322 B.C.) and Archimedes (287–212 B.C.), one has to wait until Newton (1642–1727) to find a satisfactory formulation of its fundamental principles. These principles were later expressed in a modified form by d'Alembert, Lagrange, and Hamilton. Their validity remained unchallenged, however, until Einstein formulated his *theory of relativity* (1905). While its limitations have now been recognized, *newtonian mechanics* still remains the basis of today's engineering sciences.

The basic concepts used in mechanics are *space, time, mass,* and *force*. These concepts cannot be truly defined; they should be accepted on the basis of our intuition and experience and used as a mental frame of reference for our study of mechanics.

The concept of *space* is associated with the notion of the position of a point P. The position of P can be defined by three lengths measured from a certain reference point, or *origin*, in three given directions. These lengths are known as the *coordinates* of P.

To define an event, it is not sufficient to indicate its position in space. The *time* of the event should also be given.

The concept of *mass* is used to characterize and compare bodies on the basis of certain fundamental mechanical experiments. Two bodies of the same mass, for example, will be attracted by the earth in the same manner; they will also offer the same resistance to a change in translational motion.

A *force* represents the action of one body on another. It can be exerted by actual contact or at a distance, as in the case of gravitational forces and magnetic forces. A force is characterized by its *point of application*, its *magnitude*, and its *direction;* a force is represented by a *vector* (Sec. 2.3).

In newtonian mechanics, space, time, and mass are absolute concepts, independent of each other. (This is not true in *relativistic mechanics,* where the time of an event depends upon its position, and where the mass of a body varies with its velocity.) On the other hand, the concept of force is not independent of the other three. Indeed, one of the fundamental principles of newtonian mechanics listed below indicates that the resultant force acting on a body is related to the mass of the body and to the manner in which its velocity varies with time.

You will study the conditions of rest or motion of particles and rigid bodies in terms of the four basic concepts we have introduced. By *particle* we mean a very small amount of matter which may be assumed to occupy a single point in space. A *rigid body* is a combination of a large number of particles occupying fixed positions with respect to each other. The study of the mechanics of particles is obviously a prerequisite to that of rigid bodies. Besides, the results obtained for a particle can be used directly in a large number of problems dealing with the conditions of rest or motion of actual bodies.

The study of elementary mechanics rests on six fundamental principles based on experimental evidence.

The Parallelogram Law for the Addition of Forces. This states that two forces acting on a particle may be replaced by a single force, called their *resultant,* obtained by drawing the diagonal of the parallelogram which has sides equal to the given forces (Sec. 2.2).

The Principle of Transmissibility. This states that the conditions of equilibrium or of motion of a rigid body will remain unchanged if a force acting at a given point of the rigid body is replaced by a force of the same magnitude and same direction, but acting at a different point, provided that the two forces have the same line of action (Sec. 3.3).

Newton's Three Fundamental Laws. Formulated by Sir Isaac Newton in the latter part of the seventeenth century, these laws can be stated as follows:

FIRST LAW. If the resultant force acting on a particle is zero, the particle will remain at rest (if originally at rest) or will move with constant speed in a straight line (if originally in motion) (Sec. 2.10).

SECOND LAW.   If the resultant force acting on a particle is not zero, the particle will have an acceleration proportional to the magnitude of the resultant and in the direction of this resultant force.

As you will see in Sec. 12.2, this law can be stated as

$$\mathbf{F} = m\mathbf{a} \tag{1.1}$$

where $\mathbf{F}$, m, and $\mathbf{a}$ represent, respectively, the resultant force acting on the particle, the mass of the particle, and the acceleration of the particle, expressed in a consistent system of units.

THIRD LAW.   The forces of action and reaction between bodies in contact have the same magnitude, same line of action, and opposite sense (Sec. 6.1).

Newton's Law of Gravitation.   This states that two particles of mass M and m are mutually attracted with equal and opposite forces $\mathbf{F}$ and $-\mathbf{F}$ (Fig. 1.1) of magnitude F given by the formula

$$F = G\frac{Mm}{r^2} \tag{1.2}$$

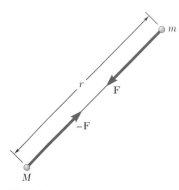

Fig. 1.1

where r = distance between the two particles

G = universal constant called the *constant of gravitation*

Newton's law of gravitation introduces the idea of an action exerted at a distance and extends the range of application of Newton's third law: the action $\mathbf{F}$ and the reaction $-\mathbf{F}$ in Fig. 1.1 are equal and opposite, and they have the same line of action.

A particular case of great importance is that of the attraction of the earth on a particle located on its surface. The force $\mathbf{F}$ exerted by the earth on the particle is then defined as the *weight* $\mathbf{W}$ of the particle. Taking M equal to the mass of the earth, m equal to the mass of the particle, and r equal to the radius R of the earth, and introducing the constant

$$g = \frac{GM}{R^2} \tag{1.3}$$

the magnitude W of the weight of a particle of mass m may be expressed as†

$$W = mg \tag{1.4}$$

Photo 1.1 When in earth orbit, people and objects are said to be *weightless* even though the gravitational force acting is approximately 90% of that experienced on the surface of the earth. This apparent contradiction will be resolved in Chapter 12 when we apply Newton's second law to the motion of particles.

The value of R in formula (1.3) depends upon the elevation of the point considered; it also depends upon its latitude, since the earth is not truly spherical. The value of g therefore varies with the position of the point considered. As long as the point actually remains on the surface of the earth, it is sufficiently accurate in most engineering computations to assume that g equals 9.81 m/s^2 or 32.2 ft/s^2.

---

†A more accurate definition of the weight $\mathbf{W}$ should take into account the rotation of the earth.

The principles we have just listed will be introduced in the course of our study of mechanics as they are needed. The study of the statics of particles carried out in Chap. 2 will be based on the parallelogram law of addition and on Newton's first law alone. The principle of transmissibility will be introduced in Chap. 3 as we begin the study of the statics of rigid bodies, and Newton's third law in Chap. 6 as we analyze the forces exerted on each other by the various members forming a structure. In the study of dynamics, Newton's second law and Newton's law of gravitation will be introduced. It will then be shown that Newton's first law is a particular case of Newton's second law (Sec. 12.2) and that the principle of transmissibility could be derived from the other principles and thus eliminated (Sec. 16.5). In the meantime, however, Newton's first and third laws, the parallelogram law of addition, and the principle of transmissibility will provide us with the necessary and sufficient foundation for the entire study of the statics of particles, rigid bodies, and systems of rigid bodies.

As noted earlier, the six fundamental principles listed above are based on experimental evidence. Except for Newton's first law and the principle of transmissibility, they are independent principles which cannot be derived mathematically from each other or from any other elementary physical principle. On these principles rests most of the intricate structure of newtonian mechanics. For more than two centuries a tremendous number of problems dealing with the conditions of rest and motion of rigid bodies, deformable bodies, and fluids were solved by applying these fundamental principles. Many of the solutions obtained could be checked experimentally, thus providing a further verification of the principles from which they were derived. It was only in the last century that Newton's mechanics was found at fault, in the study of the motion of atoms and in the study of the motion of certain planets, where it must be supplemented by the theory of relativity. But on the human or engineering scale, where velocities are small compared with the speed of light, Newton's mechanics has yet to be disproved.

1.3. SYSTEMS OF UNITS

Associated with the four fundamental concepts introduced in the preceding section are the so-called *kinetic units,* that is, the units of *length, time, mass,* and *force.* These units cannot be chosen independently if Eq. (1.1) is to be satisfied. Three of the units may be defined arbitrarily; they are then referred to as *base units.* The fourth unit, however, must be chosen in accordance with Eq. (1.1) and is referred to as a *derived unit.* Kinetic units selected in this way are said to form a *consistent system of units.*

International System of Units (SI Units†). In this system, which will be in universal use when the United States completes its conversion to SI units, the base units are the units of length, mass, and time, and they are called, respectively, the *meter* (m), the *kilogram* (kg), and the *second* (s). All three are arbitrarily defined. The second, which was originally chosen to represent 1/86 400 of the mean

†SI stands for *Système International d'Unités* (French).

solar day, is now defined as the duration of 9 192 631 770 cycles of the radiation corresponding to the transition between two levels of the fundamental state of the cesium-133 atom. The meter, originally defined as one ten-millionth of the distance from the equator to either pole, is now defined as 1 650 763.73 wavelengths of the orange-red light corresponding to a certain transition in an atom of krypton-86. The kilogram, which is approximately equal to the mass of 0.001 m^3 of water, is defined as the mass of a platinum-iridium standard kept at the International Bureau of Weights and Measures at Sèvres, near Paris, France. The unit of force is a derived unit. It is called the *newton* (N) and is defined as the force which gives an acceleration of 1 m/s^2 to a mass of 1 kg (Fig. 1.2). From Eq. (1.1) we write

$a = 1$ m/s^2

$m = 1$ kg $\quad\longrightarrow\quad F = 1$ N

Fig. 1.2

$$1 \text{ N} = (1 \text{ kg})(1 \text{ m/s}^2) = 1 \text{ kg} \cdot \text{m/s}^2 \qquad (1.5)$$

The SI units are said to form an *absolute* system of units. This means that the three base units chosen are independent of the location where measurements are made. The meter, the kilogram, and the second may be used anywhere on the earth; they may even be used on another planet. They will always have the same significance.

The *weight* of a body, or the *force of gravity* exerted on that body, should, like any other force, be expressed in newtons. From Eq. (1.4) it follows that the weight of a body of mass 1 kg (Fig. 1.3) is

$m = 1$ kg

$a = 9.81$ m/s^2

$W = 9.81$ N

Fig. 1.3

$$
\begin{aligned}
W &= mg \\
&= (1 \text{ kg})(9.81 \text{ m/s}^2) \\
&= 9.81 \text{ N}
\end{aligned}
$$

Multiples and submultiples of the fundamental SI units may be obtained through the use of the prefixes defined in Table 1.1. The multiples and submultiples of the units of length, mass, and force most frequently used in engineering are, respectively, the *kilometer* (km) and the *millimeter* (mm); the *megagram*† (Mg) and the *gram* (g); and the *kilonewton* (kN). According to Table 1.1, we have

$$
\begin{array}{ll}
1 \text{ km} = 1000 \text{ m} & 1 \text{ mm} = 0.001 \text{ m} \\
1 \text{ Mg} = 1000 \text{ kg} & 1 \text{ g} = 0.001 \text{ kg} \\
& 1 \text{ kN} = 1000 \text{ N}
\end{array}
$$

The conversion of these units into meters, kilograms, and newtons, respectively, can be effected by simply moving the decimal point three places to the right or to the left. For example, to convert 3.82 km into meters, one moves the decimal point three places to the right:

$$3.82 \text{ km} = 3820 \text{ m}$$

Similarly, 47.2 mm is converted into meters by moving the decimal point three places to the left:

$$47.2 \text{ mm} = 0.0472 \text{ m}$$

---

†Also known as a *metric ton*.

Table 1.1. **SI Prefixes**   1.3. Systems of Units **7**

Multiplication Factor	Prefix†	Symbol
$1\ 000\ 000\ 000\ 000 = 10^{12}$	tera	T
$1\ 000\ 000\ 000 = 10^{9}$	giga	G
$1\ 000\ 000 = 10^{6}$	mega	M
$1\ 000 = 10^{3}$	kilo	k
$100 = 10^{2}$	hecto‡	h
$10 = 10^{1}$	deka‡	da
$0.1 = 10^{-1}$	deci‡	d
$0.01 = 10^{-2}$	centi‡	c
$0.001 = 10^{-3}$	milli	m
$0.000\ 001 - 10^{-6}$	micro	μ
$0.000\ 000\ 001 = 10^{-9}$	nano	n
$0.000\ 000\ 000\ 001 = 10^{-12}$	pico	p
$0.000\ 000\ 000\ 000\ 001 = 10^{-15}$	femto	f
$0.000\ 000\ 000\ 000\ 000\ 001 = 10^{-18}$	atto	a

†The first syllable of every prefix is accented so that the prefix will retain its identity. Thus, the preferred pronunciation of kilometer places the accent on the first syllable, not the second.

‡The use of these prefixes should be avoided, except for the measurement of areas and volumes and for the nontechnical use of centimeter, as for body and clothing measurements.

Using scientific notation, one may also write

$$3.82\ \text{km} = 3.82 \times 10^{3}\ \text{m}$$
$$47.2\ \text{mm} = 47.2 \times 10^{-3}\ \text{m}$$

The multiples of the unit of time are the *minute* (min) and the *hour* (h). Since 1 min = 60 s and 1 h = 60 min = 3600 s, these multiples cannot be converted as readily as the others.

By using the appropriate multiple or submultiple of a given unit, one can avoid writing very large or very small numbers. For example, one usually writes 427.2 km rather than 427 200 m, and 2.16 mm rather than 0.002 16 m.†

Units of Area and Volume.   The unit of area is the *square meter* (m^2), which represents the area of a square of side 1 m; the unit of volume is the *cubic meter* (m^3), equal to the volume of a cube of side 1 m. In order to avoid exceedingly small or large numerical values in the computation of areas and volumes, one uses systems of sub-units obtained by respectively squaring and cubing not only the millimeter but also two intermediate submultiples of the meter, namely, the *decimeter* (dm) and the *centimeter* (cm). Since, by definition,

$$1\ \text{dm} = 0.1\ \text{m} = 10^{-1}\ \text{m}$$
$$1\ \text{cm} = 0.01\ \text{m} = 10^{-2}\ \text{m}$$
$$1\ \text{mm} = 0.001\ \text{m} = 10^{-3}\ \text{m}$$

†It should be noted that when more than four digits are used on either side of the decimal point to express a quantity in SI units—as in 427 200 m or 0.002 16 m—spaces, never commas, should be used to separate the digits into groups of three. This is to avoid confusion with the comma used in place of a decimal point, which is the convention in many countries.

the submultiples of the unit of area are

$$1 \text{ dm}^2 = (1 \text{ dm})^2 = (10^{-1} \text{ m})^2 = 10^{-2} \text{ m}^2$$
$$1 \text{ cm}^2 = (1 \text{ cm})^2 = (10^{-2} \text{ m})^2 = 10^{-4} \text{ m}^2$$
$$1 \text{ mm}^2 = (1 \text{ mm})^2 = (10^{-3} \text{ m})^2 = 10^{-6} \text{ m}^2$$

and the submultiples of the unit of volume are

$$1 \text{ dm}^3 = (1 \text{ dm})^3 = (10^{-1} \text{ m})^3 = 10^{-3} \text{ m}^3$$
$$1 \text{ cm}^3 = (1 \text{ cm})^3 = (10^{-2} \text{ m})^3 = 10^{-6} \text{ m}^3$$
$$1 \text{ mm}^3 = (1 \text{ mm})^3 = (10^{-3} \text{ m})^3 = 10^{-9} \text{ m}^3$$

It should be noted that when the volume of a liquid is being measured, the cubic decimeter (dm^3) is usually referred to as a *liter* (L).

Other derived SI units used to measure the moment of a force, the work of a force, etc., are shown in Table 1.2. While these units will be introduced in later chapters as they are needed, we should note an important rule at this time: When a derived unit is obtained by dividing a base unit by another base unit, a prefix may be used in the numerator of the derived unit but not in its denominator. For example, the constant k of a spring which stretches 20 mm under a load of 100 N will be expressed as

$$k = \frac{100 \text{ N}}{20 \text{ mm}} = \frac{100 \text{ N}}{0.020 \text{ m}} = 5000 \text{ N/m} \qquad \text{or} \qquad k = 5 \text{ kN/m}$$

but never as $k = 5$ N/mm.

Table 1.2. Principal SI Units Used in Mechanics

Quantity	Unit	Symbol	Formula
Acceleration	Meter per second squared	. . .	m/s^2
Angle	Radian	rad	†
Angular acceleration	Radian per second squared	. . .	rad/s^2
Angular velocity	Radian per second	. . .	rad/s
Area	Square meter	. . .	m^2
Density	Kilogram per cubic meter	. . .	kg/m^3
Energy	Joule	J	$\text{N} \cdot \text{m}$
Force	Newton	N	$\text{kg} \cdot \text{m/s}^2$
Frequency	Hertz	Hz	s^{-1}
Impulse	Newton-second	. . .	$\text{kg} \cdot \text{m/s}$
Length	Meter	m	‡
Mass	Kilogram	kg	‡
Moment of a force	Newton-meter	. . .	$\text{N} \cdot \text{m}$
Power	Watt	W	J/s
Pressure	Pascal	Pa	N/m^2
Stress	Pascal	Pa	N/m^2
Time	Second	s	‡
Velocity	Meter per second	. . .	m/s
Volume			
Solids	Cubic meter	. . .	m^3
Liquids	Liter	L	10^{-3} m^3
Work	Joule	J	$\text{N} \cdot \text{m}$

†Supplementary unit (1 revolution = 2π rad = 360°).

‡Base unit.

U.S. Customary Units.   Most practicing American engineers still commonly use a system in which the base units are the units of length, force, and time. These units are, respectively, the *foot* (ft), the *pound* (lb), and the *second* (s). The second is the same as the corresponding SI unit. The foot is defined as 0.3048 m. The pound is defined as the *weight* of a platinum standard, called the *standard pound*, which is kept at the National Institute of Standards and Technology outside Washington, the mass of which is 0.453 592 43 kg. Since the weight of a body depends upon the earth's gravitational attraction, which varies with location, it is specified that the standard pound should be placed at sea level and at a latitude of 45° to properly define a force of 1 lb. Clearly the U.S. customary units do not form an absolute system of units. Because of their dependence upon the gravitational attraction of the earth, they form a *gravitational* system of units.

While the standard pound also serves as the unit of mass in commercial transactions in the United States, it cannot be so used in engineering mechanics computations, since such a unit would not be consistent with the base units defined in the preceding paragraph. Indeed, when acted upon by a force of 1 lb, that is, when subjected to the force of gravity, the standard pound receives the acceleration of gravity, $g = 32.2$ ft/s² (Fig. 1.4), not the unit acceleration required by Eq. (1.1). The unit of mass consistent with the foot, the pound, and the second is the mass which receives an acceleration of 1 ft/s² when a force of 1 lb is applied to it (Fig. 1.5). This unit, sometimes called a *slug*, can be derived from the equation $F = ma$ after substituting 1 lb and 1 ft/s² for F and a, respectively. We write

$$F = ma \qquad 1 \text{ lb} = (1 \text{ slug})(1 \text{ ft/s}^2)$$

and obtain

$$1 \text{ slug} = \frac{1 \text{ lb}}{1 \text{ ft/s}^2} = 1 \text{ lb} \cdot \text{s}^2/\text{ft} \tag{1.6}$$

Comparing Figs. 1.4 and 1.5, we conclude that the slug is a mass 32.2 times larger than the mass of the standard pound.

The fact that in the U.S. customary system of units bodies are characterized by their weight in pounds rather than by their mass in slugs will be a convenience in the study of statics, where one constantly deals with weights and other forces and only seldom with masses. However, in the study of dynamics, where forces, masses, and accelerations are involved, the mass m of a body will be expressed in slugs when its weight W is given in pounds. Recalling Eq. (1.4), we write

$$m = \frac{W}{g} \tag{1.7}$$

where g is the acceleration of gravity ($g = 32.2$ ft/s²).

Other U.S. customary units frequently encountered in engineering problems are the *mile* (mi), equal to 5280 ft; the *inch* (in.), equal to $\frac{1}{12}$ ft; and the *kilopound* (kip), equal to a force of 1000 lb. The *ton* is often used to represent a mass of 2000 lb but, like the pound, must be converted into slugs in engineering computations.

The conversion into feet, pounds, and seconds of quantities expressed in other U.S. customary units is generally more involved

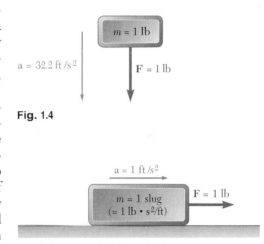

$a = 32.2$ ft/s² $m = 1$ lb $F = 1$ lb

Fig. 1.4

$a = 1$ ft/s² $m = 1$ slug ($= 1$ lb · s²/ft) $F = 1$ lb

Fig. 1.5

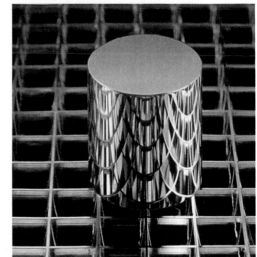

Photo 1.2   The unit of mass is the only basic unit still based on a physical standard. Work is in progress to replace this standard with one based on unchanging natural phenomena.

and requires greater attention than the corresponding operation in SI units. If, for example, the magnitude of a velocity is given as $v = 30$ mi/h, we convert it to ft/s as follows. First we write

$$v = 30\frac{\text{mi}}{\text{h}}$$

Since we want to get rid of the unit miles and introduce instead the unit feet, we should multiply the right-hand member of the equation by an expression containing miles in the denominator and feet in the numerator. But, since we do not want to change the value of the right-hand member, the expression used should have a value equal to unity. The quotient (5280 ft)/(1 mi) is such an expression. Operating in a similar way to transform the unit hour into seconds, we write

$$v = \left(30\frac{\text{mi}}{\text{h}}\right)\left(\frac{5280 \text{ ft}}{1 \text{ mi}}\right)\left(\frac{1 \text{ h}}{3600 \text{ s}}\right)$$

Carrying out the numerical computations and canceling out units which appear in both the numerator and the denominator, we obtain

$$v = 44\frac{\text{ft}}{\text{s}} = 44 \text{ ft/s}$$

1.4. CONVERSION FROM ONE SYSTEM OF UNITS TO ANOTHER

There are many instances when an engineer wishes to convert into SI units a numerical result obtained in U.S. customary units or vice versa. Because the unit of time is the same in both systems, only two kinetic base units need be converted. Thus, since all other kinetic units can be derived from these base units, only two conversion factors need be remembered.

Units of Length. By definition the U.S. customary unit of length is

$$1 \text{ ft} = 0.3048 \text{ m} \tag{1.8}$$

It follows that

$$1 \text{ mi} = 5280 \text{ ft} = 5280(0.3048 \text{ m}) = 1609 \text{ m}$$

or

$$1 \text{ mi} = 1.609 \text{ km} \tag{1.9}$$

Also

$$1 \text{ in.} = \tfrac{1}{12} \text{ ft} = \tfrac{1}{12}(0.3048 \text{ m}) = 0.0254 \text{ m}$$

or

$$1 \text{ in.} = 25.4 \text{ mm} \tag{1.10}$$

Units of Force. Recalling that the U.S. customary unit of force (pound) is defined as the weight of the standard pound (of mass 0.4536 kg) at sea level and at a latitude of 45° (where $g = 9.807$ m/s^2) and using Eq. (1.4), we write

Photo 1.3 The importance of including units in all calculations cannot be over emphasized. It was found that the $125 million Mars Climate Orbiter failed to go into orbit around Mars because the prime contractor had provided the navigation team with operating data based on U.S. units rather than the specified SI units.

$$W = mg$$
$$1 \text{ lb} = (0.4536 \text{ kg})(9.807 \text{ m/s}^2) = 4.448 \text{ kg} \cdot \text{m/s}^2$$

or, recalling Eq. (1.5),

$$1 \text{ lb} = 4.448 \text{ N} \tag{1.11}$$

Units of Mass. The U.S. customary unit of mass (slug) is a derived unit. Thus, using Eqs. (1.6), (1.8), and (1.11), we write

$$1 \text{ slug} = 1 \text{ lb} \cdot \text{s}^2/\text{ft} = \frac{1 \text{ lb}}{1 \text{ ft/s}^2} = \frac{4.448 \text{ N}}{0.3048 \text{ m/s}^2} = 14.59 \text{ N} \cdot \text{s}^2/\text{m}$$

and, recalling Eq. (1.5),

$$1 \text{ slug} = 1 \text{ lb} \cdot \text{s}^2/\text{ft} = 14.59 \text{ kg} \tag{1.12}$$

Although it cannot be used as a consistent unit of mass, we recall that the mass of the standard pound is, by definition,

$$1 \text{ pound mass} = 0.4536 \text{ kg} \tag{1.13}$$

This constant may be used to determine the *mass* in SI units (kilograms) of a body which has been characterized by its *weight* in U.S. customary units (pounds).

To convert a derived U.S. customary unit into SI units, one simply multiplies or divides by the appropriate conversion factors. For example, to convert the moment of a force which was found to be $M = 47 \text{ lb} \cdot \text{in.}$ into SI units, we use formulas (1.10) and (1.11) and write

$$M = 47 \text{ lb} \cdot \text{in.} = 47(4.448 \text{ N})(25.4 \text{ mm})$$
$$= 5310 \text{ N} \cdot \text{mm} = 5.31 \text{ N} \cdot \text{m}$$

The conversion factors given in this section may also be used to convert a numerical result obtained in SI units into U.S. customary units. For example, if the moment of a force was found to be $M = 40 \text{ N} \cdot \text{m}$, we write, following the procedure used in the last paragraph of Sec. 1.3,

$$M = 40 \text{ N} \cdot \text{m} = (40 \text{ N} \cdot \text{m})\left(\frac{1 \text{ lb}}{4.448 \text{ N}}\right)\left(\frac{1 \text{ ft}}{0.3048 \text{ m}}\right)$$

Carrying out the numerical computations and canceling out units which appear in both the numerator and the denominator, we obtain

$$M = 29.5 \text{ lb} \cdot \text{ft}$$

The U.S. customary units most frequently used in mechanics are listed in Table 1.3 with their SI equivalents.

1.5. METHOD OF PROBLEM SOLUTION

You should approach a problem in mechanics as you would approach an actual engineering situation. By drawing on your own experience and intuition, you will find it easier to understand and formulate the problem. Once the problem has been clearly stated, however, there is

Table 1.3. **U.S. Customary Units and Their SI Equivalents**

Quantity	U.S. Customary Unit	SI Equivalent
Acceleration	ft/s^2	0.3048 m/s^2
	in./s^2	0.0254 m/s^2
Area	ft^2	0.0929 m^2
	in^2	645.2 mm^2
Energy	ft · lb	1.356 J
Force	kip	4.448 kN
	lb	4.448 N
	oz	0.2780 N
Impulse	lb · s	4.448 N · s
Length	ft	0.3048 m
	in.	25.40 mm
	mi	1.609 km
Mass	oz mass	28.35 g
	lb mass	0.4536 kg
	slug	14.59 kg
	ton	907.2 kg
Moment of a force	lb · ft	1.356 N · m
	lb · in.	0.1130 N · m
Moment of inertia		
Of an area	in^4	0.4162 × 10^6 mm^4
Of a mass	lb · ft · s^2	1.356 kg · m^2
Momentum	lb · s	4.448 kg · m/s
Power	ft · lb/s	1.356 W
	hp	745.7 W
Pressure or stress	lb/ft^2	47.88 Pa
	lb/in^2 (psi)	6.895 kPa
Velocity	ft/s	0.3048 m/s
	in./s	0.0254 m/s
	mi/h (mph)	0.4470 m/s
	mi/h (mph)	1.609 km/h
Volume	ft^3	0.02832 m^3
	in^3	16.39 cm^3
Liquids	gal	3.785 L
	qt	0.9464 L
Work	ft · lb	1.356 J

no place in its solution for your particular fancy. *The solution must be based on the six fundamental principles stated in Sec. 1.2 or on theorems derived from them.* Every step taken must be justified on that basis. Strict rules must be followed, which lead to the solution in an almost automatic fashion, leaving no room for your intuition or "feeling." After an answer has been obtained, it should be checked. Here again, you may call upon your common sense and personal experience. If not completely satisfied with the result obtained, you should carefully check your formulation of the problem, the validity of the methods used for its solution, and the accuracy of your computations.

The *statement* of a problem should be clear and precise. It should contain the given data and indicate what information is required. A neat drawing showing all quantities involved should be included. Separate diagrams should be drawn for all bodies involved, indicating clearly the forces acting on each body. These diagrams are known as *free-body diagrams* and are described in detail in Secs. 2.11 and 4.2.

The *fundamental principles* of mechanics listed in Sec. 1.2 *will be used to write equations* expressing the conditions of rest or motion of the bodies considered. Each equation should be clearly related to one of the free-body diagrams. You will then proceed to solve the problem, observing strictly the usual rules of algebra and recording neatly the various steps taken.

After the answer has been obtained, it should be *carefully checked*. Mistakes in *reasoning* can often be detected by checking the units. For example, to determine the moment of a force of 50 N about a point 0.60 m from its line of action, we would have written (Sec. 3.12)

$$M = Fd = (50 \text{ N})(0.60 \text{ m}) = 30 \text{ N} \cdot \text{m}$$

The unit N · m obtained by multiplying newtons by meters is the correct unit for the moment of a force; if another unit had been obtained, we would have known that some mistake had been made.

Errors in *computation* will usually be found by substituting the numerical values obtained into an equation which has not yet been used and verifying that the equation is satisfied. The importance of correct computations in engineering cannot be overemphasized.

1.6. NUMERICAL ACCURACY

The accuracy of the solution of a problem depends upon two items: (1) the accuracy of the given data and (2) the accuracy of the computations performed.

The solution cannot be more accurate than the less accurate of these two items. For example, if the loading of a bridge is known to be 75,000 lb with a possible error of 100 lb either way, the relative error which measures the degree of accuracy of the data is

$$\frac{100 \text{ lb}}{75,000 \text{ lb}} = 0.0013 = 0.13 \text{ percent}$$

In computing the reaction at one of the bridge supports, it would then be meaningless to record it as 14,322 lb. The accuracy of the solution cannot be greater than 0.13 percent, no matter how accurate the computations are, and the possible error in the answer may be as large as $(0.13/100)(14,322 \text{ lb}) \approx 20$ lb. The answer should be properly recorded as $14,320 \pm 20$ lb.

In engineering problems, the data are seldom known with an accuracy greater than 0.2 percent. It is therefore seldom justified to write the answers to such problems with an accuracy greater than 0.2 percent. A practical rule is to use 4 figures to record numbers beginning with a "1" and 3 figures in all other cases. Unless otherwise indicated, the data given in a problem should be assumed known with a comparable degree of accuracy. A force of 40 lb, for example, should be read 40.0 lb, and a force of 15 lb should be read 15.00 lb.

Pocket electronic calculators are widely used by practicing engineers and engineering students. The speed and accuracy of these calculators facilitate the numerical computations in the solution of many problems. However, students should not record more significant figures than can be justified merely because they are easily obtained. As noted above, an accuracy greater than 0.2 percent is seldom necessary or meaningful in the solution of practical engineering problems.

CHAPTER

Statics of Particles

2

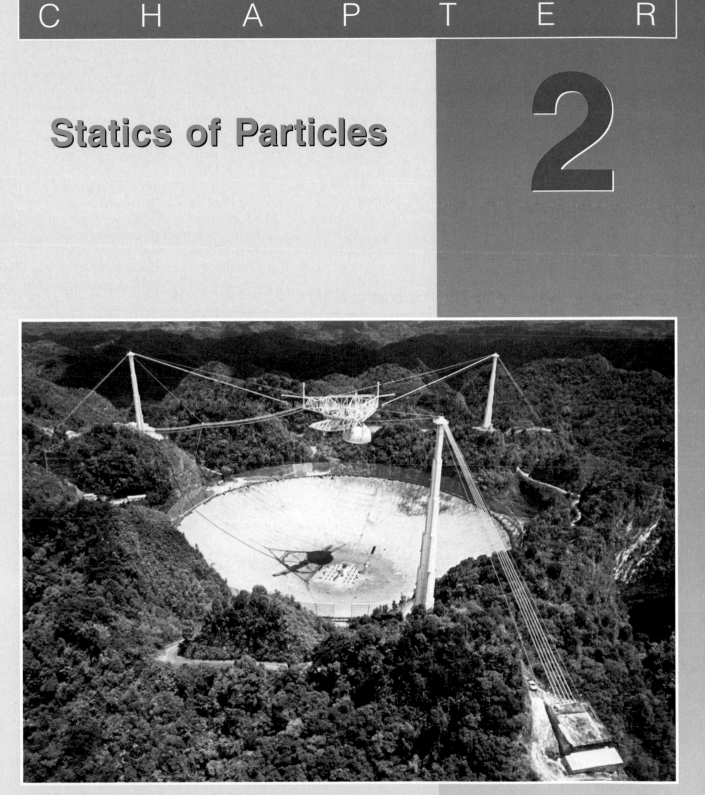

Shown above is the world's largest single-dish radio telescope, located in Arecibo, Puerto Rico. The 305-m-diameter spherical reflecting dish covers an area of approximately 20 acres. Suspended above the reflector is a 900-ton platform below which is the 328-ft-long azimuth arm. In this chapter, we will learn that by considering the platform as a *particle* in equilibrium, it will be possible to determine the tensions in the supporting cables.

STATICS OF PARTICLES

2.1 Introduction
     Forces in a Plane
2.2 Force on a Particle. Resultant of
     Two Forces
2.3 Vectors
2.4 Addition of Vectors
2.5 Resultant of Several Concurrent
     Forces
2.6 Resolution of a Force into
     Components
2.7 Rectangular Components of a
     Force. Unit Vectors
2.8 Addition of Forces by Summing *x*
     and *y* Components
2.9 Equilibrium of a Particle
2.10 Newton's First Law of Motion
2.11 Problems Involving the Equilibrium
     of a Particle. Free-Body Diagrams
     Forces in Space
2.12 Rectangular Components of a
     Force in Space
2.13 Force Defined by Its Magnitude
     and Two Points on Its Line of
     Action
2.14 Addition of Concurrent Forces in
     Space
2.15 Equilibrium of a Particle in Space

2.1. INTRODUCTION

In this chapter you will study the effect of forces acting on particles. First you will learn how to replace two or more forces acting on a given particle by a single force having the same effect as the original forces. This single equivalent force is the *resultant* of the original forces acting on the particle. Later the relations which exist among the various forces acting on a particle in a state of *equilibrium* will be derived and used to determine some of the forces acting on the particle.

The use of the word *particle* does not imply that our study will be limited to that of small corpuscles. What it means is that the size and shape of the bodies under consideration will not significantly affect the solution of the problems treated in this chapter and that all the forces acting on a given body will be assumed to be applied at the same point. Since such an assumption is verified in many practical applications, you will be able to solve a number of engineering problems in this chapter.

The first part of the chapter is devoted to the study of forces contained in a single plane, and the second part to the analysis of forces in three-dimensional space.

FORCES IN A PLANE

2.2. FORCE ON A PARTICLE. RESULTANT OF TWO FORCES

A force represents the action of one body on another and is generally characterized by its *point of application,* its *magnitude,* and its *direction.* Forces acting on a given particle, however, have the same point of application. Each force considered in this chapter will thus be completely defined by its magnitude and direction.

The magnitude of a force is characterized by a certain number of units. As indicated in Chap. 1, the SI units used by engineers to measure the magnitude of a force are the newton (N) and its multiple the kilonewton (kN), equal to 1000 N, while the U.S. customary units used for the same purpose are the pound (lb) and its multiple the kilopound (kip), equal to 1000 lb. The direction of a force is defined by the *line of action* and the *sense* of the force. The line of action is the infinite straight line along which the force acts; it is characterized by the angle it forms with some fixed axis (Fig. 2.1).

Fig. 2.1   (*a*)   (*b*)

The force itself is represented by a segment of that line; through the use of an appropriate scale, the length of this segment may be chosen to represent the magnitude of the force. Finally, the sense of the force should be indicated by an arrowhead. It is important in defining a force to indicate its sense. Two forces having the same magnitude and the same line of action but different sense, such as the forces shown in Fig. 2.1a and b, will have directly opposite effects on a particle.

Experimental evidence shows that two forces **P** and **Q** acting on a particle A (Fig. 2.2a) can be replaced by a single force **R** which has the same effect on the particle (Fig. 2.2c). This force is called the *resultant* of the forces **P** and **Q** and can be obtained, as shown in Fig. 2.2b, by constructing a parallelogram, using **P** and **Q** as two adjacent sides of the parallelogram. *The diagonal that passes through A represents the resultant.* This method for finding the resultant is known as the *parallelogram law* for the addition of two forces. This law is based on experimental evidence; it cannot be proved or derived mathematically.

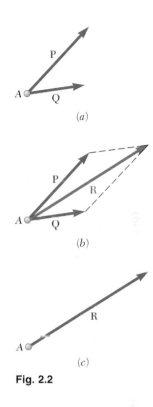

Fig. 2.2

2.3. VECTORS

It appears from the above that forces do not obey the rules of addition defined in ordinary arithmetic or algebra. For example, two forces acting at a right angle to each other, one of 4 lb and the other of 3 lb, add up to a force of 5 lb, *not* to a force of 7 lb. Forces are not the only quantities which follow the parallelogram law of addition. As you will see later, *displacements, velocities, accelerations,* and *momenta* are other examples of physical quantities possessing magnitude and direction that are added according to the parallelogram law. All these quantities can be represented mathematically by *vectors,* while those physical quantities which have magnitude but not direction, such as *volume, mass,* or *energy,* are represented by plain numbers or *scalars.*

Vectors are defined as *mathematical expressions possessing magnitude and direction, which add according to the parallelogram law.* Vectors are represented by arrows in the illustrations and will be distinguished from scalar quantities in this text through the use of boldface type (**P**). In longhand writing, a vector may be denoted by drawing a short arrow above the letter used to represent it ($\vec{P}$) or by underlining the letter ($\underline{P}$). The magnitude of a vector defines the length of the arrow used to represent the vector. In this text, italic type will be used to denote the magnitude of a vector. Thus, the magnitude of the vector **P** will be denoted by P.

A vector used to represent a force acting on a given particle has a well-defined point of application, namely, the particle itself. Such a vector is said to be a *fixed,* or *bound,* vector and cannot be moved without modifying the conditions of the problem. Other physical quantities, however, such as couples (see Chap. 3), are represented by vectors which may be freely moved in space; these vectors are called *free* vectors. Still other physical quantities, such as forces acting on a rigid body (see Chap. 3), are represented by vectors which can be moved, or slid, along their lines of action; they are known as *sliding* vectors.†

Two vectors which have the same magnitude and the same direction are said to be *equal,* whether or not they also have the same point of application (Fig. 2.4); equal vectors may be denoted by the same letter.

The *negative vector* of a given vector **P** is defined as a vector having the same magnitude as **P** and a direction opposite to that of **P** (Fig. 2.5); the negative of the vector **P** is denoted by −**P**. The vectors **P** and −**P** are commonly referred to as *equal and opposite* vectors. Clearly, we have

$$\mathbf{P} + (-\mathbf{P}) = 0$$

Fig. 2.4

Fig. 2.5

Fig. 2.6

2.4. ADDITION OF VECTORS

We saw in the preceding section that, by definition, vectors add according to the parallelogram law. Thus, the sum of two vectors **P** and **Q** is obtained by attaching the two vectors to the same point A and constructing a parallelogram, using **P** and **Q** as two sides of the parallelogram (Fig. 2.6). The diagonal that passes through A represents the sum of the vectors **P** and **Q**, and this sum is denoted by **P** + **Q**. The fact that the sign + is used to denote both vector and scalar addition should not cause any confusion if vector and scalar quantities are always carefully distinguished. Thus, we should note that the magnitude of the vector **P** + **Q** is *not,* in general, equal to the sum $P + Q$ of the magnitudes of the vectors **P** and **Q**.

Since the parallelogram constructed on the vectors **P** and **Q** does not depend upon the order in which **P** and **Q** are selected, we conclude that the addition of two vectors is *commutative,* and we write

$$\mathbf{P} + \mathbf{Q} = \mathbf{Q} + \mathbf{P} \tag{2.1}$$

---

†Some expressions have magnitude and direction, but do not add according to the parallelogram law. While these expressions may be represented by arrows, they *cannot* be considered as vectors.

A group of such expressions is the finite rotations of a rigid body. Place a closed book on a table in front of you, so that it lies in the usual fashion, with its front cover up and its binding to the left. Now rotate it through 180° about an axis parallel to the binding (Fig. 2.3a); this rotation may be represented by an arrow of length equal to 180 units and oriented as shown. Picking up the book as it lies in its new position, rotate it now through

(a)   *(b)*

Fig. 2.3 Finite rotations of rigid body

From the parallelogram law, we can derive an alternative method for determining the sum of two vectors. This method, known as the *triangle rule*, is derived as follows. Consider Fig. 2.6, where the sum of the vectors **P** and **Q** has been determined by the parallelogram law. Since the side of the parallelogram opposite **Q** is equal to **Q** in magnitude and direction, we could draw only half of the parallelogram (Fig. 2.7a). The sum of the two vectors can thus be found by *arranging* **P** *and* **Q** *in tip-to-tail fashion and then connecting the tail of* **P** *with the tip of* **Q**. In Fig. 2.7b, the other half of the parallelogram is considered, and the same result is obtained. This confirms the fact that vector addition is commutative.

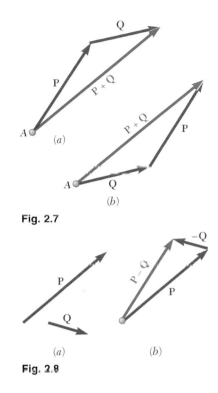

Fig. 2.7

The *subtraction* of a vector is defined as the addition of the corresponding negative vector. Thus, the vector **P** − **Q** representing the difference between the vectors **P** and **Q** is obtained by adding to **P** the negative vector −**Q** (Fig. 2.8). We write

$$\mathbf{P} - \mathbf{Q} = \mathbf{P} + (-\mathbf{Q}) \tag{2.2}$$

Here again we should observe that, while the same sign is used to denote both vector and scalar subtraction, confusion will be avoided if care is taken to distinguish between vector and scalar quantities.

We will now consider the *sum of three or more vectors*. The sum of three vectors **P**, **Q**, and **S** will, *by definition*, be obtained by first adding the vectors **P** and **Q** and then adding the vector **S** to the vector **P** + **Q**. We thus write

$$\mathbf{P} + \mathbf{Q} + \mathbf{S} = (\mathbf{P} + \mathbf{Q}) + \mathbf{S} \tag{2.3}$$

Similarly, the sum of four vectors will be obtained by adding the fourth vector to the sum of the first three. It follows that the sum of any number of vectors can be obtained by applying repeatedly the parallelogram law to successive pairs of vectors until all the given vectors are replaced by a single vector.

Fig. 2.8

180° about a horizontal axis perpendicular to the binding (Fig. 2.3*b*); this second rotation may be represented by an arrow 180 units long and oriented as shown. But the book could have been placed in this final position through a single 180° rotation about a vertical axis (Fig. 2.3*c*). We conclude that the sum of the two 180° rotations represented by arrows directed respectively along the *z* and *x* axes is a 180° rotation represented by an arrow directed along the *y* axis (Fig. 2.3*d*). Clearly, the finite rotations of a rigid body *do not* obey the parallelogram law of addition; therefore, they *cannot* be represented by vectors.

Fig. 2.9

If the given vectors are *coplanar,* that is, if they are contained in the same plane, their sum can be easily obtained graphically. For this case, the repeated application of the triangle rule is preferred to the application of the parallelogram law. In Fig. 2.9 the sum of three vectors **P**, **Q**, and **S** was obtained in that manner. The triangle rule was first applied to obtain the sum **P** + **Q** of the vectors **P** and **Q**; it was applied again to obtain the sum of the vectors **P** + **Q** and **S**. The determination of the vector **P** + **Q**, however, could have been omitted and the sum of the three vectors could have been obtained directly, as shown in Fig. 2.10, by *arranging the given vectors in tip-to-tail fashion and connecting the tail of the first vector with the tip of the last one.* This is known as the *polygon rule* for the addition of vectors.

We observe that the result obtained would have been unchanged if, as shown in Fig. 2.11, the vectors **Q** and **S** had been replaced by their sum **Q** + **S**. We may thus write

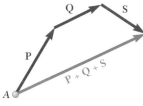

Fig. 2.10

$$\mathbf{P} + \mathbf{Q} + \mathbf{S} = (\mathbf{P} + \mathbf{Q}) + \mathbf{S} = \mathbf{P} + (\mathbf{Q} + \mathbf{S}) \qquad (2.4)$$

which expresses the fact that vector addition is *associative.* Recalling that vector addition has also been shown, in the case of two vectors, to be commutative, we write

$$\begin{aligned} \mathbf{P} + \mathbf{Q} + \mathbf{S} &= (\mathbf{P} + \mathbf{Q}) + \mathbf{S} = \mathbf{S} + (\mathbf{P} + \mathbf{Q}) \\ &= \mathbf{S} + (\mathbf{Q} + \mathbf{P}) = \mathbf{S} + \mathbf{Q} + \mathbf{P} \end{aligned} \qquad (2.5)$$

Fig. 2.11

This expression, as well as others which may be obtained in the same way, shows that the order in which several vectors are added together is immaterial (Fig. 2.12).

Product of a Scalar and a Vector. Since it is convenient to denote the sum **P** + **P** by 2**P**, the sum **P** + **P** + **P** by 3**P**, and, in general, the sum of n equal vectors **P** by the product $n\mathbf{P}$, we will define the product $n\mathbf{P}$ of a positive integer n and a vector **P** as a vector having the same direction as **P** and the magnitude $n\mathbf{P}$. Extending this definition to include all scalars, and recalling the definition of a negative vector given in Sec. 2.3, we define the product $k\mathbf{P}$ of a scalar k and a vector **P** as a vector having the same direction as **P** (if k is positive), or a direction opposite to that of **P** (if k is negative), and a magnitude equal to the product of P and of the absolute value of k (Fig. 2.13).

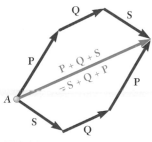

Fig. 2.12

2.5. RESULTANT OF SEVERAL CONCURRENT FORCES

Consider a particle A acted upon by several coplanar forces, that is, by several forces contained in the same plane (Fig. 2.14a). Since the forces considered here all pass through A, they are also said to be *concurrent.* The vectors representing the forces acting on A may be added by the polygon rule (Fig. 2.14b). Since the use of the polygon rule is equivalent to the repeated application of the parallelogram law, the vector **R** thus obtained represents the resultant of the given concurrent forces, that is, the single force which has the same effect on the particle A as the given forces. As indicated above, the order in which the vectors **P**, **Q**, and **S** representing the given forces are added together is immaterial.

Fig. 2.13

Fig. 2.14

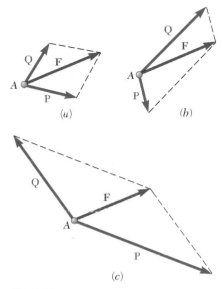

Fig. 2.15

2.6. RESOLUTION OF A FORCE INTO COMPONENTS

We have seen that two or more forces acting on a particle may be replaced by a single force which has the same effect on the particle. Conversely, a single force **F** acting on a particle may be replaced by two or more forces which, together, have the same effect on the particle. These forces are called the *components* of the original force **F**, and the process of substituting them for **F** is called *resolving the force* **F** *into components*.

Clearly, for each force **F** there exist an infinite number of possible sets of components. Sets of *two components* **P** *and* **Q** are the most important as far as practical applications are concerned. But, even then, the number of ways in which a given force **F** may be resolved into two components is unlimited (Fig. 2.15). Two cases are of particular interest:

1. *One of the Two Components,* **P**, *Is Known.* The second component, **Q**, is obtained by applying the triangle rule and joining the tip of **P** to the tip of **F** (Fig. 2.16); the magnitude and direction of **Q** are determined graphically or by trigonometry. Once **Q** has been determined, both components **P** and **Q** should be applied at *A*.

2. *The Line of Action of Each Component Is Known.* The magnitude and sense of the components are obtained by applying the parallelogram law and drawing lines, through the tip of **F**, parallel to the given lines of action (Fig. 2.17). This process leads to two well-defined components, **P** and **Q**, which can be determined graphically or computed trigonometrically by applying the law of sines.

Fig. 2.16

Many other cases can be encountered; for example, the direction of one component may be known, while the magnitude of the other component is to be as small as possible (see Sample Prob. 2.2). In all cases the appropriate triangle or parallelogram which satisfies the given conditions is drawn.

Fig. 2.17

SAMPLE PROBLEM 2.1

The two forces **P** and **Q** act on a bolt A. Determine their resultant.

SOLUTION

Graphical Solution. A parallelogram with sides equal to **P** and **Q** is drawn to scale. The magnitude and direction of the resultant are measured and found to be

$$R = 98 \text{ N} \qquad \alpha = 35° \qquad \mathbf{R} = 98 \text{ N} \; \angle 35° \quad \blacktriangleleft$$

The triangle rule may also be used. Forces **P** and **Q** are drawn in tip-to-tail fashion. Again the magnitude and direction of the resultant are measured.

$$R = 98 \text{ N} \qquad \alpha = 35° \qquad \mathbf{R} = 98 \text{ N} \; \angle 35° \quad \blacktriangleleft$$

Trigonometric Solution. The triangle rule is again used; two sides and the included angle are known. We apply the law of cosines.

$$R^2 = P^2 + Q^2 - 2PQ \cos B$$
$$R^2 = (40 \text{ N})^2 + (60 \text{ N})^2 - 2(40 \text{ N})(60 \text{ N}) \cos 155°$$
$$R = 97.73 \text{ N}$$

Now, applying the law of sines, we write

$$\frac{\sin A}{Q} = \frac{\sin B}{R} \qquad \frac{\sin A}{60 \text{ N}} = \frac{\sin 155°}{97.73 \text{ N}} \tag{1}$$

Solving Eq. (1) for sin A, we have

$$\sin A = \frac{(60 \text{ N}) \sin 155°}{97.73 \text{ N}}$$

Using a calculator, we first compute the quotient, then its arc sine, and obtain

$$A = 15.04° \qquad \alpha = 20° + A = 35.04°$$

We use 3 significant figures to record the answer (see Sec. 1.6):

$$\mathbf{R} = 97.7 \text{ N} \; \angle 35.0° \quad \blacktriangleleft$$

Alternative Trigonometric Solution. We construct the right triangle BCD and compute

$$CD = (60 \text{ N}) \sin 25° = 25.36 \text{ N}$$
$$BD = (60 \text{ N}) \cos 25° = 54.38 \text{ N}$$

Then, using triangle ACD, we obtain

$$\tan A = \frac{25.36 \text{ N}}{94.38 \text{ N}} \qquad A = 15.04°$$
$$R = \frac{25.36}{\sin A} \qquad R = 97.73 \text{ N}$$

Again, $\qquad \alpha = 20° + A = 35.04° \qquad \mathbf{R} = 97.7 \text{ N} \; \angle 35.0° \quad \blacktriangleleft$

SAMPLE PROBLEM 2.2

A barge is pulled by two tugboats. If the resultant of the forces exerted by the tugboats is a 5000-lb force directed along the axis of the barge, determine (a) the tension in each of the ropes knowing that $\alpha = 45°$, (b) the value of α for which the tension in rope 2 is minimum.

SOLUTION

a. Tension for $\alpha = 45°$. *Graphical Solution.* The parallelogram law is used; the diagonal (resultant) is known to be equal to 5000 lb and to be directed to the right. The sides are drawn parallel to the ropes. If the drawing is done to scale, we measure

$$T_1 = 3700 \text{ lb} \qquad T_2 = 2600 \text{ lb} \blacktriangleleft$$

Trigonometric Solution. The triangle rule can be used. We note that the triangle shown represents half of the parallelogram shown above. Using the law of sines, we write

$$\frac{T_1}{\sin 45°} = \frac{T_2}{\sin 30°} = \frac{5000 \text{ lb}}{\sin 105°}$$

With a calculator, we first compute and store the value of the last quotient. Multiplying this value successively by sin 45° and sin 30°, we obtain

$$T_1 = 3660 \text{ lb} \qquad T_2 = 2590 \text{ lb} \blacktriangleleft$$

b. Value of α for Minimum T_2. To determine the value of α for which the tension in rope 2 is minimum, the triangle rule is again used. In the sketch shown, line *1-1'* is the known direction of $\mathbf{T}_1$. Several possible directions of $\mathbf{T}_2$ are shown by the lines *2-2'*. We note that the minimum value of T_2 occurs when $\mathbf{T}_1$ and $\mathbf{T}_2$ are perpendicular. The minimum value of T_2 is

$$T_2 = (5000 \text{ lb}) \sin 30° = 2500 \text{ lb}$$

Corresponding values of T_1 and α are

$$T_1 = (5000 \text{ lb}) \cos 30° = 4330 \text{ lb}$$
$$\alpha = 90° - 30° \qquad\qquad \alpha = 60° \blacktriangleleft$$

SOLVING PROBLEMS ON YOUR OWN

The preceding sections were devoted to introducing and applying the *parallelogram law* for the addition of vectors.

You will now be asked to solve problems on your own. Some may resemble one of the sample problems; others may not. What all problems and sample problems in this section have in common is that they can be solved by the direct application of the parallelogram law.

Your solution of a given problem should consist of the following steps:

1. Identify which of the forces are the applied forces and which is the resultant. It is often helpful to write the vector equation which shows how the forces are related. For example, in Sample Prob. 2.1 we would have

$$\mathbf{R} = \mathbf{P} + \mathbf{Q}$$

You should keep that relation in mind as you formulate the next part of your solution.

2. Draw a parallelogram with the applied forces as two adjacent sides and the resultant as the included diagonal (Fig. 2.2). Alternatively, you can *use the triangle rule,* with the applied forces drawn in tip-to-tail fashion and the resultant extending from the tail of the first vector to the tip of the second (Fig. 2.7).

3. Indicate all dimensions. Using one of the triangles of the parallelogram, or the triangle constructed according to the triangle rule, indicate all dimensions—whether sides or angles—and determine the unknown dimensions either graphically or by trigonometry. If you use trigonometry, remember that the law of cosines should be applied first if two sides and the included angle are known [Sample Prob. 2.1], and the law of sines should be applied first if one side and all angles are known [Sample Prob. 2.2].

As is evident from the figures of Sec. 2.6, the two components of a force need not be perpendicular. Thus, when asked to resolve a force into two components, it is essential that you align the two adjacent sides of your parallelogram with the specified lines of action of the components.

If you have had prior exposure to mechanics, you might be tempted to ignore the solution techniques of this lesson in favor of resolving the forces into rectangular components. While this latter method is important and will be considered in the next section, use of the parallelogram law simplifies the solution of many problems and should be mastered at this time.

Problems†

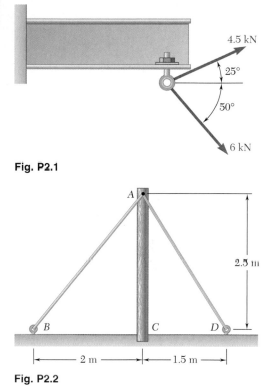

Fig. P2.1

2.1 Two forces are applied to an eye bolt fastened to a beam. Determine graphically the magnitude and direction of their resultant using (*a*) the parallelogram law, (*b*) the triangle rule.

2.2 The cable stays *AB* and *AD* help support pole *AC*. Knowing that the tension is 500 N in *AB* and 160 N in *AD*, determine graphically the magnitude and direction of the resultant of the forces exerted by the stays at *A* using (*a*) the parallelogram law, (*b*) the triangle rule.

2.3 Two forces **P** and **Q** are applied as shown at point *A* of a hook support. Knowing that *P* = 15 lb and *Q* = 25 lb, determine graphically the magnitude and direction of their resultant using (*a*) the parallelogram law, (*b*) the triangle rule.

2.4 Two forces **P** and **Q** are applied as shown at point *A* of a hook support. Knowing that *P* = 45 lb and *Q* = 15 lb, determine graphically the magnitude and direction of their resultant using (*a*) the parallelogram law, (*b*) the triangle rule.

Fig. P2.2

Fig. P2.3 and *P2.4*

2.5 Two control rods are attached at *A* to lever *AB*. Using trigonometry and knowing that the force in the left-hand rod is $F_1 = 120$ N, determine (*a*) the required force F_2 in the right-hand rod if the resultant **R** of the forces exerted by the rods on the lever is to be vertical, (*b*) the corresponding magnitude of **R**.

2.6 Two control rods are attached at *A* to lever *AB*. Using trigonometry and knowing that the force in the right-hand rod is $F_2 = 80$ N, determine (*a*) the required force F_1 in the left-hand rod if the resultant **R** of the forces exerted by the rods on the lever is to be vertical, (*b*) the corresponding magnitude of **R**.

Fig. P2.5 and P2.6

†Answers to all problems set in straight type (such as **2.1**) are given at the end of the book. Answers to problems with a number set in italic type (such as ***2.2***) are not given.

25

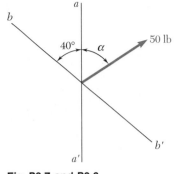

Fig. P2.7 and P2.8

2.7 The 50-lb force is to be resolved into components along lines $a-a'$ and $b-b'$. (*a*) Using trigonometry, determine the angle α knowing that the component along $a-a'$ is 35 lb. (*b*) What is the corresponding value of the component along $b-b'$?

2.8 The 50-lb force is to be resolved into components along lines $a-a'$ and $b-b'$. (*a*) Using trigonometry, determine the angle α knowing that the component along $b-b'$ is 30 lb. (*b*) What is the corresponding value of the component along $a-a'$?

2.9 To steady a sign as it is being lowered, two cables are attached to the sign at *A*. Using trigonometry and knowing that $\alpha = 25°$, determine (*a*) the required magnitude of the force **P** if the resultant **R** of the two forces applied at *A* is to be vertical, (*b*) the corresponding magnitude of **R**.

2.10 To steady a sign as it is being lowered, two cables are attached to the sign at *A*. Using trigonometry and knowing that the magnitude of **P** is 300 N, determine (*a*) the required angle α if the resultant **R** of the two forces applied at *A* is to be vertical, (*b*) the corresponding magnitude of **R**.

Fig. P2.9 and P2.10

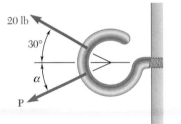

Fig. P2.11

2.11 Two forces are applied as shown to a hook support. Using trigonometry and knowing that the magnitude of **P** is 14 lb, determine (*a*) the required angle α if the resultant **R** of the two forces applied to the support is to be horizontal, (*b*) the corresponding magnitude of **R**.

2.12 For the hook support of Prob. 2.3, using trigonometry and knowing that the magnitude of **P** is 25 lb, determine (*a*) the required magnitude of the force **Q** if the resultant **R** of the two forces applied at *A* is to be vertical, (*b*) the corresponding magnitude of **R**.

2.13 For the hook support of Prob. 2.11, determine, using trigonometry, (*a*) the magnitude and direction of the smallest force **P** for which the resultant **R** of the two forces applied to the support is horizontal, (*b*) the corresponding magnitude of **R**.

2.14 As shown in Fig. P2.9, two cables are attached to a sign at *A* to steady the sign as it is being lowered. Using trigonometry, determine (*a*) the magnitude and direction of the smallest force **P** for which the resultant **R** of the two forces applied at *A* is vertical, (*b*) the corresponding magnitude of **R**.

2.15 For the hook support of Prob. 2.11, determine, using trigonometry, the magnitude and direction of the resultant of the two forces applied to the support knowing that $P = 10$ lb and $\alpha = 40°$.

2.16 Solve Prob. 2.1 using trigonometry.

2.17 Solve Prob. 2.2 using trigonometry.

2.18 Solve Prob. 2.3 using trigonometry.

2.19 Two structural members A and B are bolted to a bracket as shown. Knowing that both members are in compression and that the force is 30 kN in member A and 20 kN in member B, determine, using trigonometry, the magnitude and direction of the resultant of the forces applied to the bracket by members A and B.

2.20 Two structural members A and B are bolted to a bracket as shown. Knowing that both members are in compression and that the force is 20 kN in member A and 30 kN in member B, determine, using trigonometry, the magnitude and direction of the resultant of the forces applied to the bracket by members A and B.

Fig. *P2.19* and *P2.20*

2.7. RECTANGULAR COMPONENTS OF A FORCE. UNIT VECTORS†

In many problems it will be found desirable to resolve a force into two components which are perpendicular to each other. In Fig. 2.18, the force $\mathbf{F}$ has been resolved into a component $\mathbf{F}_x$ along the x axis and a component $\mathbf{F}_y$ along the y axis. The parallelogram drawn to obtain the two components is a *rectangle*, and $\mathbf{F}_x$ and $\mathbf{F}_y$ are called *rectangular components*.

Fig. 2.18

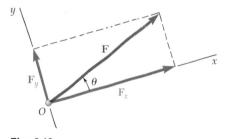

Fig. 2.19

The x and y axes are usually chosen horizontal and vertical, respectively, as in Fig. 2.18; they may, however, be chosen in any two perpendicular directions, as shown in Fig. 2.19. In determining the rectangular components of a force, the student should think of the construction lines shown in Figs. 2.18 and 2.19 as being *parallel* to the x and y axes, rather than *perpendicular* to these axes. This practice will help avoid mistakes in determining *oblique* components as in Sec. 2.6.

†The properties established in Secs. 2.7 and 2.8 may be readily extended to the rectangular components of any vector quantity.

Fig. 2.20

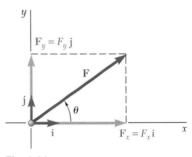

Fig. 2.21

Two vectors of unit magnitude, directed respectively along the positive x and y axes, will be introduced at this point. These vectors are called *unit vectors* and are denoted by **i** and **j**, respectively (Fig. 2.20). Recalling the definition of the product of a scalar and a vector given in Sec. 2.4, we note that the rectangular components $\mathbf{F}_x$ and $\mathbf{F}_y$ of a force **F** may be obtained by multiplying respectively the unit vectors **i** and **j** by appropriate scalars (Fig. 2.21). We write

$$\mathbf{F}_x = F_x\mathbf{i} \qquad \mathbf{F}_y = F_y\mathbf{j} \tag{2.6}$$

and

$$\mathbf{F} = F_x\mathbf{i} + F_y\mathbf{j} \tag{2.7}$$

While the scalars F_x and F_y may be positive or negative, depending upon the sense of $\mathbf{F}_x$ and of $\mathbf{F}_y$, their absolute values are respectively equal to the magnitudes of the component forces $\mathbf{F}_x$ and $\mathbf{F}_y$. The scalars F_x and F_y are called the *scalar components* of the force **F**, while the actual component forces $\mathbf{F}_x$ and $\mathbf{F}_y$ should be referred to as the *vector components* of **F**. However, when there exists no possibility of confusion, the vector as well as the scalar components of **F** may be referred to simply as the *components* of **F**. We note that the scalar component F_x is positive when the vector component $\mathbf{F}_x$ has the same sense as the unit vector **i** (that is, the same sense as the positive x axis) and is negative when $\mathbf{F}_x$ has the opposite sense. A similar conclusion may be drawn regarding the sign of the scalar component F_y.

Denoting by F the magnitude of the force **F** and by θ the angle between **F** and the x axis, measured counterclockwise from the positive x axis (Fig. 2.21), we may express the scalar components of **F** as follows:

$$F_x = F \cos \theta \qquad F_y = F \sin \theta \tag{2.8}$$

We note that the relations obtained hold for any value of the angle θ from 0° to 360° and that they define the signs as well as the absolute values of the scalar components F_x and F_y.

Example 1. A force of 800 N is exerted on a bolt A as shown in Fig. 2.22*a*. Determine the horizontal and vertical components of the force.

In order to obtain the correct sign for the scalar components F_x and F_y, the value $180° - 35° = 145°$ should be substituted for θ in Eqs. (2.8). However, it will be found more practical to determine by inspection the signs of F_x and F_y (Fig. 2.22*b*) and to use the trigonometric functions of the angle $\alpha = 35°$. We write, therefore,

$$F_x = -F \cos \alpha = -(800 \text{ N}) \cos 35° = -655 \text{ N}$$
$$F_y = +F \sin \alpha = +(800 \text{ N}) \sin 35° = +459 \text{ N}$$

The vector components of **F** are thus

$$\mathbf{F}_x = -(655 \text{ N})\mathbf{i} \qquad \mathbf{F}_y = +(459 \text{ N})\mathbf{j}$$

and we may write **F** in the form

$$\mathbf{F} = -(655 \text{ N})\mathbf{i} + (459 \text{ N})\mathbf{j}$$

Fig. 2.22

Example 2. A man pulls with a force of 300 N on a rope attached to a building, as shown in Fig. 2.23a. What are the horizontal and vertical components of the force exerted by the rope at point A?

It is seen from Fig. 2.23b that

$$F_x = +(300 \text{ N}) \cos \alpha \qquad F_y = -(300 \text{ N}) \sin \alpha$$

Observing that $AB = 10$ m, we find from Fig. 2.23a

$$\cos \alpha = \frac{8 \text{ m}}{AB} = \frac{8 \text{ m}}{10 \text{ m}} = \frac{4}{5} \qquad \sin \alpha = \frac{6 \text{ m}}{AB} = \frac{6 \text{ m}}{10 \text{ m}} = \frac{3}{5}$$

We thus obtain

$$F_x = +(300 \text{ N})\tfrac{4}{5} = +240 \text{ N} \qquad F_y = -(300 \text{ N})\tfrac{3}{5} = -180 \text{ N}$$

and write

$$\mathbf{F} = (240 \text{ N})\mathbf{i} - (180 \text{ N})\mathbf{j}$$

When a force $\mathbf{F}$ is defined by its rectangular components F_x and F_y (see Fig. 2.21), the angle θ defining its direction can be obtained by writing

$$\tan \theta = \frac{F_y}{F_x} \qquad (2.9)$$

The magnitude F of the force can be obtained by applying the Pythagorean theorem and writing

$$F = \sqrt{F_x^2 + F_y^2} \qquad (2.10)$$

or by solving for F one of the Eqs. (2.8).

Example 3. A force $\mathbf{F} = (700 \text{ lb})\mathbf{i} + (1500 \text{ lb})\mathbf{j}$ is applied to a bolt A. Determine the magnitude of the force and the angle θ it forms with the horizontal.

First we draw a diagram showing the two rectangular components of the force and the angle θ (Fig. 2.24). From Eq. (2.9), we write

$$\tan \theta = \frac{F_y}{F_x} = \frac{1500 \text{ lb}}{700 \text{ lb}}$$

Using a calculator,† we enter 1500 lb and divide by 700 lb; computing the arc tangent of the quotient, we obtain $\theta = 65.0°$. Solving the second of Eqs. (2.8) for F, we have

$$F = \frac{F_y}{\sin \theta} = \frac{1500 \text{ lb}}{\sin 65.0°} = 1655 \text{ lb}$$

The last calculation is facilitated if the value of F_y is stored when originally entered; it may then be recalled to be divided by $\sin \theta$.

---

†It is assumed that the calculator used has keys for the computation of trigonometric and inverse trigonometric functions. Some calculators also have keys for the direct conversion of rectangular coordinates into polar coordinates, and vice versa. Such calculators eliminate the need for the computation of trigonometric functions in Examples 1, 2, and 3 and in problems of the same type.

(a)

(b)

Fig. 2.23

Fig. 2.24

2.8. ADDITION OF FORCES BY SUMMING *X* AND *Y* COMPONENTS

It was seen in Sec. 2.2 that forces should be added according to the parallelogram law. From this law, two other methods, more readily applicable to the *graphical* solution of problems, were derived in Secs. 2.4 and 2.5: the triangle rule for the addition of two forces and the polygon rule for the addition of three or more forces. It was also seen that the force triangle used to define the resultant of two forces could be used to obtain a *trigonometric* solution.

When three or more forces are to be added, no practical trigonometric solution can be obtained from the force polygon which defines the resultant of the forces. In this case, an *analytic* solution of the problem can be obtained by resolving each force into two rectangular components. Consider, for instance, three forces **P**, **Q**, and **S** acting on a particle *A* (Fig. 2.25*a*). Their resultant **R** is defined by the relation

$$\mathbf{R} = \mathbf{P} + \mathbf{Q} + \mathbf{S} \tag{2.11}$$

Resolving each force into its rectangular components, we write

$$R_x\mathbf{i} + R_y\mathbf{j} = P_x\mathbf{i} + P_y\mathbf{j} + Q_x\mathbf{i} + Q_y\mathbf{j} + S_x\mathbf{i} + S_y\mathbf{j}$$
$$= (P_x + Q_x + S_x)\mathbf{i} + (P_y + Q_y + S_y)\mathbf{j}$$

from which it follows that

$$R_x = P_x + Q_x + S_x \qquad R_y = P_y + Q_y + S_y \tag{2.12}$$

or, for short,

$$R_x = \Sigma F_x \qquad R_y = \Sigma F_y \tag{2.13}$$

We thus conclude that *the scalar components R_x and R_y of the resultant **R** of several forces acting on a particle are obtained by adding algebraically the corresponding scalar components of the given forces.*†

In practice, the determination of the resultant **R** is carried out in three steps as illustrated in Fig. 2.25. First the given forces shown in Fig. 2.25*a* are resolved into their *x* and *y* components (Fig. 2.25*b*). Adding these components, we obtain the *x* and *y* components of **R** (Fig. 2.25*c*). Finally, the resultant $\mathbf{R} = R_x\mathbf{i} + R_y\mathbf{j}$ is determined by applying the parallelogram law (Fig. 2.25*d*). The procedure just described will be carried out most efficiently if the computations are arranged in a table. While it is the only practical analytic method for adding three or more forces, it is also often preferred to the trigonometric solution in the case of the addition of two forces.

(a)

(b)

(c)

(d)

Fig. 2.25

---

†Clearly, this result also applies to the addition of other vector quantities, such as velocities, accelerations, or momenta.

SAMPLE PROBLEM 2.3

Four forces act on bolt A as shown. Determine the resultant of the forces on the bolt.

SOLUTION

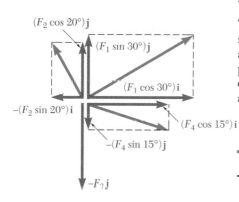

The x and y components of each force are determined by trigonometry as shown and are entered in the table below. According to the convention adopted in Sec. 2.7, the scalar number representing a force component is positive if the force component has the same sense as the corresponding coordinate axis. Thus, x components acting to the right and y components acting upward are represented by positive numbers.

Force	Magnitude, N	x Component, N	y Component, N
$\mathbf{F}_1$	150	+129.9	+75.0
$\mathbf{F}_2$	80	−27.4	+75.2
$\mathbf{F}_3$	110	0	−110.0
$\mathbf{F}_4$	100	+96.6	−25.9
		$R_x = +199.1$	$R_y = +14.3$

Thus, the resultant $\mathbf{R}$ of the four forces is

$$\mathbf{R} = R_x\mathbf{i} + R_y\mathbf{j} \qquad \mathbf{R} = (199.1\ \text{N})\mathbf{i} + (14.3\ \text{N})\mathbf{j} \blacktriangleleft$$

The magnitude and direction of the resultant may now be determined. From the triangle shown, we have

$$\tan \alpha = \frac{R_y}{R_x} = \frac{14.3\ \text{N}}{199.1\ \text{N}} \qquad \alpha = 4.1°$$

$$R = \frac{14.3\ \text{N}}{\sin \alpha} = 199.6\ \text{N} \qquad \mathbf{R} = 199.6\ \text{N} \angle 4.1° \blacktriangleleft$$

With a calculator, the last computation may be facilitated if the value of R_y is stored when originally entered; it may then be recalled to be divided by $\sin \alpha$. (Also see the footnote on p. 29.)

You saw in the preceding lesson that the resultant of two forces can be determined either graphically or from the trigonometry of an oblique triangle.

A. When three or more forces are involved, the determination of their resultant **R** is best carried out by first resolving each force into *rectangular components*. Two cases may be encountered, depending upon the way in which each of the given forces is defined:

*Case 1. The force **F** is defined by its magnitude F and the angle α it forms with the x axis.* The x and y components of the force can be obtained by multiplying F by cos α and sin α, respectively [Example 1].

*Case 2. The force **F** is defined by its magnitude F and the coordinates of two points A and B on its line of action* (Fig. 2.23). The angle α that **F** forms with the x axis may first be determined by trigonometry. However, the components of **F** may also be obtained directly from proportions among the various dimensions involved, without actually determining α [Example 2].

B. Rectangular components of the resultant. The components R_x and R_y of the resultant can be obtained by adding algebraically the corresponding components of the given forces [Sample Prob. 2.3].

You can express the resultant in *vectorial form* using the unit vectors **i** and **j**, which are directed along the x and y axes, respectively:

$$\mathbf{R} = R_x\mathbf{i} + R_y\mathbf{j}$$

Alternatively, you can determine the *magnitude and direction* of the resultant by solving the right triangle of sides R_x and R_y for R and for the angle that **R** forms with the x axis.

In the examples and sample problem of this lesson, the x and y axes were horizontal and vertical, respectively. You should remember, however, that for some problems it will be more efficient to rotate the axes to align them with one or more of the applied forces.

Problems

2.21 Determine the x and y components of each of the forces shown.

2.22 Determine the x and y components of each of the forces shown.

Fig. P2.21

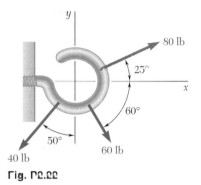

Fig. P2.22

2.23 and *2.24* Determine the x and y components of each of the forces shown.

Fig. P2.23

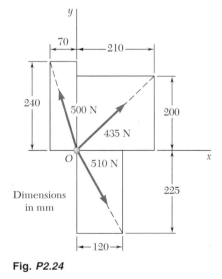

Fig. *P2.24*

2.25 While emptying a wheelbarrow, a gardener exerts on each handle AB a force **P** directed along line CD. Knowing that **P** must have a 135-N horizontal component, determine (*a*) the magnitude of the force **P**, (*b*) its vertical component.

Fig. P2.25

Fig. P2.26

Fig. P2.27

Fig. P2.29 and P2.30

Fig. P2.28

2.26 Member *BD* exerts on member *ABC* a force **P** directed along line *BD*. Knowing that **P** must have a 960-N vertical component, determine (*a*) the magnitude of the force **P**, (*b*) its horizontal component.

2.27 Member *CB* of the vise shown exerts on block *B* a force **P** directed along line *CB*. Knowing that **P** must have a 260-lb horizontal component, determine (*a*) the magnitude of the force **P**, (*b*) its vertical component.

2.28 Activator rod *AB* exerts on crank *BCD* a force **P** directed along line *AB*. Knowing that **P** must have a 25-lb component perpendicular to arm *BC* of the crank, determine (*a*) the magnitude of the force **P**, (*b*) its component along line *BC*.

2.29 The guy wire *BD* exerts on the telephone pole *AC* a force **P** directed along *BD*. Knowing that **P** has a 450-N component along line *AC*, determine (*a*) the magnitude of the force **P**, (*b*) its component in a direction perpendicular to *AC*.

2.30 The guy wire *BD* exerts on the telephone pole *AC* a force **P** directed along *BD*. Knowing that **P** must have a 200-N component perpendicular to the pole *AC*, determine (*a*) the magnitude of the force **P**, (*b*) its component along line *AC*.

2.31 Determine the resultant of the three forces of Prob. 2.24.

2.32 Determine the resultant of the three forces of Prob. 2.21.

2.33 Determine the resultant of the three forces of Prob. 2.22.

2.34 Determine the resultant of the three forces of Prob. 2.23.

2.35 Knowing that $\alpha = 35°$, determine the resultant of the three forces shown.

2.36 Knowing that $\alpha = 65°$, determine the resultant of the three forces shown.

Fig. P2.35 and P2.36

Fig. *P2.37*

2.37 Knowing that the tension in cable *BC* is 145 lb, determine the resultant of the three forces exerted at point *B* of beam *AB*.

2.38 Knowing that $\alpha = 50°$, determine the resultant of the three forces shown.

2.39 Determine (*a*) the required value of α if the resultant of the three forces shown is to be vertical, (*b*) the corresponding magnitude of the resultant.

2.40 For the beam of Prob. 2.37, determine (*a*) the required tension in cable *BC* if the resultant of the three forces exerted at point *B* is to be vertical, (*b*) the corresponding magnitude of the resultant.

2.41 Boom *AB* is held in the position shown by three cables. Knowing that the tensions in cables *AC* and *AD* are 4 kN and 5.2 kN, respectively, determine (*a*) the tension in cable *AE* if the resultant of the tensions exerted at point *A* of the boom must be directed along *AB*, (*b*) the corresponding magnitude of the resultant.

2.42 For the block of Probs. 2.35 and 2.36, determine (*a*) the required value of α if the resultant of the three forces shown is to be parallel to the incline, (*b*) the corresponding magnitude of the resultant.

Fig. *P2.38* and P2.39

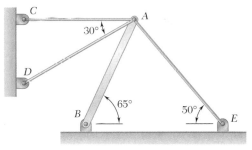

Fig. P2.41

2.9. EQUILIBRIUM OF A PARTICLE

In the preceding sections, we discussed the methods for determining the resultant of several forces acting on a particle. Although it has not occurred in any of the problems considered so far, it is quite possible for the resultant to be zero. In such a case, the net effect of the given forces is zero, and the particle is said to be in equilibrium. We thus have the following definition: *When the resultant of all the forces acting on a particle is zero, the particle is in equilibrium.*

A particle which is acted upon by two forces will be in equilibrium if the two forces have the same magnitude and the same line of action but opposite sense. The resultant of the two forces is then zero. Such a case is shown in Fig. 2.26.

Fig. 2.26

Fig. 2.27

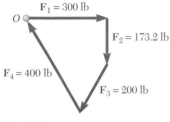

Fig. 2.28

Another case of equilibrium of a particle is represented in Fig. 2.27, where four forces are shown acting on A. In Fig. 2.28, the resultant of the given forces is determined by the polygon rule. Starting from point O with $\mathbf{F}_1$ and arranging the forces in tip-to-tail fashion, we find that the tip of $\mathbf{F}_4$ coincides with the starting point O. Thus the resultant $\mathbf{R}$ of the given system of forces is zero, and the particle is in equilibrium.

The closed polygon drawn in Fig. 2.28 provides a *graphical* expression of the equilibrium of A. To express *algebraically* the conditions for the equilibrium of a particle, we write

$$\mathbf{R} = \Sigma\mathbf{F} = 0 \qquad\qquad (2.14)$$

Resolving each force $\mathbf{F}$ into rectangular components, we have

$$\Sigma(F_x\mathbf{i} + F_y\mathbf{j}) = 0 \qquad \text{or} \qquad (\Sigma F_x)\mathbf{i} + (\Sigma F_y)\mathbf{j} = 0$$

We conclude that the necessary and sufficient conditions for the equilibrium of a particle are

$$\Sigma F_x = 0 \qquad \Sigma F_y = 0 \qquad\qquad (2.15)$$

Returning to the particle shown in Fig. 2.27, we check that the equilibrium conditions are satisfied. We write

$$\Sigma F_x = 300 \text{ lb} - (200 \text{ lb}) \sin 30° - (400 \text{ lb}) \sin 30°$$
$$= 300 \text{ lb} - 100 \text{ lb} - 200 \text{ lb} = 0$$
$$\Sigma F_y = -173.2 \text{ lb} - (200 \text{ lb}) \cos 30° + (400 \text{ lb}) \cos 30°$$
$$= -173.2 \text{ lb} - 173.2 \text{ lb} + 346.4 \text{ lb} = 0$$

2.10. NEWTON'S FIRST LAW OF MOTION

In the latter part of the seventeenth century, Sir Isaac Newton formulated three fundamental laws upon which the science of mechanics is based. The first of these laws can be stated as follows:

If the resultant force acting on a particle is zero, the particle will remain at rest (if originally at rest) or will move with constant speed in a straight line (if originally in motion).

From this law and from the definition of equilibrium given in Sec. 2.9, it is seen that a particle in equilibrium either is at rest or is moving in a straight line with constant speed. In the following section, various problems concerning the equilibrium of a particle will be considered.

2.11. PROBLEMS INVOLVING THE EQUILIBRIUM OF A PARTICLE. FREE-BODY DIAGRAMS

In practice, a problem in engineering mechanics is derived from an actual physical situation. A sketch showing the physical conditions of the problem is known as a *space diagram.*

The methods of analysis discussed in the preceding sections apply to a system of forces acting on a particle. A large number of problems involving actual structures, however, can be reduced to problems concerning the equilibrium of a particle. This is done by choosing a significant particle and drawing a separate diagram showing this particle and all the forces acting on it. Such a diagram is called a *free-body diagram.*

As an example, consider the 75-kg crate shown in the space diagram of Fig. 2.29a. This crate was lying between two buildings, and it is now being lifted onto a truck, which will remove it. The crate is supported by a vertical cable, which is joined at A to two ropes which pass over pulleys attached to the buildings at B and C. It is desired to determine the tension in each of the ropes AB and AC.

In order to solve this problem, a free-body diagram showing a particle in equilibrium must be drawn. Since we are interested in the rope tensions, the free-body diagram should include at least one of these tensions or, if possible, both tensions. Point A is seen to be a good free body for this problem. The free-body diagram of point A is shown in Fig. 2.29b. It shows point A and the forces exerted on A by the vertical cable and the two ropes. The force exerted by the cable is directed downward, and its magnitude is equal to the weight W of the crate. Recalling Eq. (1.4), we write

$$W - mg = (75 \text{ kg})(9.81 \text{ m/s}^2) = 736 \text{ N}$$

and indicate this value in the free-body diagram. The forces exerted by the two ropes are not known. Since they are respectively equal in magnitude to the tensions in rope AB and rope AC, we denote them by $\mathbf{T}_{AB}$ and $\mathbf{T}_{AC}$ and draw them away from A in the directions shown in the space diagram. No other detail is included in the free-body diagram.

Since point A is in equilibrium, the three forces acting on it must form a closed triangle when drawn in tip-to-tail fashion. This *force triangle* has been drawn in Fig. 2.29c. The values T_{AB} and T_{AC} of the tension in the ropes may be found graphically if the triangle is drawn to scale, or they may be found by trigonometry. If the latter method of solution is chosen, we use the law of sines and write

$$\frac{T_{AB}}{\sin 60°} = \frac{T_{AC}}{\sin 40°} = \frac{736 \text{ N}}{\sin 80°}$$

$$T_{AB} = 647 \text{ N} \qquad T_{AC} = 480 \text{ N}$$

When a particle is in *equilibrium under three forces*, the problem can be solved by drawing a force triangle. When a particle is in *equilibrium under more than three forces*, the problem can be solved graphically by drawing a force polygon. If an analytic solution is desired, the *equations of equilibrium* given in Sec. 2.9 should be solved:

$$\Sigma F_x = 0 \qquad \Sigma F_y = 0 \qquad (2.15)$$

These equations can be solved for no more than *two unknowns;* similarly, the force triangle used in the case of equilibrium under three forces can be solved for two unknowns.

The more common types of problems are those in which the two unknowns represent (1) the two components (or the magnitude and direction) of a single force, (2) the magnitudes of two forces, each of known direction. Problems involving the determination of the maximum or minimum value of the magnitude of a force are also encountered (see Probs. 2.59 through 2.64).

(a) Space diagram

(b) Free-body diagram   (c) Force triangle

Fig. 2.29

Photo 2.1 As illustrated in the above example, it is possible to determine the tensions in the cables supporting the shaft shown by treating the hook as a particle and then applying the equations of equilibrium to the forces acting on the hook.

SAMPLE PROBLEM 2.4

In a ship-unloading operation, a 3500-lb automobile is supported by a cable. A rope is tied to the cable at A and pulled in order to center the automobile over its intended position. The angle between the cable and the vertical is $2°$, while the angle between the rope and the horizontal is $30°$. What is the tension in the rope?

SOLUTION

Free-Body Diagram. Point A is chosen as a free body, and the complete free-body diagram is drawn. T_{AB} is the tension in the cable AB, and T_{AC} is the tension in the rope.

Equilibrium Condition. Since only three forces act on the free body, we draw a force triangle to express that it is in equilibrium. Using the law of sines, we write

$$\frac{T_{AB}}{\sin 120°} = \frac{T_{AC}}{\sin 2°} = \frac{3500 \text{ lb}}{\sin 58°}$$

With a calculator, we first compute and store the value of the last quotient. Multiplying this value successively by $\sin 120°$ and $\sin 2°$, we obtain

$$T_{AB} = 3570 \text{ lb} \qquad T_{AC} = 144 \text{ lb} \quad \blacktriangleleft$$

SAMPLE PROBLEM 2.5

Determine the magnitude and direction of the smallest force **F** which will maintain the package shown in equilibrium. Note that the force exerted by the rollers on the package is perpendicular to the incline.

SOLUTION

Free-Body Diagram. We choose the package as a free body, assuming that it can be treated as a particle. We draw the corresponding free-body diagram.

Equilibrium Condition. Since only three forces act on the free body, we draw a force triangle to express that it is in equilibrium. Line *1-1'* represents the known direction of **P**. In order to obtain the minimum value of the force **F**, we choose the direction of **F** perpendicular to that of **P**. From the geometry of the triangle obtained, we find

$$F = (294 \text{ N}) \sin 15° = 76.1 \text{ N} \qquad \alpha = 15°$$
$$\mathbf{F} = 76.1 \text{ N} \searrow 15° \quad \blacktriangleleft$$

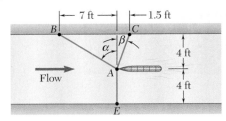

SAMPLE PROBLEM 2.6

As part of the design of a new sailboat, it is desired to determine the drag force which may be expected at a given speed. To do so, a model of the proposed hull is placed in a test channel and three cables are used to keep its bow on the centerline of the channel. Dynamometer readings indicate that for a given speed, the tension is 40 lb in cable AB and 60 lb in cable AE. Determine the drag force exerted on the hull and the tension in cable AC.

SOLUTION

Determination of the Angles. First, the angles α and β defining the direction of cables AB and AC are determined. We write

$$\tan \alpha = \frac{7 \text{ ft}}{4 \text{ ft}} = 1.75 \qquad \tan \beta = \frac{1.5 \text{ ft}}{4 \text{ ft}} = 0.375$$

$$\alpha = 60.26° \qquad \beta = 20.56°$$

Free-Body Diagram. Choosing the hull as a free body, we draw the free-body diagram shown. It includes the forces exerted by the three cables on the hull, as well as the drag force $\mathbf{F}_D$ exerted by the flow.

Equilibrium Condition. We express that the hull is in equilibrium by writing that the resultant of all forces is zero:

$$\mathbf{R} = \mathbf{T}_{AB} + \mathbf{T}_{AC} + \mathbf{T}_{AE} + \mathbf{F}_D = 0 \tag{1}$$

Since more than three forces are involved, we resolve the forces into x and y components:

$$\mathbf{T}_{AB} = -(40 \text{ lb}) \sin 60.26°\mathbf{i} + (40 \text{ lb}) \cos 60.26°\mathbf{j}$$
$$= -(34.73 \text{ lb})\mathbf{i} + (19.84 \text{ lb})\mathbf{j}$$
$$\mathbf{T}_{AC} = T_{AC} \sin 20.56°\mathbf{i} + T_{AC} \cos 20.56°\mathbf{j}$$
$$= 0.3512 T_{AC}\mathbf{i} + 0.9363 T_{AC}\mathbf{j}$$
$$\mathbf{T}_{AE} = -(60 \text{ lb})\mathbf{j}$$
$$\mathbf{F}_D = F_D\mathbf{i}$$

Substituting the expressions obtained into Eq. (1) and factoring the unit vectors $\mathbf{i}$ and $\mathbf{j}$, we have

$$(-34.73 \text{ lb} + 0.3512 T_{AC} + F_D)\mathbf{i} + (19.84 \text{ lb} + 0.9363 T_{AC} - 60 \text{ lb})\mathbf{j} = 0$$

This equation will be satisfied if, and only if, the coefficients of $\mathbf{i}$ and $\mathbf{j}$ are equal to zero. We thus obtain the following two equilibrium equations, which express, respectively, that the sum of the x components and the sum of the y components of the given forces must be zero.

$$\Sigma F_x = 0: \qquad -34.73 \text{ lb} + 0.3512 T_{AC} + F_D = 0 \tag{2}$$
$$\Sigma F_y = 0: \qquad 19.84 \text{ lb} + 0.9363 T_{AC} - 60 \text{ lb} = 0 \tag{3}$$

From Eq. (3) we find $\qquad\qquad\qquad\qquad T_{AC} = +42.9 \text{ lb} \;\blacktriangleleft$

and, substituting this value into Eq. (2), $\qquad\qquad F_D = +19.66 \text{ lb} \;\blacktriangleleft$

In drawing the free-body diagram, we assumed a sense for each unknown force. A positive sign in the answer indicates that the assumed sense is correct. The complete force polygon may be drawn to check the results.

SOLVING PROBLEMS
ON YOUR OWN

When a particle is in *equilibrium,* the resultant of the forces acting on the particle must be zero. Expressing this fact in the case of a particle under *coplanar forces* will provide you with two relations among these forces. As you saw in the preceding sample problems, these relations can be used to determine two unknowns—such as the magnitude and direction of one force or the magnitudes of two forces.

Drawing a free-body diagram is the first step in the solution of a problem involving the equilibrium of a particle. This diagram shows the particle and all the forces acting on it. Indicate on your free-body diagram the magnitudes of known forces as well as any angle or dimensions that define the direction of a force. Any unknown magnitude or angle should be denoted by an appropriate symbol. Nothing else should be included on the free-body diagram.

Drawing a clear and accurate free-body diagram is a must in the solution of any equilibrium problem. Skipping this step might save you pencil and paper, but is very likely to lead you to a wrong solution.

Case 1.   If only three forces are involved in the free-body diagram, the rest of the solution is best carried out by drawing these forces in tip-to-tail fashion to form a *force triangle.* This triangle can be solved graphically or using trigonometry for no more than two unknowns [Sample Probs. 2.4 and 2.5].

Case 2.   If more than three forces are involved, it is to your advantage to use an *analytic solution.* Begin by selecting appropriate x and y axes and resolve each of the forces shown in the free-body diagram into x and y components. Expressing that the sum of the x components and the sum of the y components of all the forces are both zero, you will obtain two equations which you can solve for no more than two unknowns [Sample Prob. 2.6].

It is strongly recommended that when using an analytic solution the equations of equilibrium be written in the same form as Eqs. (2) and (3) of Sample Prob. 2.6. The practice adopted by some students of initially placing the unknowns on the left side of the equation and the known quantities on the right side may lead to confusion in assigning the appropriate sign to each term.

We have noted that regardless of the method used to solve a two-dimensional equilibrium problem we can determine at most two unknowns. If a two-dimensional problem involves more than two unknowns, one or more additional relations must be obtained from the information contained in the statement of the problem.

Some of the following problems contain small pulleys. We will assume that the pulleys are frictionless, so the tension in the rope or cable passing over a pulley is the same on each side of the pulley. In Chap. 4 we will discuss why the tension is the same.

Problems

2.43 Two cables are tied together at *C* and are loaded as shown. Determine the tension (*a*) in cable *AC*, (*b*) in cable *BC*.

Fig. P2.43

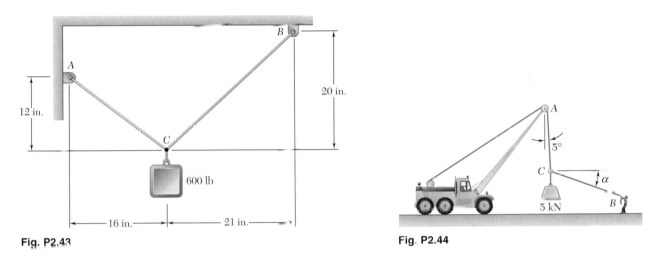

Fig. P2.44

2.44 Knowing that $\alpha = 25°$, determine the tension (*a*) in cable *AC*, (*b*) in rope *BC*.

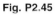

Fig. P2.45

2.45 Knowing that $\alpha = 50°$ and that boom *AC* exerts on pin *C* a force directed along line *AC*, determine (*a*) the magnitude of that force, (*b*) the tension in cable *BC*.

2.46 Two cables are tied together at *C* and are loaded as shown. Knowing that $\alpha = 30°$, determine the tension (*a*) in cable *AC*, (*b*) in cable *BC*.

Fig. P2.46

2.47 A chairlift has been stopped in the position shown. Knowing that each chair weighs 300 N and that the skier in chair E weighs 890 N, determine the weight of the skier in chair F.

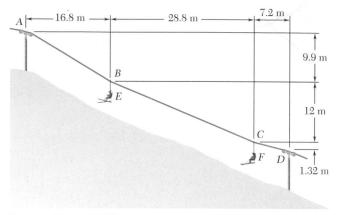

Fig. *P2.47* and *P2.48*

2.48 A chairlift has been stopped in the position shown. Knowing that each chair weighs 300 N and that the skier in chair F weighs 800 N, determine the weight of the skier in chair E.

2.49 Four wooden members are joined with metal plate connectors and are in equilibrium under the action of the four forces shown. Knowing that F_A = 510 lb and F_B = 480 lb, determine the magnitudes of the other two forces.

Fig. P2.49 and P2.50

2.50 Four wooden members are joined with metal plate connectors and are in equilibrium under the action of the four forces shown. Knowing that F_A = 420 lb and F_C = 540 lb, determine the magnitudes of the other two forces.

2.51 Two forces **P** and **Q** are applied as shown to an aircraft connection. Knowing that the connection is in equilibrium and that P = 400 lb and Q = 520 lb, determine the magnitudes of the forces exerted on the rods A and B.

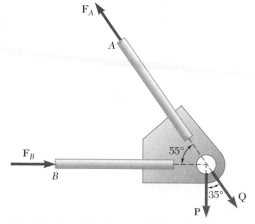

Fig. *P2.51* and *P2.52*

2.52 Two forces **P** and **Q** are applied as shown to an aircraft connection. Knowing that the connection is in equilibrium and that the magnitudes of the forces exerted on rods A and B are F_A = 600 lb and F_B = 320 lb, determine the magnitudes of **P** and **Q**.

2.53 Two cables tied together at C are loaded as shown. Knowing that $W = 840$ N, determine the tension (a) in cable AC, (b) in cable BC.

2.54 Two cables tied together at C are loaded as shown. Determine the range of values of W for which the tension will not exceed 1050 N in either cable.

Fig. P2.53 and P2.54

2.55 The cabin of an aerial tramway is suspended from a set of wheels that can roll freely on the support cable ACB and is being pulled at a constant speed by cable DE. Knowing that $\alpha = 40°$ and $\beta = 35°$, that the combined weight of the cabin, its support system, and its passengers is 24.8 kN, and assuming the tension in cable DF to be negligible, determine the tension (a) in the support cable ACB, (b) in the traction cable DE.

2.56 The cabin of an aerial tramway is suspended from a set of wheels that can roll freely on the support cable ACB and is being pulled at a constant speed by cable DE. Knowing that $\alpha = 42°$ and $\beta = 32°$, that the tension in cable DE is 20 kN, and assuming the tension in cable DF to be negligible, determine (a) the combined weight of the cabin, its support system, and its passengers, (h) the tension in the support cable ACB.

2.57 A block of weight W is suspended from a 500-mm long cord and two springs of which the unstretched lengths are 450 mm. Knowing that the constants of the springs are $k_{AB} = 1500$ N/m and $k_{AD} = 500$ N/m, determine (a) the tension in the cord, (b) the weight of the block.

Fig. P2.55 and P2.56

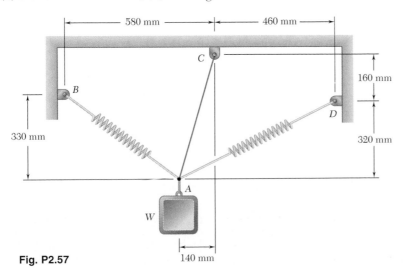

Fig. P2.57

2.58 A load of weight 400 N is suspended from a spring and two cords which are attached to blocks of weights $3W$ and W as shown. Knowing that the constant of the spring is 800 N/m, determine (a) the value of W, (b) the unstretched length of the spring.

Fig. P2.58

2.59 For the cables and loading of Prob. 2.46, determine (a) the value of α for which the tension in cable BC is as small as possible, (b) the corresponding value of the tension.

2.60 Knowing that portions AC and BC of cable ACB must be equal, determine the shortest length of cable which can be used to support the load shown if the tension in the cable is not to exceed 725 N.

2.61 Two cables tied together at C are loaded as shown. Knowing that the maximum allowable tension in each cable is 200 lb, determine (a) the magnitude of the largest force $\mathbf{P}$ which may be applied at C, (b) the corresponding value of α.

2.62 Two cables tied together at C are loaded as shown. Knowing that the maximum allowable tension is 300 lb in cable AC and 150 lb in cable BC, determine (a) the magnitude of the largest force $\mathbf{P}$ which may be applied at C, (b) the corresponding value of α.

2.63 For the structure and loading of Prob. 2.45, determine (a) the value of α for which the tension in cable BC is as small as possible, (b) the corresponding value of the tension.

2.64 Boom AB is supported by cable BC and a hinge at A. Knowing that the boom exerts on pin B a force directed along the boom and that the tension in rope BD is 70 lb, determine (a) the value of α for which the tension in cable BC is as small as possible, (b) the corresponding value of the tension.

2.65 Collar A shown in Fig. P2.65 and P2.66 can slide on a frictionless vertical rod and is attached as shown to a spring. The constant of the spring is 660 N/m, and the spring is unstretched when $h = 300$ mm. Knowing that the system is in equilibrium when $h = 400$ mm, determine the weight of the collar.

Fig. P2.60

Fig. P2.61 and *P2.62*

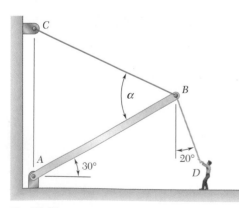

Fig. P2.64

2.66 The 40-N collar A can slide on a frictionless vertical rod and is attached as shown to a spring. The spring is unstretched when $h = 300$ mm. Knowing that the constant of the spring is 560 N/m, determine the value of h for which the system is in equilibrium.

2.67 A 280-kg crate is supported by several rope-and-pulley arrangements as shown. Determine for each arrangement the tension in the rope. (*Hint:* The tension in the rope is the same on each side of a simple pulley. This can be proved by the methods of Chap. 4.)

Fig. P2.65 and P2.66

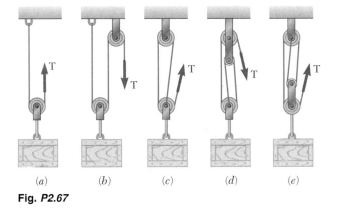

Fig. **P2.67**

2.68 Solve parts b and d of Prob. 2.67 assuming that the free end of the rope is attached to the crate.

2.69 A 350-lb load is supported by the rope-and-pulley arrangement shown. Knowing that $\beta = 25°$, determine the magnitude and direction of the force **P** which should be exerted on the free end of the rope to maintain equilibrium. (See the hint of Prob. 2.67.)

2.70 A 350-lb load is supported by the rope-and-pulley arrangement shown. Knowing that $\alpha = 35°$, determine (*a*) the angle β, (*b*) the magnitude of the force **P** which should be exerted on the free end of the rope to maintain equilibrium. (See the hint of Prob. 2.67.)

2.71 A load **Q** is applied to the pulley C, which can roll on the cable ACB. The pulley is held in the position shown by a second cable CAD, which passes over the pulley A and supports a load **P**. Knowing that $P = 800$ N, determine (*a*) the tension in cable ACB, (*b*) the magnitude of load **Q**.

2.72 A 2000-N load **Q** is applied to the pulley C, which can roll on the cable ACB. The pulley is held in the position shown by a second cable CAD, which passes over the pulley A and supports a load **P**. Determine (*a*) the tension in cable ACB, (*b*) the magnitude of load **P**.

Fig. P2.69 and P2.70

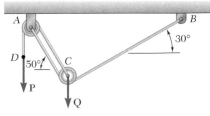

Fig. P2.71 and *P2.72*

FORCES IN SPACE

2.12. RECTANGULAR COMPONENTS OF A FORCE IN SPACE

The problems considered in the first part of this chapter involved only two dimensions; they could be formulated and solved in a single plane. In this section and in the remaining sections of the chapter, we will discuss problems involving the three dimensions of space.

Consider a force **F** acting at the origin O of the system of rectangular coordinates x, y, z. To define the direction of **F**, we draw the

Fig. 2.30

Fig. 2.31

vertical plane $OBAC$ containing $\mathbf{F}$ (Fig. 2.30*a*). This plane passes through the vertical y axis; its orientation is defined by the angle ϕ it forms with the xy plane. The direction of $\mathbf{F}$ within the plane is defined by the angle θ_y that $\mathbf{F}$ forms with the y axis. The force $\mathbf{F}$ may be resolved into a vertical component $\mathbf{F}_y$ and a horizontal component $\mathbf{F}_h$; this operation, shown in Fig. 2.30*b*, is carried out in plane $OBAC$ according to the rules developed in the first part of the chapter. The corresponding scalar components are

$$F_y = F \cos \theta_y \qquad F_h = F \sin \theta_y \qquad (2.16)$$

But $\mathbf{F}_h$ may be resolved into two rectangular components $\mathbf{F}_x$ and $\mathbf{F}_z$ along the x and z axes, respectively. This operation, shown in Fig. 2.30*c*, is carried out in the xz plane. We obtain the following expressions for the corresponding scalar components:

$$F_x = F_h \cos \phi = F \sin \theta_y \cos \phi$$
$$F_z = F_h \sin \phi = F \sin \theta_y \sin \phi \qquad (2.17)$$

The given force $\mathbf{F}$ has thus been resolved into three rectangular vector components $\mathbf{F}_x$, $\mathbf{F}_y$, $\mathbf{F}_z$, which are directed along the three coordinate axes.

Applying the Pythagorean theorem to the triangles OAB and OCD of Fig. 2.30, we write

$$F^2 = (OA)^2 = (OB)^2 + (BA)^2 = F_y^2 + F_h^2$$
$$F_h^2 = (OC)^2 = (OD)^2 + (DC)^2 = F_x^2 + F_z^2$$

Eliminating F_h^2 from these two equations and solving for F, we obtain the following relation between the magnitude of $\mathbf{F}$ and its rectangular scalar components:

$$F = \sqrt{F_x^2 + F_y^2 + F_z^2} \qquad (2.18)$$

The relationship existing between the force $\mathbf{F}$ and its three components $\mathbf{F}_x$, $\mathbf{F}_y$, $\mathbf{F}_z$ is more easily visualized if a "box" having $\mathbf{F}_x$, $\mathbf{F}_y$, $\mathbf{F}_z$ for edges is drawn as shown in Fig. 2.31. The force $\mathbf{F}$ is then represented by the diagonal OA of this box. Figure 2.31*b* shows the right triangle OAB used to derive the first of the formulas (2.16): $F_y = F \cos \theta_y$. In Fig. 2.31*a* and *c*, two other right triangles have also been drawn: OAD and OAE. These triangles are seen to occupy in the box positions comparable with that of triangle OAB. Denoting by θ_x and θ_z, respectively, the angles that $\mathbf{F}$ forms with the x and z axes, we can derive two formulas similar to $F_y = F \cos \theta_y$. We thus write

$$F_x = F \cos \theta_x \qquad F_y = F \cos \theta_y \qquad F_z = F \cos \theta_z \qquad (2.19)$$

The three angles θ_x, θ_y, θ_z define the direction of the force $\mathbf{F}$; they are more commonly used for this purpose than the angles θ_y and ϕ introduced at the beginning of this section. The cosines of θ_x, θ_y, θ_z are known as the *direction cosines* of the force $\mathbf{F}$.

Introducing the unit vectors $\mathbf{i}$, $\mathbf{j}$, and $\mathbf{k}$, directed respectively along the x, y, and z axes (Fig. 2.32), we can express $\mathbf{F}$ in the form

$$\mathbf{F} = F_x\mathbf{i} + F_y\mathbf{j} + F_z\mathbf{k} \qquad (2.20)$$

where the scalar components F_x, F_y, F_z are defined by the relations (2.19).

Example 1. A force of 500 N forms angles of 60°, 45°, and 120°, respectively, with the x, y, and z axes. Find the components F_x, F_y, and F_z of the force.

Substituting $F = 500$ N, $\theta_x = 60°$, $\theta_y = 45°$, $\theta_z = 120°$ into formulas (2.19), we write

$$F_x = (500 \text{ N}) \cos 60° = +250 \text{ N}$$
$$F_y = (500 \text{ N}) \cos 45° = +354 \text{ N}$$
$$F_z = (500 \text{ N}) \cos 120° = -250 \text{ N}$$

Carrying into Eq. (2.20) the values obtained for the scalar components of **F**, we have

$$\mathbf{F} = (250 \text{ N})\mathbf{i} + (354 \text{ N})\mathbf{j} - (250 \text{ N})\mathbf{k}$$

As in the case of two-dimensional problems, a plus sign indicates that the component has the same sense as the corresponding axis, and a minus sign indicates that it has the opposite sense.

The angle a force **F** forms with an axis should be measured from the positive side of the axis and will always be between 0 and 180°. An angle θ_x smaller than 90° (acute) indicates that **F** (assumed attached to O) is on the same side of the yz plane as the positive x axis; $\cos \theta_x$ and F_x will then be positive. An angle θ_x larger than 90° (obtuse) indicates that **F** is on the other side of the yz plane, $\cos \theta_x$ and F_x will then be negative. In Example 1 the angles θ_x and θ_y are acute, while θ_z is obtuse; consequently, F_x and F_y are positive, while F_z is negative.

Substituting into (2.20) the expressions obtained for F_x, F_y, F_z in (2.19), we write

$$\mathbf{F} = F(\cos \theta_x \mathbf{i} + \cos \theta_y \mathbf{j} + \cos \theta_z \mathbf{k}) \qquad (2.21)$$

which shows that the force **F** can be expressed as the product of the scalar F and the vector

$$\boldsymbol{\lambda} = \cos \theta_x \mathbf{i} + \cos \theta_y \mathbf{j} + \cos \theta_z \mathbf{k} \qquad (2.22)$$

Clearly, the vector $\boldsymbol{\lambda}$ is a vector whose magnitude is equal to 1 and whose direction is the same as that of **F** (Fig. 2.33). The vector $\boldsymbol{\lambda}$ is referred to as the *unit vector* along the line of action of **F**. It follows from (2.22) that the components of the unit vector $\boldsymbol{\lambda}$ are respectively equal to the direction cosines of the line of action of **F**:

$$\lambda_x = \cos \theta_x \qquad \lambda_y = \cos \theta_y \qquad \lambda_z = \cos \theta_z \qquad (2.23)$$

We should observe that the values of the three angles θ_x, θ_y, θ_z are not independent. Recalling that the sum of the squares of the components of a vector is equal to the square of its magnitude, we write

$$\lambda_x^2 + \lambda_y^2 + \lambda_z^2 = 1$$

or, substituting for λ_x, λ_y, λ_z from (2.23),

$$\cos^2 \theta_x + \cos^2 \theta_y + \cos^2 \theta_z = 1 \qquad (2.24)$$

In Example 1, for instance, once the values $\theta_x = 60°$ and $\theta_y = 45°$ have been selected, the value of θ_z *must* be equal to 60° or 120° in order to satisfy identity (2.24).

When the components F_x, F_y, F_z of a force **F** are given, the magnitude F of the force is obtained from (2.18). The relations (2.19) can then be solved for the direction cosines,

$$\cos \theta_x = \frac{F_x}{F} \qquad \cos \theta_y = \frac{F_y}{F} \qquad \cos \theta_z = \frac{F_z}{F} \qquad (2.25)$$

and the angles θ_x, θ_y, θ_z characterizing the direction of **F** can be found.

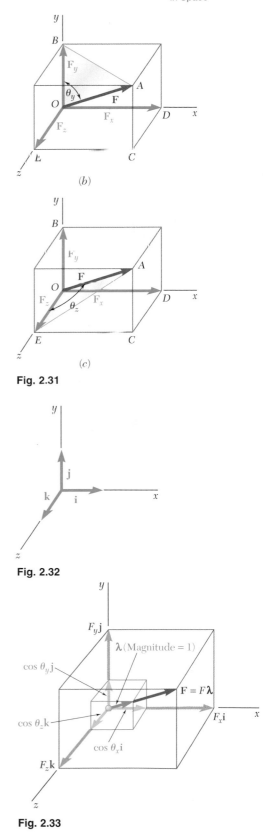

Fig. 2.31

Fig. 2.32

Fig. 2.33

Example 2. A force $\mathbf{F}$ has the components $F_x = 20$ lb, $F_y = -30$ lb, $F_z = 60$ lb. Determine its magnitude F and the angles θ_x, θ_y, θ_z it forms with the coordinate axes.

From formula (2.18) we obtain†

$$F = \sqrt{F_x^2 + F_y^2 + F_z^2}$$
$$= \sqrt{(20 \text{ lb})^2 + (-30 \text{ lb})^2 + (60 \text{ lb})^2}$$
$$= \sqrt{4900} \text{ lb} = 70 \text{ lb}$$

Substituting the values of the components and magnitude of $\mathbf{F}$ into Eqs. (2.25), we write

$$\cos \theta_x = \frac{F_x}{F} = \frac{20 \text{ lb}}{70 \text{ lb}} \qquad \cos \theta_y = \frac{F_y}{F} = \frac{-30 \text{ lb}}{70 \text{ lb}} \qquad \cos \theta_z = \frac{F_z}{F} = \frac{60 \text{ lb}}{70 \text{ lb}}$$

Calculating successively each quotient and its arc cosine, we obtain

$$\theta_x = 73.4° \qquad \theta_y = 115.4° \qquad \theta_z = 31.0°$$

These computations can be carried out easily with a calculator.

2.13. FORCE DEFINED BY ITS MAGNITUDE AND TWO POINTS ON ITS LINE OF ACTION

In many applications, the direction of a force $\mathbf{F}$ is defined by the coordinates of two points, $M(x_1, y_1, z_1)$ and $N(x_2, y_2, z_2)$, located on its line of action (Fig. 2.34). Consider the vector $\overrightarrow{MN}$ joining M and N

Fig. 2.34

and of the same sense as $\mathbf{F}$. Denoting its scalar components by d_x, d_y, d_z, respectively, we write

$$\overrightarrow{MN} = d_x\mathbf{i} + d_y\mathbf{j} + d_z\mathbf{k} \tag{2.26}$$

The unit vector $\boldsymbol{\lambda}$ along the line of action of $\mathbf{F}$ (i.e., along the line MN) may be obtained by dividing the vector $\overrightarrow{MN}$ by its magnitude MN. Substituting for $\overrightarrow{MN}$ from (2.26) and observing that MN is equal to the distance d from M to N, we write

$$\boldsymbol{\lambda} = \frac{\overrightarrow{MN}}{MN} = \frac{1}{d}(d_x\mathbf{i} + d_y\mathbf{j} + d_z\mathbf{k}) \tag{2.27}$$

---

†With a calculator programmed to convert rectangular coordinates into polar coordinates, the following procedure will be found more expeditious for computing F: First determine F_h from its two rectangular components F_x and F_z (Fig. 2.30c), then determine F from its two rectangular components F_h and F_y (Fig. 2.30b). The actual order in which the three components F_x, F_y, F_z are entered is immaterial.

Recalling that $\mathbf{F}$ is equal to the product of F and λ, we have

$$\mathbf{F} = F\lambda = \frac{F}{d}(d_x\mathbf{i} + d_y\mathbf{j} + d_z\mathbf{k}) \qquad (2.28)$$

from which it follows that the scalar components of $\mathbf{F}$ are, respectively,

$$F_x = \frac{Fd_x}{d} \qquad F_y = \frac{Fd_y}{d} \qquad F_z = \frac{Fd_z}{d} \qquad (2.29)$$

The relations (2.29) considerably simplify the determination of the components of a force $\mathbf{F}$ of given magnitude F when the line of action of $\mathbf{F}$ is defined by two points M and N. Subtracting the coordinates of M from those of N, we first determine the components of the vector $\overrightarrow{MN}$ and the distance d from M to N:

$$d_x = x_2 - x_1 \qquad d_y = y_2 - y_1 \qquad d_z = z_2 - z_1$$
$$d = \sqrt{d_x^2 + d_y^2 + d_z^2}$$

Substituting for F and for d_x, d_y, d_z, and d into the relations (2.29), we obtain the components F_x, F_y, F_z of the force.

The angles θ_x, θ_y, θ_z that $\mathbf{F}$ forms with the coordinate axes can then be obtained from Eqs. (2.25). Comparing Eqs. (2.22) and (2.27), we can also write

$$\cos\theta_x = \frac{d_x}{d} \qquad \cos\theta_y = \frac{d_y}{d} \qquad \cos\theta_z = \frac{d_z}{d} \qquad (2.30)$$

and determine the angles θ_x, θ_y, θ_z directly from the components and magnitude of the vector $\overrightarrow{MN}$.

2.14. ADDITION OF CONCURRENT FORCES IN SPACE

The resultant $\mathbf{R}$ of two or more forces in space will be determined by summing their rectangular components. Graphical or trigonometric methods are generally not practical in the case of forces in space.

The method followed here is similar to that used in Sec. 2.8 with coplanar forces. Setting

$$\mathbf{R} = \Sigma\mathbf{F}$$

we resolve each force into its rectangular components and write

$$R_x\mathbf{i} + R_y\mathbf{j} + R_z\mathbf{k} = \Sigma(F_x\mathbf{i} + F_y\mathbf{j} + F_z\mathbf{k})$$
$$= (\Sigma F_x)\mathbf{i} + (\Sigma F_y)\mathbf{j} + (\Sigma F_z)\mathbf{k}$$

from which it follows that

$$R_x = \Sigma F_x \qquad R_y = \Sigma F_y \qquad R_z = \Sigma F_z \qquad (2.31)$$

The magnitude of the resultant and the angles θ_x, θ_y, θ_z that the resultant forms with the coordinate axes are obtained using the method discussed in Sec. 2.12. We write

$$R = \sqrt{R_x^2 + R_y^2 + R_z^2} \qquad (2.32)$$

$$\cos\theta_x = \frac{R_x}{R} \qquad \cos\theta_y = \frac{R_y}{R} \qquad \cos\theta_z = \frac{R_z}{R} \qquad (2.33)$$

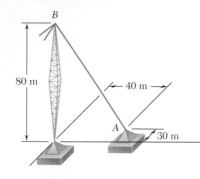

SAMPLE PROBLEM 2.7

A tower guy wire is anchored by means of a bolt at A. The tension in the wire is 2500 N. Determine (*a*) the components F_x, F_y, F_z of the force acting on the bolt, (*b*) the angles θ_x, θ_y, θ_z defining the direction of the force.

SOLUTION

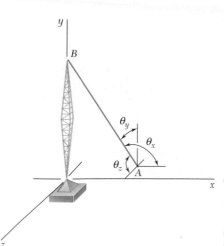

a. Components of the Force. The line of action of the force acting on the bolt passes through A and B, and the force is directed from A to B. The components of the vector $\overrightarrow{AB}$, which has the same direction as the force, are

$$d_x = -40 \text{ m} \qquad d_y = +80 \text{ m} \qquad d_z = +30 \text{ m}$$

The total distance from A to B is

$$AB = d = \sqrt{d_x^2 + d_y^2 + d_z^2} = 94.3 \text{ m}$$

Denoting by **i**, **j**, **k** the unit vectors along the coordinate axes, we have

$$\overrightarrow{AB} = -(40 \text{ m})\mathbf{i} + (80 \text{ m})\mathbf{j} + (30 \text{ m})\mathbf{k}$$

Introducing the unit vector $\boldsymbol{\lambda} = \overrightarrow{AB}/AB$, we write

$$\mathbf{F} = F\boldsymbol{\lambda} = F\frac{\overrightarrow{AB}}{AB} = \frac{2500 \text{ N}}{94.3 \text{ m}}\overrightarrow{AB}$$

Substituting the expression found for $\overrightarrow{AB}$, we obtain

$$\mathbf{F} = \frac{2500 \text{ N}}{94.3 \text{ m}}[-(40 \text{ m})\mathbf{i} + (80 \text{ m})\mathbf{j} + (30 \text{ m})\mathbf{k}]$$

$$\mathbf{F} = -(1060 \text{ N})\mathbf{i} + (2120 \text{ N})\mathbf{j} + (795 \text{ N})\mathbf{k}$$

The components of $\mathbf{F}$, therefore, are

$$F_x = -1060 \text{ N} \qquad F_y = +2120 \text{ N} \qquad F_z = +795 \text{ N} \quad \blacktriangleleft$$

b. Direction of the Force. Using Eqs. (2.25), we write

$$\cos \theta_x = \frac{F_x}{F} = \frac{-1060 \text{ N}}{2500 \text{ N}} \qquad \cos \theta_y = \frac{F_y}{F} = \frac{+2120 \text{ N}}{2500 \text{ N}}$$

$$\cos \theta_z = \frac{F_z}{F} = \frac{+795 \text{ N}}{2500 \text{ N}}$$

Calculating successively each quotient and its arc cosine, we obtain

$$\theta_x = 115.1° \qquad \theta_y = 32.0° \qquad \theta_z = 71.5° \quad \blacktriangleleft$$

(*Note:* This result could have been obtained by using the components and magnitude of the vector $\overrightarrow{AB}$ rather than those of the force $\mathbf{F}$.)

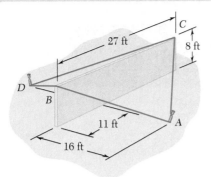

SAMPLE PROBLEM 2.8

A wall section of precast concrete is temporarily held by the cables shown. Knowing that the tension is 840 lb in cable AB and 1200 lb in cable AC, determine the magnitude and direction of the resultant of the forces exerted by cables AB and AC on stake A.

SOLUTION

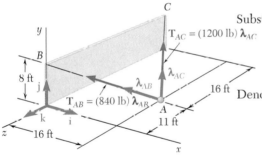

Components of the Forces.   The force exerted by each cable on stake A will be resolved into x, y, and z components. We first determine the components and magnitude of the vectors $\overrightarrow{AB}$ and $\overrightarrow{AC}$, measuring them from A toward the wall section. Denoting by $\mathbf{i}$, $\mathbf{j}$, $\mathbf{k}$ the unit vectors along the coordinate axes, we write

$$\overrightarrow{AB} = -(16 \text{ ft})\mathbf{i} + (8 \text{ ft})\mathbf{j} + (11 \text{ ft})\mathbf{k} \qquad AB = 21 \text{ ft}$$
$$\overrightarrow{AC} = -(16 \text{ ft})\mathbf{i} + (8 \text{ ft})\mathbf{j} - (16 \text{ ft})\mathbf{k} \qquad AC = 24 \text{ ft}$$

Denoting by $\boldsymbol{\lambda}_{AB}$ the unit vector along AB, we have

$$\mathbf{T}_{AB} = T_{AB}\boldsymbol{\lambda}_{AB} = T_{AB}\frac{\overrightarrow{AB}}{AB} = \frac{840 \text{ lb}}{21 \text{ ft}}\overrightarrow{AB}$$

Substituting the expression found for $\overrightarrow{AB}$, we obtain

$$\mathbf{T}_{AB} = \frac{840 \text{ lb}}{21 \text{ ft}}[-(16 \text{ ft})\mathbf{i} + (8 \text{ ft})\mathbf{j} + (11 \text{ ft})\mathbf{k}]$$
$$\mathbf{T}_{AB} = -(640 \text{ lb})\mathbf{i} + (320 \text{ lb})\mathbf{j} + (440 \text{ lb})\mathbf{k}$$

Denoting by $\boldsymbol{\lambda}_{AC}$ the unit vector along AC, we obtain in a similar way

$$\mathbf{T}_{AC} = T_{AC}\boldsymbol{\lambda}_{AC} = T_{AC}\frac{\overrightarrow{AC}}{AC} = \frac{1200 \text{ lb}}{24 \text{ ft}}\overrightarrow{AC}$$
$$\mathbf{T}_{AC} = -(800 \text{ lb})\mathbf{i} + (400 \text{ lb})\mathbf{j} - (800 \text{ lb})\mathbf{k}$$

Resultant of the Forces.   The resultant $\mathbf{R}$ of the forces exerted by the two cables is

$$\mathbf{R} = \mathbf{T}_{AB} + \mathbf{T}_{AC} = -(1440 \text{ lb})\mathbf{i} + (720 \text{ lb})\mathbf{j} - (360 \text{ lb})\mathbf{k}$$

The magnitude and direction of the resultant are now determined:

$$R = \sqrt{R_x^2 + R_y^2 + R_z^2} = \sqrt{(-1440)^2 + (720)^2 + (-360)^2}$$
$$R = 1650 \text{ lb} \quad \blacktriangleleft$$

From Eqs. (2.33) we obtain

$$\cos \theta_x = \frac{R_x}{R} = \frac{-1440 \text{ lb}}{1650 \text{ lb}} \qquad \cos \theta_y = \frac{R_y}{R} = \frac{+720 \text{ lb}}{1650 \text{ lb}}$$

$$\cos \theta_z = \frac{R_z}{R} = \frac{-360 \text{ lb}}{1650 \text{ lb}}$$

Calculating successively each quotient and its arc cosine, we have

$$\theta_x = 150.8° \qquad \theta_y = 64.1° \qquad \theta_z = 102.6° \quad \blacktriangleleft$$

SOLVING PROBLEMS
ON YOUR OWN

In this lesson we saw that a *force in space* can be defined by its magnitude and direction or by its three rectangular components F_x, F_y, and F_z.

A. When a force is defined by its magnitude and direction, its rectangular components F_x, F_y, and F_z can be found as follows:

Case 1. If the direction of the force **F** is defined by the angles θ_y and ϕ shown in Fig. 2.30, projections of **F** through these angles or their complements will yield the components of **F** [Eqs. (2.17)]. Note that the x and z components of **F** are found by first projecting **F** onto the horizontal plane; the projection $\mathbf{F}_h$ obtained in this way is then resolved into the components $\mathbf{F}_x$ and $\mathbf{F}_z$ (Fig. 2.30c).

When solving problems of this type, we strongly encourage you first to sketch the force **F** and then its projection $\mathbf{F}_h$ and components F_x, F_y, and F_z before beginning the mathematical part of the solution.

Case 2. If the direction of the force **F** is defined by the angles θ_x, θ_y, θ_z that **F** forms with the coordinate axes, each component can be obtained by multiplying the magnitude F of the force by the cosine of the corresponding angle [Example 1]:

$$F_x = F \cos \theta_x \qquad F_y = F \cos \theta_y \qquad F_z = F \cos \theta_z$$

Case 3. If the direction of the force **F** is defined by two points M and N located on its line of action (Fig. 2.34), you will first express the vector $\overrightarrow{MN}$ drawn from M to N in terms of its components d_x, d_y, d_z and the unit vectors **i**, **j**, **k**:

$$\overrightarrow{MN} = d_x\mathbf{i} + d_y\mathbf{j} + d_z\mathbf{k}$$

Next, you will determine the unit vector $\boldsymbol{\lambda}$ along the line of action of **F** by dividing the vector $\overrightarrow{MN}$ by its magnitude MN. Multiplying $\boldsymbol{\lambda}$ by the magnitude of **F**, you will obtain the desired expression for **F** in terms of its rectangular components [Sample Prob. 2.7]:

$$\mathbf{F} = F\boldsymbol{\lambda} = \frac{F}{d}(d_x\mathbf{i} + d_y\mathbf{j} + d_z\mathbf{k})$$

It is advantageous to use a consistent and meaningful system of notation when determining the rectangular components of a force. The method used in this text is illustrated in Sample Prob. 2.8 where, for example, the force $\mathbf{T}_{AB}$ acts from stake A toward point B. Note that the subscripts have been ordered to agree with the direction of the force. It is recommended that you adopt the same notation, as it will help you identify point 1 (the first subscript) and point 2 (the second subscript).

When forming the vector defining the line of action of a force, you may think of its scalar components as the number of steps you must take in each coordinate direction to go from point 1 to point 2. It is essential that you always remember to assign the correct sign to each of the components.

B.   When a force is defined by its rectangular components F_x, $\mathbf{F}_y$, $\mathbf{F}_z$, you can obtain its magnitude F by writing

$$F = \sqrt{F_x^2 + F_y^2 + F_z^2}$$

You can determine the direction cosines of the line of action of $\mathbf{F}$ by dividing the components of the force by F:

$$\cos \theta_x = \frac{F_x}{F} \qquad \cos \theta_y = \frac{F_y}{F} \qquad \cos \theta_z = \frac{F_z}{F}$$

From the direction cosines you can obtain the angles θ_x, θ_y, θ_z that $\mathbf{F}$ forms with the coordinate axes [Example 2].

C.   To determine the resultant R of two or more forces in three-dimensional space, first determine the rectangular components of each force by one of the procedures described above. Adding these components will yield the components R_x, R_y, R_z of the resultant. The magnitude and direction of the resultant can then be obtained as indicated above for the force $\mathbf{F}$ [Sample Prob. 2.8].

Fig. P2.73 and *P2.74*

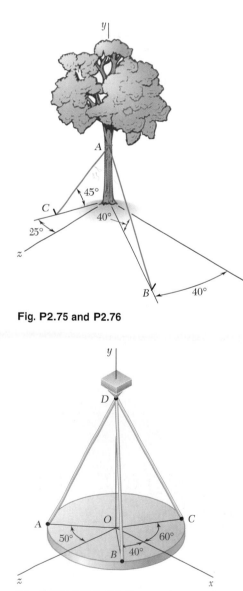

Fig. P2.75 and P2.76

2.73 Determine (*a*) the *x*, *y*, and *z* components of the 200-lb force, (*b*) the angles θ_x, θ_y, and θ_z that the force forms with the coordinate axes.

2.74 Determine (*a*) the *x*, *y*, and *z* components of the 420-lb force, (*b*) the angles θ_x, θ_y, and θ_z that the force forms with the coordinate axes.

2.75 To stabilize a tree partially uprooted in a storm, cables *AB* and *AC* are attached to the upper trunk of the tree and then are fastened to steel rods anchored in the ground. Knowing that the tension in cable *AB* is 4.2 kN, determine (*a*) the components of the force exerted by this cable on the tree, (*b*) the angles θ_x, θ_y, and θ_z that the force forms with axes at *A* which are parallel to the coordinate axes.

2.76 To stabilize a tree partially uprooted in a storm, cables *AB* and *AC* are attached to the upper trunk of the tree and then are fastened to steel rods anchored in the ground. Knowing that the tension in cable *AC* is 3.6 kN, determine (*a*) the components of the force exerted by this cable on the tree, (*b*) the angles θ_x, θ_y, and θ_z that the force forms with axes at *A* which are parallel to the coordinate axes.

2.77 A horizontal circular plate is suspended as shown from three wires which are attached to a support at *D* and form 30° angles with the vertical. Knowing that the *x* component of the force exerted by wire *AD* on the plate is 220.6 N, determine (*a*) the tension in wire *AD*, (*b*) the angles θ_x, θ_y, and θ_z that the force exerted at *A* forms with the coordinate axes.

2.78 A horizontal circular plate is suspended as shown from three wires which are attached to a support at *D* and form 30° angles with the vertical. Knowing that the *z* component of the force exerted by wire *BD* on the plate is −64.28 N, determine (*a*) the tension in wire *BD*, (*b*) the angles θ_x, θ_y, and θ_z that the force exerted at *B* forms with the coordinate axes.

2.79 A horizontal circular plate is suspended as shown from three wires which are attached to a support at *D* and form 30° angles with the vertical. Knowing that the tension in wire *CD* is 120 lb, determine (*a*) the components of the force exerted by this wire on the plate, (*b*) the angles θ_x, θ_y, and θ_z that the force forms with the coordinate axes.

2.80 A horizontal circular plate is suspended as shown from three wires which are attached to a support at *D* and form 30° angles with the vertical. Knowing that the *x* component of the force exerted by wire *CD* on the plate is −40 lb, determine (*a*) the tension in wire *CD*, (*b*) the angles θ_x, θ_y, and θ_z that the force exerted at *C* forms with the coordinate axes.

2.81 Determine the magnitude and direction of the force $\mathbf{F} = (800\ \text{lb})\mathbf{i} + (260\ \text{lb})\mathbf{j} - (320\ \text{lb})\mathbf{k}$.

Fig. P2.77, P2.78, P2.79, and *P2.80*

2.82 Determine the magnitude and direction of the force $\mathbf{F} = (400 \text{ N})\mathbf{i} - (1200 \text{ N})\mathbf{j} + (300 \text{ N})\mathbf{k}$.

2.83 A force acts at the origin of a coordinate system in a direction defined by the angles $\theta_x = 64.5°$ and $\theta_z = 55.9°$. Knowing that the y component of the force is -200 N, determine (a) the angle θ_y, (b) the other components and the magnitude of the force.

2.84 A force acts at the origin of a coordinate system in a direction defined by the angles $\theta_x = 75.4°$ and $\theta_y = 132.6°$. Knowing that the z component of the force is -60 N, determine (a) the angle θ_z, (b) the other components and the magnitude of the force.

2.85 A force $\mathbf{F}$ of magnitude 400 N acts at the origin of a coordinate system. Knowing that $\theta_x = 28.5°$, $F_y = -80$ N, and $F_z > 0$, determine (a) the components F_x and F_z, (b) the angles θ_y and θ_z.

2.86 A force $\mathbf{F}$ of magnitude 600 lb acts at the origin of a coordinate system. Knowing that $F_x = 200$ lb, $\theta_z = 136.8°$, and $F_y < 0$, determine (a) the components F_y and F_z, (b) the angles θ_x and θ_y.

2.87 A transmission tower is held by three guy wires anchored by bolts at B, C, and D. If the tension in wire AB is 2100 N, determine the components of the force exerted by the wire on the bolt at B.

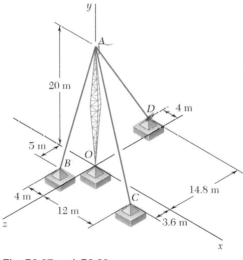

Fig. P2.87 and *P2.88*

2.88 A transmission tower is held by three guy wires anchored by bolts at B, C, and D. If the tension in wire AD is 1260 N, determine the components of the force exerted by the wire on the bolt at D.

2.89 A rectangular plate is supported by three cables as shown. Knowing that the tension in cable AB is 204 lb, determine the components of the force exerted on the plate at B.

2.90 A rectangular plate is supported by three cables as shown. Knowing that the tension in cable AD is 195 lb, determine the components of the forces exerted on the plate at D.

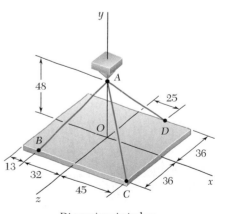

Dimensions in inches

Fig. P2.89 and P2.90

2.91 A steel rod is bent into a semicircular ring of radius 0.96 m and is supported in part by cables *BD* and *BE* which are attached to the ring at *B*. Knowing that the tension in cable *BD* is 220 N, determine the components of this force exerted by the cable on the support at *D*.

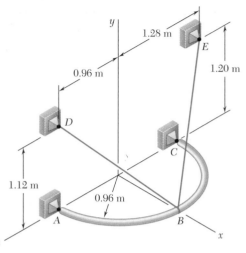

Fig. P2.91 and P2.92

2.92 A steel rod is bent into a semicircular ring of radius 0.96 m and is supported in part by cables *BD* and *BE* which are attached to the ring at *B*. Knowing that the tension in cable *BE* is 250 N, determine the components of this force exerted by the cable on the support at *E*.

2.93 Find the magnitude and direction of the resultant of the two forces shown knowing that $P = 500$ lb and $Q = 600$ lb.

2.94 Find the magnitude and direction of the resultant of the two forces shown knowing that $P = 600$ N and $Q = 400$ N.

2.95 Knowing that the tension is 850 N in cable *AB* and 1020 N in cable *AC*, determine the magnitude and direction of the resultant of the forces exerted at *A* by the two cables.

Fig. *P2.93* and P2.94

Fig. P2.95

2.96 Assuming that in Prob. 2.95 the tension is 1020 N in cable *AB* and 850 N in cable *AC*, determine the magnitude and direction of the resultant of the forces exerted at *A* by the two cables.

2.97 For the semicircular ring of Prob. 2.91, determine the magnitude and direction of the resultant of the forces exerted by the cables at *B* knowing that the tensions in cables *BD* and *BE* are 220 N and 250 N, respectively.

2.98 To stabilize a tree partially uprooted in a storm, cables *AB* and *AC* are attached to the upper trunk of the tree and then are fastened to steel rods anchored in the ground. Knowing that the tension in *AB* is 920 lb and that the resultant of the forces exerted at *A* by cables *AB* and *AC* lies in the *yz* plane, determine (*a*) the tension in *AC*, (*b*) the magnitude and direction of the resultant of the two forces.

2.99 To stabilize a tree partially uprooted in a storm, cables *AB* and *AC* are attached to the upper trunk of the tree and then are fastened to steel rods anchored in the ground. Knowing that the tension in *AC* is 850 lb and that the resultant of the forces exerted at *A* by cables *AB* and *AC* lies in the *yz* plane, determine (*a*) the tension in *AB*, (*b*) the magnitude and direction of the resultant of the two forces.

2.100 For the plate of Prob. 2.89, determine the tension in cables *AB* and *AD* knowing that the tension in cable *AC* is 27 lb and that the resultant of the forces exerted by the three cables at *A* must be vertical.

Fig. P2.98 and *P2.99*

2.15. EQUILIBRIUM OF A PARTICLE IN SPACE

According to the definition given in Sec. 2.9, a particle *A* is in equilibrium if the resultant of all the forces acting on *A* is zero. The components R_x, R_y, R_z of the resultant are given by the relations (2.31); expressing that the components of the resultant are zero, we write

$$\Sigma F_x = 0 \qquad \Sigma F_y = 0 \qquad \Sigma F_z = 0 \qquad (2.34)$$

Equations (2.34) represent the necessary and sufficient conditions for the equilibrium of a particle in space. They can be used to solve problems dealing with the equilibrium of a particle involving no more than three unknowns.

To solve such problems, you first should draw a free-body diagram showing the particle in equilibrium and *all* the forces acting on it. You can then write the equations of equilibrium (2.34) and solve them for three unknowns. In the more common types of problems, these unknowns will represent (1) the three components of a single force or (2) the magnitude of three forces, each of known direction.

Photo 2.2 While the tension in the *four* cables supporting the shipping container cannot be found using the *three* equations of (2.34), a relation between the tensions can be obtained by considering the equilibrium of the hook.

SAMPLE PROBLEM 2.9

A 200-kg cylinder is hung by means of two cables AB and AC, which are attached to the top of a vertical wall. A horizontal force **P** perpendicular to the wall holds the cylinder in the position shown. Determine the magnitude of **P** and the tension in each cable.

SOLUTION

Free-body Diagram. Point A is chosen as a free body; this point is subjected to four forces, three of which are of unknown magnitude.

Introducing the unit vectors **i**, **j**, **k**, we resolve each force into rectangular components.

$$\mathbf{P} = P\mathbf{i}$$
$$\mathbf{W} = -mg\mathbf{j} = -(200 \text{ kg})(9.81 \text{ m/s}^2)\mathbf{j} = -(1962 \text{ N})\mathbf{j} \tag{1}$$

In the case of $\mathbf{T}_{AB}$ and $\mathbf{T}_{AC}$, it is necessary first to determine the components and magnitudes of the vectors $\overrightarrow{AB}$ and $\overrightarrow{AC}$. Denoting by $\boldsymbol{\lambda}_{AB}$ the unit vector along AB, we write

$$\overrightarrow{AB} = -(1.2 \text{ m})\mathbf{i} + (10 \text{ m})\mathbf{j} + (8 \text{ m})\mathbf{k} \qquad AB = 12.862 \text{ m}$$
$$\boldsymbol{\lambda}_{AB} = \frac{\overrightarrow{AB}}{12.862 \text{ m}} = -0.09330\mathbf{i} + 0.7775\mathbf{j} + 0.6220\mathbf{k}$$
$$\mathbf{T}_{AB} = T_{AB}\boldsymbol{\lambda}_{AB} = -0.09330T_{AB}\mathbf{i} + 0.7775T_{AB}\mathbf{j} + 0.6220T_{AB}\mathbf{k} \tag{2}$$

Denoting by $\boldsymbol{\lambda}_{AC}$ the unit vector along AC, we write in a similar way

$$\overrightarrow{AC} = -(1.2 \text{ m})\mathbf{i} + (10 \text{ m})\mathbf{j} - (10 \text{ m})\mathbf{k} \qquad AC = 14.193 \text{ m}$$
$$\boldsymbol{\lambda}_{AC} = \frac{\overrightarrow{AC}}{14.193 \text{ m}} = -0.08455\mathbf{i} + 0.7046\mathbf{j} - 0.7046\mathbf{k}$$
$$\mathbf{T}_{AC} = T_{AC}\boldsymbol{\lambda}_{AC} = -0.08455T_{AC}\mathbf{i} + 0.7046T_{AC}\mathbf{j} - 0.7046T_{AC}\mathbf{k} \tag{3}$$

Equilibrium Condition. Since A is in equilibrium, we must have

$$\Sigma\mathbf{F} = 0: \qquad \mathbf{T}_{AB} + \mathbf{T}_{AC} + \mathbf{P} + \mathbf{W} = 0$$

or, substituting from (1), (2), (3) for the forces and factoring **i**, **j**, **k**,

$$(-0.09330T_{AB} - 0.08455T_{AC} + P)\mathbf{i}$$
$$+ (0.7775T_{AB} + 0.7046T_{AC} - 1962 \text{ N})\mathbf{j}$$
$$+ (0.6220T_{AB} - 0.7046T_{AC})\mathbf{k} = 0$$

Setting the coefficients of **i**, **j**, **k** equal to zero, we write three scalar equations, which express that the sums of the x, y, and z components of the forces are respectively equal to zero.

$$\Sigma F_x = 0: \qquad -0.09330T_{AB} - 0.08455T_{AC} + P = 0$$
$$\Sigma F_y = 0: \qquad +0.7775T_{AB} + 0.7046T_{AC} - 1962 \text{ N} = 0$$
$$\Sigma F_z = 0: \qquad +0.6220T_{AB} - 0.7046T_{AC} = 0$$

Solving these equations, we obtain

$$P = 235 \text{ N} \qquad T_{AB} = 1402 \text{ N} \qquad T_{AC} = 1238 \text{ N} \qquad \blacktriangleleft$$

We saw earlier that when a particle is in *equilibrium*, the resultant of the forces acting on the particle must be zero. Expressing this fact in the case of the equilibrium of a *particle in three-dimensional space* will provide you with three relations among the forces acting on the particle. These relations can be used to determine three unknowns—usually the magnitudes of three forces.

Your solution will consist of the following steps.

1. Draw a free-body diagram of the particle. This diagram shows the particle and all the forces acting on it. Indicate on the diagram the magnitudes of known forces, as well as any angles or dimensions that define the direction of a force. Any unknown magnitude or angle should be denoted by an appropriate symbol. Nothing else should be included on your free-body diagram.

2. Resolve each of the forces into rectangular components. Following the method used in the preceding lesson, you will determine for each force **F** the unit vector **λ** defining the direction of that force and express **F** as the product of its magnitude F and the unit vector **λ**. When two points on the line of action of **F** are known, you will obtain an expression of the form

$$\mathbf{F} = F\boldsymbol{\lambda} = \frac{F}{d}(d_x\mathbf{i} + d_y\mathbf{j} + d_z\mathbf{k})$$

where d, d_x, d_y, and d_z are dimensions obtained from the free-body diagram of the particle. We also showed the direction of **F** can be defined in terms of the angles θ_y and ϕ. If a force is known in magnitude as well as in direction, then F is known and the expression obtained for **F** is completely defined; otherwise F is one of the three unknowns that should be determined.

3. Set the resultant, or sum, of the forces exerted on the particle equal to zero. You will obtain a vectorial equation consisting of terms containing the unit vectors **i**, **j**, or **k**. You will group the terms containing the same unit vector and factor that vector. For the vectorial equation to be satisfied, the coefficient of each of the unit vectors must be equal to zero. Thus, setting each coefficient equal to zero will yield three scalar equations that you can solve for no more than three unknowns [Sample Prob. 2.9].

2.101 The support assembly shown is bolted in place at B, C, and D and supports a downward force **P** at A. Knowing that the forces in members AB, AC, and AD are directed along the respective members and that the force in member AB is 146 N, determine the magnitude of **P**.

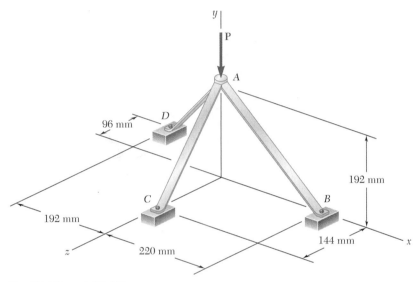

Fig. P2.101 and *P2.102*

2.102 The support assembly shown is bolted in place at B, C, and D and supports a downward force **P** at A. Knowing that the forces in members AB, AC, and AD are directed along the respective members and that $P = 200$ N, determine the forces in the members.

2.103 Three cables are used to tether a balloon as shown. Determine the vertical force **P** exerted by the balloon at A knowing that the tension in cable AB is 60 lb.

2.104 Three cables are used to tether a balloon as shown. Determine the vertical force **P** exerted by the balloon at A knowing that the tension in cable AC is 100 lb.

2.105 The crate shown in Fig. P2.105 and P2.108 is supported by three cables. Determine the weight of the crate knowing that the tension in cable AB is 3 kN.

2.106 For the crate of Prob. 2.105, determine the weight of the crate knowing that the tension in cable AD is 2.8 kN.

2.107 For the crate of Prob. 2.105, determine the weight of the crate knowing that the tension in cable AC is 2.4 kN.

Fig. P2.103 and P2.104

Fig. P2.105 and P2.108

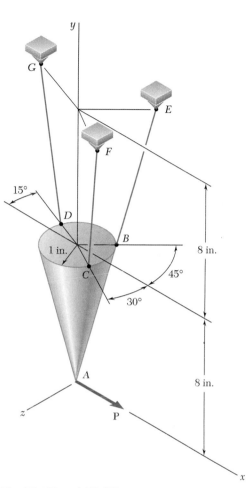

Fig. P2.109 and *P2.110*

2.108 A 750-kg crate is supported by three cables as shown. Determine the tension in each cable.

2.109 A force **P** is applied as shown to a uniform cone which is supported by three cords, where the lines of action of the cords pass through the vertex A of the cone. Knowing that P = 0 and that the tension in cord BE is 0.2 lb, determine the weight W of the cone.

2.110 A force **P** is applied as shown to a uniform cone which is supported by three cords, where the lines of action of the cords pass through the vertex A of the cone. Knowing that the cone weighs 1.6 lb, determine the range of values of P for which cord CF is taut.

2.111 A transmission tower is held by three guy wires attached to a pin at A and anchored by bolts at B, C, and D. If the tension in wire AB is 3.6 kN, determine the vertical force **P** exerted by the tower on the pin at A.

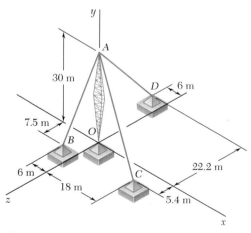

Fig. P2.111 and P2.112

2.112 A transmission tower is held by three guy wires attached to a pin at A and anchored by bolts at B, C, and D. If the tension in wire AC is 2.6 kN, determine the vertical force **P** exerted by the tower on the pin at A.

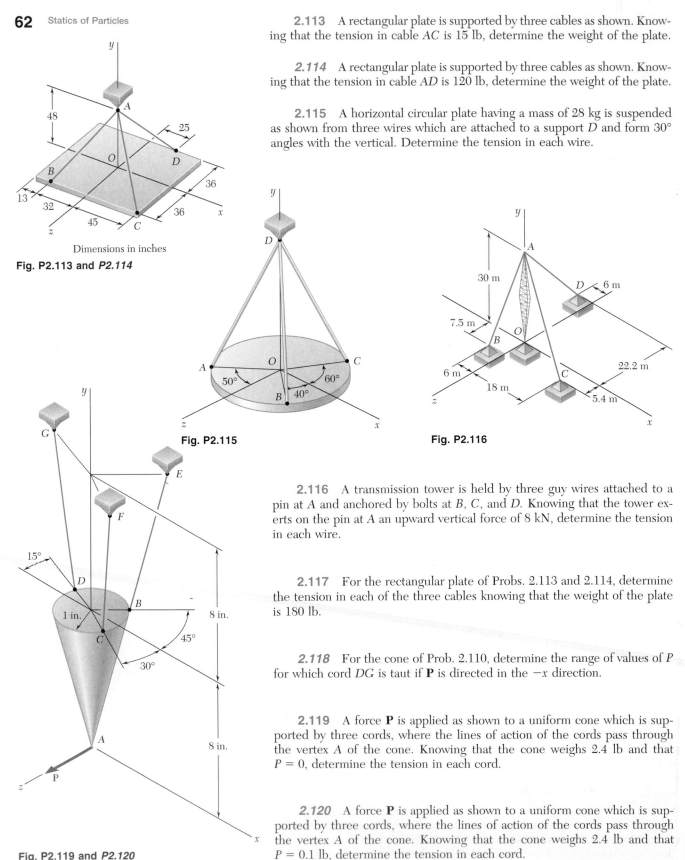

Dimensions in inches

Fig. P2.113 and *P2.114*

Fig. P2.115

Fig. P2.116

Fig. P2.119 and *P2.120*

2.113 A rectangular plate is supported by three cables as shown. Knowing that the tension in cable AC is 15 lb, determine the weight of the plate.

2.114 A rectangular plate is supported by three cables as shown. Knowing that the tension in cable AD is 120 lb, determine the weight of the plate.

2.115 A horizontal circular plate having a mass of 28 kg is suspended as shown from three wires which are attached to a support D and form 30° angles with the vertical. Determine the tension in each wire.

2.116 A transmission tower is held by three guy wires attached to a pin at A and anchored by bolts at B, C, and D. Knowing that the tower exerts on the pin at A an upward vertical force of 8 kN, determine the tension in each wire.

2.117 For the rectangular plate of Probs. 2.113 and 2.114, determine the tension in each of the three cables knowing that the weight of the plate is 180 lb.

2.118 For the cone of Prob. 2.110, determine the range of values of P for which cord DG is taut if **P** is directed in the −x direction.

2.119 A force **P** is applied as shown to a uniform cone which is supported by three cords, where the lines of action of the cords pass through the vertex A of the cone. Knowing that the cone weighs 2.4 lb and that P = 0, determine the tension in each cord.

2.120 A force **P** is applied as shown to a uniform cone which is supported by three cords, where the lines of action of the cords pass through the vertex A of the cone. Knowing that the cone weighs 2.4 lb and that P = 0.1 lb, determine the tension in each cord.

2.121 Using two ropes and a roller chute, two workers are unloading a 200-kg cast-iron counterweight from a truck. Knowing that at the instant shown the counterweight is kept from moving and that the positions of points *A*, *B*, and *C* are, respectively, *A*(0, −0.5 m, 1 m), *B*(−0.6 m, 0.8 m, 0), and *C*(0.7 m, 0.9 m, 0), and assuming that no friction exists between the counterweight and the chute, determine the tension in each rope. (*Hint:* Since there is no friction, the force exerted by the chute on the counterweight must be perpendicular to the chute.)

2.122 Solve Prob. 2.121 assuming that a third worker is exerting a force **P** = −(180 N)**i** on the counterweight.

2.123 A piece of machinery of weight *W* is temporarily supported by cables *AB*, *AC*, and *ADE*. Cable *ADE* is attached to the ring at *A*, passes over the pulley at *D* and back through the ring, and is attached to the support at *E*. Knowing that *W* = 320 lb, determine the tension in each cable. (*Hint:* The tension is the same in all portions of cable *ADE*.)

Fig. P2.121

Fig. P2.123 and P2.124

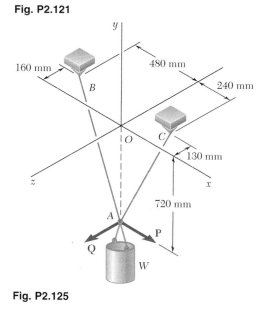

Fig. P2.125

2.124 A piece of machinery of weight *W* is temporarily supported by cables *AB*, *AC*, and *ADE*. Cable *ADE* is attached to the ring at *A*, passes over the pulley at *D* and back through the ring, and is attached to the support at *E*. Knowing that the tension in cable *AB* is 68 lb, determine (*a*) the tension in *AC*, (*b*) the tension in *ADE*, (*c*) the weight *W*. (*Hint:* The tension is the same in all portions of cable *ADE*.)

2.125 A container of weight *W* is suspended from ring *A*. Cable *BAC* passes through the ring and is attached to fixed supports at *B* and *C*. Two forces **P** = *P***i** and **Q** = *Q***k** are applied to the ring to maintain the container in the position shown. Knowing that *W* = 1200 N, determine *P* and *Q*. (*Hint:* The tension is the same in both portions of cable *BAC*.)

2.126 For the system of Prob. 2.125, determine *W* and *P* knowing that *Q* = 160 N.

2.127 Collars *A* and *B* are connected by a 1-m-long wire and can slide freely on frictionless rods. If a force **P** = (680 N)**j** is applied at *A*, determine (*a*) the tension in the wire when *y* = 300 mm, (*b*) the magnitude of the force **Q** required to maintain the equilibrium of the system.

2.128 Solve Prob. 2.127 assuming *y* = 550 mm.

Fig. P2.127

In this chapter we have studied the effect of forces on particles, that is, on bodies of such shape and size that all forces acting on them may be assumed applied at the same point.

Forces are *vector quantities*; they are characterized by a *point of application*, a *magnitude*, and a *direction*, and they add according to the *parallelogram law* (Fig. 2.35). The magnitude and direction of the resultant **R** of two forces **P** and **Q** can be determined either graphically or by trigonometry, using successively the law of cosines and the law of sines [Sample Prob. 2.1].

Any given force acting on a particle can be resolved into two or more *components*, that is, it can be replaced by two or more forces which have the same effect on the particle. A force **F** can be resolved into two components **P** and **Q** by drawing a parallelogram which has **F** for its diagonal; the components **P** and **Q** are then represented by the two adjacent sides of the parallelogram (Fig. 2.36) and can be determined either graphically or by trigonometry [Sec. 2.6].

A force **F** is said to have been resolved into two *rectangular components* if its components F_x and F_y are perpendicular to each other and are directed along the coordinate axes (Fig. 2.37). Introducing the *unit vectors* **i** and **j** along the x and y axes, respectively, we write [Sec. 2.7]

$$\mathbf{F}_x = F_x\mathbf{i} \qquad \mathbf{F}_y = F_y\mathbf{j} \tag{2.6}$$

and

$$\mathbf{F} = F_x\mathbf{i} + F_y\mathbf{j} \tag{2.7}$$

where F_x and F_y are the *scalar components* of **F**. These components, which can be positive or negative, are defined by the relations

$$F_x = F \cos \theta \qquad F_y = F \sin \theta \tag{2.8}$$

When the rectangular components F_x and F_y of a force **F** are given, the angle θ defining the direction of the force can be obtained by writing

$$\tan \theta = \frac{F_y}{F_x} \tag{2.9}$$

The magnitude F of the force can then be obtained by solving one of the equations (2.8) for F or by applying the Pythagorean theorem and writing

$$F = \sqrt{F_x^2 + F_y^2} \tag{2.10}$$

Resultant of two forces

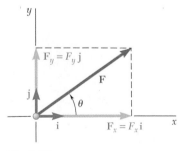

Fig. 2.35

Components of a force

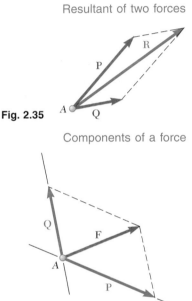

Fig. 2.36

Rectangular components
Unit vectors

Fig. 2.37

When *three or more coplanar forces* act on a particle, the rectangular components of their resultant **R** can be obtained by adding algebraically the corresponding components of the given forces [Sec. 2.8]. We have

$$R_x = \Sigma F_x \qquad R_y = \Sigma F_y \qquad (2.13)$$

The magnitude and direction of **R** can then be determined from relations similar to Eqs. (2.9) and (2.10) [Sample Prob. 2.3].

A force **F** in *three-dimensional space* can be resolved into rectangular components F_x, F_y, and F_z [Sec. 2.12]. Denoting by θ_x, θ_y, and θ_z, respectively, the angles that **F** forms with the x, y, and z axes (Fig. 2.38), we have

$$F_x = F \cos \theta_x \qquad F_y = F \cos \theta_y \qquad F_z = F \cos \theta_z \quad (2.19)$$

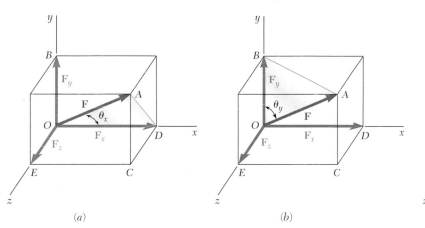

Fig. 2.38   (a)   (b)   (c)

The cosines of θ_x, θ_y, θ_z are known as the *direction cosines* of the force **F**. Introducing the unit vectors **i**, **j**, **k** along the coordinate axes, we write

$$\mathbf{F} = F_x\mathbf{i} + F_y\mathbf{j} + F_z\mathbf{k} \qquad (2.20)$$

or

$$\mathbf{F} = F(\cos \theta_x\mathbf{i} + \cos \theta_y\mathbf{j} + \cos \theta_z\mathbf{k}) \qquad (2.21)$$

which shows (Fig. 2.39) that **F** is the product of its magnitude F and the unit vector

$$\boldsymbol{\lambda} = \cos \theta_x\mathbf{i} + \cos \theta_y\mathbf{j} + \cos \theta_z\mathbf{k}$$

Since the magnitude of **λ** is equal to unity, we must have

$$\cos^2 \theta_x + \cos^2 \theta_y + \cos^2 \theta_z = 1 \qquad (2.24)$$

When the rectangular components F_x, F_y, F_z of a force **F** are given, the magnitude F of the force is found by writing

$$F = \sqrt{F_x^2 + F_y^2 + F_z^2} \qquad (2.18)$$

and the direction cosines of **F** are obtained from Eqs. (2.19). We have

$$\cos \theta_x = \frac{F_x}{F} \qquad \cos \theta_y = \frac{F_y}{F} \qquad \cos \theta_z = \frac{F_z}{F} \quad (2.25)$$

Fig. 2.39

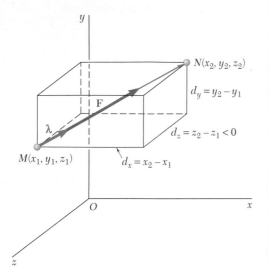

Fig. 2.40

When a force **F** is defined in three-dimensional space by its magnitude F and two points M and N on its line of action [Sec. 2.13], its rectangular components can be obtained as follows. We first express the vector $\overrightarrow{MN}$ joining points M and N in terms of its components d_x, d_y, and d_z (Fig. 2.40); we write

$$\overrightarrow{MN} = d_x\mathbf{i} + d_y\mathbf{j} + d_z\mathbf{k} \qquad (2.26)$$

We next determine the unit vector $\boldsymbol{\lambda}$ along the line of action of **F** by dividing $\overrightarrow{MN}$ by its magnitude $MN = d$:

$$\boldsymbol{\lambda} = \frac{\overrightarrow{MN}}{MN} = \frac{1}{d}(d_x\mathbf{i} + d_y\mathbf{j} + d_z\mathbf{k}) \qquad (2.27)$$

Recalling that **F** is equal to the product of F and $\boldsymbol{\lambda}$, we have

$$\mathbf{F} = F\boldsymbol{\lambda} = \frac{F}{d}(d_x\mathbf{i} + d_y\mathbf{j} + d_z\mathbf{k}) \qquad (2.28)$$

from which it follows [Sample Probs. 2.7 and 2.8] that the scalar components of **F** are, respectively,

$$F_x = \frac{Fd_x}{d} \qquad F_y = \frac{Fd_y}{d} \qquad F_z = \frac{Fd_z}{d} \qquad (2.29)$$

Resultant of forces in space

When *two or more forces* act on a particle in *three-dimensional space*, the rectangular components of their resultant **R** can be obtained by adding algebraically the corresponding components of the given forces [Sec. 2.14]. We have

$$R_x = \Sigma F_x \qquad R_y = \Sigma F_y \qquad R_z = \Sigma F_z \qquad (2.31)$$

The magnitude and direction of **R** can then be determined from relations similar to Eqs. (2.18) and (2.25) [Sample Prob. 2.8].

Equilibrium of a particle

A particle is said to be in *equilibrium* when the resultant of all the forces acting on it is zero [Sec. 2.9]. The particle will then remain at rest (if originally at rest) or move with constant speed in a straight line (if originally in motion) [Sec. 2.10].

Free-body diagram

To solve a problem involving a particle in equilibrium, one first should draw a *free-body diagram* of the particle showing all the forces acting on it [Sec. 2.11]. If *only three coplanar forces* act on the particle, a *force triangle* may be drawn to express that the particle is in equilibrium. Using graphical methods or trigonometry, this triangle can be solved for no more than two unknowns [Sample Prob. 2.4]. If *more than three coplanar forces* are involved, the equations of equilibrium

$$\Sigma F_x = 0 \qquad \Sigma F_y = 0 \qquad (2.15)$$

should be used. These equations can be solved for no more than two unknowns [Sample Prob. 2.6].

Equilibrium in space

When a particle is in *equilibrium in three-dimensional space* [Sec. 2.15], the three equations of equilibrium

$$\Sigma F_x = 0 \qquad \Sigma F_y = 0 \qquad \Sigma F_z = 0 \qquad (2.34)$$

should be used. These equations can be solved for no more than three unknowns [Sample Prob. 2.9].

Review Problems

2.129 Member *BD* exerts on member *ABC* a force **P** directed along line *BD*. Knowing that **P** must have a 300-lb horizontal component, determine (*a*) the magnitude of the force **P**, (*b*) its vertical component.

2.130 A container of weight *W* is suspended from ring *A*, to which cables *AC* and *AE* are attached. A force **P** is applied to the end *F* of a third cable which passes over a pulley at *B* and through ring *A* and which is attached to a support at *D*. Knowing that *W* = 1000 N, determine the magnitude of **P**. (*Hint:* The tension is the same in all portions of cable *FBAD*.)

Fig. P2.129

Fig. *P2.130* and P2.131

2.131 A container of weight *W* is suspended from ring *A*, to which cables *AC* and *AE* are attached. A force **P** is applied to the end *F* of a third cable which passes over a pulley at *B* and through ring *A* and which is attached to a support at *D*. Knowing that the tension in cable *AC* is 150 N, determine (*a*) the magnitude of the force **P**, (*b*) the weight *W* of the container. (See the hint for Prob. 2.130.)

2.132 Two cables tied together at *C* are loaded as shown. Knowing that *Q* = 60 lb, determine the tension (*a*) in cable *AC*, (*b*) in cable *BC*.

2.133 Two cables tied together at *C* are loaded as shown. Determine the range of values of *Q* for which the tension will not exceed 60 lb in either cable.

Fig. P2.132 and *P2.133*

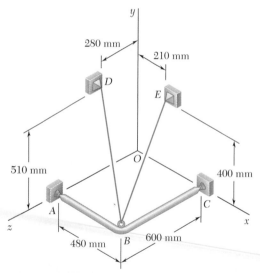

Fig. P2.134 and P2.135

Fig. *P2.138* and P2.139

2.134 A welded connection is in equilibrium under the action of the four forces shown. Knowing that $F_A = 8$ kN and $F_B = 16$ kN, determine the magnitudes of the other two forces.

2.135 A welded connection is in equilibrium under the action of the four forces shown. Knowing that $F_A = 5$ kN and $F_D = 6$ kN, determine the magnitudes of the other two forces.

2.136 Collar A is connected as shown to a 50-lb load and can slide on a frictionless horizontal rod. Determine the magnitude of the force **P** required to maintain the equilibrium of the collar when (*a*) $x = 4.5$ in., (*b*) $x = 15$ in.

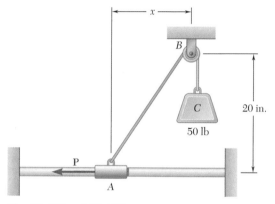

Fig. *P2.136* and P2.137

2.137 Collar A is connected as shown to a 50-lb load and can slide on a frictionless horizontal rod. Determine the distance x for which the collar is in equilibrium when $P = 48$ lb.

2.138 A frame ABC is supported in part by cable DBE which passes through a frictionless ring at B. Knowing that the tension in the cable is 385 N, determine the components of the force exerted by the cable on the support at D.

2.139 A frame ABC is supported in part by cable DBE which passes through a frictionless ring at B. Determine the magnitude and direction of the resultant of the forces exerted by the cable at B knowing that the tension in the cable is 385 N.

2.140 A steel tank is to be positioned in an excavation. Using trigonometry, determine (*a*) the magnitude and direction of the smallest force **P** for which the resultant **R** of the two forces applied at A is vertical, (*b*) the corresponding magnitude of **R**.

Fig. P2.140

Computer Problems

2.C1 Using computational software, determine the magnitude and direction of the resultant of n coplanar forces applied at a point A. Use this software to solve Probs. 2.32, 2.33, 2.34, and 2.36.

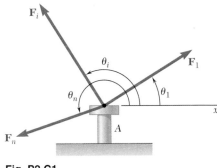

Fig. P2.C1

2.C2 A worker plans to lift a 60-lb 5-gallon bucket of paint by tying a rope to the scaffold at A and then passing the rope through the bail of the bucket at B and finally over the pulley at C. (*a*) Plot the tension in the rope as a function of the height y for 2 ft $\leq y \leq$ 18 ft. (*b*) Evaluate the worker's plan.

2.C3 The collar A can slide freely on the horizontal frictionless rod shown. The spring attached to the collar has a spring constant k and is undeformed when the collar is directly below support B. Express in terms of k and the distance c the magnitude of the force **P** required to maintain equilibrium of the system. Plot P as a function of c for values of c from 0 to 600 mm when (*a*) $k = 2$ N/mm, (*b*) $k = 3$ N/mm, (*c*) $k = 4$ N/mm.

Fig. P2.C2

Fig. P2.C3

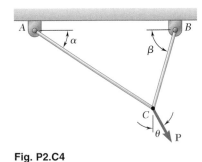

Fig. P2.C4

2.C4 A load **P** is supported by two cables as shown. Using computational software, determine the tension in each cable as a function of P and θ. For the following three sets of numerical values, plot the tensions for values of θ ranging from $\theta_1 = \beta - 90°$ to $\theta_2 = 90°$ and then determine from the graphs (a) the value of θ for which the tension in the two cables is as small as possible, (b) the corresponding value of the tension.

(1) $\alpha = 35°$, $\beta = 75°$, $P = 1.6$ kN
(2) $\alpha = 50°$, $\beta = 30°$, $P = 2.4$ kN
(3) $\alpha = 40°$, $\beta = 60°$, $P = 1.0$ kN

2.C5 Cables AC and BC are tied together at C and are loaded as shown. Knowing that $P = 100$ lb, (a) express the tension in each cable as a function of θ. (b) Plot the tension in each cable for $0 \le \theta \le 90°$. (c) From the graph obtained in part a, determine the smallest value of θ for which both cables are in tension.

Fig. P2.C5

Fig. P2.C6

2.C6 A container of weight W is suspended from ring A to which cable AB of length 5 m and spring AC are attached. The constant of the spring is 100 N/m, and its unstretched length is 3 m. Determine the tension in the cable when (a) $W = 120$ N, (b) $W = 160$ N.

2.C7 An acrobat is walking on a tightrope of length $L = 80.3$ ft attached to supports A and B at a distance of 80.0 ft from each other. The combined weight of the acrobat and his balancing pole is 200 lb, and the friction between his shoes and the rope is large enough to prevent him from slipping. Neglecting the weight of the rope and any elastic deformation, use computational software to determine the deflection y and the tension in portions AC and BC of the rope for values of x from 0.5 ft to 40 ft using 0.5-ft increments. From the results obtained, determine (a) the maximum deflection of the rope, (b) the maximum tension in the rope, (c) the minimum values of the tension in portions AC and BC of the rope.

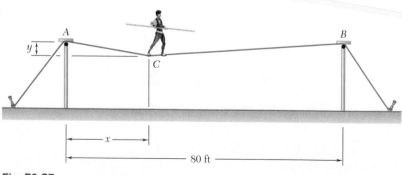

Fig. P2.C7

2.C8 The transmission tower shown is guyed by three cables attached to a pin at *A* and anchored at points *B*, *C*, and *D*. Cable *AD* is 21 m long and the tension in that cable is 20 kN. (*a*) Express the *x*, *y*, and *z* components of the force exerted by cable *AD* on the anchor at *D* and the corresponding angles θ_x, θ_y, and θ_z in terms of α. (*b*) Plot the components of the force and the angles θ_x, θ_y, and θ_z for $0 \le \alpha \le 60°$.

2.C9 A tower is guyed by cables *AB* and *AC*. A worker ties a rope of length 12 m to the tower at *A* and exerts a constant force of 160 N on the rope. (*a*) Express the tension in each cable as a function of θ knowing that the resultant of the tensions in the cables and the rope is directed downward. (*b*) Plot the tension in each cable as a function of θ for $0 \le \theta \le 180°$, and determine from the graph the range of values of θ for which both cables are taut.

Fig. P2.C8

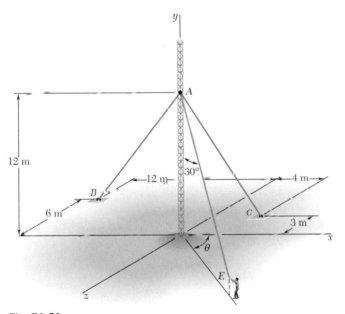

Fig. P2.C9

2.C10 Collars *A* and *B* are connected by a 10-in.-long wire and can slide freely on frictionless rods. If a force **Q** of magnitude 25 lb is applied to collar *B* as shown, determine the tension in the wire and the corresponding magnitude of the force **P** required to maintain equilibrium. Plot the tension in the wire and the magnitude of the force **P** for $0 \le x \le 5$ in.

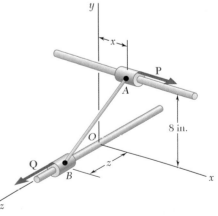

Fig. P2.C10

Rigid Bodies: Equivalent Systems of Forces

3

It will be shown in this chapter that the forces exerted by the four tugboats on the USS *Pasadena* could be replaced with a single equivalent force exerted by one tugboat.

RIGID BODIES: EQUIVALENT SYSTEMS OF FORCES

3.1 Introduction
3.2 External and Internal Forces
3.3 Principle of Transmissibility. Equivalent Forces
3.4 Vector Product of Two Vectors
3.5 Vector Products Expressed in Terms of Rectangular Components
3.6 Moment of a Force about a Point
3.7 Varignon's Theorem
3.8 Rectangular Components of the Moment of a Force
3.9 Scalar Product of Two Vectors
3.10 Mixed Triple Product of Three Vectors
3.11 Moment of a Force about a Given Axis
3.12 Moment of a Couple
3.13 Equivalent Couples
3.14 Addition of Couples
3.15 Couples Can Be Represented by Vectors
3.16 Resolution of a Given Force Into a Force at O and a Couple
3.17 Reduction of a System of Forces to One Force and One Couple
3.18 Equivalent Systems of Forces
3.19 Equipollent Systems of Vectors
3.20 Further Reduction of a System of Forces
3.21 Reduction of a System of Forces to a Wrench

3.1. INTRODUCTION

In the preceding chapter it was assumed that each of the bodies considered could be treated as a single particle. Such a view, however, is not always possible, and a body, in general, should be treated as a combination of a large number of particles. The size of the body will have to be taken into consideration, as well as the fact that forces will act on different particles and thus will have different points of application.

Most of the bodies considered in elementary mechanics are assumed to be *rigid*, a *rigid body* being defined as one which does not deform. Actual structures and machines, however, are never absolutely rigid and deform under the loads to which they are subjected. But these deformations are usually small and do not appreciably affect the conditions of equilibrium or motion of the structure under consideration. They are important, though, as far as the resistance of the structure to failure is concerned and are considered in the study of mechanics of materials.

In this chapter you will study the effect of forces exerted on a rigid body, and you will learn how to replace a given system of forces by a simpler equivalent system. This analysis will rest on the fundamental assumption that the effect of a given force on a rigid body remains unchanged if that force is moved along its line of action (*principle of transmissibility*). It follows that forces acting on a rigid body can be represented by *sliding vectors*, as indicated earlier in Sec. 2.3.

Two important concepts associated with the effect of a force on a rigid body are the *moment of a force about a point* (Sec. 3.6) and the *moment of a force about an axis* (Sec. 3.11). Since the determination of these quantities involves the computation of vector products and scalar products of two vectors, the fundamentals of vector algebra will be introduced in this chapter and applied to the solution of problems involving forces acting on rigid bodies.

Another concept introduced in this chapter is that of a *couple*, that is, the combination of two forces which have the same magnitude, parallel lines of action, and opposite sense (Sec. 3.12). As you will see, any system of forces acting on a rigid body can be replaced by an equivalent system consisting of one force acting at a given point and one couple. This basic system is called a *force-couple system*. In the case of concurrent, coplanar, or parallel forces, the equivalent force-couple system can be further reduced to a single force, called the *resultant* of the system, or to a single couple, called the *resultant couple* of the system.

3.2. EXTERNAL AND INTERNAL FORCES

Forces acting on rigid bodies can be separated into two groups: (1) *external forces* and (2) *internal forces*.

1. The *external forces* represent the action of other bodies on the rigid body under consideration. They are entirely responsible for the external behavior of the rigid body. They will either cause it to move or ensure that it remains at rest. We shall be concerned only with external forces in this chapter and in Chaps. 4 and 5.

2. The *internal forces* are the forces which hold together the particles forming the rigid body. If the rigid body is structurally composed of several parts, the forces holding the component parts together are also defined as internal forces. Internal forces will be considered in Chaps. 6 and 7.

As an example of external forces, let us consider the forces acting on a disabled truck that three people are pulling forward by means of a rope attached to the front bumper (Fig. 3.1). The external forces acting on the truck are shown in a *free-body diagram* (Fig. 3.2). Let us first consider the *weight* of the truck. Although it embodies the effect of the earth's pull on each of the particles forming the truck, the weight can be represented by the single force **W** The *point of application* of this force, that is, the point at which the force acts, is defined as the *center of gravity* of the truck. It will be seen in Chap. 5 how centers of gravity can be determined. The weight **W** tends to make the truck move vertically downward. In fact, it would actually cause the truck to move downward, that is, to fall, if it were not for the presence of the ground. The ground opposes the downward motion of the truck by means of the reactions **R**₁ and **R**₂. These forces are exerted *by* the ground *on* the truck and must therefore be included among the external forces acting on the truck.

The people pulling on the rope exert the force **F**. The point of application of **F** is on the front bumper. The force **F** tends to make the truck move forward in a straight line and does actually make it move, since no external force opposes this motion. (Rolling resistance has been neglected here for simplicity.) This forward motion of the truck, during which each straight line keeps its original orientation (the floor of the truck remains horizontal, and the walls remain vertical), is known as a *translation*. Other forces might cause the truck to move differently. For example, the force exerted by a jack placed under the front axle would cause the truck to pivot about its rear axle. Such a motion is a *rotation*. It can be concluded, therefore, that each of the *external forces* acting on a *rigid body* can, if unopposed, impart to the rigid body a motion of translation or rotation, or both.

Fig. 3.1

Fig. 3.2

3.3. PRINCIPLE OF TRANSMISSIBILITY. EQUIVALENT FORCES

The *principle of transmissibility* states that the conditions of equilibrium or motion of a rigid body will remain unchanged if a force **F** acting at a given point of the rigid body is replaced by a force **F**′ of the same magnitude and same direction, but acting at a different point, *provided that the two forces have the same line of action* (Fig. 3.3). The two forces **F** and **F**′ have the same effect on the rigid body and are said to be *equivalent*. This principle, which states that the action of a force may be *transmitted* along its line of action, is based on experimental evidence. It *cannot* be derived from the properties established so far in this text and must therefore be accepted as an experimental law. However, as you will see in Sec. 16.5, the principle of transmissibility can be derived from the study of the dynamics of rigid bodies, but this study requires the introduction of Newton's

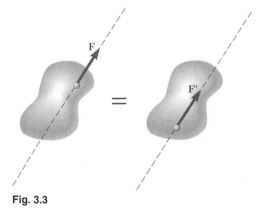

Fig. 3.3

second and third laws and of a number of other concepts as well. Therefore, our study of the statics of rigid bodies will be based on the three principles introduced so far, that is, the parallelogram law of addition, Newton's first law, and the principle of transmissibility.

It was indicated in Chap. 2 that the forces acting on a particle could be represented by vectors. These vectors had a well-defined point of application, namely, the particle itself, and were therefore fixed, or bound, vectors. In the case of forces acting on a rigid body, however, the point of application of the force does not matter, as long as the line of action remains unchanged. Thus, forces acting on a rigid body must be represented by a different kind of vector, known as a *sliding vector*, since forces may be allowed to slide along their lines of action. We should note that all the properties which will be derived in the following sections for the forces acting on a rigid body will be valid more generally for any system of sliding vectors. In order to keep our presentation more intuitive, however, we will carry it out in terms of physical forces rather than in terms of mathematical sliding vectors.

Fig. 3.4

Returning to the example of the truck, we first observe that the line of action of the force **F** is a horizontal line passing through both the front and the rear bumpers of the truck (Fig. 3.4). Using the principle of transmissibility, we can therefore replace **F** by an *equivalent force* **F′** acting on the rear bumper. In other words, the conditions of motion are unaffected, and all the other external forces acting on the truck (**W**, **R**$_1$, **R**$_2$) remain unchanged if the people push on the rear bumper instead of pulling on the front bumper.

The principle of transmissibility and the concept of equivalent forces have limitations, however. Consider, for example, a short bar *AB* acted upon by equal and opposite axial forces **P**$_1$ and **P**$_2$, as shown in Fig. 3.5*a*. According to the principle of transmissibility, the force **P**$_2$ can be replaced by a force **P**$_2'$ having the same magnitude, the same direction, and the same line of action but acting at *A* instead of *B* (Fig. 3.5*b*). The forces **P**$_1$ and **P**$_2'$ acting on the same particle can

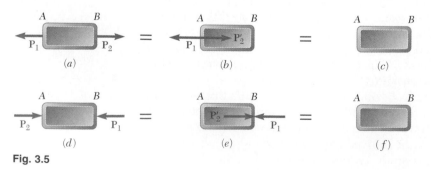

Fig. 3.5

be added according to the rules of Chap. 2, and, as these forces are equal and opposite, their sum is equal to zero. Thus, in terms of the external behavior of the bar, the original system of forces shown in Fig. 3.5a is equivalent to no force at all (Fig. 3.5c).

Consider now the two equal and opposite forces $\mathbf{P}_1$ and $\mathbf{P}_2$ acting on the bar AB as shown in Fig. 3.5d. The force $\mathbf{P}_2$ can be replaced by a force $\mathbf{P}_2'$ having the same magnitude, the same direction, and the same line of action but acting at B instead of at A (Fig. 3.5e). The forces $\mathbf{P}_1$ and $\mathbf{P}_2'$ can then be added, and their sum is again zero (Fig. 3.5f). From the point of view of the mechanics of rigid bodies, the systems shown in Fig. 3.5a and d are thus equivalent. But the *internal forces* and *deformations* produced by the two systems are clearly different. The bar of Fig. 3.5a is in *tension* and, if not absolutely rigid, will increase in length slightly; the bar of Fig. 3.5d is in *compression* and, if not absolutely rigid, will decrease in length slightly. Thus, while the principle of transmissibility may be used freely to determine the conditions of motion or equilibrium of rigid bodies and to compute the external forces acting on these bodies, it should be avoided, or at least used with care, in determining internal forces and deformations.

3.4. VECTOR PRODUCT OF TWO VECTORS

In order to gain a better understanding of the effect of a force on a rigid body, a new concept, the concept of *a moment of a force about a point*, will be introduced at this time. This concept will be more clearly understood, and applied more effectively, if we first add to the mathematical tools at our disposal the *vector product* of two vectors.

The vector product of two vectors $\mathbf{P}$ and $\mathbf{Q}$ is defined as the vector $\mathbf{V}$ which satisfies the following conditions.

1. The line of action of $\mathbf{V}$ is perpendicular to the plane containing $\mathbf{P}$ and $\mathbf{Q}$ (Fig. 3.6a).
2. The magnitude of $\mathbf{V}$ is the product of the magnitudes of $\mathbf{P}$ and $\mathbf{Q}$ and of the sine of the angle θ formed by $\mathbf{P}$ and $\mathbf{Q}$ (the measure of which will always be 180° or less); we thus have

$$V = PQ \sin \theta \qquad (3.1)$$

3. The direction of $\mathbf{V}$ is obtained from the *right-hand rule*. Close your right hand and hold it so that your fingers are curled in the same sense as the rotation through θ which brings the vector $\mathbf{P}$ in line with the vector $\mathbf{Q}$; your thumb will then indicate the direction of the vector $\mathbf{V}$ (Fig. 3.6b). Note that if $\mathbf{P}$ and $\mathbf{Q}$ do not have a common point of application, they should first be redrawn from the same point. The three vectors $\mathbf{P}$, $\mathbf{Q}$, and $\mathbf{V}$—taken in that order—are said to form a *right-handed triad*.†

(b)

Fig. 3.6

---

†We should note that the x, y, and z axes used in Chap. 2 form a right-handed system of orthogonal axes and that the unit vectors $\mathbf{i}$, $\mathbf{j}$, $\mathbf{k}$ defined in Sec. 2.12 form a right-handed orthogonal triad.

Fig. 3.7

As stated above, the vector **V** satisfying these three conditions (which define it uniquely) is referred to as the vector product of **P** and **Q**; it is represented by the mathematical expression

$$\mathbf{V} = \mathbf{P} \times \mathbf{Q} \qquad (3.2)$$

Because of the notation used, the vector product of two vectors **P** and **Q** is also referred to as the *cross product* of **P** and **Q**.

It follows from Eq. (3.1) that, when two vectors **P** and **Q** have either the same direction or opposite directions, their vector product is zero. In the general case when the angle θ formed by the two vectors is neither $0°$ nor $180°$, Eq. (3.1) can be given a simple geometric interpretation: The magnitude V of the vector product of **P** and **Q** is equal to the area of the parallelogram which has **P** and **Q** for sides (Fig. 3.7). The vector product **P** × **Q** will therefore remain unchanged if we replace **Q** by a vector **Q′** which is coplanar with **P** and **Q** and such that the line joining the tips of **Q** and **Q′** is parallel to **P**. We write

$$\mathbf{V} = \mathbf{P} \times \mathbf{Q} = \mathbf{P} \times \mathbf{Q}' \qquad (3.3)$$

From the third condition used to define the vector product **V** of **P** and **Q**, namely, the condition stating that **P**, **Q**, and **V** must form a right-handed triad, it follows that vector products *are not commutative*, that is, **Q** × **P** is not equal to **P** × **Q**. Indeed, we can easily check that **Q** × **P** is represented by the vector −**V**, which is equal and opposite to **V**. We thus write

$$\mathbf{Q} \times \mathbf{P} = -(\mathbf{P} \times \mathbf{Q}) \qquad (3.4)$$

Example. Let us compute the vector product **V** = **P** × **Q** where the vector **P** is of magnitude 6 and lies in the zx plane at an angle of $30°$ with the x axis, and where the vector **Q** is of magnitude 4 and lies along the x axis (Fig. 3.8).

It follows immediately from the definition of the vector product that the vector **V** must lie along the y axis, have the magnitude

$$V = PQ \sin \theta = (6)(4) \sin 30° = 12$$

and be directed upward.

Fig. 3.8

We saw that the commutative property does not apply to vector products. We may wonder whether the *distributive* property holds, that is, whether the relation

$$\mathbf{P} \times (\mathbf{Q}_1 + \mathbf{Q}_2) = \mathbf{P} \times \mathbf{Q}_1 + \mathbf{P} \times \mathbf{Q}_2 \qquad (3.5)$$

is valid. The answer is *yes*. Many readers are probably willing to accept without formal proof an answer which they intuitively feel is correct. However, since the entire structure of both vector algebra and statics depends upon the relation (3.5), we should take time out to derive it.

We can, without any loss of generality, assume that **P** is directed along the y axis (Fig. 3.9a). Denoting by **Q** the sum of **Q**$_1$ and **Q**$_2$, we drop perpendiculars from the tips of **Q**, **Q**$_1$, and **Q**$_2$ onto the zx plane, defining in this way the vectors **Q′**, **Q′**$_1$, and **Q′**$_2$. These vectors will be referred to, respectively, as the *projections* of **Q**, **Q**$_1$, and **Q**$_2$ on the zx plane. Recalling the property expressed by Eq. (3.3), we note that the left-hand member of Eq. (3.5) can be replaced by $\mathbf{P} \times \mathbf{Q}'$ and that, similarly, the vector products $\mathbf{P} \times \mathbf{Q}_1$ and $\mathbf{P} \times \mathbf{Q}_2$ can respectively be replaced by $\mathbf{P} \times \mathbf{Q}'_1$ and $\mathbf{P} \times \mathbf{Q}'_2$. Thus, the relation to be proved can be written in the form

$$\mathbf{P} \times \mathbf{Q}' = \mathbf{P} \times \mathbf{Q}'_1 + \mathbf{P} \times \mathbf{Q}'_2 \qquad (3.5')$$

We now observe that $\mathbf{P} \times \mathbf{Q}'$ can be obtained from $\mathbf{Q}'$ by multiplying this vector by the scalar P and rotating it counterclockwise through 90° in the zx plane (Fig. 3.9b); the other two vector products

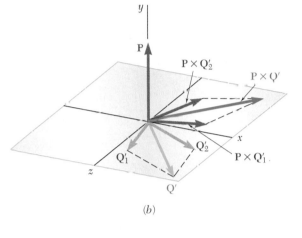

(a)   (b)

Fig. 3.9

in (3.5') can be obtained in the same manner from $\mathbf{Q}'_1$ and $\mathbf{Q}'_2$, respectively. Now, since the projection of a parallelogram onto an arbitrary plane is a parallelogram, the projection $\mathbf{Q}'$ of the sum $\mathbf{Q}$ of $\mathbf{Q}_1$ and $\mathbf{Q}_2$ must be the sum of the projections $\mathbf{Q}'_1$ and $\mathbf{Q}'_2$ of $\mathbf{Q}_1$ and $\mathbf{Q}_2$ on the same plane (Fig. 3.9a). This relation between the vectors $\mathbf{Q}'$, $\mathbf{Q}'_1$, and $\mathbf{Q}'_2$ will still hold after the three vectors have been multiplied by the scalar P and rotated through 90° (Fig. 3.9b). Thus, the relation (3.5') has been proved, and we can now be sure that the distributive property holds for vector products.

A third property, the associative property, does not apply to vector products; we have in general

$$(\mathbf{P} \times \mathbf{Q}) \times \mathbf{S} \neq \mathbf{P} \times (\mathbf{Q} \times \mathbf{S}) \qquad (3.6)$$

3.5. VECTOR PRODUCTS EXPRESSED IN TERMS OF RECTANGULAR COMPONENTS

Let us now determine the vector product of any two of the unit vectors $\mathbf{i}$, $\mathbf{j}$, and $\mathbf{k}$, which were defined in Chap. 2. Consider first the product $\mathbf{i} \times \mathbf{j}$ (Fig. 3.10a). Since both vectors have a magnitude equal to 1 and since they are at a right angle to each other, their vector product will also be a unit vector. This unit vector must be $\mathbf{k}$, since the vectors $\mathbf{i}$, $\mathbf{j}$, and $\mathbf{k}$ are mutually perpendicular and form a right-handed triad. On the other hand, it follows from the right-hand rule given on page 77 that the product $\mathbf{j} \times \mathbf{i}$ will be equal to $-\mathbf{k}$ (Fig. 3.10b). Finally, it should be observed that the vector product of a unit

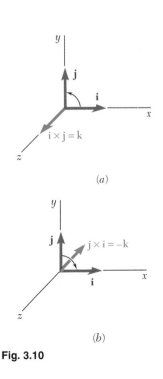

(a)

(b)

Fig. 3.10

vector with itself, such as $\mathbf{i} \times \mathbf{i}$, is equal to zero, since both vectors have the same direction. The vector products of the various possible pairs of unit vectors are

$$
\begin{array}{lll}
\mathbf{i} \times \mathbf{i} = 0 & \mathbf{j} \times \mathbf{i} = -\mathbf{k} & \mathbf{k} \times \mathbf{i} = \mathbf{j} \\
\mathbf{i} \times \mathbf{j} = \mathbf{k} & \mathbf{j} \times \mathbf{j} = 0 & \mathbf{k} \times \mathbf{j} = -\mathbf{i} \\
\mathbf{i} \times \mathbf{k} = -\mathbf{j} & \mathbf{j} \times \mathbf{k} = \mathbf{i} & \mathbf{k} \times \mathbf{k} = 0
\end{array} \tag{3.7}
$$

Fig. 3.11

By arranging in a circle and in counterclockwise order the three letters representing the unit vectors (Fig. 3.11), we can simplify the determination of the sign of the vector product of two unit vectors: The product of two unit vectors will be positive if they follow each other in counterclockwise order and will be negative if they follow each other in clockwise order.

We can now easily express the vector product $\mathbf{V}$ of two given vectors $\mathbf{P}$ and $\mathbf{Q}$ in terms of the rectangular components of these vectors. Resolving $\mathbf{P}$ and $\mathbf{Q}$ into components, we first write

$$
\mathbf{V} = \mathbf{P} \times \mathbf{Q} = (P_x\mathbf{i} + P_y\mathbf{j} + P_z\mathbf{k}) \times (Q_x\mathbf{i} + Q_y\mathbf{j} + Q_z\mathbf{k})
$$

Making use of the distributive property, we express $\mathbf{V}$ as the sum of vector products, such as $P_x\mathbf{i} \times Q_y\mathbf{j}$. Observing that each of the expressions obtained is equal to the vector product of two unit vectors, such as $\mathbf{i} \times \mathbf{j}$, multiplied by the product of two scalars, such as P_xQ_y, and recalling the identities (3.7), we obtain, after factoring out $\mathbf{i}$, $\mathbf{j}$, and $\mathbf{k}$,

$$
\mathbf{V} = (P_yQ_z - P_zQ_y)\mathbf{i} + (P_zQ_x - P_xQ_z)\mathbf{j} + (P_xQ_y - P_yQ_x)\mathbf{k} \tag{3.8}
$$

The rectangular components of the vector product $\mathbf{V}$ are thus found to be

$$
\begin{aligned}
V_x &= P_yQ_z - P_zQ_y \\
V_y &= P_zQ_x - P_xQ_z \\
V_z &= P_xQ_y - P_yQ_x
\end{aligned} \tag{3.9}
$$

Returning to Eq. (3.8), we observe that its right-hand member represents the expansion of a determinant. The vector product $\mathbf{V}$ can thus be expressed in the following form, which is more easily memorized:†

$$
\mathbf{V} = \begin{vmatrix} \mathbf{i} & \mathbf{j} & \mathbf{k} \\ P_x & P_y & P_z \\ Q_x & Q_y & Q_z \end{vmatrix} \tag{3.10}
$$

†Any determinant consisting of three rows and three columns can be evaluated by repeating the first and second columns and forming products along each diagonal line. The sum of the products obtained along the red lines is then subtracted from the sum of the products obtained along the black lines.

3.6. MOMENT OF A FORCE ABOUT A POINT

Let us now consider a force **F** acting on a rigid body (Fig. 3.12a). As we know, the force **F** is represented by a vector which defines its magnitude and direction. However, the effect of the force on the rigid body depends also upon its point of application A. The position of A can be conveniently defined by the vector **r** which joins the fixed reference point O with A; this vector is known as the *position vector* of A.† The position vector **r** and the force **F** define the plane shown in Fig. 3.12a.

We will define the *moment of* **F** *about* O as the vector product of **r** and **F**:

$$\mathbf{M}_O = \mathbf{r} \times \mathbf{F} \tag{3.11}$$

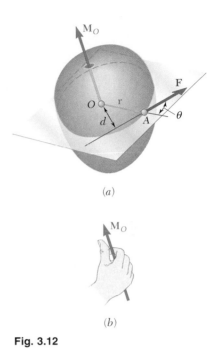

(a)

(b)

Fig. 3.12

According to the definition of the vector product given in Sec. 3.4, the moment $\mathbf{M}_O$ must be perpendicular to the plane containing O and the force **F**. The sense of $\mathbf{M}_O$ is defined by the sense of the rotation which will bring the vector **r** in line with the vector **F**; this rotation will be observed as *counterclockwise* by an observer located at the tip of $\mathbf{M}_O$. Another way of defining the sense of $\mathbf{M}_O$ is furnished by a variation of the right-hand rule: Close your right hand and hold it so that your fingers are curled in the sense of the rotation that **F** would impart to the rigid body about a fixed axis directed along the line of action of $\mathbf{M}_O$; your thumb will indicate the sense of the moment $\mathbf{M}_O$ (Fig. 3.12b).

Finally, denoting by θ the angle between the lines of action of the position vector **r** and the force **F**, we find that the magnitude of the moment of **F** about O is

$$M_O = rF \sin \theta = Fd \tag{3.12}$$

where d represents the perpendicular distance from O to the line of action of **F**. Since the tendency of a force **F** to make a rigid body rotate about a fixed axis perpendicular to the force depends upon the distance of **F** from that axis as well as upon the magnitude of **F**, we note that *the magnitude of* $\mathbf{M}_O$ *measures the tendency of the force* **F** *to make the rigid body rotate about a fixed axis directed along* $\mathbf{M}_O$.

In the SI system of units, where a force is expressed in newtons (N) and a distance in meters (m), the moment of a force is expressed in newton-meters (N · m). In the U.S. customary system of units, where a force is expressed in pounds and a distance in feet or inches, the moment of a force is expressed in lb · ft or lb · in.

We can observe that although the moment $\mathbf{M}_O$ of a force about a point depends upon the magnitude, the line of action, and the sense of the force, it does *not* depend upon the actual position of the point of application of the force along its line of action. Conversely, the moment $\mathbf{M}_O$ of a force **F** does not characterize the position of the point of application of **F**.

---

†We can easily verify that position vectors obey the law of vector addition and, thus, are truly vectors. Consider, for example, the position vectors **r** and **r**′ of A with respect to two reference points O and O' and the position vector **s** of O with respect to O' (Fig. 3.40a, Sec. 3.16). We verify that the position vector $\mathbf{r}' = \overrightarrow{O'A}$ can be obtained from the position vectors $\mathbf{s} = \overrightarrow{O'O}$ and $\mathbf{r} = \overrightarrow{OA}$ by applying the triangle rule for the addition of vectors.

However, as it will be seen presently, the moment $\mathbf{M}_O$ of a force $\mathbf{F}$ of given magnitude and direction *completely defines the line of action of* $\mathbf{F}$. Indeed, the line of action of $\mathbf{F}$ must lie in a plane through O perpendicular to the moment $\mathbf{M}_O$; its distance d from O must be equal to the quotient M_O/F of the magnitudes of $\mathbf{M}_O$ and $\mathbf{F}$; and the sense of $\mathbf{M}_O$ determines whether the line of action of $\mathbf{F}$ is to be drawn on one side or the other of the point O.

We recall from Sec. 3.3 that the principle of transmissibility states that two forces $\mathbf{F}$ and $\mathbf{F}'$ are equivalent (that is, have the same effect on a rigid body) if they have the same magnitude, same direction, and same line of action. This principle can now be restated as follows: *Two forces* $\mathbf{F}$ *and* $\mathbf{F}'$ *are equivalent if, and only if, they are equal* (that is, have the same magnitude and same direction) *and have equal moments about a given point* O. The necessary and sufficient conditions for two forces $\mathbf{F}$ and $\mathbf{F}'$ to be equivalent are thus

$$\mathbf{F} = \mathbf{F}' \qquad \text{and} \qquad \mathbf{M}_O = \mathbf{M}'_O \qquad (3.13)$$

We should observe that it follows from this statement that if the relations (3.13) hold for a given point O, they will hold for any other point.

Problems Involving Only Two Dimensions. Many applications deal with two-dimensional structures, that is, structures which have length and breadth but only negligible depth and which are subjected to forces contained in the plane of the structure. Two-dimensional structures and the forces acting on them can be readily represented on a sheet of paper or on a blackboard. Their analysis is therefore considerably simpler than that of three-dimensional structures and forces.

(a) $M_O = + Fd$

(b) $M_O = - Fd$

Fig. 3.13

Consider, for example, a rigid slab acted upon by a force $\mathbf{F}$ (Fig. 3.13). The moment of $\mathbf{F}$ about a point O chosen in the plane of the figure is represented by a vector $\mathbf{M}_O$ perpendicular to that plane and of magnitude Fd. In the case of Fig. 3.13a the vector $\mathbf{M}_O$ points *out of* the paper, while in the case of Fig. 3.13b it points *into* the paper. As we look at the figure, we observe in the first case that $\mathbf{F}$ tends to rotate the slab counterclockwise and in the second case that it tends to rotate the slab clockwise. Therefore, it is natural to refer to the sense of the moment of $\mathbf{F}$ about O in Fig. 3.13a as counterclockwise ↺, and in Fig. 3.13b as clockwise ↻.

Since the moment of a force $\mathbf{F}$ acting in the plane of the figure must be perpendicular to that plane, we need only specify the *magnitude* and the *sense* of the moment of $\mathbf{F}$ about O. This can be done by assigning to the magnitude M_O of the moment a positive or negative sign according to whether the vector $\mathbf{M}_O$ points out of or into the paper.

3.7. VARIGNON'S THEOREM

The distributive property of vector products can be used to determine the moment of the resultant of several *concurrent forces*. If several forces $\mathbf{F}_1$, $\mathbf{F}_2$, . . . are applied at the same point A (Fig. 3.14), and if we denote by $\mathbf{r}$ the position vector of A, it follows immediately from Eq. (3.5) of Sec. 3.4 that

$$\mathbf{r} \times (\mathbf{F}_1 + \mathbf{F}_2 + \cdots) = \mathbf{r} \times \mathbf{F}_1 + \mathbf{r} \times \mathbf{F}_2 + \cdots \qquad (3.14)$$

In words, *the moment about a given point O of the resultant of several concurrent forces is equal to the sum of the moments of the various forces about the same point O.* This property, which was originally established by the French mathematician Pierre Varignon (1654–1722) long before the introduction of vector algebra, is known as *Varignon's theorem.*

The relation (3.14) makes it possible to replace the direct determination of the moment of a force $\mathbf{F}$ by the determination of the moments of two or more component forces. As you will see in the next section, $\mathbf{F}$ will generally be resolved into components parallel to the coordinate axes. However, it may be more expeditious in some instances to resolve $\mathbf{F}$ into components which are not parallel to the coordinate axes (see Sample Prob. 3.3).

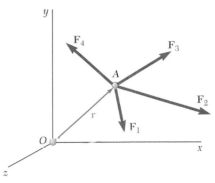

Fig. 3.14

3.8. RECTANGULAR COMPONENTS OF THE MOMENT OF A FORCE

In general, the determination of the moment of a force in space will be considerably simplified if the force and the position vector of its point of application are resolved into rectangular x, y, and z components. Consider, for example, the moment $\mathbf{M}_O$ about O of a force $\mathbf{F}$ whose components are F_x, F_y, and F_z and which is applied at a point A of coordinates x, y, and z (Fig. 3.15). Observing that the components of the position vector $\mathbf{r}$ are respectively equal to the coordinates x, y, and z of the point A, we write

$$\mathbf{r} = x\mathbf{i} + y\mathbf{j} + z\mathbf{k} \qquad (3.15)$$
$$\mathbf{F} = F_x\mathbf{i} + F_y\mathbf{j} + F_z\mathbf{k} \qquad (3.16)$$

Substituting for $\mathbf{r}$ and $\mathbf{F}$ from (3.15) and (3.16) into

$$\mathbf{M}_O = \mathbf{r} \times \mathbf{F} \qquad (3.11)$$

and recalling the results obtained in Sec. 3.5, we write the moment $\mathbf{M}_O$ of $\mathbf{F}$ about O in the form

$$\mathbf{M}_O = M_x\mathbf{i} + M_y\mathbf{j} + M_z\mathbf{k} \qquad (3.17)$$

where the components M_x, M_y, and M_z are defined by the relations

$$\begin{aligned} M_x &= yF_z - zF_y \\ M_y &= zF_x - xF_z \\ M_z &= xF_y - yF_x \end{aligned} \qquad (3.18)$$

Fig. 3.15

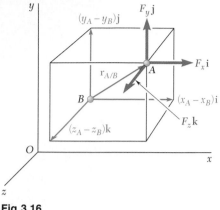

Fig 3.16

As you will see in Sec. 3.11, the scalar components M_x, M_y, and M_z of the moment $\mathbf{M}_O$ measure the tendency of the force $\mathbf{F}$ to impart to a rigid body a motion of rotation about the x, y, and z axes, respectively. Substituting from (3.18) into (3.17), we can also write $\mathbf{M}_O$ in the form of the determinant

$$\mathbf{M}_O = \begin{vmatrix} \mathbf{i} & \mathbf{j} & \mathbf{k} \\ x & y & z \\ F_x & F_y & F_z \end{vmatrix} \quad (3.19)$$

To compute the moment $\mathbf{M}_B$ about an arbitrary point B of a force $\mathbf{F}$ applied at A (Fig. 3.16), we must replace the position vector $\mathbf{r}$ in Eq. (3.11) by a vector drawn from B to A. This vector is the *position vector of A relative to B* and will be denoted by $\mathbf{r}_{A/B}$. Observing that $\mathbf{r}_{A/B}$ can be obtained by subtracting $\mathbf{r}_B$ from $\mathbf{r}_A$, we write

$$\mathbf{M}_B = \mathbf{r}_{A/B} \times \mathbf{F} = (\mathbf{r}_A - \mathbf{r}_B) \times \mathbf{F} \quad (3.20)$$

or, using the determinant form,

$$\mathbf{M}_B = \begin{vmatrix} \mathbf{i} & \mathbf{j} & \mathbf{k} \\ x_{A/B} & y_{A/B} & z_{A/B} \\ F_x & F_y & F_z \end{vmatrix} \quad (3.21)$$

where $x_{A/B}$, $y_{A/B}$, and $z_{A/B}$ denote the components of the vector $\mathbf{r}_{A/B}$:

$$x_{A/B} = x_A - x_B \qquad y_{A/B} = y_A - y_B \qquad z_{A/B} = z_A - z_B$$

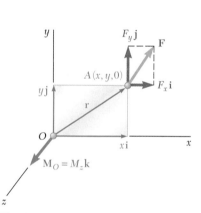

Fig. 3.17

In the case of *problems involving only two dimensions,* the force $\mathbf{F}$ can be assumed to lie in the xy plane (Fig. 3.17). Setting $z = 0$ and $F_z = 0$ in Eq. (3.19), we obtain

$$\mathbf{M}_O = (xF_y - yF_x)\mathbf{k}$$

We verify that the moment of $\mathbf{F}$ about O is perpendicular to the plane of the figure and that it is completely defined by the scalar

$$M_O = M_z = xF_y - yF_x \quad (3.22)$$

As noted earlier, a positive value for M_O indicates that the vector $\mathbf{M}_O$ points out of the paper (the force $\mathbf{F}$ tends to rotate the body counterclockwise about O), and a negative value indicates that the vector $\mathbf{M}_O$ points into the paper (the force $\mathbf{F}$ tends to rotate the body clockwise about O).

To compute the moment about $B(x_B, y_B)$ of a force lying in the xy plane and applied at $A(x_A, y_A)$ (Fig. 3.18), we set $z_{A/B} = 0$ and $F_z = 0$ in the relations (3.21) and note that the vector $\mathbf{M}_B$ is perpendicular to the xy plane and is defined in magnitude and sense by the scalar

$$M_B = (x_A - x_B)F_y - (y_A - y_B)F_x \quad (3.23)$$

Fig. 3.18

SAMPLE PROBLEM 3.1

A 100-lb vertical force is applied to the end of a lever which is attached to a shaft at O. Determine (a) the moment of the 100-lb force about O; (b) the horizontal force applied at A which creates the same moment about O; (c) the smallest force applied at A which creates the same moment about O; (d) how far from the shaft a 240-lb vertical force must act to create the same moment about O; (e) whether any one of the forces obtained in parts b, c, and d is equivalent to the original force.

SOLUTION

a. Moment about O. The perpendicular distance from O to the line of action of the 100-lb force is

$$d = (24 \text{ in.}) \cos 60° = 12 \text{ in.}$$

The magnitude of the moment about O of the 100-lb force is

$$M_O = Fd = (100 \text{ lb})(12 \text{ in.}) = 1200 \text{ lb} \cdot \text{in.}$$

Since the force tends to rotate the lever clockwise about O, the moment will be represented by a vector $\mathbf{M}_O$ perpendicular to the plane of the figure and pointing *into* the paper. We express this fact by writing

$$\mathbf{M}_O = 1200 \text{ lb} \cdot \text{in.} \ \downarrow \quad \blacktriangleleft$$

b. Horizontal Force. In this case, we have

$$d = (24 \text{ in.}) \sin 60° = 20.8 \text{ in.}$$

Since the moment about O must be 1200 lb · in., we write

$$M_O = Fd$$
$$1200 \text{ lb} \cdot \text{in.} = F(20.8 \text{ in.})$$
$$F = 57.7 \text{ lb} \qquad \mathbf{F} = 57.7 \text{ lb} \rightarrow \quad \blacktriangleleft$$

c. Smallest Force. Since $M_O = Fd$, the smallest value of F occurs when d is maximum. We choose the force perpendicular to OA and note that $d = 24$ in.; thus

$$M_O = Fd$$
$$1200 \text{ lb} \cdot \text{in.} = F(24 \text{ in.})$$
$$F = 50 \text{ lb} \qquad \mathbf{F} = 50 \text{ lb} \ ⦨30° \quad \blacktriangleleft$$

d. 240-lb Vertical Force. In this case $M_O = Fd$ yields

$$1200 \text{ lb} \cdot \text{in.} = (240 \text{ lb})d \qquad d = 5 \text{ in.}$$
but
$$OB \cos 60° = d \qquad\qquad OB = 10 \text{ in.} \quad \blacktriangleleft$$

e. None of the forces considered in parts b, c, and d is equivalent to the original 100-lb force. Although they have the same moment about O, they have different x and y components. In other words, although each force tends to rotate the shaft in the same manner, each causes the lever to pull on the shaft in a different way.

SAMPLE PROBLEM 3.2

A force of 800 N acts on a bracket as shown. Determine the moment of the force about B.

SOLUTION

The moment $\mathbf{M}_B$ of the force $\mathbf{F}$ about B is obtained by forming the vector product

$$\mathbf{M}_B = \mathbf{r}_{A/B} \times \mathbf{F}$$

where $\mathbf{r}_{A/B}$ is the vector drawn from B to A. Resolving $\mathbf{r}_{A/B}$ and $\mathbf{F}$ into rectangular components, we have

$$\mathbf{r}_{A/B} = -(0.2 \text{ m})\mathbf{i} + (0.16 \text{ m})\mathbf{j}$$
$$\mathbf{F} = (800 \text{ N}) \cos 60°\mathbf{i} + (800 \text{ N}) \sin 60°\mathbf{j}$$
$$= (400 \text{ N})\mathbf{i} + (693 \text{ N})\mathbf{j}$$

Recalling the relations (3.7) for the vector products of unit vectors (Sec. 3.5), we obtain

$$\mathbf{M}_B = \mathbf{r}_{A/B} \times \mathbf{F} = [-(0.2)\mathbf{i} + (0.16 \text{ m})\mathbf{j}] \times [(400 \text{ N})\mathbf{i} + (693 \text{ N})\mathbf{j}]$$
$$= -(138.6 \text{ N} \cdot \text{m})\mathbf{k} - (64.0 \text{ N} \cdot \text{m})\mathbf{k}$$
$$= -(202.6 \text{ N} \cdot \text{m})\mathbf{k} \qquad\qquad \mathbf{M}_B = 203 \text{ N} \cdot \text{m} \downarrow \quad \blacktriangleleft$$

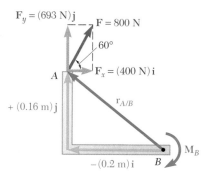

The moment $\mathbf{M}_B$ is a vector perpendicular to the plane of the figure and pointing *into* the paper.

SAMPLE PROBLEM 3.3

A 30-lb force acts on the end of the 3-ft lever as shown. Determine the moment of the force about O.

SOLUTION

The force is replaced by two components, one component $\mathbf{P}$ in the direction of OA and one component $\mathbf{Q}$ perpendicular to OA. Since O is on the line of action of $\mathbf{P}$, the moment of $\mathbf{P}$ about O is zero and the moment of the 30-lb force reduces to the moment of $\mathbf{Q}$, which is clockwise and, thus, is represented by a negative scalar.

$$Q = (30 \text{ lb}) \sin 20° = 10.26 \text{ lb}$$
$$M_O = -Q(3 \text{ ft}) = -(10.26 \text{ lb})(3 \text{ ft}) = -30.8 \text{ lb} \cdot \text{ft}$$

Since the value obtained for the scalar M_O is negative, the moment $\mathbf{M}_O$ points *into* the paper. We write

$$\mathbf{M}_O = 30.8 \text{ lb} \cdot \text{ft} \downarrow \quad \blacktriangleleft$$

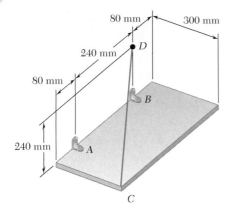

SAMPLE PROBLEM 3.4

A rectangular plate is supported by brackets at A and B and by a wire CD. Knowing that the tension in the wire is 200 N, determine the moment about A of the force exerted by the wire on point C.

SOLUTION

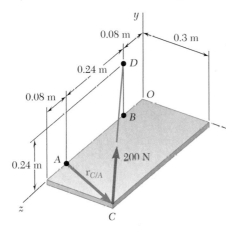

The moment $\mathbf{M}_A$ about A of the force $\mathbf{F}$ exerted by the wire on point C is obtained by forming the vector product

$$\mathbf{M}_A = \mathbf{r}_{C/A} \times \mathbf{F} \tag{1}$$

where $\mathbf{r}_{C/A}$ is the vector drawn from A to C,

$$\mathbf{r}_{C/A} = \overrightarrow{AC} = (0.3 \text{ m})\mathbf{i} + (0.08 \text{ m})\mathbf{k} \tag{2}$$

and $\mathbf{F}$ is the 200-N force directed along CD. Introducing the unit vector $\boldsymbol{\lambda} = \overrightarrow{CD}/CD$, we write

$$\mathbf{F} = F\boldsymbol{\lambda} = (200 \text{ N})\frac{\overrightarrow{CD}}{CD} \tag{3}$$

Resolving the vector $\overrightarrow{CD}$ into rectangular components, we have

$$\overrightarrow{CD} = -(0.3 \text{ m})\mathbf{i} + (0.24 \text{ m})\mathbf{j} - (0.32 \text{ m})\mathbf{k} \qquad CD = 0.50 \text{ m}$$

Substituting into (3), we obtain

$$\mathbf{F} = \frac{200 \text{ N}}{0.50 \text{ m}}[-(0.3 \text{ m})\mathbf{i} + (0.24 \text{ m})\mathbf{j} - (0.32 \text{ m})\mathbf{k}]$$
$$= -(120 \text{ N})\mathbf{i} + (96 \text{ N})\mathbf{j} - (128 \text{ N})\mathbf{k} \tag{4}$$

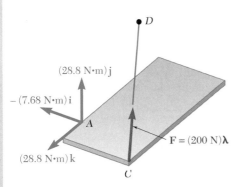

Substituting for $\mathbf{r}_{C/A}$ and $\mathbf{F}$ from (2) and (4) into (1) and recalling the relations (3.7) of Sec. 3.5, we obtain

$$\mathbf{M}_A = \mathbf{r}_{C/A} \times \mathbf{F} = (0.3\mathbf{i} + 0.08\mathbf{k}) \times (-120\mathbf{i} + 96\mathbf{j} - 128\mathbf{k})$$
$$= (0.3)(96)\mathbf{k} + (0.3)(-128)(-\mathbf{j}) + (0.08)(-120)\mathbf{j} + (0.08)(96)(-\mathbf{i})$$
$$\mathbf{M}_A = -(7.68 \text{ N} \cdot \text{m})\mathbf{i} + (28.8 \text{ N} \cdot \text{m})\mathbf{j} + (28.8 \text{ N} \cdot \text{m})\mathbf{k} \blacktriangleleft$$

Alternative Solution. As indicated in Sec. 3.8, the moment $\mathbf{M}_A$ can be expressed in the form of a determinant:

$$\mathbf{M}_A = \begin{vmatrix} \mathbf{i} & \mathbf{j} & \mathbf{k} \\ x_C - x_A & y_C - y_A & z_C - z_A \\ F_x & F_y & F_z \end{vmatrix} = \begin{vmatrix} \mathbf{i} & \mathbf{j} & \mathbf{k} \\ 0.3 & 0 & 0.08 \\ -120 & 96 & -128 \end{vmatrix}$$
$$\mathbf{M}_A = -(7.68 \text{ N} \cdot \text{m})\mathbf{i} + (28.8 \text{ N} \cdot \text{m})\mathbf{j} + (28.8 \text{ N} \cdot \text{m})\mathbf{k} \blacktriangleleft$$

SOLVING PROBLEMS ON YOUR OWN

In this lesson we introduced the *vector product* or *cross product* of two vectors. In the following problems, you will use the vector product to compute the *moment of a force about a point* and also to determine the *perpendicular distance* from a point to a line.

We defined the moment of the force **F** about the point O of a rigid body as

$$\mathbf{M}_O = \mathbf{r} \times \mathbf{F} \tag{3.11}$$

where **r** is the position vector *from O to any point* on the line of action of **F**. Since the vector product is not commutative, it is absolutely necessary when computing such a product that you place the vectors in the proper order and that each vector have the correct sense. The moment $\mathbf{M}_O$ is important because its magnitude is a measure of the tendency of the force **F** to cause the rigid body to rotate about an axis directed along $\mathbf{M}_O$.

1. **Computing the moment $\mathbf{M}_O$ of a force in two dimensions.** You can use one of the following procedures:

   a. Use Eq. (3.12), $M_O = Fd$, which expresses the magnitude of the moment as the product of the magnitude of **F** and the *perpendicular distance d* from O to the line of action of **F** (Sample Prob. 3.1).

   b. Express **r** and **F** in component form and formally evaluate the vector product $\mathbf{M}_O = \mathbf{r} \times \mathbf{F}$ [Sample Prob. 3.2].

   c. Resolve **F** into components respectively parallel and perpendicular to the position vector **r**. Only the perpendicular component contributes to the moment of **F** [Sample Prob. 3.3].

   d. Use Eq. (3.22), $M_O = M_z = xF_y - yF_x$. When applying this method, the simplest approach is to treat the scalar components of **r** and **F** as positive and then to assign, by observation, the proper sign to the moment produced by each force component. For example, applying this method to solve Sample Prob. 3.2, we observe that both force components tend to produce a clockwise rotation about B. Therefore, the moment of each force about B should be represented by a negative scalar. We then have for the total moment

$$M_B = -(0.16 \text{ m})(400 \text{ N}) - (0.20 \text{ m})(693 \text{ N}) = -202.6 \text{ N} \cdot \text{m}$$

2. **Computing the moment $\mathbf{M}_O$ of a force F in three dimensions.** Following the method of Sample Prob. 3.4, the first step in the process is to select the most convenient (simplest) position vector **r**. You should next express **F** in terms of its rectangular components. The final step is to evaluate the vector product $\mathbf{r} \times \mathbf{F}$ to determine the moment. In most three-dimensional problems you will find it easiest to calculate the vector product using a determinant.

3. **Determining the perpendicular distance d from a point A to a given line.** First assume that a force **F** of known magnitude F lies along the given line. Next determine its moment about A by forming the vector product $\mathbf{M}_A = \mathbf{r} \times \mathbf{F}$, and calculate this product as indicated above. Then compute its magnitude M_A. Finally, substitute the values of F and M_A into the equation $M_A = Fd$ and solve for d.

Problems

3.1 A 13.2-N force **P** is applied to the lever which controls the auger of a snowblower. Determine the moment of **P** about *A* when *α* is equal to 30°.

3.2 The force **P** is applied to the lever which controls the auger of a snowblower. Determine the magnitude and the direction of the smallest force **P** which has a 2.20-N · m counterclockwise moment about *A*.

3.3 A 13.1-N force **P** is applied to the lever which controls the auger of a snowblower. Determine the value of *α* knowing that the moment of **P** about *A* is counterclockwise and has a magnitude of 1.95 N · m.

3.4 A foot valve for a pneumatic system is hinged at *B*. Knowing that *α* = 28°, determine the moment of the 4-lb force about point *B* by resolving the force into horizontal and vertical components.

Fig. P3.1, P3.2, and P3.3

Fig. P3.4 and *P3.5*

3.5 A foot valve for a pneumatic system is hinged at *B*. Knowing that *α* = 28°, determine the moment of the 4-lb force about point *B* by resolving the force into components along *ABC* and in a direction perpendicular to *ABC*.

3.6 It is known that a vertical force of 800 N is required to remove the nail at *C* from the board. As the nail first starts moving, determine (*a*) the moment about *B* of the force exerted on the nail, (*b*) the magnitude of the force **P** which creates the same moment about *B* if *α* = 10°, (*c*) the smallest force **P** which creates the same moment about *B*.

Fig. P3.6

Fig. P3.7 and *P3.8*

3.7 A sign is suspended from two chains *AE* and *BF*. Knowing that the tension in *BF* is 45 lb, determine (*a*) the moment about *A* of the force exerted by the chain at *B*, (*b*) the smallest force applied at *C* which creates the same moment about *A*.

3.8 A sign is suspended from two chains *AE* and *BF*. Knowing that the tension in *BF* is 45 lb, determine (*a*) the moment about *A* of the force exerted by the chain at *B*, (*b*) the magnitude and sense of the vertical force applied at *C* which creates the same moment about *A*, (*c*) the smallest force applied at *B* which creates the same moment about *A*.

3.9 and 3.10 The tailgate of a car is supported by the hydraulic lift *BC*. If the lift exerts a 125-N force directed along its center line on the ball and socket at *B*, determine the moment of the force about *A*.

Fig. P3.9

Fig. P3.10

3.11 A winch puller *AB* is used to straighten a fence post. Knowing that the tension in cable *BC* is 260 lb, length *a* is 8 in., length *b* is 35 in., and length *d* is 76 in., determine the moment about *D* of the force exerted by the cable at *C* by resolving that force into horizontal and vertical components applied (*a*) at point *C*, (*b*) at point *E*.

3.12 It is known that a force with a moment of 7840 lb · in. about *D* is required to straighten the fence post *CD*. If *a* = 8 in., *b* = 35 in., and *d* = 112 in., determine the tension that must be developed in the cable of winch puller *AB* to create the required moment about point *D*.

Fig. P3.11, P3.12, and *P3.13*

3.13 It is known that a force with a moment of 1152 N · m about *D* is required to straighten the fence post *CD*. If the capacity of the winch puller *AB* is 2880 N, determine the minimum value of distance *d* to create the specified moment about point *D* knowing that *a* = 0.24 m and *b* = 1.05 m.

3.14 A mechanic uses a piece of pipe *AB* as a lever when tightening an alternator belt. When he pushes down at *A*, a force of 580 N is exerted on the alternator *B*. Determine the moment of that force about bolt *C* if its line of action passes through *O*.

144 mm

108 mm

86 mm

78 mm

Fig. *P3.14*

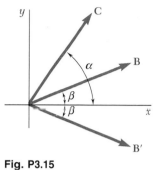

Fig. P3.15

3.15 Form the vector products **B** × **C** and **B'** × **C**, where *B* = *B'*, and use the results obtained to prove the identity

$$\sin \alpha \cos \beta = \tfrac{1}{2}\sin(\alpha + \beta) + \tfrac{1}{2}\sin(\alpha - \beta).$$

3.16 A line passes through the points (420 mm, −150 mm) and (−140 mm, 180 mm). Determine the perpendicular distance *d* from the line to the origin *O* of the system of coordinates.

3.17 A plane contains the vectors **A** and **B**. Determine the unit vector normal to the plane when **A** and **B** are equal to, respectively, (*a*) 4**i** − 2**j** + 3**k** and −2**i** + 6**j** − 5**k**, (*b*) 7**i** + **j** − 4**k** and −6**i** − 3**j** + 2**k**.

3.18 The vectors **P** and **Q** are two adjacent sides of a parallelogram. Determine the area of the parallelogram when (*a*) **P** = (8 in.)**i** + (2 in.)**j** − (1 in.)**k** and **Q** = −(3 in.)**i** + (4 in.)**j** + (2 in.)**k**, (*b*) **P** = −(3 in.)**i** + (6 in.)**j** + (4 in.)**k** and **Q** = (2 in.)**i** + (5 in.)**j** − (3 in.)**k**.

3.19 Determine the moment about the origin *O* of the force **F** = −(5 N)**i** − (2 N)**j** + (3 N)**k** which acts at a point *A*. Assume that the position vector of *A* is (*a*) **r** = (4 m)**i** − (2 m)**j** − (1 m)**k**, (*b*) **r** = −(8 m)**i** + (3 m)**j** + (4 m)**k**, (*c*) **r** = (7.5 m)**i** + (3 m)**j** − (4.5 m)**k**.

3.20 Determine the moment about the origin *O* of the force **F** = −(1.5 lb)**i** + (3 lb)**j** − (2 lb)**k** which acts at a point *A*. Assume that the position vector of *A* is (*a*) **r** = (2.5 ft)**i** − (1 ft)**j** + (2 ft)**k**, (*b*) **r** = (4.5 ft)**i** − (9 ft)**j** + (6 ft)**k**, (*c*) **r** = (4 ft)**i** − (1 ft)**j** + (7 ft)**k**.

3.21 Before the trunk of a large tree is felled, cables *AB* and *BC* are attached as shown. Knowing that the tension in cables *AB* and *BC* are 777 N and 990 N, respectively, determine the moment about *O* of the resultant force exerted on the tree by the cables at *B*.

3.22 Before a telephone cable is strung, rope *BAC* is tied to a stake at *B* and is passed over a pulley at *A*. Knowing that portion *AC* of the rope lies in a plane parallel to the *xy* plane and that the tension **T** in the rope is 124 N, determine the moment about *O* of the resultant force exerted on the pulley by the rope.

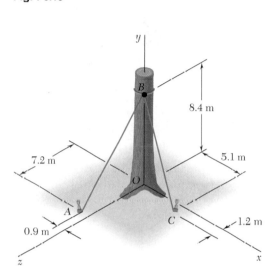

8.4 m

7.2 m

5.1 m

0.9 m

1.2 m

Fig. P3.21

10°

1 m

9 m

1.5 m

2.8 m

Fig. *P3.22*

Fig. P3.23

Fig. *P3.24*

3.23   An 8-lb force is applied to a wrench to tighten a showerhead. Knowing that the centerline of the wrench is parallel to the *x* axis, determine the moment of the force about *A*.

3.24   A wooden board *AB*, which is used as a temporary prop to support a small roof, exerts at point *A* of the roof a 228 N force directed along *BA*. Determine the moment about *C* of that force.

3.25   The ramp *ABCD* is supported by cables at corners *C* and *D*. The tension in each of the cables is 360 lb. Determine the moment about *A* of the force exerted by (*a*) the cable at *D*, (*b*) the cable at *C*.

Fig. P3.25

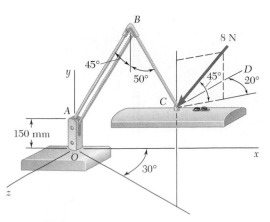

Fig. P3.26

3.26   The arms *AB* and *BC* of a desk lamp lie in a vertical plane that forms an angle of 30° with the *xy* plane. To reposition the light, a force of magnitude 8 N is applied at *C* as shown. Determine the moment of the force about *O* knowing that *AB* = 450 mm, *BC* = 325 mm, and line *CD* is parallel to the *z* axis.

3.27 In Prob. 3.21, determine the perpendicular distance from point O to cable AB.

3.28 In Prob. 3.21, determine the perpendicular distance from point O to cable BC.

3.29 In Prob. 3.24, determine the perpendicular distance from point D to a line drawn through points A and B.

3.30 In Prob. 3.24, determine the perpendicular distance from point C to a line drawn through points A and B.

3.31 In Prob. 3.25, determine the perpendicular distance from point A to portion DE of cable DEF.

3.32 In Prob. 3.25, determine the perpendicular distance from point A to a line drawn through points C and G.

3.33 In Prob. 3.25, determine the perpendicular distance from point B to a line drawn through points D and E.

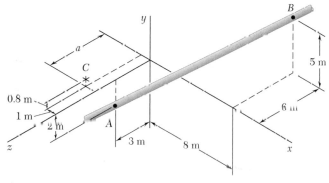

Fig. P3.34

3.34 Determine the value of a which minimizes the perpendicular distance from point C to a section of pipeline that passes through points A and B.

3.9. SCALAR PRODUCT OF TWO VECTORS

The *scalar product* of two vectors $\mathbf{P}$ and $\mathbf{Q}$ is defined as the product of the magnitudes of $\mathbf{P}$ and $\mathbf{Q}$ and of the cosine of the angle θ formed by $\mathbf{P}$ and $\mathbf{Q}$ (Fig. 3.19). The scalar product of $\mathbf{P}$ and $\mathbf{Q}$ is denoted by $\mathbf{P} \cdot \mathbf{Q}$. We write therefore

$$\mathbf{P} \cdot \mathbf{Q} = PQ \cos \theta \qquad (3.24)$$

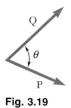

Fig. 3.19

Note that the expression just defined is not a vector but a *scalar*, which explains the name *scalar product*; because of the notation used, $\mathbf{P} \cdot \mathbf{Q}$ is also referred to as the *dot product* of the vectors $\mathbf{P}$ and $\mathbf{Q}$.

It follows from its very definition that the scalar product of two vectors is *commutative*, that is, that

$$\mathbf{P} \cdot \mathbf{Q} = \mathbf{Q} \cdot \mathbf{P} \qquad (3.25)$$

To prove that the scalar product is also *distributive*, we must prove the relation

$$\mathbf{P} \cdot (\mathbf{Q}_1 + \mathbf{Q}_2) = \mathbf{P} \cdot \mathbf{Q}_1 + \mathbf{P} \cdot \mathbf{Q}_2 \qquad (3.26)$$

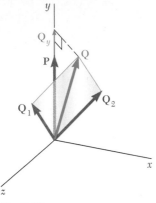

Fig. 3.20

We can, without any loss of generality, assume that $\mathbf{P}$ is directed along the y axis (Fig. 3.20). Denoting by $\mathbf{Q}$ the sum of $\mathbf{Q}_1$ and $\mathbf{Q}_2$ and by θ_y the angle $\mathbf{Q}$ forms with the y axis, we express the left-hand member of (3.26) as follows:

$$\mathbf{P} \cdot (\mathbf{Q}_1 + \mathbf{Q}_2) = \mathbf{P} \cdot \mathbf{Q} = PQ \cos \theta_y = PQ_y \qquad (3.27)$$

where Q_y is the y component of $\mathbf{Q}$. We can, in a similar way, express the right-hand member of (3.26) as

$$\mathbf{P} \cdot \mathbf{Q}_1 + \mathbf{P} \cdot \mathbf{Q}_2 = P(Q_1)_y + P(Q_2)_y \qquad (3.28)$$

Since $\mathbf{Q}$ is the sum of $\mathbf{Q}_1$ and $\mathbf{Q}_2$, its y component must be equal to the sum of the y components of $\mathbf{Q}_1$ and $\mathbf{Q}_2$. Thus, the expressions obtained in (3.27) and (3.28) are equal, and the relation (3.26) has been proved.

As far as the third property—the associative property—is concerned, we note that this property cannot apply to scalar products. Indeed, $(\mathbf{P} \cdot \mathbf{Q}) \cdot \mathbf{S}$ has no meaning, since $\mathbf{P} \cdot \mathbf{Q}$ is not a vector but a scalar.

The scalar product of two vectors $\mathbf{P}$ and $\mathbf{Q}$ can be expressed in terms of their rectangular components. Resolving $\mathbf{P}$ and $\mathbf{Q}$ into components, we first write

$$\mathbf{P} \cdot \mathbf{Q} = (P_x\mathbf{i} + P_y\mathbf{j} + P_z\mathbf{k}) \cdot (Q_x\mathbf{i} + Q_y\mathbf{j} + Q_z\mathbf{k})$$

Making use of the distributive property, we express $\mathbf{P} \cdot \mathbf{Q}$ as the sum of scalar products, such as $P_x\mathbf{i} \cdot Q_x\mathbf{i}$ and $P_x\mathbf{i} \cdot Q_y\mathbf{j}$. However, from the definition of the scalar product it follows that the scalar products of the unit vectors are either zero or one.

$$\begin{array}{lll} \mathbf{i} \cdot \mathbf{i} = 1 & \mathbf{j} \cdot \mathbf{j} = 1 & \mathbf{k} \cdot \mathbf{k} = 1 \\ \mathbf{i} \cdot \mathbf{j} = 0 & \mathbf{j} \cdot \mathbf{k} = 0 & \mathbf{k} \cdot \mathbf{i} = 0 \end{array} \qquad (3.29)$$

Thus, the expression obtained for $\mathbf{P} \cdot \mathbf{Q}$ reduces to

$$\mathbf{P} \cdot \mathbf{Q} = P_xQ_x + P_yQ_y + P_zQ_z \qquad (3.30)$$

In the particular case when $\mathbf{P}$ and $\mathbf{Q}$ are equal, we note that

$$\mathbf{P} \cdot \mathbf{P} = P_x^2 + P_y^2 + P_z^2 = P^2 \qquad (3.31)$$

Applications

1. *Angle formed by two given vectors.* Let two vectors be given in terms of their components:

$$\mathbf{P} = P_x\mathbf{i} + P_y\mathbf{j} + P_z\mathbf{k}$$
$$\mathbf{Q} = Q_x\mathbf{i} + Q_y\mathbf{j} + Q_z\mathbf{k}$$

To determine the angle formed by the two vectors, we equate the expressions obtained in (3.24) and (3.30) for their scalar product and write

$$PQ \cos \theta = P_xQ_x + P_yQ_y + P_zQ_z$$

Solving for $\cos \theta$, we have

$$\cos \theta = \frac{P_xQ_x + P_yQ_y + P_zQ_z}{PQ} \qquad (3.32)$$

2. *Projection of a vector on a given axis.* Consider a vector **P** forming an angle θ with an axis, or directed line, OL (Fig. 3.21). The *projection of* **P** *on the axis* OL is defined as the scalar

$$P_{OL} = P \cos \theta \qquad (3.33)$$

We note that the projection P_{OL} is equal in absolute value to the length of the segment OA; it will be positive if OA has the same sense as the axis OL, that is, if θ is acute, and negative otherwise. If **P** and OL are at a right angle, the projection of **P** on OL is zero.

Consider now a vector **Q** directed along OL and of the same sense as OL (Fig. 3.22). The scalar product of **P** and **Q** can be expressed as

$$\mathbf{P} \cdot \mathbf{Q} = PQ \cos \theta = P_{OL}Q \qquad (3.34)$$

from which it follows that

$$P_{OL} = \frac{\mathbf{P} \cdot \mathbf{Q}}{Q} = \frac{P_xQ_x + P_yQ_y + P_zQ_z}{Q} \qquad (3.35)$$

In the particular case when the vector selected along OL is the unit vector $\boldsymbol{\lambda}$ (Fig. 3.23), we write

$$P_{OL} = \mathbf{P} \cdot \boldsymbol{\lambda} \qquad (3.36)$$

Resolving **P** and $\boldsymbol{\lambda}$ into rectangular components and recalling from Sec. 2.12 that the components of $\boldsymbol{\lambda}$ along the coordinate axes are respectively equal to the direction cosines of OL, we express the projection of **P** on OL as

$$P_{OL} = P_x \cos \theta_x + P_y \cos \theta_y + P_z \cos \theta_z \qquad (3.37)$$

where θ_x, θ_y, and θ_z denote the angles that the axis OL forms with the coordinate axes.

Fig. 3.21

Fig. 3.22

Fig. 3.23

3.10. MIXED TRIPLE PRODUCT OF THREE VECTORS

We define the *mixed triple product* of the three vectors **S**, **P**, and **Q** as the scalar expression

$$\mathbf{S} \cdot (\mathbf{P} \times \mathbf{Q}) \qquad (3.38)$$

obtained by forming the scalar product of **S** with the vector product of **P** and **Q**.†

---

†Another kind of triple product will be introduced later (Chap. 15): the *vector triple product* $\mathbf{S} \times (\mathbf{P} \times \mathbf{Q})$.

Fig. 3.24

Fig. 3.25

Fig. 3.26

A simple geometrical interpretation can be given for the mixed triple product of **S**, **P**, and **Q** (Fig. 3.24). We first recall from Sec. 3.4 that the vector **P** × **Q** is perpendicular to the plane containing **P** and **Q** and that its magnitude is equal to the area of the parallelogram which has **P** and **Q** for sides. On the other hand, Eq. (3.34) indicates that the scalar product of **S** and **P** × **Q** can be obtained by multiplying the magnitude of **P** × **Q** (that is, the area of the parallelogram defined by **P** and **Q**) by the projection of **S** on the vector **P** × **Q** (that is, by the projection of **S** on the normal to the plane containing the parallelogram). The mixed triple product is thus equal, in absolute value, to the volume of the parallelepiped having the vectors **S**, **P**, and **Q** for sides (Fig. 3.25). We note that the sign of the mixed triple product will be positive if **S**, **P**, and **Q** form a right-handed triad and negative if they form a left-handed triad [that is, **S** · (**P** × **Q**) will be negative if the rotation which brings **P** into line with **Q** is observed as clockwise from the tip of **S**]. The mixed triple product will be zero if **S**, **P**, and **Q** are coplanar.

Since the parallelepiped defined in the preceding paragraph is independent of the order in which the three vectors are taken, the six mixed triple products which can be formed with **S**, **P**, and **Q** will all have the same absolute value, although not the same sign. It is easily shown that

$$
\begin{aligned}
\mathbf{S} \cdot (\mathbf{P} \times \mathbf{Q}) &= \mathbf{P} \cdot (\mathbf{Q} \times \mathbf{S}) = \mathbf{Q} \cdot (\mathbf{S} \times \mathbf{P}) \\
&= -\mathbf{S} \cdot (\mathbf{Q} \times \mathbf{P}) = -\mathbf{P} \cdot (\mathbf{S} \times \mathbf{Q}) = -\mathbf{Q} \cdot (\mathbf{P} \times \mathbf{S})
\end{aligned} \tag{3.39}
$$

Arranging in a circle and in counterclockwise order the letters representing the three vectors (Fig. 3.26), we observe that the sign of the mixed triple product remains unchanged if the vectors are permuted in such a way that they are still read in counterclockwise order. Such a permutation is said to be a *circular permutation*. It also follows from Eq. (3.39) and from the commutative property of scalar products that the mixed triple product of **S**, **P**, and **Q** can be defined equally well as **S** · (**P** × **Q**) or (**S** × **P**) · **Q**.

The mixed triple product of the vectors **S**, **P**, and **Q** can be expressed in terms of the rectangular components of these vectors. Denoting **P** × **Q** by **V** and using formula (3.30) to express the scalar product of **S** and **V**, we write

$$
\mathbf{S} \cdot (\mathbf{P} \times \mathbf{Q}) = \mathbf{S} \cdot \mathbf{V} = S_x V_x + S_y V_y + S_z V_z
$$

Substituting from the relations (3.9) for the components of **V**, we obtain

$$
\begin{aligned}
\mathbf{S} \cdot (\mathbf{P} \times \mathbf{Q}) = S_x(P_y Q_z - P_z Q_y) &+ S_y(P_z Q_x - P_x Q_z) \\
&+ S_z(P_x Q_y - P_y Q_x)
\end{aligned} \tag{3.40}
$$

This expression can be written in a more compact form if we observe that it represents the expansion of a determinant:

$$
\mathbf{S} \cdot (\mathbf{P} \times \mathbf{Q}) = \begin{vmatrix} S_x & S_y & S_z \\ P_x & P_y & P_z \\ Q_x & Q_y & Q_z \end{vmatrix} \tag{3.41}
$$

By applying the rules governing the permutation of rows in a determinant, we could easily verify the relations (3.39) which were derived earlier from geometrical considerations.

3.11. MOMENT OF A FORCE ABOUT A GIVEN AXIS

Now that we have further increased our knowledge of vector algebra, we can introduce a new concept, the concept of *moment of a force about an axis*. Consider again a force **F** acting on a rigid body and the moment $\mathbf{M}_O$ of that force about O (Fig. 3.27). Let OL be an axis through O; *we define the moment M_{OL} of **F** about OL as the projection OC of the moment $\mathbf{M}_O$ onto the axis OL.* Denoting by $\boldsymbol{\lambda}$ the unit vector along OL and recalling from Secs. 3.9 and 3.6, respectively, the expressions (3.36) and (3.11) obtained for the projection of a vector on a given axis and for the moment $\mathbf{M}_O$ of a force **F**, we write

$$M_{OL} = \boldsymbol{\lambda} \cdot \mathbf{M}_O = \boldsymbol{\lambda} \cdot (\mathbf{r} \times \mathbf{F}) \qquad (3.42)$$

which shows that the moment M_{OL} of **F** about the axis OL is the scalar obtained by forming the mixed triple product of $\boldsymbol{\lambda}$, **r**, and **F**. Expressing M_{OL} in the form of a determinant, we write

$$M_{OL} = \begin{vmatrix} \lambda_x & \lambda_y & \lambda_z \\ x & y & z \\ F_x & F_y & F_z \end{vmatrix} \qquad (3.43)$$

where $\lambda_x, \lambda_y, \lambda_z$ = direction cosines of axis OL
$\qquad x, y, z$ = coordinates of point of application of **F**
$\qquad F_x, F_y, F_z$ = components of force **F**

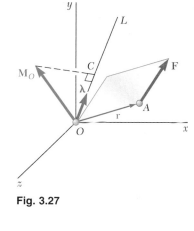

Fig. 3.27

The physical significance of the moment M_{OL} of a force **F** about a fixed axis OL becomes more apparent if we resolve **F** into two rectangular components $\mathbf{F}_1$ and $\mathbf{F}_2$, with $\mathbf{F}_1$ parallel to OL and $\mathbf{F}_2$ lying in a plane P perpendicular to OL (Fig. 3.28). Resolving **r** similarly into two components $\mathbf{r}_1$ and $\mathbf{r}_2$ and substituting for **F** and **r** into (3.42), we write

$$\begin{aligned} M_{OL} &= \boldsymbol{\lambda} \cdot [(\mathbf{r}_1 + \mathbf{r}_2) \times (\mathbf{F}_1 + \mathbf{F}_2)] \\ &= \boldsymbol{\lambda} \cdot (\mathbf{r}_1 \times \mathbf{F}_1) + \boldsymbol{\lambda} \cdot (\mathbf{r}_1 \times \mathbf{F}_2) + \boldsymbol{\lambda} \cdot (\mathbf{r}_2 \times \mathbf{F}_1) + \boldsymbol{\lambda} \cdot (\mathbf{r}_2 \times \mathbf{F}_2) \end{aligned}$$

Fig. 3.28

Noting that all of the mixed triple products except the last one are equal to zero, since they involve vectors which are coplanar when drawn from a common origin (Sec. 3.10), we have

$$M_{OL} = \boldsymbol{\lambda} \cdot (\mathbf{r}_2 \times \mathbf{F}_2) \qquad (3.44)$$

The vector product $\mathbf{r}_2 \times \mathbf{F}_2$ is perpendicular to the plane P and represents the moment of the component $\mathbf{F}_2$ of **F** about the point Q where OL intersects P. Therefore, the scalar M_{OL}, which will be positive if $\mathbf{r}_2 \times \mathbf{F}_2$ and OL have the same sense and negative otherwise, measures the tendency of $\mathbf{F}_2$ to make the rigid body rotate about the fixed axis OL. Since the other component $\mathbf{F}_1$ of **F** does not tend to make the body rotate about OL, we conclude that *the moment M_{OL} of **F** about OL measures the tendency of the force **F** to impart to the rigid body a motion of rotation about the fixed axis OL.*

It follows from the definition of the moment of a force about an axis that the moment of **F** about a coordinate axis is equal to the component of $\mathbf{M}_O$ along that axis. Substituting successively each of the unit vectors **i**, **j**, and **k** for $\boldsymbol{\lambda}$ in (3.42), we observe that the expressions thus obtained for the *moments of* **F** *about the coordinate axes* are respectively equal to the expressions obtained in Sec. 3.8 for the components of the moment $\mathbf{M}_O$ of **F** about O:

$$
\begin{aligned}
M_x &= yF_z - zF_y \\
M_y &= zF_x - xF_z \\
M_z &= xF_y - yF_x
\end{aligned}
\tag{3.18}
$$

We observe that just as the components F_x, F_y, and F_z of a force **F** acting on a rigid body measure, respectively, the tendency of **F** to move the rigid body in the x, y, and z directions, the moments M_x, M_y, and M_z of **F** about the coordinate axes measure the tendency of **F** to impart to the rigid body a motion of rotation about the x, y, and z axes, respectively.

More generally, the moment of a force **F** applied at A about an axis which does not pass through the origin is obtained by choosing an arbitrary point B on the axis (Fig. 3.29) and determining the projection on the axis BL of the moment $\mathbf{M}_B$ of **F** about B. We write

$$
M_{BL} = \boldsymbol{\lambda} \cdot \mathbf{M}_B = \boldsymbol{\lambda} \cdot (\mathbf{r}_{A/B} \times \mathbf{F})
\tag{3.45}
$$

where $\mathbf{r}_{A/B} = \mathbf{r}_A - \mathbf{r}_B$ represents the vector drawn from B to A. Expressing M_{BL} in the form of a determinant, we have

$$
M_{BL} =
\begin{vmatrix}
\lambda_x & \lambda_y & \lambda_z \\
x_{A/B} & y_{A/B} & z_{A/B} \\
F_x & F_y & F_z
\end{vmatrix}
\tag{3.46}
$$

where λ_x, λ_y, λ_z = direction cosines of axis BL
$$x_{A/B} = x_A - x_B \qquad y_{A/B} = y_A - y_B \qquad z_{A/B} = z_A - z_B$$
F_x, F_y, F_z = components of force **F**

It should be noted that the result obtained is independent of the choice of the point B on the given axis. Indeed, denoting by M_{CL} the result obtained with a different point C, we have

$$
\begin{aligned}
M_{CL} &= \boldsymbol{\lambda} \cdot [(\mathbf{r}_A - \mathbf{r}_C) \times \mathbf{F}] \\
&= \boldsymbol{\lambda} \cdot [(\mathbf{r}_A - \mathbf{r}_B) \times \mathbf{F}] + \boldsymbol{\lambda} \cdot [(\mathbf{r}_B - \mathbf{r}_C) \times \mathbf{F}]
\end{aligned}
$$

But, since the vectors $\boldsymbol{\lambda}$ and $\mathbf{r}_B - \mathbf{r}_C$ lie in the same line, the volume of the parallelepiped having the vectors $\boldsymbol{\lambda}$, $\mathbf{r}_B - \mathbf{r}_C$, and **F** for sides is zero, as is the mixed triple product of these three vectors (Sec. 3.10). The expression obtained for M_{CL} thus reduces to its first term, which is the expression used earlier to define M_{BL}. In addition, it follows from Sec. 3.6 that, when computing the moment of **F** about the given axis, A can be any point on the line of action of **F**.

Fig. 3.29

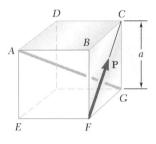

SAMPLE PROBLEM 3.5

A cube of side a is acted upon by a force $\mathbf{P}$ as shown. Determine the moment of $\mathbf{P}$ (a) about A, (b) about the edge AB, (c) about the diagonal AG of the cube. (d) Using the result of part c, determine the perpendicular distance between AG and FC.

SOLUTION

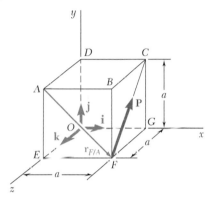

a. Moment about A. Choosing x, y, and z axes as shown, we resolve into rectangular components the force $\mathbf{P}$ and the vector $\mathbf{r}_{F/A} = \overrightarrow{AF}$ drawn from A to the point of application F of $\mathbf{P}$.

$$\mathbf{r}_{F/A} = a\mathbf{i} - a\mathbf{j} = a(\mathbf{i} - \mathbf{j})$$
$$\mathbf{P} = (P/\sqrt{2})\mathbf{j} - (P/\sqrt{2})\mathbf{k} = (P/\sqrt{2})(\mathbf{j} - \mathbf{k})$$

The moment of $\mathbf{P}$ about A is

$$\mathbf{M}_A = \mathbf{r}_{F/A} \times \mathbf{P} = a(\mathbf{i} - \mathbf{j}) \times (P/\sqrt{2})(\mathbf{j} - \mathbf{k})$$
$$\mathbf{M}_A = (aP/\sqrt{2})(\mathbf{i} + \mathbf{j} + \mathbf{k}) \quad \blacktriangleleft$$

b. Moment about AB. Projecting $\mathbf{M}_A$ on AB, we write

$$M_{AB} = \mathbf{i} \cdot \mathbf{M}_A = \mathbf{i} \cdot (aP/\sqrt{2})(\mathbf{i} + \mathbf{j} + \mathbf{k})$$
$$M_{AB} = aP/\sqrt{2} \quad \blacktriangleleft$$

We verify that, since AB is parallel to the x axis, M_{AB} is also the x component of the moment $\mathbf{M}_A$.

c. Moment about Diagonal AG. The moment of $\mathbf{P}$ about AG is obtained by projecting $\mathbf{M}_A$ on AG. Denoting by $\boldsymbol{\lambda}$ the unit vector along AG, we have

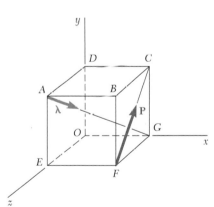

$$\boldsymbol{\lambda} = \frac{\overrightarrow{AG}}{AG} = \frac{a\mathbf{i} - a\mathbf{j} - a\mathbf{k}}{a\sqrt{3}} = (1/\sqrt{3})(\mathbf{i} - \mathbf{j} - \mathbf{k})$$
$$M_{AG} = \boldsymbol{\lambda} \cdot \mathbf{M}_A = (1/\sqrt{3})(\mathbf{i} - \mathbf{j} - \mathbf{k}) \cdot (aP/\sqrt{2})(\mathbf{i} + \mathbf{j} + \mathbf{k})$$
$$M_{AG} = (aP/\sqrt{6})(1 - 1 - 1) \quad M_{AG} = -aP/\sqrt{6} \quad \blacktriangleleft$$

Alternative Method. The moment of $\mathbf{P}$ about AG can also be expressed in the form of a determinant:

$$M_{AG} = \begin{vmatrix} \lambda_x & \lambda_y & \lambda_z \\ x_{F/A} & y_{F/A} & z_{F/A} \\ F_x & F_y & F_z \end{vmatrix} = \begin{vmatrix} 1/\sqrt{3} & -1/\sqrt{3} & -1/\sqrt{3} \\ a & -a & 0 \\ 0 & P/\sqrt{2} & -P/\sqrt{2} \end{vmatrix} = -aP/\sqrt{6}$$

d. Perpendicular Distance between AG and FC. We first observe that $\mathbf{P}$ is perpendicular to the diagonal AG. This can be checked by forming the scalar product $\mathbf{P} \cdot \boldsymbol{\lambda}$ and verifying that it is zero:

$$\mathbf{P} \cdot \boldsymbol{\lambda} = (P/\sqrt{2})(\mathbf{j} - \mathbf{k}) \cdot (1/\sqrt{3})(\mathbf{i} - \mathbf{j} - \mathbf{k}) = (P\sqrt{6})(0 - 1 + 1) = 0$$

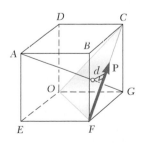

The moment M_{AG} can then be expressed as $-Pd$, where d is the perpendicular distance from AG to FC. (The negative sign is used since the rotation imparted to the cube by $\mathbf{P}$ appears as clockwise to an observer at G.) Recalling the value found for M_{AG} in part c,

$$M_{AG} = -Pd = -aP/\sqrt{6} \qquad d = a/\sqrt{6} \quad \blacktriangleleft$$

SOLVING PROBLEMS ON YOUR OWN

In the problems for this lesson you will apply the *scalar product* or *dot product* of two vectors to determine the *angle formed by two given vectors* and the *projection of a force on a given axis*. You will also use the *mixed triple product* of three vectors to find the *moment of a force about a given axis* and the *perpendicular distance between two lines*.

1. Calculating the angle formed by two given vectors.   First express the vectors in terms of their components and determine the magnitudes of the two vectors. The cosine of the desired angle is then obtained by dividing the scalar product of the two vectors by the product of their magnitudes [Eq. (3.32)].

2. Computing the projection of a vector P on a given axis OL.   In general, begin by expressing $\mathbf{P}$ and the unit vector $\boldsymbol{\lambda}$, that defines the direction of the axis, in component form. Take care that $\boldsymbol{\lambda}$ has the correct sense (that is, $\boldsymbol{\lambda}$ is directed from O to L). The required projection is then equal to the scalar product $\mathbf{P} \cdot \boldsymbol{\lambda}$. However, if you know the angle θ formed by $\mathbf{P}$ and $\boldsymbol{\lambda}$, the projection is also given by $P \cos \theta$.

3. Determining the moment M_{OL} of a force about a given axis OL.   We defined M_{OL} as

$$M_{OL} = \boldsymbol{\lambda} \cdot \mathbf{M}_O = \boldsymbol{\lambda} \cdot (\mathbf{r} \times \mathbf{F}) \tag{3.42}$$

where $\boldsymbol{\lambda}$ is the unit vector along OL and $\mathbf{r}$ is a position vector *from any point* on the line OL *to any point* on the line of action of $\mathbf{F}$. As was the case for the moment of a force about a point, choosing the most convenient position vector will simplify your calculations. Also, recall the warning of the previous lesson: the vectors $\mathbf{r}$ and $\mathbf{F}$ must have the correct sense, and they must be placed in the proper order. The procedure you should follow when computing the moment of a force about an axis is illustrated in part *c* of Sample Prob. 3.5. The two essential steps in this procedure are to first express $\boldsymbol{\lambda}$, $\mathbf{r}$, and $\mathbf{F}$ in terms of their rectangular components and to then evaluate the mixed triple product $\boldsymbol{\lambda} \cdot (\mathbf{r} \times \mathbf{F})$ to determine the moment about the axis. In most three-dimensional problems the most convenient way to compute the mixed triple product is by using a determinant.

As noted in the text, when $\boldsymbol{\lambda}$ is directed along one of the coordinate axes, M_{OL} is equal to the scalar component of $\mathbf{M}_O$ along that axis.

4. Determining the perpendicular distance between two lines. You should remember that it is the perpendicular component $\mathbf{F}_2$ of the force $\mathbf{F}$ that tends to make a body rotate about a given axis OL (Fig. 3.28). It then follows that

$$M_{OL} = F_2 d$$

where M_{OL} is the moment of $\mathbf{F}$ about axis OL and d is the perpendicular distance between OL and the line of action of $\mathbf{F}$. This last equation gives us a simple technique for determining d. First assume that a force $\mathbf{F}$ of known magnitude F lies along one of the given lines and that the unit vector $\boldsymbol{\lambda}$ lies along the other line. Next compute the moment M_{OL} of the force $\mathbf{F}$ about the second line using the method discussed above. The magnitude of the parallel component, F_1, of $\mathbf{F}$ is obtained using the scalar product:

$$F_1 = \mathbf{F} \cdot \boldsymbol{\lambda}$$

The value of F_2 is then determined from

$$F_2 = \sqrt{F^2 - F_1^2}$$

Finally, substitute the values of M_{OL} and F_2 into the equation $M_{OL} = F_2 d$ and solve for d.

You should now realize that the calculation of the perpendicular distance in part d of Sample Prob. 3.5 was simplified by $\mathbf{P}$ being perpendicular to the diagonal AG. In general, the two given lines will not be perpendicular, so that the technique just outlined will have to be used when determining the perpendicular distance between them.

Problems

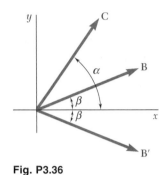

Fig. P3.36

3.35 Given the vectors $\mathbf{P} = 7\mathbf{i} - 2\mathbf{j} + 5\mathbf{k}$, $\mathbf{Q} = -3\mathbf{i} - 4\mathbf{j} + 6\mathbf{k}$, and $\mathbf{S} = 8\mathbf{i} + \mathbf{j} - 9\mathbf{k}$, compute the scalar products $\mathbf{P} \cdot \mathbf{Q}$, $\mathbf{P} \cdot \mathbf{S}$, and $\mathbf{Q} \cdot \mathbf{S}$.

3.36 Form the scalar products $\mathbf{B} \cdot \mathbf{C}$ and $\mathbf{B}' \cdot \mathbf{C}$, where $B = B'$, and use the results obtained to prove the identity

$$\cos \alpha \cos \beta = \tfrac{1}{2} \cos(\alpha + \beta) + \tfrac{1}{2} \cos(\alpha - \beta).$$

3.37 Consider the volleyball net shown. Determine the angle formed by guy wires AB and AC.

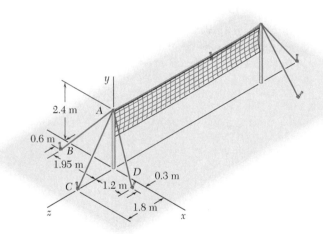

Fig. P3.37 and *P3.38*

3.38 Consider the volleyball net shown. Determine the angle formed by guy wires AC and AD.

3.39 Steel framing members AB, BC, and CD are joined at B and C and are braced using cables EF and EG. Knowing that E is at the midpoint of BC and that the tension in cable EF is 330 N, determine (*a*) the angle between EF and member BC, (*b*) the projection on BC of the force exerted by cable EF at point E.

3.40 Steel framing members AB, BC, and CD are joined at B and C and are braced using cables EF and EG. Knowing that E is at the midpoint of BC and that the tension in cable EG is 445 N, determine (*a*) the angle between EG and member BC, (*b*) the projection on BC of the force exerted by cable EG at point E.

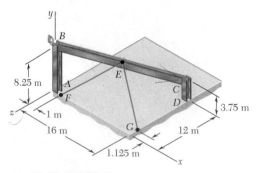

Fig. P3.39 and P3.40

3.41 Slider *P* can move along rod *OA*. An elastic cord *PC* is attached to the slider and to the vertical member *BC*. Knowing that the distance from *O* to *P* is 0.12 m and the tension in the cord is 30 N, determine (*a*) the angle between the elastic cord and the rod *OA*, (*b*) the projection on *OA* of the force exerted by cord *PC* at point *P*.

3.42 Slider *P* can move along rod *OA*. An elastic cord *PC* is attached to the slider and to the vertical member *BC*. Determine the distance from *O* to *P* for which cord *PC* and rod *OA* are perpendicular.

3.43 Determine the volume of the parallelepiped of Fig. 3.25 when (*a*) $\mathbf{P} = -(7\text{ in.})\mathbf{i} - (1\text{ in.})\mathbf{j} + (2\text{ in.})\mathbf{k}$, $\mathbf{Q} = (3\text{ in.})\mathbf{i} - (2\text{ in.})\mathbf{j} + (4\text{ in.})\mathbf{k}$, and $\mathbf{S} = -(5\text{ in.})\mathbf{i} + (6\text{ in.})\mathbf{j} - (1\text{ in.})\mathbf{k}$, (*b*) $\mathbf{P} = (1\text{ in.})\mathbf{i} + (2\text{ in.})\mathbf{j} - (1\text{ in.})\mathbf{k}$, $\mathbf{Q} = -(8\text{ in.})\mathbf{i} - (1\text{ in.})\mathbf{j} + (9\text{ in.})\mathbf{k}$, and $\mathbf{S} = (2\text{ in.})\mathbf{i} + (3\text{ in.})\mathbf{j} + (1\text{ in.})\mathbf{k}$.

3.44 Given the vectors $\mathbf{P} = 4\mathbf{i} - 2\mathbf{j} + P_z\mathbf{k}$, $\mathbf{Q} = \mathbf{i} + 3\mathbf{j} - 5\mathbf{k}$, and $\mathbf{S} = -6\mathbf{i} + 2\mathbf{j} - \mathbf{k}$, determine the value of P_z for which the three vectors are coplanar.

3.45 The 0.732×1.2-m lid *ABCD* of a storage bin is hinged along side *AB* and is held open by looping cord *DEC* over a frictionless hook at *E*. If the tension in the cord is 54 N, determine the moment about each of the coordinate axes of the force exerted by the cord at *D*.

3.46 The 0.732×1.2 m lid *ABCD* of a storage bin is hinged along side *AB* and is held open by looping cord *DEC* over a frictionless hook at *E*. If the tension in the cord is 54 N, determine the moment about each of the coordinate axes of the force exerted by the cord at *C*.

3.47 A fence consists of wooden posts and a steel cable fastened to each post and anchored in the ground at *A* and *D*. Knowing that the sum of the moments about the *z* axis of the forces exerted by the cable on the posts at *B* and *C* is -66 N · m, determine the magnitude of $\mathbf{T}_{CD}$ when $T_{BA} = 56$ N.

Dimensions in mm

Fig. P3.41 and *P3.42*

Fig. P3.45 and P3.46

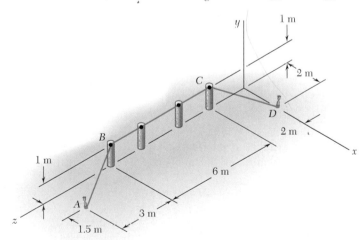

Fig. P3.47 and P3.48

3.48 A fence consists of wooden posts and a steel cable fastened to each post and anchored in the ground at *A* and *D*. Knowing that the sum of the moments about the *y* axis of the forces exerted by the cable on the posts at *B* and *C* is 212 N · m, determine the magnitude of $\mathbf{T}_{BA}$ when $T_{CD} = 33$ N.

Fig. P3.49 and *P3.50*

3.49 To lift a heavy crate, a man uses a block and tackle attached to the bottom of an I-beam at hook B. Knowing that the moments about the y and z axes of the force exerted at B by portion AB of the rope are, respectively, 100 lb · ft and −400 lb · ft, determine the distance a.

3.50 To lift a heavy crate, a man uses a block and tackle attached to the bottom of an I-beam at hook B. Knowing that the man applies a 200-lb force to end A of the rope and that the moment of that force about the y axis is 175 lb · ft, determine the distance a.

3.51 A force P is applied to the lever of an arbor press. Knowing that P lies in a plane parallel to the yz plane and that $M_x = 230$ lb · in., $M_y = -200$ lb · in., and $M_z = -35$ lb · in., determine the magnitude of P and the values of ϕ and θ.

Fig. P3.51 and *P3.52*

3.52 A force P is applied to the lever of an arbor press. Knowing that P lies in a plane parallel to the yz plane and that $M_y = -180$ lb · in. and $M_z = -30$ lb · in., determine the moment M_x of P about the x axis when $\theta = 60°$.

3.53 The triangular plate ABC is supported by ball-and-socket joints at B and D and is held in the position shown by cables AE and CF. If the force exerted by cable AE at A is 220 lb, determine the moment of that force about the line joining points D and B.

3.54 The triangular plate ABC is supported by ball-and-socket joints at B and D and is held in the position shown by cables AE and CF. If the force exerted by cable CF at C is 132 lb, determine the moment of that force about the line joining points D and B.

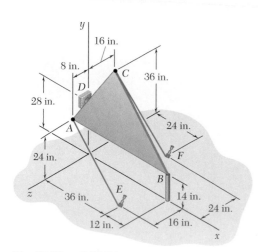

Fig. P3.53 and *P3.54*

3.55 A mast is mounted on the roof of a house using bracket *ABCD* and is guyed by cables *EF*, *EG*, and *EH*. Knowing that the force exerted by cable *EF* at *E* is 66 N, determine the moment of that force about the line joining points *D* and *I*.

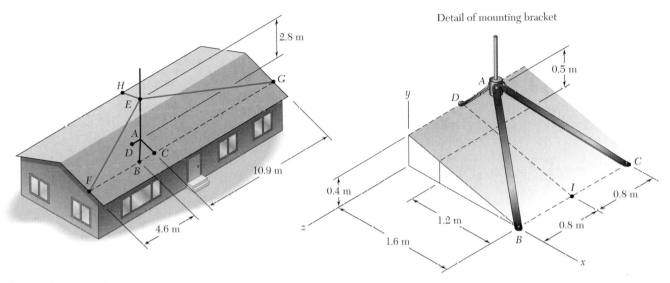

Fig. P3.55 and P3.56

3.56 A mast is mounted on the roof of a house using bracket *ABCD* and is guyed by cables *EF*, *EG*, and *EH*. Knowing that the force exerted by cable *EG* at *E* is 61.5 N, determine the moment of that force about the line joining points *D* and *I*.

3.57 A rectangular tetrahedron has six edges of length *a*. A force **P** is directed as shown along edge *BC*. Determine the moment of **P** about edge *OA*.

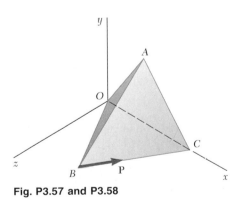

Fig. P3.57 and P3.58

3.58 A rectangular tetrahedron has six edges of length *a*. (*a*) Show that two opposite edges, such as *OA* and *BC*, are perpendicular to each other. (*b*) Use this property and the result obtained in Prob. 3.57 to determine the perpendicular distance between edges *OA* and *BC*.

3.59 The 8-ft-wide portion *ABCD* of an inclined, cantilevered walkway is partially supported by members *EF* and *GH*. Knowing that the compressive force exerted by member *EF* on the walkway at *F* is 5400 lb, determine the moment of that force about edge *AD*.

Fig. P3.59 and *P3.60*

3.60 The 8-ft-wide portion *ABCD* of an inclined, cantilevered walkway is partially supported by members *EF* and *GH*. Knowing that the compressive force exerted by member *GH* on the walkway at *H* is 4800 lb, determine the moment of that force about edge *AD*.

3.61 Two forces $\mathbf{F}_1$ and $\mathbf{F}_2$ in space have the same magnitude *F*. Prove that the moment of $\mathbf{F}_1$ about the line of action of $\mathbf{F}_2$ is equal to the moment of $\mathbf{F}_2$ about the line of action of $\mathbf{F}_1$.

***3.62** In Prob. 3.53, determine the perpendicular distance between cable *AE* and the line joining points *D* and *B*.

***3.63** In Prob. 3.54, determine the perpendicular distance between cable *CF* and the line joining points *D* and *B*.

***3.64** In Prob. 3.55, determine the perpendicular distance between cable *EF* and the line joining points *D* and *I*.

***3.65* In Prob. 3.56, determine the perpendicular distance between cable *EG* and the line joining points *D* and *I*.

***3.66** In Prob. 3.41, determine the perpendicular distance between post *BC* and the line connecting points *O* and *A*.

***3.67* In Prob. 3.45, determine the perpendicular distance between cord *DE* and the *y* axis.

3.12. MOMENT OF A COUPLE

*Two forces **F** and −**F** having the same magnitude, parallel lines of action, and opposite sense are said to form a couple.* (Fig. 3.30). Clearly, the sum of the components of the two forces in any direction is zero. The sum of the moments of the two forces about a given point, however, is not zero. While the two forces will not translate the body on which they act, they will tend to make it rotate.

Denoting by $\mathbf{r}_A$ and $\mathbf{r}_B$, respectively, the position vectors of the points of application of **F** and −**F** (Fig. 3.31), we find that the sum of the moments of the two forces about O is

$$\mathbf{r}_A \times \mathbf{F} + \mathbf{r}_B \times (-\mathbf{F}) = (\mathbf{r}_A - \mathbf{r}_B) \times \mathbf{F}$$

Setting $\mathbf{r}_A - \mathbf{r}_B = \mathbf{r}$, where **r** is the vector joining the points of application of the two forces, we conclude that the sum of the moments of **F** and −**F** about O is represented by the vector

$$\mathbf{M} = \mathbf{r} \times \mathbf{F} \qquad (3.47)$$

The vector **M** is called the *moment of the couple;* it is a vector perpendicular to the plane containing the two forces, and its magnitude is

$$M = rF \sin \theta = Fd \qquad (3.48)$$

where d is the perpendicular distance between the lines of action of **F** and −**F**. The sense of **M** is defined by the right-hand rule.

Since the vector **r** in (3.47) is independent of the choice of the origin O of the coordinate axes, we note that the same result would have been obtained if the moments of **F** and −**F** had been computed about a different point O'. Thus, the moment **M** of a couple is a *free vector* (Sec. 2.3) which can be applied at any point (Fig. 3.32).

From the definition of the moment of a couple, it also follows that two couples, one consisting of the forces $\mathbf{F}_1$ and $-\mathbf{F}_1$, the other of the forces $\mathbf{F}_2$ and $-\mathbf{F}_2$ (Fig. 3.33), will have equal moments if

$$F_1 d_1 = F_2 d_2 \qquad (3.49)$$

and if the two couples lie in parallel planes (or in the same plane) and have the same sense.

Fig. 3.30

Fig. 3.31

Fig. 3.32

Fig. 3.33

Photo 3.1 The parallel upward and downward forces of equal magnitude exerted on the arms of the lug nut wrench are an example of a couple.

3.13. EQUIVALENT COUPLES

Figure 3.34 shows three couples which act successively on the same rectangular box. As seen in the preceding section, the only motion a couple can impart to a rigid body is a rotation. Since each of the three couples shown has the same moment **M** (same direction and same magnitude $M = 120$ lb · in.), we can expect the three couples to have the same effect on the box.

(a)

(b)   (c)

Fig. 3.34

As reasonable as this conclusion appears, we should not accept it hastily. While intuitive feeling is of great help in the study of mechanics, it should not be accepted as a substitute for logical reasoning. Before stating that two systems (or groups) of forces have the same effect on a rigid body, we should prove that fact on the basis of the experimental evidence introduced so far. This evidence consists of the parallelogram law for the addition of two forces (Sec. 2.2) and the principle of transmissibility (Sec. 3.3). Therefore, we will state that *two systems of forces are equivalent* (that is, they have the same effect on a rigid body) *if we can transform one of them into the other by means of one or several of the following operations:* (1) replacing two forces acting on the same particle by their resultant; (2) resolving a force into two components; (3) canceling two equal and opposite forces acting on the same particle; (4) attaching to the same particle two equal and opposite forces; (5) moving a force along its line of action. Each of these operations is easily justified on the basis of the parallelogram law or the principle of transmissibility.

Let us now prove that *two couples having the same moment* **M** *are equivalent*. First consider two couples contained in the same plane, and assume that this plane coincides with the plane of the figure (Fig. 3.35). The first couple consists of the forces $\mathbf{F}_1$ and $-\mathbf{F}_1$ of magnitude F_1, which are located at a distance d_1 from each other (Fig. 3.35a), and the second couple consists of the forces $\mathbf{F}_2$ and $-\mathbf{F}_2$ of magnitude F_2, which are located at a distance d_2 from each other (Fig. 3.35d). Since the two couples have the same moment **M**, which is perpendicular to the plane of the figure, they must have the same sense (assumed here to be counterclockwise), and the relation

$$F_1 d_1 = F_2 d_2 \tag{3.49}$$

must be satisfied. To prove that they are equivalent, we shall show that the first couple can be transformed into the second by means of the operations listed above.

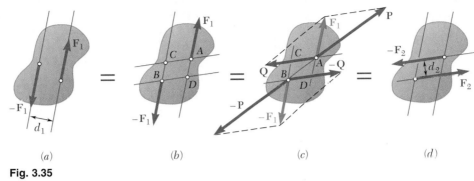

Fig. 3.35

Denoting by A, B, C, D the points of intersection of the lines of action of the two couples, we first slide the forces $\mathbf{F}_1$ and $-\mathbf{F}_1$ until they are attached, respectively, at A and B, as shown in Fig. 3.35b. The force $\mathbf{F}_1$ is then resolved into a component $\mathbf{P}$ along line AB and a component $\mathbf{Q}$ along AC (Fig. 3.35c); similarly, the force $-\mathbf{F}_1$ is resolved into $-\mathbf{P}$ along AB and $-\mathbf{Q}$ along BD. The forces $\mathbf{P}$ and $-\mathbf{P}$ have the same magnitude, the same line of action, and opposite sense; they can be moved along their common line of action until they are applied at the same point and may then be canceled. Thus the couple formed by $\mathbf{F}_1$ and $-\mathbf{F}_1$ reduces to a couple consisting of $\mathbf{Q}$ and $-\mathbf{Q}$.

We will now show that the forces $\mathbf{Q}$ and $-\mathbf{Q}$ are respectively equal to the forces $-\mathbf{F}_2$ and $\mathbf{F}_2$. The moment of the couple formed by $\mathbf{Q}$ and $-\mathbf{Q}$ can be obtained by computing the moment of $\mathbf{Q}$ about B; similarly, the moment of the couple formed by $\mathbf{F}_1$ and $-\mathbf{F}_1$ is the moment of $\mathbf{F}_1$ about B. But, by Varignon's theorem, the moment of $\mathbf{F}_1$ is equal to the sum of the moments of its components $\mathbf{P}$ and $\mathbf{Q}$. Since the moment of $\mathbf{P}$ about B is zero, the moment of the couple formed by $\mathbf{Q}$ and $-\mathbf{Q}$ must be equal to the moment of the couple formed by $\mathbf{F}_1$ and $-\mathbf{F}_1$. Recalling (3.49), we write

$$Qd_2 = F_1 d_1 = F_2 d_2 \qquad \text{and} \qquad Q = F_2$$

Thus the forces $\mathbf{Q}$ and $-\mathbf{Q}$ are respectively equal to the forces $-\mathbf{F}_2$ and $\mathbf{F}_2$, and the couple of Fig. 3.35a is equivalent to the couple of Fig. 3.35d.

Next consider two couples contained in parallel planes P_1 and P_2; we will prove that they are equivalent if they have the same moment. In view of the foregoing, we can assume that the couples consist of forces of the same magnitude F acting along parallel lines (Fig. 3.36a and d). We propose to show that the couple contained in plane P_1 can be transformed into the couple contained in plane P_2 by means of the standard operations listed above.

Let us consider the two planes defined respectively by the lines of action of $\mathbf{F}_1$ and $-\mathbf{F}_2$ and by those of $-\mathbf{F}_1$ and $\mathbf{F}_2$ (Fig. 3.36b). At a point on their line of intersection we attach two forces $\mathbf{F}_3$ and $-\mathbf{F}_3$, respectively equal to $\mathbf{F}_1$ and $-\mathbf{F}_1$. The couple formed by $\mathbf{F}_1$ and $-\mathbf{F}_3$ can be replaced by a couple consisting of $\mathbf{F}_3$ and $-\mathbf{F}_2$ (Fig. 3.36c), since both couples clearly have the same moment and are contained in the same plane. Similarly, the couple formed by $-\mathbf{F}_1$ and $\mathbf{F}_3$ can be replaced by a couple consisting of $-\mathbf{F}_3$ and $\mathbf{F}_2$. Canceling the two

(a)

(b)

(c)

(d)

Fig. 3.36

equal and opposite forces $\mathbf{F}_3$ and $-\mathbf{F}_3$, we obtain the desired couple in plane P_2 (Fig. 3.36d). Thus, we conclude that two couples having the same moment $\mathbf{M}$ are equivalent, whether they are contained in the same plane or in parallel planes.

The property we have just established is very important for the correct understanding of the mechanics of rigid bodies. It indicates that when a couple acts on a rigid body, it does not matter where the two forces forming the couple act or what magnitude and direction they have. The only thing which counts is the *moment* of the couple (magnitude and direction). Couples with the same moment will have the same effect on the rigid body.

3.14. ADDITION OF COUPLES

Consider two intersecting planes P_1 and P_2 and two couples acting respectively in P_1 and P_2. We can, without any loss of generality, assume that the couple in P_1 consists of two forces $\mathbf{F}_1$ and $-\mathbf{F}_1$ perpendicular to the line of intersection of the two planes and acting respectively at A and B (Fig. 3.37a). Similarly, we assume that the couple in P_2 consists of two forces $\mathbf{F}_2$ and $-\mathbf{F}_2$ perpendicular to AB and acting respectively at A and B. It is clear that the resultant $\mathbf{R}$ of $\mathbf{F}_1$ and $\mathbf{F}_2$ and the resultant $-\mathbf{R}$ of $-\mathbf{F}_1$ and $-\mathbf{F}_2$ form a couple. Denoting by $\mathbf{r}$ the vector joining B to A and recalling the definition of the moment of a couple (Sec. 3.12), we express the moment $\mathbf{M}$ of the resulting couple as follows:

$$\mathbf{M} = \mathbf{r} \times \mathbf{R} = \mathbf{r} \times (\mathbf{F}_1 + \mathbf{F}_2)$$

and, by Varignon's theorem,

$$\mathbf{M} = \mathbf{r} \times \mathbf{F}_1 + \mathbf{r} \times \mathbf{F}_2$$

But the first term in the expression obtained represents the moment $\mathbf{M}_1$ of the couple in P_1, and the second term represents the moment $\mathbf{M}_2$ of the couple in P_2. We have

$$\mathbf{M} = \mathbf{M}_1 + \mathbf{M}_2 \tag{3.50}$$

and we conclude that the sum of two couples of moments $\mathbf{M}_1$ and $\mathbf{M}_2$ is a couple of moment $\mathbf{M}$ equal to the vector sum of $\mathbf{M}_1$ and $\mathbf{M}_2$ (Fig. 3.37b).

(a)

(b)

Fig. 3.37

3.15. COUPLES CAN BE REPRESENTED BY VECTORS

As we saw in Sec. 3.13, couples which have the same moment, whether they act in the same plane or in parallel planes, are equivalent. There is therefore no need to draw the actual forces forming a given couple in order to define its effect on a rigid body (Fig. 3.38a). It is sufficient to draw an arrow equal in magnitude and direction to the moment $\mathbf{M}$ of the couple (Fig. 3.38b). On the other hand, we saw in Sec. 3.14 that the sum of two couples is itself a couple and that the moment $\mathbf{M}$ of the resultant couple can be obtained by forming the vector sum of the moments $\mathbf{M}_1$ and $\mathbf{M}_2$ of the given couples. Thus, couples obey the law of addition of vectors, and the arrow used in Fig. 3.38b to represent the couple defined in Fig. 3.38a can truly be considered a vector.

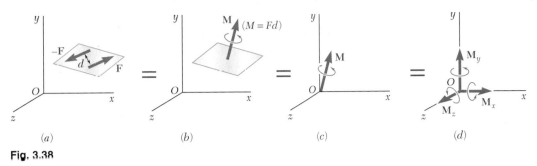

Fig. 3.38

The vector representing a couple is called a *couple vector*. Note that, in Fig. 3.38, a red arrow is used to distinguish the couple vector, *which represents the couple itself*, from the *moment* of the couple, which was represented by a green arrow in earlier figures. Also note that the symbol ↺ is added to this red arrow to avoid any confusion with vectors representing forces. A couple vector, like the moment of a couple, is a free vector. Its point of application, therefore, can be chosen at the origin of the system of coordinates, if so desired (Fig. 3.38c). Furthermore, the couple vector $\mathbf{M}$ can be resolved into component vectors $\mathbf{M}_x$, $\mathbf{M}_y$, and $\mathbf{M}_z$, which are directed along the coordinate axes (Fig. 3.38d). These component vectors represent couples acting, respectively, in the yz, zx, and xy planes.

3.16. RESOLUTION OF A GIVEN FORCE INTO A FORCE AT *O* AND A COUPLE

Consider a force $\mathbf{F}$ acting on a rigid body at a point A defined by the position vector $\mathbf{r}$ (Fig. 3.39a). Suppose that for some reason we would rather have the force act at point O. While we can move $\mathbf{F}$ along its line of action (principle of transmissibility), we cannot move it to a point O which does not lie on the original line of action without modifying the action of $\mathbf{F}$ on the rigid body.

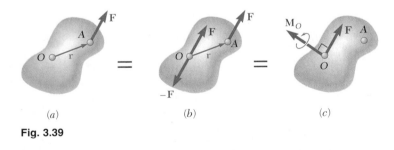

(a)   (b)   (c)

Fig. 3.39

We can, however, attach two forces at point O, one equal to $\mathbf{F}$ and the other equal to $-\mathbf{F}$, without modifying the action of the original force on the rigid body (Fig. 3.39b). As a result of this transformation, a force $\mathbf{F}$ is now applied at O; the other two forces form a

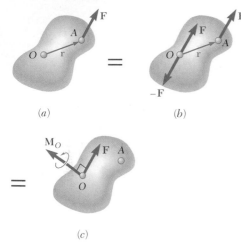

(a)   *(b)*

(c)

Fig. 3.39 (*repeated*)

couple of moment $\mathbf{M}_O = \mathbf{r} \times \mathbf{F}$. Thus, *any force* $\mathbf{F}$ *acting on a rigid body can be moved to an arbitrary point* O *provided that a couple is added whose moment is equal to the moment of* $\mathbf{F}$ *about* O. The couple tends to impart to the rigid body the same rotational motion about O that the force $\mathbf{F}$ tended to produce before it was transferred to O. The couple is represented by a couple vector $\mathbf{M}_O$ perpendicular to the plane containing $\mathbf{r}$ and $\mathbf{F}$. Since $\mathbf{M}_O$ is a free vector, it may be applied anywhere; for convenience, however, the couple vector is usually attached at O, together with $\mathbf{F}$, and the combination obtained is referred to as a *force-couple system* (Fig. 3.39c).

If the force $\mathbf{F}$ had been moved from A to a different point O' (Fig. 3.40a and c), the moment $\mathbf{M}_{O'} = \mathbf{r}' \times \mathbf{F}$ of $\mathbf{F}$ about O' should have been computed, and a new force-couple system, consisting of $\mathbf{F}$ and of the couple vector $\mathbf{M}_{O'}$, would have been attached at O'. The relation existing between the moments of $\mathbf{F}$ about O and O' is obtained by writing

$$\mathbf{M}_{O'} = \mathbf{r}' \times \mathbf{F} = (\mathbf{r} + \mathbf{s}) \times \mathbf{F} = \mathbf{r} \times \mathbf{F} + \mathbf{s} \times \mathbf{F}$$

$$\mathbf{M}_{O'} = \mathbf{M}_O + \mathbf{s} \times \mathbf{F} \tag{3.51}$$

where $\mathbf{s}$ is the vector joining O' to O. Thus, the moment $\mathbf{M}_{O'}$ of $\mathbf{F}$ about O' is obtained by adding to the moment $\mathbf{M}_O$ of $\mathbf{F}$ about O the vector product $\mathbf{s} \times \mathbf{F}$ representing the moment about O' of the force $\mathbf{F}$ applied at O.

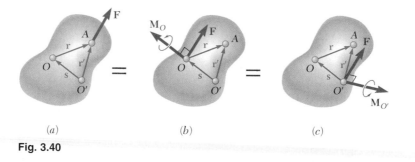

(a)   *(b)*   *(c)*

Fig. 3.40

Photo 3.2 The force exerted by each hand on the wrench could be replaced with an equivalent force-couple system acting on the nut.

This result could also have been established by observing that, in order to transfer to O' the force-couple system attached at O (Fig. 3.40b and c), the couple vector $\mathbf{M}_O$ can be freely moved to O'; to move the force $\mathbf{F}$ from O to O', however, it is necessary to add to $\mathbf{F}$ a couple vector whose moment is equal to the moment about O' of the force $\mathbf{F}$ applied at O. Thus, the couple vector $\mathbf{M}_{O'}$ must be the sum of $\mathbf{M}_O$ and the vector $\mathbf{s} \times \mathbf{F}$.

As noted above, the force-couple system obtained by transferring a force $\mathbf{F}$ from a point A to a point O consists of $\mathbf{F}$ and a couple vector $\mathbf{M}_O$ perpendicular to $\mathbf{F}$. Conversely, any force-couple system consisting of a force $\mathbf{F}$ and a couple vector $\mathbf{M}_O$ which are *mutually perpendicular* can be replaced by a single equivalent force. This is done by moving the force $\mathbf{F}$ in the plane perpendicular to $\mathbf{M}_O$ until its moment about O is equal to the moment of the couple to be eliminated.

SAMPLE PROBLEM 3.6

Determine the components of the single couple equivalent to the two couples shown.

SOLUTION

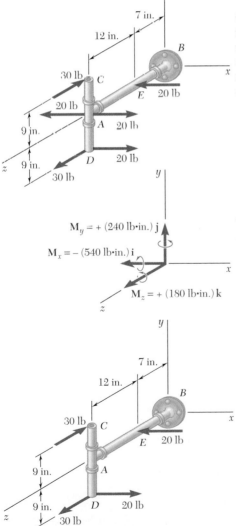

Our computations will be simplified if we attach two equal and opposite 20-lb forces at A. This enables us to replace the original 20-lb-force couple by two new 20-lb-force couples, one of which lies in the zx plane and the other in a plane parallel to the xy plane. The three couples shown in the adjoining sketch can be represented by three couple vectors $\mathbf{M}_x$, $\mathbf{M}_y$, and $\mathbf{M}_z$ directed along the coordinate axes. The corresponding moments are

$$M_x = -(30 \text{ lb})(18 \text{ in.}) = -540 \text{ lb} \cdot \text{in.}$$
$$M_y = +(20 \text{ lb})(12 \text{ in.}) = +240 \text{ lb} \cdot \text{in.}$$
$$M_z = +(20 \text{ lb})(9 \text{ in.}) = +180 \text{ lb} \cdot \text{in.}$$

These three moments represent the components of the single couple $\mathbf{M}$ equivalent to the two given couples. We write

$$\mathbf{M} = -(540 \text{ lb} \cdot \text{in.})\mathbf{i} + (240 \text{ lb} \cdot \text{in.})\mathbf{j} + (180 \text{ lb} \cdot \text{in.})\mathbf{k} \quad \blacktriangleleft$$

Alternative Solution. The components of the equivalent single couple $\mathbf{M}$ can also be obtained by computing the sum of the moments of the four given forces about an arbitrary point. Selecting point D, we write

$$\mathbf{M} = \mathbf{M}_D = (18 \text{ in.})\mathbf{j} \times (-30 \text{ lb})\mathbf{k} + [(9 \text{ in.})\mathbf{j} - (12 \text{ in.})\mathbf{k}] \times (-20 \text{ lb})\mathbf{i}$$

and, after computing the various vector products,

$$\mathbf{M} = -(540 \text{ lb} \cdot \text{in.})\mathbf{i} + (240 \text{ lb} \cdot \text{in.})\mathbf{j} + (180 \text{ lb} \cdot \text{in.})\mathbf{k} \quad \blacktriangleleft$$

113

SAMPLE PROBLEM 3.7

Replace the couple and force shown by an equivalent single force applied to the lever. Determine the distance from the shaft to the point of application of this equivalent force.

SOLUTION

First the given force and couple are replaced by an equivalent force-couple system at O. We move the force $\mathbf{F} = -(400 \text{ N})\mathbf{j}$ to O and at the same time add a couple of moment $\mathbf{M}_O$ equal to the moment about O of the force in its original position.

$$\mathbf{M}_O = \overrightarrow{OB} \times \mathbf{F} = [(0.150 \text{ m})\mathbf{i} + (0.260 \text{ m})\mathbf{j}] \times (-400 \text{ N})\mathbf{j}$$
$$= -(60 \text{ N} \cdot \text{m})\mathbf{k}$$

This couple is added to the couple of moment $-(24 \text{ N} \cdot \text{m})\mathbf{k}$ formed by the two 200-N forces, and a couple of moment $-(84 \text{ N} \cdot \text{m})\mathbf{k}$ is obtained. This last couple can be eliminated by applying $\mathbf{F}$ at a point C chosen in such a way that

$$-(84 \text{ N} \cdot \text{m})\mathbf{k} = \overrightarrow{OC} \times \mathbf{F}$$
$$= [(OC) \cos 60°\mathbf{i} + (OC) \sin 60°\mathbf{j}] \times (-400 \text{ N})\mathbf{j}$$
$$= -(OC) \cos 60°(400 \text{ N})\mathbf{k}$$

We conclude that

$$(OC) \cos 60° = 0.210 \text{ m} = 210 \text{ mm} \qquad OC = 420 \text{ mm} \quad \blacktriangleleft$$

Alternative Solution.   Since the effect of a couple does not depend on its location, the couple of moment $-(24 \text{ N} \cdot \text{m})\mathbf{k}$ can be moved to B; we thus obtain a force-couple system at B. The couple can now be eliminated by applying $\mathbf{F}$ at a point C chosen in such a way that

$$-(24 \text{ N} \cdot \text{m})\mathbf{k} = \overrightarrow{BC} \times \mathbf{F}$$
$$= -(BC) \cos 60°(400 \text{ N})\mathbf{k}$$

We conclude that

$$(BC) \cos 60° = 0.060 \text{ m} = 60 \text{ mm} \qquad BC = 120 \text{ mm}$$
$$OC = OB + BC = 300 \text{ mm} + 120 \text{ mm} \qquad OC = 420 \text{ mm} \quad \blacktriangleleft$$

In this lesson we discussed the properties of *couples*. To solve the problems which follow, you will need to remember that the net effect of a couple is to produce a moment **M**. Since this moment is independent of the point about which it is computed, **M** is a *free vector* and thus remains unchanged as it is moved from point to point. Also, two couples are *equivalent* (that is, they have the same effect on a given rigid body) if they produce the same moment.

When determining the moment of a couple, all previous techniques for computing moments apply. Also, since the moment of a couple is a free vector, it should be computed relative to the most convenient point.

Because the only effect of a couple is to produce a moment, it is possible to represent a couple with a vector, the *couple vector*, which is equal to the moment of the couple. The couple vector is a free vector and will be represented by a special symbol, ↗, to distinguish it from force vectors.

In solving the problems in this lesson, you will be called upon to perform the following operations:

1. Adding two or more couples. This results in a new couple, the moment of which is obtained by adding vectorially the moments of the given couples [Sample Prob. 3.6].

2. Replacing a force with an equivalent force-couple system at a specified point. As explained in Sec. 3.16, the force of the force-couple system is equal to the original force, while the required couple vector is equal to the moment of the original force about the given point. In addition, it is important to observe that the force and the couple vector are perpendicular to each other. Conversely, it follows that a force-couple system can be reduced to a single force only if the force and couple vector are mutually perpendicular (see the next paragraph).

3. Replacing a force-couple system (with F perpendicular to M) with a single equivalent force. Note that the requirement that **F** and **M** be mutually perpendicular will be satisfied in all two-dimensional problems. The single equivalent force is equal to **F** and is applied in such a way that its moment about the original point of application is equal to **M** [Sample Prob. 3.7].

Fig. P3.68

Problems

3.68 A plate in the shape of a parallelogram is acted upon by two couples. Determine (*a*) the moment of the couple formed by the two 21-N forces, (*b*) the perpendicular distance between the 12-N forces if the resultant of the two couples is zero, (*c*) the value of α if the resultant couple is 1.8 N · m clockwise and *d* is 1.05 m.

3.69 A couple **M** of magnitude 10 lb · ft is applied to the handle of a screwdriver to tighten a screw into a block of wood. Determine the magnitudes of the two smallest horizontal forces that are equivalent to **M** if they are applied (*a*) at corners *A* and *D*, (*b*) at corners *B* and *C*, (*c*) anywhere on the block.

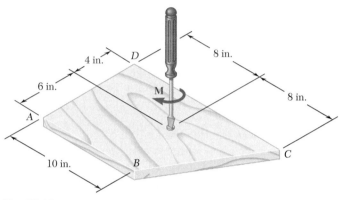

Fig. P3.69

3.70 Two 60-mm-diameter pegs are mounted on a steel plate at *A* and *C*, and two rods are attached to the plate at *B* and *D*. A cord is passed around the pegs and pulled as shown, while the rods exert on the plate 10-N forces as indicated. (*a*) Determine the resulting couple acting on the plate when $T = 36$ N. (*b*) If only the cord is used, in what direction should it be pulled to create the same couple with the minimum tension in the cord? (*c*) Determine the value of that minimum tension.

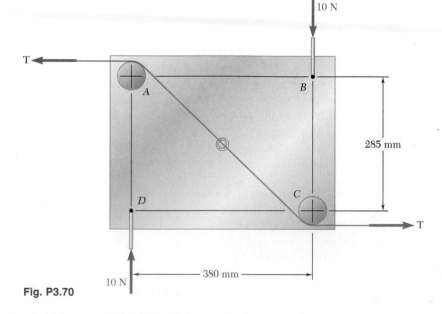

Fig. P3.70

3.71 The steel plate shown will support six 50-mm-diameter idler rollers mounted on the plate as shown. Two flat belts pass around the rollers, and rollers A and D will be adjusted so that the tension in each belt is 45 N. Determine (a) the resultant couple acting on the plate if $a = 0.2$ m, (b) the value of a so that the resultant couple acting on the plate is 54 N · m clockwise.

Fig. P3.71

Fig. P3.72

3.72 The shafts of an angle drive are acted upon by the two couples shown. Replace the two couples with a single equivalent couple, specifying its magnitude and the direction of its axis.

3.73 and 3.74 Knowing that $P = 0$, replace the two remaining couples with a single equivalent couple, specifying its magnitude and the direction of its axis.

Fig. P3.73 and P3.76

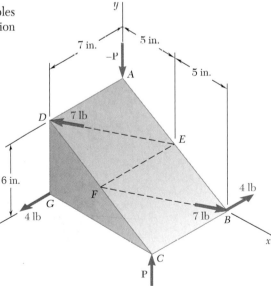

Fig. P3.74 and *P3.75*

3.75 Knowing that $P = 5$ lb, replace the three couples with a single equivalent couple, specifying its magnitude and the direction of its axis.

3.76 Knowing that $P = 210$ N, replace the three couples with a single equivalent couple, specifying its magnitude and the direction of its axis.

Fig. P3.77

3.77 In a manufacturing operation, three holes are drilled simultaneously in a workpiece. Knowing that the holes are perpendicular to the surfaces of the workpiece, replace the couples applied to the drills with a single equivalent couple, specifying its magnitude and the direction of its axis.

3.78 The tension in the cable attached to the end C of an adjustable boom ABC is 1000 N. Replace the force exerted by the cable at C with an equivalent force-couple system (a) at A, (b) at B.

Fig. P3.78

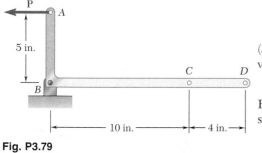

Fig. P3.79

3.79 The 20-lb horizontal force P acts on a bell crank as shown. (a) Replace P with an equivalent force-couple system at B. (b) Find the two vertical forces at C and D which are equivalent to the couple found in part a.

3.80 A 700-N force P is applied at point A of a structural member. Replace P with (a) an equivalent force-couple system at C, (b) an equivalent system consisting of a vertical force at B and a second force at D.

Fig. P3.80

Fig. P3.81 and P3.82

3.81 A landscaper tries to plumb a tree by applying a 240-N force as shown. Two helpers then attempt to plumb the same tree, with one pulling at B and the other pushing with a parallel force at C. Determine these two forces so that they are equivalent to the single 240-N force shown in the figure.

3.82 A landscaper tries to plumb a tree by applying a 240-N force as shown. (a) Replace that force with an equivalent force-couple system at C. (b) Two helpers attempt to plumb the same tree, with one applying a horizontal force at C and the other pulling at B. Determine these two forces if they are to be equivalent to the single force of part a.

3.83   A dirigible is tethered by a cable attached to its cabin at *B*. If the tension in the cable is 250 lb, replace the force exerted by the cable at *B* with an equivalent system formed by two parallel forces applied at *A* and *C*.

3.84   Three workers trying to move a 3 × 3 × 4-ft crate apply to the crate the three horizontal forces shown. (*a*) If *P* = 60 lb, replace the three forces with an equivalent force-couple system at *A*. (*b*) Replace the force-couple system of part *a* with a single force, and determine where it should be applied to side *AB*. (*c*) Determine the magnitude of **P** so that the three forces can be replaced with a single equivalent force applied at *B*.

Fig. P3.83

Fig. P3.84

3.85   A force and a couple are applied to a beam. (*a*) Replace this system with a single force **F** applied at point *G*, and determine the distance *d*. (*b*) Solve part *a* assuming that the directions of the two 600-N forces are reversed.

3.86   Three cables attached to a disk exert on it the forces shown. (*a*) Replace the three forces with an equivalent force-couple system at *A*. (*b*) Determine the single force which is equivalent to the force-couple system obtained in part *a*, and specify its point of application on a line drawn through points *A* and *D*.

Fig. P3.85

Fig. P3.86

Fig. P3.87

Fig. P3.89

3.87   While tapping a hole, a machinist applies the horizontal forces shown to the handle of the tap wrench. Show that these forces are equivalent to a single force, and specify, if possible, the point of application of the single force on the handle.

3.88   A rectangular plate is acted upon by the force and couple shown. This system is to be replaced with a single equivalent force. (*a*) For $\alpha = 40°$, specify the magnitude and the line of action of the equivalent force. (*b*) Specify the value of α if the line of action of the equivalent force is to intersect line *CD* 12 in. to the right of *D*.

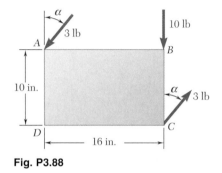

Fig. P3.88

3.89   A hexagonal plate is acted upon by the force **P** and the couple shown. Determine the magnitude and the direction of the smallest force **P** for which this system can be replaced with a single force at *E*.

3.90   An eccentric, compressive 270-lb force **P** is applied to the end of a cantilever beam. Replace **P** with an equivalent force-couple system at *G*.

Fig. P3.90

3.91   Two workers use blocks and tackles attached to the bottom of an I-beam to lift a large cylindrical tank. Knowing that the tension in rope *AB* is 324 N, replace the force exerted at *A* by rope *AB* with an equivalent force-couple system at *E*.

3.92   Two workers use blocks and tackles attached to the bottom of an I-beam to lift a large cylindrical tank. Knowing that the tension in rope *CD* is 366 N, replace the force exerted at *C* by rope *CD* with an equivalent force-couple system at *O*.

Fig. P3.91 and P3.92

3.93 To keep a door closed, a wooden stick is wedged between the floor and the doorknob. The stick exerts at B a 45-lb force directed along line AB. Replace that force with an equivalent force-couple system at C.

3.94 A 25-lb force acting in a vertical plane parallel to the yz plane is applied to the 8-in.-long horizontal handle AB of a socket wrench. Replace the force with an equivalent force-couple system at the origin O of the coordinate system.

Fig. P3.93

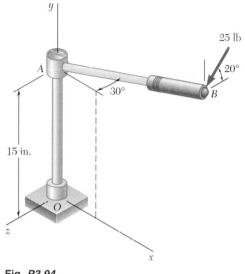

Fig. *P3.94*

3.95 A 315-N force $\mathbf{F}$ and 70-N · m couple $\mathbf{M}$ are applied to corner A of the block shown. Replace the given force-couple system with an equivalent force-couple system at corner D.

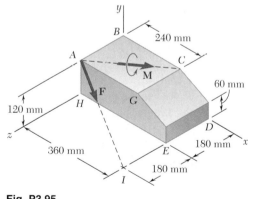

Fig. P3.95

3.96 The handpiece of a miniature industrial grinder weighs 2.4 N, and its center of gravity is located on the y axis. The head of the handpiece is offset in the xz plane in such a way that line BC forms an angle of 25° with the x direction. Show that the weight of the handpiece and the two couples $\mathbf{M}_1$ and $\mathbf{M}_2$ can be replaced with a single equivalent force. Further assuming that $M_1 = 0.068$ N · m and $M_2 = 0.065$ N · m, determine (*a*) the magnitude and the direction of the equivalent force, (*b*) the point where its line of action intersects the xz plane.

Fig. P3.96

3.97 A 20-lb force $\mathbf{F}_1$ and a 40-lb $\cdot$ ft couple $\mathbf{M}_1$ are applied to corner E of the bent plate shown. If $\mathbf{F}_1$ and $\mathbf{M}_1$ are to be replaced with an equivalent force-couple system $(\mathbf{F}_2, \mathbf{M}_2)$ at corner B and if $(M_2)_z = 0$, determine (a) the distance d, (b) $\mathbf{F}_2$ and $\mathbf{M}_2$.

Fig. P3.97

3.17. REDUCTION OF A SYSTEM OF FORCES TO ONE FORCE AND ONE COUPLE

Consider a system of forces $\mathbf{F}_1$, $\mathbf{F}_2$, $\mathbf{F}_3$, . . . , acting on a rigid body at the points A_1, A_2, A_3, . . . , *defined by the position vectors* $\mathbf{r}_1$, $\mathbf{r}_2$, $\mathbf{r}_3$, *etc.* (Fig. 3.41a). As seen in the preceding section, $\mathbf{F}_1$ can be moved from A_1 to a given point O if a couple of moment $\mathbf{M}_1$ equal to the moment $\mathbf{r}_1 \times \mathbf{F}_1$ of $\mathbf{F}_1$ about O is added to the original system of forces. Repeating this procedure with $\mathbf{F}_2$, $\mathbf{F}_3$, . . . , we obtain the system shown in Fig. 3.41b, which consists of the original forces, now acting at O, and the added couple vectors. Since the forces are now concurrent, they can be added vectorially and replaced by their resultant $\mathbf{R}$. Similarly, the couple vectors $\mathbf{M}_1$, $\mathbf{M}_2$, $\mathbf{M}_3$, . . . , can be added vectorially and replaced by a single couple vector $\mathbf{M}_O^R$. Any system of forces, however complex, can thus be reduced to an *equivalent force-couple system acting at a given point O* (Fig. 3.41c). We should note that while each of the couple vectors $\mathbf{M}_1$, $\mathbf{M}_2$, $\mathbf{M}_3$, . . . , in Fig. 3.41b is perpendicular to its corresponding force, the resultant force $\mathbf{R}$ and the resultant couple vector $\mathbf{M}_O^R$ in Fig. 3.41c will not, in general, be perpendicular to each other.

Fig. 3.41

The equivalent force-couple system is defined by the equations

$$\mathbf{R} = \Sigma\mathbf{F} \qquad \mathbf{M}_O^R = \Sigma\mathbf{M}_O = \Sigma(\mathbf{r} \times \mathbf{F}) \qquad (3.52)$$

which express that the force $\mathbf{R}$ is obtained by adding all the forces of the system, while the moment of the resultant couple vector $\mathbf{M}_O^R$, called the *moment resultant* of the system, is obtained by adding the moments about O of all the forces of the system.

Once a given system of forces has been reduced to a force and a couple at a point O, it can easily be reduced to a force and a couple at another point O'. While the resultant force $\mathbf{R}$ will remain unchanged, the new moment resultant $\mathbf{M}_{O'}^R$ will be equal to the sum of $\mathbf{M}_O^R$ and the moment about O' of the force $\mathbf{R}$ attached at O (Fig. 3.42). We have

$$\mathbf{M}_{O'}^R = \mathbf{M}_O^R + \mathbf{s} \times \mathbf{R} \qquad (3.53)$$

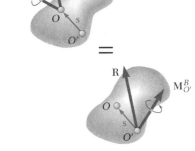

Fig. 3.42

In practice, the reduction of a given system of forces to a single force $\mathbf{R}$ at O and a couple vector $\mathbf{M}_O^R$ will be carried out in terms of components. Resolving each position vector $\mathbf{r}$ and each force $\mathbf{F}$ of the system into rectangular components, we write

$$\mathbf{r} = x\mathbf{i} + y\mathbf{j} + z\mathbf{k} \qquad (3.54)$$
$$\mathbf{F} = F_x\mathbf{i} + F_y\mathbf{j} + F_z\mathbf{k} \qquad (3.55)$$

Substituting for $\mathbf{r}$ and $\mathbf{F}$ in (3.52) and factoring out the unit vectors $\mathbf{i}$, $\mathbf{j}$, $\mathbf{k}$, we obtain $\mathbf{R}$ and $\mathbf{M}_O^R$ in the form

$$\mathbf{R} = R_x\mathbf{i} + R_y\mathbf{j} + R_z\mathbf{k} \qquad \mathbf{M}_O^R = M_x^R\mathbf{i} + M_y^R\mathbf{j} + M_z^R\mathbf{k} \qquad (3.56)$$

The components R_x, R_y, R_z represent, respectively, the sums of the x, y, and z components of the given forces and measure the tendency of the system to impart to the rigid body a motion of translation in the x, y, or z direction. Similarly, the components M_x^R, M_y^R, M_z^R represent, respectively, the sum of the moments of the given forces about the x, y, and z axes and measure the tendency of the system to impart to the rigid body a motion of rotation about the x, y, or z axis.

If the magnitude and direction of the force $\mathbf{R}$ are desired, they can be obtained from the components R_x, R_y, R_z by means of the relations (2.18) and (2.19) of Sec. 2.12; similar computations will yield the magnitude and direction of the couple vector $\mathbf{M}_O^R$.

3.18. EQUIVALENT SYSTEMS OF FORCES

We saw in the preceding section that any system of forces acting on a rigid body can be reduced to a force-couple system at a given point O. This equivalent force-couple system characterizes completely the effect of the given force system on the rigid body. *Two systems of forces are equivalent, therefore, if they can be reduced to the same*

Photo 3.3 The forces exerted by the tow rope and the rear tugboat on the ship can be replaced with an equivalent force-couple system when analyzing the motion of the ship.

force-couple system at a given point O Recalling that the force-couple system at O is defined by the relations (3.52), we state that *two systems of forces,* $\mathbf{F}_1$, $\mathbf{F}_2$, $\mathbf{F}_3$, . . . , *and* $\mathbf{F}'_1$, $\mathbf{F}'_2$, $\mathbf{F}'_3$, . . . , *which act on the same rigid body are equivalent if, and only if, the sums of the forces and the sums of the moments about a given point O of the forces of the two systems are, respectively, equal.* Expressed mathematically, the necessary and sufficient conditions for the two systems of forces to be equivalent are

$$\Sigma \mathbf{F} = \Sigma \mathbf{F}' \qquad \text{and} \qquad \Sigma \mathbf{M}_O = \Sigma \mathbf{M}'_O \qquad (3.57)$$

Note that to prove that two systems of forces are equivalent, the second of the relations (3.57) must be established with respect to *only one point O.* It will hold, however, with respect to *any point* if the two systems are equivalent.

Resolving the forces and moments in (3.57) into their rectangular components, we can express the necessary and sufficient conditions for the equivalence of two systems of forces acting on a rigid body as follows:

$$\begin{array}{ccc} \Sigma F_x = \Sigma F'_x & \Sigma F_y = \Sigma F'_y & \Sigma F_z = \Sigma F'_z \\ \Sigma M_x = \Sigma M'_x & \Sigma M_y = \Sigma M'_y & \Sigma M_z = \Sigma M'_z \end{array} \qquad (3.58)$$

These equations have a simple physical significance. They express that two systems of forces are equivalent if they tend to impart to the rigid body (1) the same translation in the x, y, and z directions, respectively, and (2) the same rotation about the x, y, and z axes, respectively.

3.19. EQUIPOLLENT SYSTEMS OF VECTORS

In general, when two systems of vectors satisfy Eqs. (3.57) or (3.58), that is, when their resultants and their moment resultants about an arbitrary point O are respectively equal, the two systems are said to be *equipollent.* The result established in the preceding section can thus be restated as follows: *If two systems of forces acting on a rigid body are equipollent, they are also equivalent.*

It is important to note that this statement does not apply to *any* system of vectors. Consider, for example, a system of forces acting on a set of independent particles which do *not* form a rigid body. A different system of forces acting on the same particles may happen to be equipollent to the first one; that is, it may have the same resultant and the same moment resultant. Yet, since different forces will now act on the various particles, their effects on these particles will be different; the two systems of forces, while equipollent, are *not equivalent.*

3.20. FURTHER REDUCTION OF A SYSTEM OF FORCES

We saw in Sec. 3.17 that any given system of forces acting on a rigid body can be reduced to an equivalent force-couple system at O consisting of a force $\mathbf{R}$ equal to the sum of the forces of the system and a couple vector $\mathbf{M}_O^R$ of moment equal to the moment resultant of the system.

When $\mathbf{R} = 0$, the force-couple system reduces to the couple vector $\mathbf{M}_O^R$. The given system of forces can then be reduced to a single couple, called the *resultant couple* of the system.

Let us now investigate the conditions under which a given system of forces can be reduced to a single force. It follows from Sec. 3.16 that the force-couple system at O can be replaced by a single force $\mathbf{R}$ acting along a new line of action if $\mathbf{R}$ and $\mathbf{M}_O^R$ are mutually perpendicular. The systems of forces which can be reduced to a single force, or *resultant,* are therefore the systems for which the force $\mathbf{R}$ and the couple vector $\mathbf{M}_O^R$ are mutually perpendicular. While this condition *is generally not satisfied* by systems of forces in space, it *will be satisfied* by systems consisting of (1) concurrent forces, (2) coplanar forces, or (3) parallel forces. These three cases will be discussed separately.

1. *Concurrent forces* are applied at the same point and can therefore be added directly to obtain their resultant $\mathbf{R}$. Thus, they always reduce to a single force. Concurrent forces were discussed in detail in Chap. 2.

2. *Coplanar forces* act in the same plane, which may be assumed to be the plane of the figure (Fig. 3.43*a*). The sum $\mathbf{R}$ of the forces of the system will also lie in the plane of the figure, while the moment of each force about O, and thus the moment resultant $\mathbf{M}_O^R$, will be perpendicular to that plane. The force-couple system at O consists, therefore, of a force $\mathbf{R}$ and a couple vector $\mathbf{M}_O^R$ which are mutually perpendicular (Fig. 3.43*b*).† They can be reduced to a single force $\mathbf{R}$ by moving $\mathbf{R}$ in the plane of the figure until its moment about O becomes equal to $\mathbf{M}_O^R$. The distance from O to the line of action of $\mathbf{R}$ is $d = M_O^R/R$ (Fig. 3.43*c*).

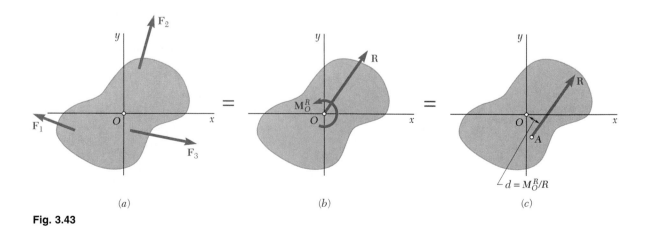

(a)   (b)   (c)

Fig. 3.43

†Since the couple vector $\mathbf{M}_O^R$ is perpendicular to the plane of the figure, it has been represented by the symbol ↺. A counterclockwise couple ↺ represents a vector pointing out of the paper, and a clockwise couple ↻ represents a vector pointing into the paper.

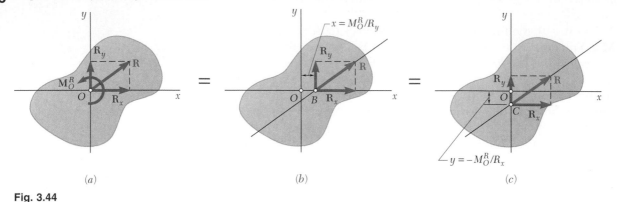

Fig. 3.44

As noted in Sec. 3.17, the reduction of a system of forces is considerably simplified if the forces are resolved into rectangular components. The force-couple system at O is then characterized by the components (Fig. 3.44a)

$$R_x = \Sigma F_x \qquad R_y = \Sigma F_y \qquad M_z^R = M_O^R = \Sigma M_O \qquad (3.59)$$

To reduce the system to a single force $\mathbf{R}$, we express that the moment of $\mathbf{R}$ about O must be equal to $\mathbf{M}_O^R$. Denoting by x and y the coordinates of the point of application of the resultant and recalling formula (3.22) of Sec. 3.8, we write

$$xR_y - yR_x = M_O^R$$

which represents the equation of the line of action of $\mathbf{R}$. We can also determine directly the x and y intercepts of the line of action of the resultant by noting that $\mathbf{M}_O^R$ must be equal to the moment about O of the y component of $\mathbf{R}$ when $\mathbf{R}$ is attached at B (Fig. 3.44b) and to the moment of its x component when $\mathbf{R}$ is attached at C (Fig. 3.44c).

3. *Parallel forces* have parallel lines of action and may or may not have the same sense. Assuming here that the forces are parallel to the y axis (Fig. 3.45a), we note that their sum $\mathbf{R}$ will also be parallel to the y axis. On the other hand, since the moment of a given force must be perpendicular to that force, the moment about O of each force of the system, and thus the moment resultant $\mathbf{M}_O^R$, will lie in the zx plane. The force-couple system at O consists, therefore, of a force $\mathbf{R}$ and

Fig. 3.45

a couple vector $\mathbf{M}_O^R$ which are mutually perpendicular (Fig. 3.45b). They can be reduced to a single force $\mathbf{R}$ (Fig. 3.45c) or, if $\mathbf{R} = 0$, to a single couple of moment $\mathbf{M}_O^R$.

In practice, the force-couple system at O will be characterized by the components

$$R_y = \Sigma F_y \qquad M_x^R = \Sigma M_x \qquad M_z^R = \Sigma M_z \qquad (3.60)$$

The reduction of the system to a single force can be carried out by moving $\mathbf{R}$ to a new point of application $A(x, 0, z)$ chosen so that the moment of $\mathbf{R}$ about O is equal to $\mathbf{M}_O^R$. We write

$$\mathbf{r} \times \mathbf{R} = \mathbf{M}_O^R$$
$$(x\mathbf{i} + z\mathbf{k}) \times R_y\mathbf{j} = M_x^R\mathbf{i} + M_z^R\mathbf{k}$$

By computing the vector products and equating the coefficients of the corresponding unit vectors in both members of the equation, we obtain two scalar equations which define the coordinates of A:

$$-zR_y = M_x^R \qquad xR_y = M_z^R$$

These equations express that the moments of $\mathbf{R}$ about the x and z axes must, respectively, be equal to M_x^R and M_z^R.

Photo 3.4 The parallel wind forces acting on the highway signs can be reduced to a single equivalent force. Determining this force can simplify the calculation of the forces acting on the supports of the frame to which the signs are attached.

*3.21. REDUCTION OF A SYSTEM OF FORCES TO A WRENCH

In the general case of a system of forces in space, the equivalent force-couple system at O consists of a force $\mathbf{R}$ and a couple vector $\mathbf{M}_O^R$ which are not perpendicular, and neither of which is zero (Fig. 3.46a). Thus, the system of forces *cannot* be reduced to a single force or to a single couple. The couple vector, however, can be replaced by two other couple vectors obtained by resolving $\mathbf{M}_O^R$ into a component $\mathbf{M}_1$ along $\mathbf{R}$ and a component $\mathbf{M}_2$ in a plane perpendicular to $\mathbf{R}$ (Fig. 3.46b). The couple vector $\mathbf{M}_2$ and the force $\mathbf{R}$ can then be replaced by a single force $\mathbf{R}$ acting along a new line of action. The original system of forces thus reduces to $\mathbf{R}$ and to the couple vector $\mathbf{M}_1$ (Fig. 3.46c), that is, to $\mathbf{R}$ and a couple acting in the plane perpendicular to $\mathbf{R}$. This particular force-couple system is called a *wrench*, and the resulting combination of push and twist is found in drilling and tapping operations

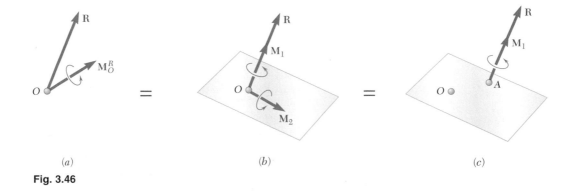

(a)   (b)   (c)

Fig. 3.46

and in the tightening and the loosening of screws. The line of action of **R** is known as the *axis of the wrench*, and the ratio $p = M_1/R$ is called the *pitch* of the wrench. A wrench, therefore, consists of two collinear vectors, namely, a force **R** and a couple vector

$$\mathbf{M}_1 = p\mathbf{R} \tag{3.61}$$

Recalling the expression (3.35) obtained in Sec. 3.9 for the projection of a vector on the line of action of another vector, we note that the projection of $\mathbf{M}_O^R$ on the line of action of **R** is

$$M_1 = \frac{\mathbf{R} \cdot \mathbf{M}_O^R}{R}$$

Thus, the pitch of the wrench can be expressed as†

$$p = \frac{M_1}{R} = \frac{\mathbf{R} \cdot \mathbf{M}_O^R}{R^2} \tag{3.62}$$

To define the axis of the wrench, we can write a relation involving the position vector **r** of an arbitrary point P located on that axis. Attaching the resultant force **R** and couple vector $\mathbf{M}_1$ at P (Fig. 3.47) and expressing that the moment about O of this force-couple system is equal to the moment resultant $\mathbf{M}_O^R$ of the original force system, we write

$$\mathbf{M}_1 + \mathbf{r} \times \mathbf{R} = \mathbf{M}_O^R \tag{3.63}$$

or, recalling Eq. (3.61),

$$p\mathbf{R} + \mathbf{r} \times \mathbf{R} = \mathbf{M}_O^R \tag{3.64}$$

Photo 3.5 The pushing-turning action associated with the tightening of a screw illustrates the collinear lines of action of the force and couple vector that constitute a wrench.

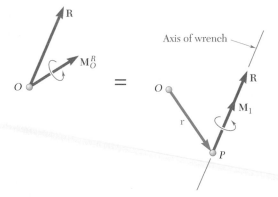

Fig. 3.47

†The expressions obtained for the projection of the couple vector on the line of action of **R** and for the pitch of the wrench are independent of the choice of point O. Using the relation (3.53) of Sec. 3.17, we note that if a different point O' had been used, the numerator in (3.62) would have been

$$\mathbf{R} \cdot \mathbf{M}_{O'}^R = \mathbf{R} \cdot (\mathbf{M}_O^R + \mathbf{s} \times \mathbf{R}) = \mathbf{R} \cdot \mathbf{M}_O^R + \mathbf{R} \cdot (\mathbf{s} \times \mathbf{R})$$

Since the mixed triple product $\mathbf{R} \cdot (\mathbf{s} \times \mathbf{R})$ is identically equal to zero, we have

$$\mathbf{R} \cdot \mathbf{M}_{O'}^R = \mathbf{R} \cdot \mathbf{M}_O^R$$

Thus, the scalar product $\mathbf{R} \cdot \mathbf{M}_O^R$ is independent of the choice of point O.

SAMPLE PROBLEM 3.8

A 4.80-m-long beam is subjected to the forces shown. Reduce the given system of forces to (a) an equivalent force-couple system at A, (b) an equivalent force-couple system at B, (c) a single force or resultant.

Note. Since the reactions at the supports are not included in the given system of forces, the given system will not maintain the beam in equilibrium.

SOLUTION

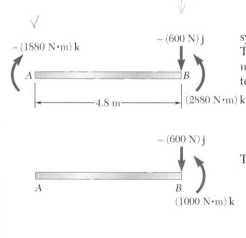

a. **Force-Couple System at A.** The force-couple system at A equivalent to the given system of forces consists of a force **R** and a couple $\mathbf{M}_A^R$ defined as follows:

$$\mathbf{R} = \Sigma\mathbf{F}$$
$$= (150\text{ N})\mathbf{j} - (600\text{ N})\mathbf{j} + (100\text{ N})\mathbf{j} - (250\text{ N})\mathbf{j} = -(600\text{ N})\mathbf{j}$$
$$\mathbf{M}_A^R = \Sigma(\mathbf{r} \times \mathbf{F})$$
$$= (1.6\mathbf{i}) \times (-600\mathbf{j}) + (2.8\mathbf{i}) \times (100\mathbf{j}) + (4.8\mathbf{i}) \times (-250\mathbf{j})$$
$$= -(1880\text{ N}\cdot\text{m})\mathbf{k}$$

The equivalent force-couple system at A is thus

$$\mathbf{R} = 600\text{ N} \downarrow \qquad \mathbf{M}_A^R = 1880\text{ N}\cdot\text{m} \; \downarrow \quad \blacktriangleleft$$

b. **Force-Couple System at B.** We propose to find a force-couple system at B equivalent to the force-couple system at A determined in part *a*. The force **R** is unchanged, but a new couple $\mathbf{M}_B^R$ must be determined, the moment of which is equal to the moment about B of the force-couple system determined in part *a*. Thus, we have

$$\mathbf{M}_B^R = \mathbf{M}_A^R + \overrightarrow{BA} \times \mathbf{R}$$
$$= -(1880\text{ N}\cdot\text{m})\mathbf{k} + (-4.8\text{ m})\mathbf{i} \times (-600\text{ N})\mathbf{j}$$
$$= -(1880\text{ N}\cdot\text{m})\mathbf{k} + (2880\text{ N}\cdot\text{m})\mathbf{k} = +(1000\text{ N}\cdot\text{m})\mathbf{k}$$

The equivalent force-couple system at B is thus

$$\mathbf{R} = 600\text{ N} \downarrow \qquad \mathbf{M}_B^R = 1000\text{ N}\cdot\text{m} \; \uparrow \quad \blacktriangleleft$$

c. **Single Force or Resultant.** The resultant of the given system of forces is equal to **R**, and its point of application must be such that the moment of **R** about A is equal to $\mathbf{M}_A^R$. We write

$$\mathbf{r} \times \mathbf{R} = \mathbf{M}_A^R$$
$$x\mathbf{i} \times (-600\text{ N})\mathbf{j} = -(1880\text{ N}\cdot\text{m})\mathbf{k}$$
$$-x(600\text{ N})\mathbf{k} = -(1880\text{ N}\cdot\text{m})\mathbf{k}$$

and conclude that $x = 3.13$ m. Thus, the single force equivalent to the given system is defined as

$$\mathbf{R} = 600\text{ N} \downarrow \qquad x = 3.13\text{ m} \quad \blacktriangleleft$$

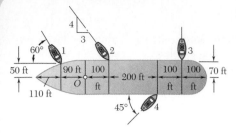

SAMPLE PROBLEM 3.9

Four tugboats are used to bring an ocean liner to its pier. Each tugboat exerts a 5000-lb force in the direction shown. Determine (a) the equivalent force-couple system at the foremast O, (b) the point on the hull where a single, more powerful tugboat should push to produce the same effect as the original four tugboats.

SOLUTION

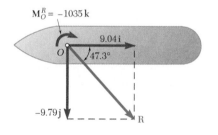

a. **Force-Couple System at O.** Each of the given forces is resolved into components in the diagram shown (kip units are used). The force-couple system at O equivalent to the given system of forces consists of a force **R** and a couple $\mathbf{M}_O^R$ defined as follows:

$$\mathbf{R} = \Sigma\mathbf{F}$$
$$= (2.50\mathbf{i} - 4.33\mathbf{j}) + (3.00\mathbf{i} - 4.00\mathbf{j}) + (-5.00\mathbf{j}) + (3.54\mathbf{i} + 3.54\mathbf{j})$$
$$= 9.04\mathbf{i} - 9.79\mathbf{j}$$

$$\mathbf{M}_O^R = \Sigma(\mathbf{r} \times \mathbf{F})$$
$$= (-90\mathbf{i} + 50\mathbf{j}) \times (2.50\mathbf{i} - 4.33\mathbf{j})$$
$$+ (100\mathbf{i} + 70\mathbf{j}) \times (3.00\mathbf{i} - 4.00\mathbf{j})$$
$$+ (400\mathbf{i} + 70\mathbf{j}) \times (-5.00\mathbf{j})$$
$$+ (300\mathbf{i} - 70\mathbf{j}) \times (3.54\mathbf{i} + 3.54\mathbf{j})$$
$$= (390 - 125 - 400 - 210 - 2000 + 1062 + 248)\mathbf{k}$$
$$= -1035\mathbf{k}$$

The equivalent force-couple system at O is thus

$$\mathbf{R} = (9.04 \text{ kips})\mathbf{i} - (9.79 \text{ kips})\mathbf{j} \qquad \mathbf{M}_O^R = -(1035 \text{ kip} \cdot \text{ft})\mathbf{k}$$

or $\qquad\qquad \mathbf{R} = 13.33 \text{ kips} \;\diagdown 47.3° \qquad \mathbf{M}_O^R = 1035 \text{ kip} \cdot \text{ft} \downarrow$ ◄

Remark. Since all the forces are contained in the plane of the figure, we could have expected the sum of their moments to be perpendicular to that plane. Note that the moment of each force component could have been obtained directly from the diagram by first forming the product of its magnitude and perpendicular distance to O and then assigning to this product a positive or a negative sign depending upon the sense of the moment.

b. **Single Tugboat.** The force exerted by a single tugboat must be equal to **R**, and its point of application A must be such that the moment of **R** about O is equal to $\mathbf{M}_O^R$. Observing that the position vector of A is

$$\mathbf{r} = x\mathbf{i} + 70\mathbf{j}$$

we write

$$\mathbf{r} \times \mathbf{R} = \mathbf{M}_O^R$$
$$(x\mathbf{i} + 70\mathbf{j}) \times (9.04\mathbf{i} - 9.79\mathbf{j}) = -1035\mathbf{k}$$
$$-x(9.79)\mathbf{k} - 633\mathbf{k} = -1035\mathbf{k} \qquad x = 41.1 \text{ ft} ◄$$

SAMPLE PROBLEM 3.10

Three cables are attached to a bracket as shown. Replace the forces exerted by the cables with an equivalent force-couple system at A.

SOLUTION

We first determine the relative position vectors drawn from point A to the points of application of the various forces and resolve the forces into rectangular components. Observing that $\mathbf{F}_B = (700 \text{ N})\boldsymbol{\lambda}_{BE}$ where

$$\boldsymbol{\lambda}_{BE} = \frac{\overrightarrow{BE}}{BE} = \frac{75\mathbf{i} - 150\mathbf{j} + 50\mathbf{k}}{175}$$

we have, using meters and newtons,

$$\mathbf{r}_{B/A} = \overrightarrow{AB} = 0.075\mathbf{i} + 0.050\mathbf{k} \qquad \mathbf{F}_B = 300\mathbf{i} - 600\mathbf{j} + 200\mathbf{k}$$
$$\mathbf{r}_{C/A} = \overrightarrow{AC} = 0.075\mathbf{i} - 0.050\mathbf{k} \qquad \mathbf{F}_C = 707\mathbf{i} \qquad - 707\mathbf{k}$$
$$\mathbf{r}_{D/A} = \overrightarrow{AD} = 0.100\mathbf{i} - 0.100\mathbf{j} \qquad \mathbf{F}_D = 600\mathbf{i} + 1039\mathbf{j}$$

The force-couple system at A equivalent to the given forces consists of a force $\mathbf{R} = \Sigma\mathbf{F}$ and a couple $\mathbf{M}_A^R = \Sigma(\mathbf{r} \times \mathbf{F})$. The force $\mathbf{R}$ is readily obtained by adding respectively the x, y, and z components of the forces:

$$\mathbf{R} = \Sigma\mathbf{F} = (1607 \text{ N})\mathbf{i} + (439 \text{ N})\mathbf{j} - (507 \text{ N})\mathbf{k} \quad \blacktriangleleft$$

The computation of $\mathbf{M}_A^R$ will be facilitated if we express the moments of the forces in the form of determinants (Sec. 3.8):

$$\mathbf{r}_{B/A} \times \mathbf{F}_B = \begin{vmatrix} \mathbf{i} & \mathbf{j} & \mathbf{k} \\ 0.075 & 0 & 0.050 \\ 300 & -600 & 200 \end{vmatrix} = 30\mathbf{i} \qquad -45\mathbf{k}$$

$$\mathbf{r}_{C/A} \times \mathbf{F}_C = \begin{vmatrix} \mathbf{i} & \mathbf{j} & \mathbf{k} \\ 0.075 & 0 & -0.050 \\ 707 & 0 & -707 \end{vmatrix} = \qquad 17.68\mathbf{j}$$

$$\mathbf{r}_{D/A} \times \mathbf{F}_D = \begin{vmatrix} \mathbf{i} & \mathbf{j} & \mathbf{k} \\ 0.100 & -0.100 & 0 \\ 600 & 1039 & 0 \end{vmatrix} = \qquad 163.9\mathbf{k}$$

Adding the expressions obtained, we have

$$\mathbf{M}_A^R = \Sigma(\mathbf{r} \times \mathbf{F}) = (30 \text{ N} \cdot \text{m})\mathbf{i} + (17.68 \text{ N} \cdot \text{m})\mathbf{j} + (118.9 \text{ N} \cdot \text{m})\mathbf{k} \quad \blacktriangleleft$$

The rectangular components of the force $\mathbf{R}$ and the couple $\mathbf{M}_A^R$ are shown in the adjoining sketch.

SAMPLE PROBLEM 3.11

A square foundation mat supports the four columns shown. Determine the magnitude and point of application of the resultant of the four loads.

SOLUTION

We first reduce the given system of forces to a force-couple system at the origin O of the coordinate system. This force-couple system consists of a force $\mathbf{R}$ and a couple vector $\mathbf{M}_O^R$ defined as follows:

$$\mathbf{R} = \Sigma\mathbf{F} \qquad \mathbf{M}_O^R = \Sigma(\mathbf{r} \times \mathbf{F})$$

The position vectors of the points of application of the various forces are determined, and the computations are arranged in tabular form.

$\mathbf{r}$, ft	$\mathbf{F}$, kips	$\mathbf{r} \times \mathbf{F}$, kip · ft
0	$-40\mathbf{j}$	0
$10\mathbf{i}$	$-12\mathbf{j}$	$-120\mathbf{k}$
$10\mathbf{i} + 5\mathbf{k}$	$-8\mathbf{j}$	$40\mathbf{i} - 80\mathbf{k}$
$4\mathbf{i} + 10\mathbf{k}$	$-20\mathbf{j}$	$200\mathbf{i} - 80\mathbf{k}$
	$\mathbf{R} = -80\mathbf{j}$	$\mathbf{M}_O^R = 240\mathbf{i} - 280\mathbf{k}$

Since the force $\mathbf{R}$ and the couple vector $\mathbf{M}_O^R$ are mutually perpendicular, the force-couple system obtained can be reduced further to a single force $\mathbf{R}$. The new point of application of $\mathbf{R}$ will be selected in the plane of the mat and in such a way that the moment of $\mathbf{R}$ about O will be equal to $\mathbf{M}_O^R$. Denoting by $\mathbf{r}$ the position vector of the desired point of application, and by x and z its coordinates, we write

$$\mathbf{r} \times \mathbf{R} = \mathbf{M}_O^R$$
$$(x\mathbf{i} + z\mathbf{k}) \times (-80\mathbf{j}) = 240\mathbf{i} - 280\mathbf{k}$$
$$-80x\mathbf{k} + 80z\mathbf{i} = 240\mathbf{i} - 280\mathbf{k}$$

from which it follows that

$$-80x = -280 \qquad 80z = 240$$
$$x = 3.50 \text{ ft} \qquad z = 3.00 \text{ ft}$$

We conclude that the resultant of the given system of forces is

$$\mathbf{R} = 80 \text{ kips} \downarrow \qquad \text{at } x = 3.50 \text{ ft}, z = 3.00 \text{ ft} \blacktriangleleft$$

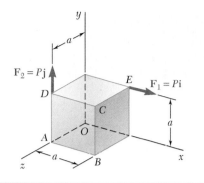

SAMPLE PROBLEM 3.12

Two forces of the same magnitude P act on a cube of side a as shown. Replace the two forces by an equivalent wrench, and determine (a) the magnitude and direction of the resultant force $\mathbf{R}$, (b) the pitch of the wrench, (c) the point where the axis of the wrench intersects the yz plane.

SOLUTION

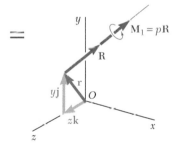

Equivalent Force-Couple System at O. We first determine the equivalent force-couple system at the origin O. We observe that the position vectors of the points of application E and D of the two given forces are $\mathbf{r}_E = a\mathbf{i} + a\mathbf{j}$ and $\mathbf{r}_D = a\mathbf{j} + a\mathbf{k}$. The resultant $\mathbf{R}$ of the two forces and their moment resultant $\mathbf{M}_O^R$ about O are

$$\mathbf{R} = \mathbf{F}_1 + \mathbf{F}_2 = P\mathbf{i} + P\mathbf{j} = P(\mathbf{i} + \mathbf{j}) \tag{1}$$
$$\mathbf{M}_O^R = \mathbf{r}_E \times \mathbf{F}_1 + \mathbf{r}_D \times \mathbf{F}_2 = (a\mathbf{i} + a\mathbf{j}) \times P\mathbf{i} + (a\mathbf{j} + a\mathbf{k}) \times P\mathbf{j}$$
$$= -Pa\mathbf{k} - Pa\mathbf{i} = -Pa(\mathbf{i} + \mathbf{k}) \tag{2}$$

a. Resultant Force $\mathbf{R}$. It follows from Eq. (1) and the adjoining sketch that the resultant force $\mathbf{R}$ has the magnitude $R = P\sqrt{2}$, lies in the xy plane, and forms angles of $45°$ with the x and y axes. Thus

$$R = P\sqrt{2} \qquad \theta_x = \theta_y = 45° \qquad \theta_z = 90° \quad \blacktriangleleft$$

b. Pitch of Wrench. Recalling formula (3.62) of Sec. 3.21 and Eqs. (1) and (2) above, we write

$$p = \frac{\mathbf{R} \cdot \mathbf{M}_O^R}{R^2} = \frac{P(\mathbf{i} + \mathbf{j}) \cdot (-Pa)(\mathbf{i} + \mathbf{k})}{(P\sqrt{2})^2} = \frac{-P^2a(1 + 0 + 0)}{2P^2} \qquad p = -\frac{a}{2} \quad \blacktriangleleft$$

c. Axis of Wrench. It follows from the above and from Eq. (3.61) that the wrench consists of the force $\mathbf{R}$ found in (1) and the couple vector

$$\mathbf{M}_1 = p\mathbf{R} = -\frac{a}{2}P(\mathbf{i} + \mathbf{j}) = -\frac{Pa}{2}(\mathbf{i} + \mathbf{j}) \tag{3}$$

To find the point where the axis of the wrench intersects the yz plane, we express that the moment of the wrench about O is equal to the moment resultant $\mathbf{M}_O^R$ of the original system:

$$\mathbf{M}_1 + \mathbf{r} \times \mathbf{R} = \mathbf{M}_O^R$$

or, noting that $\mathbf{r} = y\mathbf{j} + z\mathbf{k}$ and substituting for $\mathbf{R}$, $\mathbf{M}_O^R$, and $\mathbf{M}_1$ from Eqs. (1), (2), and (3),

$$-\frac{Pa}{2}(\mathbf{i} + \mathbf{j}) + (y\mathbf{j} + z\mathbf{k}) \times P(\mathbf{i} + \mathbf{j}) = -Pa(\mathbf{i} + \mathbf{k})$$

$$-\frac{Pa}{2}\mathbf{i} - \frac{Pa}{2}\mathbf{j} - Py\mathbf{k} + Pz\mathbf{j} - Pz\mathbf{i} = -Pa\mathbf{i} - Pa\mathbf{k}$$

Equating the coefficients of $\mathbf{k}$, and then the coefficients of $\mathbf{j}$, we find

$$y = a \qquad z = a/2 \quad \blacktriangleleft$$

This lesson was devoted to the reduction and simplification of force systems. In solving the problems which follow, you will be asked to perform the operations discussed below.

1. Reducing a force system to a force and a couple at a given point A. The force is the *resultant* **R** of the system and is obtained by adding the various forces; the moment of the couple is the *moment resultant* of the system and is obtained by adding the moments about A of the various forces. We have

$$\mathbf{R} = \Sigma\mathbf{F} \qquad \mathbf{M}_A^R = \Sigma(\mathbf{r} \times \mathbf{F})$$

where the position vector **r** is drawn from A to *any point* on the line of action of **F**.

2. Moving a force-couple system from point A to point B. If you wish to reduce a given force system to a force-couple system at point B after you have reduced it to a force-couple system at point A, you need not recompute the moments of the forces about B. The resultant **R** remains unchanged, and the new moment resultant $\mathbf{M}_B^R$ can be obtained by adding to $\mathbf{M}_A^R$ the moment about B of the force **R** applied at A [Sample Prob. 3.8]. Denoting by **s** the vector drawn from B to A, you can write

$$\mathbf{M}_B^R = \mathbf{M}_A^R + \mathbf{s} \times \mathbf{R}$$

3. Checking whether two force systems are equivalent. First reduce each force system to a force-couple system *at the same, but arbitrary, point A* (as explained in paragraph 1). The two systems are equivalent (that is, they have the same effect on the given rigid body) if the two force-couple systems you have obtained are identical, that is, if

$$\Sigma\mathbf{F} = \Sigma\mathbf{F}' \qquad \text{and} \qquad \Sigma\mathbf{M}_A = \Sigma\mathbf{M}_A'$$

You should recognize that if the first of these equations is not satisfied, that is, if the two systems do not have the same resultant **R**, the two systems cannot be equivalent and there is then no need to check whether or not the second equation is satisfied.

4. Reducing a given force system to a single force. First reduce the given system to a force-couple system consisting of the resultant **R** and the couple vector $\mathbf{M}_A^R$ at some convenient point A (as explained in paragraph 1). You will recall from the previous lesson that further reduction to a single force is possible *only if the force* **R** *and the couple vector* $\mathbf{M}_A^R$ *are mutually perpendicular.* This will certainly be the case for systems of forces which are either *concurrent, coplanar,* or *parallel.* The required single force can then be obtained by moving **R** until its moment about A is equal to $\mathbf{M}_A^R$, as you did in several problems of the preceding lesson. More formally, you can write that the position vector **r** drawn from A to any point on the line of action of the single force **R** must satisfy the equation

$$\mathbf{r} \times \mathbf{R} = \mathbf{M}_A^R$$

This procedure was used in Sample Probs. 3.8, 3.9, and 3.11.

5. *Reducing a given force system to a wrench.* If the given system is comprised of forces which are not concurrent, coplanar, or parallel, the equivalent force-couple system at a point A will consist of a force **R** and a couple vector $\mathbf{M}_A^R$ which, in general, *are not mutually perpendicular.* (To check whether **R** and $\mathbf{M}_A^R$ are mutually perpendicular, form their scalar product. If this product is zero, they are mutually perpendicular; otherwise, they are not.) If **R** and $\mathbf{M}_A^R$ are not mutually perpendicular, the force-couple system (and thus the given system of forces) *cannot be reduced to a single force.* However, the system can be reduced to a *wrench*— the combination of a force **R** and a couple vector $\mathbf{M}_1$ directed along a common line of action called the *axis of the wrench* (Fig. 3.47). The ratio $p = M_1/R$ is called the *pitch* of the wrench.

To reduce a given force system to a wrench, you should follow these steps:

    a. Reduce the given system to an equivalent force-couple system ($\mathbf{R}$, $\mathbf{M}_O^R$), typically located at the origin O.

    b. Determine the pitch p from Eq. (3.62)

$$p = \frac{M_1}{R} = \frac{\mathbf{R} \cdot \mathbf{M}_O^R}{R^2} \tag{3.62}$$

and the couple vector from $\mathbf{M}_1 = p\mathbf{R}$.

    c. Express that the moment about O of the wrench is equal to the moment resultant $\mathbf{M}_O^R$ of the force-couple system at O:

$$\mathbf{M}_1 + \mathbf{r} \times \mathbf{R} = \mathbf{M}_O^R \tag{3.63}$$

This equation allows you to determine the point where the line of action of the wrench intersects a specified plane, since the position vector **r** is directed from O to that point.

These steps are illustrated in Sample Prob. 3.12. Although the determination of a wrench and the point where its axis intersects a plane may appear difficult, the process is simply the application of several of the ideas and techniques developed in this chapter. Thus, once you have mastered the wrench, you can feel confident that you understand much of Chap. 3.

Fig. P3.98

Fig. P3.99

3.98 A 4-ft-long beam is subjected to a variety of loadings. (*a*) Replace each loading with an equivalent force-couple system at end *A* of the beam. (*b*) Which of the loadings are equivalent?

3.99 A 4-ft-long beam is loaded as shown. Determine the loading of Prob. 3.98 which is equivalent to this loading.

3.100 Determine the single equivalent force and the distance from point *A* to its line of action for the beam and loading of (*a*) Prob. 3.98*b*, (*b*) Prob. 3.98*d*, (*c*) Prob. 3.98*e*.

3.101 Five separate force-couple systems act at the corners of a metal block, which has been machined into the shape shown. Determine which of these systems is equivalent to a force $\mathbf{F} = (10\ \text{N})\mathbf{j}$ and a couple of moment $\mathbf{M} = (6\ \text{N} \cdot \text{m})\mathbf{i} + (4\ \text{N} \cdot \text{m})\mathbf{k}$ located at point *A*.

Fig. P3.101

3.102 The masses of two children sitting at ends A and B of a seesaw are 38 kg and 29 kg, respectively. Where should a third child sit so that the resultant of the weights of the three children will pass through C if she has a mass of (a) 27 kg, (b) 24 kg.

3.103 Three stage lights are mounted on a pipe as shown. The mass of each light is $m_A = m_B = 1.8$ kg and $m_C = 1.6$ kg. (a) If $d = 0.75$ m, determine the distance from D to the line of action of the resultant of the weights of the three lights. (b) Determine the value of d so that the resultant of the weights passes through the midpoint of the pipe.

Fig. P3.102

Fig. P3.103

3.104 Three hikers are shown crossing a footbridge. Knowing that the weights of the hikers at points C, D, and E are 800 N, 700 N, and 540 N, respectively, determine (a) the horizontal distance from A to the line of action of the resultant of the three weights when $a = 1.1$ m, (b) the value of a so that the loads on the bridge supports at A and B are equal.

3.105 Gear C is rigidly attached to arm AB. If the forces and couple shown can be reduced to a single equivalent force at A, determine the equivalent force and the magnitude of the couple **M**.

Fig. P3.104

Fig. P3.105

Fig. P3.106

3.106 To test the strength of a 25 × 20-in. suitcase, forces are applied as shown. Knowing that $P = 18$ lb, (*a*) determine the resultant of the applied forces, (*b*) locate the two points where the line of action of the resultant intersects the edge of the suitcase.

3.107 Solve Prob. 3.106 assuming that $P = 28$ lb.

3.108 As four holes are punched simultaneously in a piece of aluminum sheet metal, the punches exert on the piece the forces shown. Knowing that the forces are perpendicular to the surfaces of the piece, determine (*a*) the resultant of the applied forces when $\alpha = 45°$ and the point of intersection of the line of action of that resultant with a line drawn through points *A* and *B*, (*b*) the value of α so that the line of action of the resultant passes through fold *EF*.

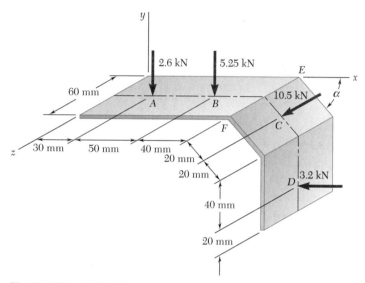

Fig. P3.108 and P3.109

3.109 As four holes are punched simultaneously in a piece of aluminum sheet metal, the punches exert on the piece the forces shown. Knowing that the forces are perpendicular to the surfaces of the piece, determine (*a*) the value of α so that the resultant of the applied forces is parallel to the 10.5 N force, (*b*) the corresponding resultant of the applied forces and the point of intersection of its line of action with a line drawn through points *A* and *B*.

3.110 A truss supports the loading shown. Determine the equivalent force acting on the truss and the point of intersection of its line of action with a line through points *A* and *G*.

3.111 Three forces and a couple act on crank *ABC*. For $P = 5$ lb and $\alpha = 40°$, (*a*) determine the resultant of the given system of forces, (*b*) locate the point where the line of action of the resultant intersects a line drawn through points *B* and *C*, (*c*) locate the point where the line of action of the resultant intersects a line drawn through points *A* and *B*.

3.112 Three forces and a couple act on crank *ABC*. Determine the value of *d* so that the given system of forces is equivalent to zero at (*a*) point *B*, (*b*) point *D*.

Fig. P3.110

Fig. P3.111 and *P3.112*

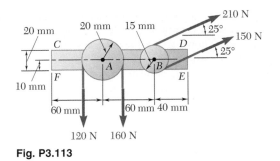

Fig. P3.113

3.113 Pulleys A and B are mounted on bracket $CDEF$. The tension on each side of the two belts is as shown. Replace the four forces with a single equivalent force, and determine where its line of action intersects the bottom edge of the bracket.

3.114 As follower AB rolls along the surface of member C, it exerts a constant force $\mathbf{F}$ perpendicular to the surface. (*a*) Replace $\mathbf{F}$ with an equivalent force-couple system at the point D obtained by drawing the perpendicular from the point of contact to the x axis. (*b*) For $a = 1$ m and $b = 2$ m, determine the value of x for which the moment of the equivalent force-couple system at D is maximum.

3.115 As plastic bushings are inserted into a 3-in.-diameter cylindrical sheet metal container, the insertion tool exerts the forces shown on the enclosure. Each of the forces is parallel to one of the coordinate axes. Replace these forces with an equivalent force-couple system at C.

3.116 Two 300-mm-diameter pulleys are mounted on line shaft AD. The belts B and C lie in vertical planes parallel to the yz plane. Replace the belt forces shown with an equivalent force-couple system at A.

Fig. P3.114

Fig. P3.115

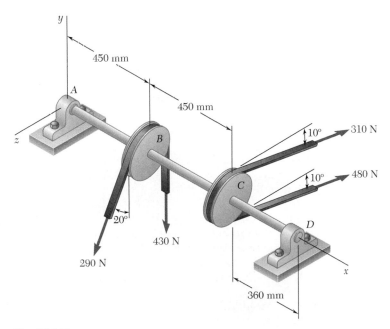

Fig. P3.116

3.117 A mechanic uses a crowfoot wrench to loosen a bolt at C. The mechanic holds the socket wrench handle at points A and B and applies forces at these points. Knowing that these forces are equivalent to a force-couple system at C consisting of the force $\mathbf{C} = -(40 \text{ N})\mathbf{i} + (20 \text{ N})\mathbf{k}$ and the couple $\mathbf{M}_C = (40 \text{ N} \cdot \text{m})\mathbf{i}$, determine the forces applied at A and B when $A_z = 10$ N.

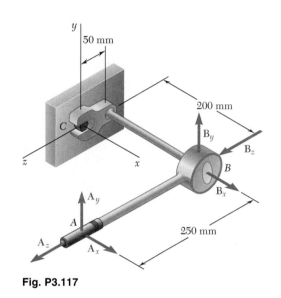

Fig. P3.117

3.118 While using a pencil sharpener, a student applies the forces and couple shown. (*a*) Determine the forces exerted at B and C knowing that these forces and the couple are equivalent to a force-couple system at A consisting of the force $\mathbf{R} = (3.9 \text{ lb})\mathbf{i} + R_y\mathbf{j} - (1.1 \text{ lb})\mathbf{k}$ and the couple $\mathbf{M}_A^R = M_x\mathbf{i} + (1.5 \text{ lb} \cdot \text{ft})\mathbf{j} - (1.1 \text{ lb} \cdot \text{ft})\mathbf{k}$. (*b*) Find the corresponding values of R_y and M_x.

Fig. *P3.118*

3.119 A portion of the flue for a furnace is attached to the ceiling at A. While supporting the free end of the flue at F, a worker pushes in at E and pulls out at F to align end E with the furnace. Knowing that the 10-lb force at F lies in a plane parallel to the yz plane, determine (a) the angle α the force at F should form with the horizontal if duct AB is not to tend to rotate about the vertical, (b) the force-couple system at B equivalent to the given force system when this condition is satisfied.

Fig. P3.119 and P3.120

3.120 A portion of the flue for a furnace is attached to the ceiling at A. While supporting the free end of the flue at F, a worker pushes in at E and pulls out at F to align end E with the furnace. Knowing that the 10-lb force at F lies in a plane parallel to the yz plane and that $\alpha = 60°$, (a) replace the given force system with an equivalent force-couple system at C, (b) determine whether duct CD will tend to rotate clockwise or counterclockwise relative to elbow C, as viewed from D to C.

3.121 The head-and-motor assembly of a radial drill press was originally positioned with arm AB parallel to the z axis and the axis of the chuck and bit parallel to the y axis. The assembly was then rotated 25° about the y axis and 20° about the centerline of the horizontal arm AB, bringing it into the position shown. The drilling process was started by switching on the motor and rotating the handle to bring the bit into contact with the workpiece. Replace the force and couple exerted by the drill press with an equivalent force-couple system at the center O of the base of the vertical column.

Fig. P3.121

Fig. P3.122

3.122 While a sagging porch is leveled and repaired, a screw jack is used to support the front of the porch. As the jack is expanded, it exerts on the porch the force-couple system shown, where $R = 300$ N and $M = 37.5$ N · m. Replace this force-couple system with an equivalent force-couple system at C.

3.123 Three children are standing on a 15×15-ft raft. If the weights of the children at points A, B, and C are 85 lb, 60 lb, and 90 lb, respectively, determine the magnitude and the point of application of the resultant of the three weights.

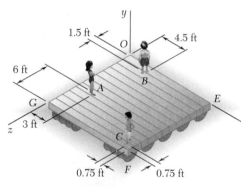

Fig. P3.123 and P3.124

3.124 Three children are standing on a 15×15-ft raft. The weights of the children at points A, B, and C are 85 lb, 60 lb, and 90 lb, respectively. If a fourth child of weight 95 lb climbs onto the raft, determine where she should stand if the other children remain in the positions shown and the line of action of the resultant of the four weights is to pass through the center of the raft.

3.125 The forces shown are the resultant downward loads on sections of the flat roof of a building because of accumulated snow. Determine the magnitude and the point of application of the resultant of these four loads.

Fig. P3.125 and P3.126

3.126 The forces shown are the resultant downward loads on sections of the flat roof of a building because of accumulated snow. If the snow represented by the 580-kN force is shoveled so that this load acts at E, determine a and b knowing that the point of application of the resultant of the four loads is then at B.

*3.127 A group of students loads a 2 × 4-m flatbed trailer with two 0.6 × 0.6 × 0.6-m boxes and one 0.6 × 0.6 × 1.2-m box. Each of the boxes at the rear of the trailer is positioned so that it is aligned with both the back and a side of the trailer. Determine the smallest load the students should place in a second 0.6 × 0.6 × 1.2-m box and where on the trailer they should secure it, without any part of the box overhanging the sides of the trailer, if each box is uniformly loaded and the line of action of the resultant of the weights of the four boxes is to pass through the point of intersection of the centerlines of the trailer and the axle. (*Hint:* Keep in mind that the box may be placed either on its side or on its end.)

200 N

400 N

180 N

1.6 m

A

C

B

1.8 m

2.2 m

Fig. _P3.127_

*3.128 Solve Prob. 3.127 if the students want to place as much weight as possible in the fourth box and that at least one side of the box must coincide with a side of the trailer.

*3.129 A block of wood is acted upon by three forces of the same magnitude *P* and having the directions shown. Replace the three forces with an equivalent wrench and determine (*a*) the magnitude and direction of the resultant **R**, (*b*) the pitch of the wrench, (*c*) the point where the axis of the wrench intersects the *xy* plane.

*3.130 A piece of sheet metal is bent into the shape shown and is acted upon by three forces. Replace the three forces with an equivalent wrench and determine (*a*) the magnitude and direction of the resultant **R**, (*b*) the pitch of the wrench, (*c*) the point where the axis of the wrench intersects the *yz* plane.

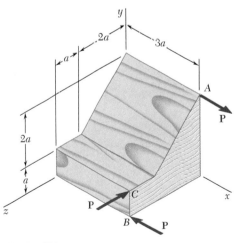

y

2*a*

3*a*

a

A

P

2*a*

a

C

P

z

B

P

x

Fig. P3.129

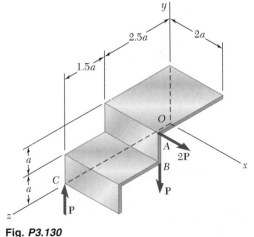

y

2.5*a*

2*a*

1.5*a*

O

A

2P

B

a

C

a

P

P

x

z

Fig. _P3.130_

*3.131 and *3.132   The forces and couples shown are applied to two screws as a piece of sheet metal is fastened to a block of wood. Reduce the forces and the couples to an equivalent wrench and determine (*a*) the resultant force **R**, (*b*) the pitch of the wrench, (*c*) the point where the axis of the wrench intersects the *xz* plane.

Fig. P3.131

Fig. *P3.132*

*3.133 and *3.134   Two bolts *A* and *B* are tightened by applying the forces and couple shown. Replace the two wrenches with a single equivalent wrench and determine (*a*) the resultant **R**, (*b*) the pitch of the single equivalent wrench, (*c*) the point where the axis of the wrench intersects the *xz* plane.

Fig. P3.133

Fig. *P3.134*

*3.135 A flagpole is guyed by three cables. If the tensions in the cables have the same magnitude P, replace the forces exerted on the pole with an equivalent wrench and determine (a) the resultant force $\mathbf{R}$, (b) the pitch of the wrench, (c) the point where the axis of the wrench intersects the xz plane.

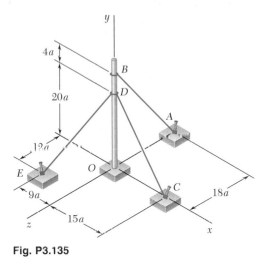

Fig. P3.135

*3.136 and *3.137 Determine whether the force-and-couple system shown can be reduced to a single equivalent force $\mathbf{R}$. If it can, determine $\mathbf{R}$ and the point where the line of action of $\mathbf{R}$ intersects the yz plane. If it cannot be so reduced, replace the given system with an equivalent wrench and determine its resultant, its pitch, and the point where its axis intersects the yz plane.

Fig. *P3.136*

Fig. P3.137

*3.138 Replace the wrench shown with an equivalent system consisting of two forces perpendicular to the y axis and applied respectively at A and B.

*3.139 Show that, in general, a wrench can be replaced with two forces chosen in such a way that one force passes through a given point while the other force lies in a given plane.

*3.140 Show that a wrench can be replaced with two perpendicular forces, one of which is applied at a given point.

*3.141 Show that a wrench can be replaced with two forces, one of which has a prescribed line of action.

Fig. P3.138

Principle of transmissibility

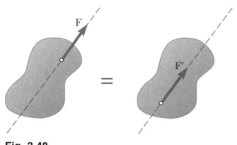

Fig. 3.48

Vector product of two vectors

(a)

(b)

Fig. 3.49

Fig. 3.50

In this chapter we studied the effect of forces exerted on a rigid body. We first learned to distinguish between *external* and *internal* forces [Sec. 3.2] and saw that, according to the *principle of transmissibility,* the effect of an external force on a rigid body remains unchanged if that force is moved along its line of action [Sec. 3.3]. In other words, two forces **F** and **F′** acting on a rigid body at two different points have the same effect on that body if they have the same magnitude, same direction, and same line of action (Fig. 3.48). Two such forces are said to be *equivalent*.

Before proceeding with the discussion of *equivalent systems of forces,* we introduced the concept of the *vector product of two vectors* [Sec. 3.4]. The vector product

$$\mathbf{V} = \mathbf{P} \times \mathbf{Q}$$

of the vectors **P** and **Q** was defined as a vector perpendicular to the plane containing **P** and **Q** (Fig. 3.49), of magnitude

$$V = PQ \sin \theta \qquad (3.1)$$

and directed in such a way that a person located at the tip of **V** will observe as counterclockwise the rotation through θ which brings the vector **P** in line with the vector **Q**. The three vectors **P**, **Q**, and **V**—taken in that order—are said to form a *right-handed triad.* It follows that the vector products **Q** × **P** and **P** × **Q** are represented by equal and opposite vectors. We have

$$\mathbf{Q} \times \mathbf{P} = -(\mathbf{P} \times \mathbf{Q}) \qquad (3.4)$$

It also follows from the definition of the vector product of two vectors that the vector products of the unit vectors **i**, **j**, and **k** are

$$\mathbf{i} \times \mathbf{i} = 0 \qquad \mathbf{i} \times \mathbf{j} = \mathbf{k} \qquad \mathbf{j} \times \mathbf{i} = -\mathbf{k}$$

and so on. The sign of the vector product of two unit vectors can be obtained by arranging in a circle and in counterclockwise order the three letters representing the unit vectors (Fig. 3.50): The vector product of two unit vectors will be positive if they follow each other in counterclockwise order and negative if they follow each other in clockwise order.

The *rectangular components of the vector product* **V** of two vectors **P** and **Q** were expressed [Sec. 3.5] as

$$V_x = P_y Q_z - P_z Q_y$$
$$V_y = P_z Q_x - P_x Q_z \qquad (3.9)$$
$$V_z = P_x Q_y - P_y Q_x$$

Rectangular components of vector product

Using a determinant, we also wrote

$$\mathbf{V} = \begin{vmatrix} \mathbf{i} & \mathbf{j} & \mathbf{k} \\ P_x & P_y & P_z \\ Q_x & Q_y & Q_z \end{vmatrix} \qquad (3.10)$$

The *moment of a force* **F** *about a point* O was defined [Sec. 3.6] as the vector product

Moment of a force about a point

$$\mathbf{M}_O = \mathbf{r} \times \mathbf{F} \qquad (3.11)$$

where **r** is the *position vector* drawn from O to the point of application A of the force **F** (Fig. 3.51). Denoting by θ the angle between the lines of action of **r** and **F**, we found that the magnitude of the moment of **F** about O can be expressed as

$$M_O = rF \sin \theta = Fd \qquad (3.12)$$

where d represents the perpendicular distance from O to the line of action of **F**.

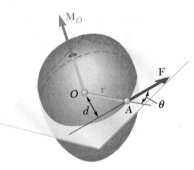

Fig. 3.51

The *rectangular components of the moment* $\mathbf{M}_O$ *of a force* **F** were expressed [Sec. 3.8] as

Rectangular components of moment

$$M_x = yF_z - zF_y$$
$$M_y = zF_x - xF_z \qquad (3.18)$$
$$M_z = xF_y - yF_x$$

where x, y, z are the components of the position vector **r** (Fig. 3.52). Using a determinant form, we also wrote

$$\mathbf{M}_O = \begin{vmatrix} \mathbf{i} & \mathbf{j} & \mathbf{k} \\ x & y & z \\ F_x & F_y & F_z \end{vmatrix} \qquad (3.19)$$

In the more general case of the moment about an arbitrary point B of a force **F** applied at A, we had

$$\mathbf{M}_B = \begin{vmatrix} \mathbf{i} & \mathbf{j} & \mathbf{k} \\ x_{A/B} & y_{A/B} & z_{A/B} \\ F_x & F_y & F_z \end{vmatrix} \qquad (3.21)$$

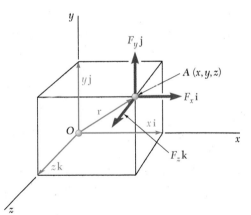

Fig. 3.52

where $x_{A/B}$, $y_{A/B}$, and $z_{A/B}$ denote the components of the vector $\mathbf{r}_{A/B}$:

$$x_{A/B} = x_A - x_B \qquad y_{A/B} = y_A - y_B \qquad z_{A/B} = z_A - z_B$$

In the case of *problems involving only two dimensions,* the force **F** can be assumed to lie in the *xy* plane. Its moment **M**$_B$ about a point *B* in the same plane is perpendicular to that plane (Fig. 3.53) and is completely defined by the scalar

$$M_B = (x_A - x_B)F_y - (y_A - y_B)F_x \qquad (3.23)$$

Fig. 3.53

Various methods for the computation of the moment of a force about a point were illustrated in Sample Probs. 3.1 through 3.4.

Scalar product of two vectors

Fig. 3.54

The *scalar product* of two vectors **P** and **Q** [Sec. 3.9] was denoted by **P · Q** and was defined as the scalar quantity

$$\mathbf{P} \cdot \mathbf{Q} = PQ \cos \theta \qquad (3.24)$$

where θ is the angle between **P** and **Q** (Fig. 3.54). By expressing the scalar product of **P** and **Q** in terms of the rectangular components of the two vectors, we determined that

$$\mathbf{P} \cdot \mathbf{Q} = P_x Q_x + P_y Q_y + P_z Q_z \qquad (3.30)$$

Projection of a vector on an axis

Fig. 3.55

The *projection of a vector* **P** *on an axis OL* (Fig. 3.55) can be obtained by forming the scalar product of **P** and the unit vector **λ** along *OL*. We have

$$P_{OL} = \mathbf{P} \cdot \boldsymbol{\lambda} \qquad (3.36)$$

or, using rectangular components,

$$P_{OL} = P_x \cos \theta_x + P_y \cos \theta_y + P_z \cos \theta_z \qquad (3.37)$$

where θ_x, θ_y, and θ_z denote the angles that the axis *OL* forms with the coordinate axes.

Mixed triple product of three vectors

The *mixed triple product* of the three vectors **S**, **P**, and **Q** was defined as the scalar expression

$$\mathbf{S} \cdot (\mathbf{P} \times \mathbf{Q}) \qquad (3.38)$$

obtained by forming the scalar product of **S** with the vector product of **P** and **Q** [Sec. 3.10]. It was shown that

$$\mathbf{S} \cdot (\mathbf{P} \times \mathbf{Q}) = \begin{vmatrix} S_x & S_y & S_z \\ P_x & P_y & P_z \\ Q_x & Q_y & Q_z \end{vmatrix} \quad (3.41)$$

where the elements of the determinant are the rectangular components of the three vectors.

The *moment of a force* **F** *about an axis* OL [Sec. 3.11] was defined as the projection OC on OL of the moment $\mathbf{M}_O$ of the force **F** (Fig. 3.56), that is, as the mixed triple product of the unit vector **λ**, the position vector **r**, and the force **F**:

$$M_{OL} = \mathbf{\lambda} \cdot \mathbf{M}_O = \mathbf{\lambda} \cdot (\mathbf{r} \times \mathbf{F}) \quad (3.42)$$

Using the determinant form for the mixed triple product, we have

$$M_{OL} = \begin{vmatrix} \lambda_x & \lambda_y & \lambda_z \\ x & y & z \\ F_x & F_y & F_z \end{vmatrix} \quad (3.43)$$

where $\lambda_x, \lambda_y, \lambda_z$ = direction cosines of axis OL
x, y, z = components of **r**
F_x, F_y, F_z = components of **F**

An example of the determination of the moment of a force about a skew axis was given in Sample Prob. 3.5.

Two forces **F** *and* $-\mathbf{F}$ *having the same magnitude, parallel lines of action, and opposite sense are said to form a couple* [Sec. 3.12]. It was shown that the moment of a couple is independent of the point about which it is computed; it is a vector **M** perpendicular to the plane of the couple and equal in magnitude to the product of the common magnitude F of the forces and the perpendicular distance d between their lines of action (Fig. 3.57).

Two couples having the same moment **M** are *equivalent*, that is, they have the same effect on a given rigid body [Sec. 3.13]. The sum of two couples is itself a couple [Sec. 3.14], and the moment **M** of the resultant couple can be obtained by adding vectorially the moments $\mathbf{M}_1$ and $\mathbf{M}_2$ of the original couples [Sample Prob. 3.6]. It follows that a couple can be represented by a vector, called a *couple vector*, equal in magnitude and direction to the moment **M** of the couple [Sec. 3.15]. A couple vector is a *free vector* which can be attached to the origin O if so desired and resolved into components (Fig. 3.58).

Moment of a force about an axis

Fig. 3.56

Couples

Fig. 3.57

Fig. 3.58

Fig. 3.59

Force-couple system

Any force $\mathbf{F}$ acting at a point A of a rigid body can be replaced by a *force-couple system* at an arbitrary point O, consisting of the force $\mathbf{F}$ applied at O and a couple of moment $\mathbf{M}_O$ equal to the moment about O of the force $\mathbf{F}$ in its original position [Sec. 3.16]; it should be noted that the force $\mathbf{F}$ and the couple vector $\mathbf{M}_O$ are always perpendicular to each other (Fig. 3.59).

Reduction of a system of forces to a force-couple system

It follows [Sec. 3.17] that *any system of forces can be reduced to a force-couple system at a given point O* by first replacing each of the forces of the system by an equivalent force-couple system at O (Fig. 3.60) and then adding all the forces and all the couples determined in this manner to obtain a resultant force $\mathbf{R}$ and a resultant couple vector $\mathbf{M}_O^R$ [Sample Probs. 3.8 through 3.11]. Note that, in general, the resultant $\mathbf{R}$ and the couple vector $\mathbf{M}_O^R$ will not be perpendicular to each other.

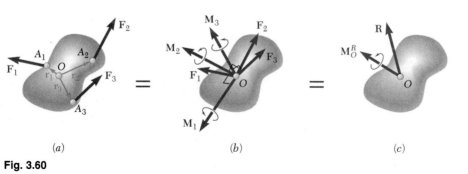

(a)   (b)   (c)

Fig. 3.60

Equivalent systems of forces

We concluded from the above [Sec. 3.18] that, as far as a rigid body is concerned, *two systems of forces, $\mathbf{F}_1$, $\mathbf{F}_2$, $\mathbf{F}_3$, . . . and $\mathbf{F}_1'$, $\mathbf{F}_2'$, $\mathbf{F}_3'$, . . . , are equivalent if, and only if,*

$$\Sigma\mathbf{F} = \Sigma\mathbf{F}' \quad \text{and} \quad \Sigma\mathbf{M}_O = \Sigma\mathbf{M}_O' \quad (3.57)$$

Further reduction of a system of forces

If the resultant force $\mathbf{R}$ and the resultant couple vector $\mathbf{M}_O^R$ are perpendicular to each other, the force-couple system at O can be further reduced to a single resultant force [Sec. 3.20]. This will be the case for systems consisting either of (a) concurrent forces (cf. Chap. 2), (b) coplanar forces [Sample Probs. 3.8 and 3.9], or (c) parallel forces [Sample Prob. 3.11]. If the resultant $\mathbf{R}$ and the couple vector $\mathbf{M}_O^R$ are *not* perpendicular to each other, the system *cannot* be reduced to a single force. It can, however, be reduced to a special type of force-couple system called a *wrench*, consisting of the resultant $\mathbf{R}$ and a couple vector $\mathbf{M}_1$ directed along $\mathbf{R}$ [Sec. 3.21 and Sample Prob. 3.12].

Review Problems

3.142 A worker tries to move a rock by applying a 360-N force to a steel bar as shown. (*a*) Replace that force with an equivalent force-couple system at *D*. (*b*) Two workers attempt to move the same rock by applying a vertical force at *A* and another force at *D*. Determine these two forces if they are to be equivalent to the single force of part *a*.

Fig. P3.142 and *P3.143*

3.143 A worker tries to move a rock by applying a 360-N force to a steel bar as shown. If two workers attempt to move the same rock by applying a force at *A* and a parallel force at *C*, determine these two forces so that they will be equivalent to the single 360-N force shown in the figure.

3.144 A force and a couple are applied as shown to the end of a cantilever beam. (*a*) Replace this system with a single force **F** applied at point *C*, and determine the distance *d* from *C* to a line drawn through points *D* and *E*. (*b*) Solve part *a* if the directions of the two 360-N forces are reversed.

3.145 A crate of mass 80 kg is held in the position shown. Determine (*a*) the moment produced by the weight **W** of the crate about *E*, (*b*) the smallest force applied at *B* which creates a moment of equal magnitude and opposite sense about *E*.

3.146 A crate of mass 80 kg is held in the position shown. Determine (*a*) the moment produced by the weight **W** of the crate about *E*, (*b*) the smallest force applied at *A* which creates a moment of equal magnitude and opposite sense about *E*, (*c*) the magnitude, sense, and point of application on the bottom of the crate of the smallest vertical force which creates a moment of equal magnitude and opposite sense about *E*.

3.147 A farmer uses cables and winch pullers *B* and *E* to plumb one side of a small barn. Knowing that the sum of the moments about the *x* axis of the forces exerted by the cables on the barn at points *A* and *D* is equal to 4728 lb · ft, determine the magnitude of $\mathbf{T}_{DE}$ when $T_{AB} = 255$ lb.

3.148 Solve Prob. 3.147 when the tension in cable *AB* is 306 lb.

Fig. P3.144

Fig. P3.145 and *P3.146*

Fig. P3.147

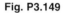

Fig. P3.149

3.149 As an adjustable brace BC is used to bring a wall into plumb, the force-couple system shown is exerted on the wall. Replace this force-couple system with an equivalent force-couple system at A knowing that $R = 21.2$ lb and $M = 13.25$ lb $\cdot$ ft.

3.150 Two parallel 60-N forces are applied to a lever as shown. Determine the moment of the couple formed by the two forces (a) by resolving each force into horizontal and vertical components and adding the moments of the two resulting couples, (b) by using the perpendicular distance between the two forces, (c) by summing the moments of the two forces about point A.

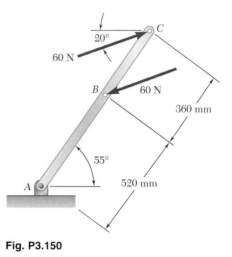

Fig. P3.150

3.151 A 32-lb motor is mounted on the floor. Find the resultant of the weight and the forces exerted on the belt, and determine where the line of action of the resultant intersects the floor.

Fig. P3.151

Fig. P3.152 and *P3.153*

3.152 To loosen a frozen valve, a force $\mathbf{F}$ of magnitude 70 lb is applied to the handle of the valve. Knowing that $\theta = 25°$, $M_x = -61$ lb $\cdot$ ft, and $M_z = -43$ lb $\cdot$ ft, determine ϕ and d.

3.153 When a force $\mathbf{F}$ is applied to the handle of the valve shown, its moments about the x and z axes are, respectively, $M_x = -77$ lb $\cdot$ ft and $M_z = -81$ lb $\cdot$ ft. For $d = 27$ in., determine the moment M_y of $\mathbf{F}$ about the y axis.

Computer Problems

3.C1 Rod *AB* is held in place by a cord *AC* that has a tension *T*. Using computational software, determine the moment about *B* of the force exerted by the cord at point *A* in terms of the tension *T* and the distance *c*. Plot the moment about *B* for 320 mm $\leq c \leq$ 960 mm when (*a*) *T* = 50 N, (*b*) *T* = 75 N, (*c*) *T* = 100 N.

Fig. P3.C1

3.C2 A 400-mm tube *AB* can slide freely along a horizontal rod. The ends *A* and *B* of the tube are connected by elastic cords fixed to point *C*. (*a*) Determine the angle θ between the two cords *AC* and *BC* as a function of *x*. (*b*) Plot the angle θ between the cords *AC* and *BC* as a function of *x* for −400 mm $\leq x \leq$ 400 mm.

Fig. P3.C2

3.C3 Using computational software, determine the perpendicular distance between the line of action of a force **F** and the line *OL*. Use this software to solve (*a*) Prob. 3.62, (*b*) Prob. 3.63, (*c*) Prob. 3.65.

Fig. P3.C3

153

Fig. P3.C4

Fig. P3.C6

3.C4 A friend asks for your help in designing flower planter boxes. The boxes are to have 4, 5, 6, or 8 sides, which are to tilt outward at 10°, 20°, or 30°. Using computational software, determine the bevel angle α for each of the twelve planter designs. (*Hint:* The bevel angle α is equal to one-half of the angle formed by the inward normals of two adjacent sides.)

3.C5 A crane is oriented so that the end of the 50-ft boom *OA* lies in the *yz* plane as shown. The tension in cable *AB* is 875 lb. Determine the moment about each coordinate axes of the force exerted on *A* by cable *AB* as a function of *x*. Plot each moment as a function of *x* for -15 ft $\leq x \leq 15$ ft.

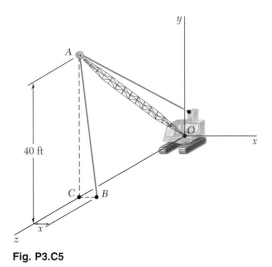

Fig. P3.C5

3.C6 A physical therapy patient lying on her right side holds a 1.5-kg dumbbell *A* in her left hand and slowly raises it along a circular path. Knowing that the center of the path is at *B* and that the path lies in the *xy* plane, plot the scalar components of the moment about *C* of the weight of the dumbbell as functions of θ for $-70° \leq \theta \leq 40°$.

3.C7 The right angle horizontal speed-reducer shown weighs 40 lb, and its center of gravity is located on the *y* axis. Knowing that $M_1 = An^2$ and $M_2 = An^3$, determine the single force that is equivalent to the weight of the unit and the two couples acting on it and the point of coordinates $(x, 0, z)$ where the line of action of the single force intersects the floor. Plot *x* and *z* as functions of *n* for $2 \leq n \leq 6$ when (*a*) $A = 0.75$ lb · ft, (*b*) $A = 1.5$ lb · ft, (*c*) $A = 2$ lb · ft.

Fig. P3.C7

3.C8 A beam *AB* is subjected to several vertical forces as shown. Using computational software, determine the magnitude of the resultant of the forces and the distance x_c to point *C*, the point where the line of action of the resultant intersects *AB*. Use this software to solve (*a*) Prob. 3.103*a*, (*b*) Prob. 3.104*a*.

Fig. P3.C8

3.C9 Using computational software, determine the magnitude and the point of application of the resultant of the vertical forces $\mathbf{P}_1$, $\mathbf{P}_2$, . . . , $\mathbf{P}_n$ which act at points $\mathbf{A}_1$, $\mathbf{A}_2$, . . . , $\mathbf{A}_n$ that are located in the *xz* plane. Use this software to solve (*a*) Prob. 3.123, (*b*) Prob. 3.125.

Fig. P3.C9

3.C10 Using computational software, determine whether a system of forces and couples can be reduced to a single equivalent force $\mathbf{R}$, and, if it can, determine $\mathbf{R}$ and the point where the line of action of $\mathbf{R}$ intersects the *yz* plane. Use this software to solve (*a*) Prob. 3.136, (*b*) Prob. 3.137.

Fig. P3.C10

C H A P T E R

4

Equilibrium of Rigid Bodies

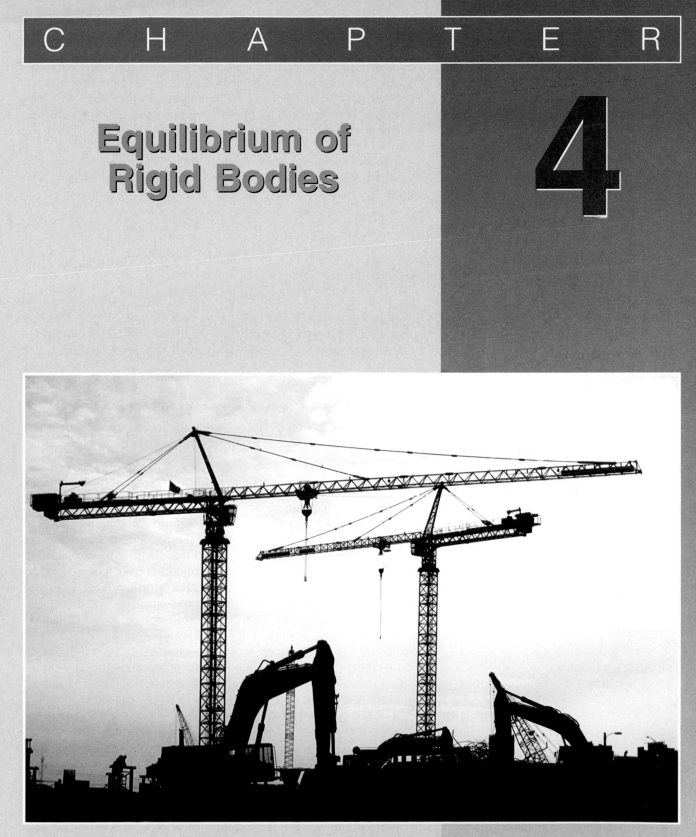

On many construction projects, large cranes such as those shown are used. The sum of the moments of the loads and the counterweights about the base of a crane must be adjusted so that the couple of the reaction at the base does not cause failure of the tower.

EQUILIBRIUM OF RIGID BODIES

4.1 Introduction
4.2 Free-Body Diagram
     Equilibrium in Two Dimensions
4.3 Reactions at Supports and Connections for a Two-Dimensional Structure
4.4 Equilibrium of a Rigid Body in Two Dimensions
4.5 Statically Indeterminate Reactions. Partial Constraints
4.6 Equilibrium of a Two-Force Body
4.7 Equilibrium of a Three-Force Body
     Equilibrium in Three Dimensions
4.8 Equilibrium of a Rigid Body in Three Dimensions
4.9 Reactions at Supports and Connections for a Three-Dimensional Structure

4.1. INTRODUCTION

We saw in the preceding chapter that the external forces acting on a rigid body can be reduced to a force-couple system at some arbitrary point O. When the force and the couple are both equal to zero, the external forces form a system equivalent to zero, and the rigid body is said to be in *equilibrium*.

The necessary and sufficient conditions for the equilibrium of a rigid body, therefore, can be obtained by setting $\mathbf{R}$ and $\mathbf{M}_O^R$ equal to zero in the relations (3.52) of Sec. 3.17:

$$\Sigma\mathbf{F} = 0 \qquad \Sigma\mathbf{M}_O = \Sigma(\mathbf{r} \times \mathbf{F}) = 0 \tag{4.1}$$

Resolving each force and each moment into its rectangular components, we can express the necessary and sufficient conditions for the equilibrium of a rigid body with the following six scalar equations:

$$\Sigma F_x = 0 \qquad \Sigma F_y = 0 \qquad \Sigma F_z = 0 \tag{4.2}$$
$$\Sigma M_x = 0 \qquad \Sigma M_y = 0 \qquad \Sigma M_z = 0 \tag{4.3}$$

The equations obtained can be used to determine unknown forces applied to the rigid body or unknown reactions exerted on it by its supports. We note that Eqs. (4.2) express the fact that the components of the external forces in the x, y, and z directions are balanced; Eqs. (4.3) express the fact that the moments of the external forces about the x, y, and z axes are balanced. Therefore, for a rigid body in equilibrium, the system of the external forces will impart no translational or rotational motion to the body considered.

In order to write the equations of equilibrium for a rigid body, it is essential to first identify all of the forces acting on that body and then to draw the corresponding *free-body diagram*. In this chapter we first consider the equilibrium of *two-dimensional structures* subjected to forces contained in their planes and learn how to draw their free-body diagrams. In addition to the forces *applied* to a structure, the *reactions* exerted on the structure by its supports will be considered. A specific reaction will be associated with each type of support. You will learn how to determine whether the structure is properly supported, so that you can know in advance whether the equations of equilibrium can be solved for the unknown forces and reactions.

Later in the chapter, the equilibrium of three-dimensional structures will be considered, and the same kind of analysis will be given to these structures and their supports.

4.2. FREE-BODY DIAGRAM

In solving a problem concerning the equilibrium of a rigid body, it is essential to consider *all* of the forces acting on the body; it is equally important to exclude any force which is not directly applied to the body. Omitting a force or adding an extraneous one would destroy the conditions of equilibrium. Therefore, the first step in the solution of the problem should be to draw a *free-body diagram* of the rigid body under consideration. Free-body diagrams have already been used on many occasions in Chap. 2. However, in view of their importance to the solution of equilibrium problems, we summarize here the various steps which must be followed in drawing a free-body diagram.

1. A clear decision should be made regarding the choice of the free body to be used. This body is then detached from the ground and is separated from all other bodies. The contour of the body thus isolated is sketched.

2. All external forces should be indicated on the free-body diagram. These forces represent the actions exerted *on* the free body *by* the ground and *by* the bodies which have been detached; they should be applied at the various points where the free body was supported by the ground or was connected to the other bodies. The *weight* of the free body should also be included among the external forces, since it represents the attraction exerted by the earth on the various particles forming the free body. As will be seen in Chap. 5, the weight should be applied at the center of gravity of the body. When the free body is made of several parts, the forces the various parts exert on each other should *not* be included among the external forces. These forces are internal forces as far as the free body is concerned.

3. The magnitudes and directions of the *known external forces* should be clearly marked on the free-body diagram. When indicating the directions of these forces, it must be remembered that the forces shown on the free-body diagram must be those which are exerted *on*, and not *by*, the free body. Known external forces generally include the *weight* of the free body and *forces applied* for a given purpose.

4. *Unknown external forces* usually consist of the *reactions*, through which the ground and other bodies oppose a possible motion of the free body. The reactions constrain the free body to remain in the same position and, for that reason, are sometimes called *constraining forces*. Reactions are exerted at the points where the free body is *supported by* or *connected to* other bodies and should be clearly indicated. Reactions are discussed in detail in Secs. 4.3 and 4.8.

5. The free-body diagram should also include dimensions, since these may be needed in the computation of moments of forces. Any other detail, however, should be omitted.

Photo 4.1 A free-body diagram of the tractor shown would include all of the external forces acting on the tractor: the weight of the tractor, the weight of the load in the bucket, and the forces exerted by the ground on the tires.

Photo 4.2 In Chap. 6, we will discuss how to determine the internal forces in structures made of several connected pieces, such as the forces in the members that support the bucket of the tractor of Photo 4.1.

EQUILIBRIUM IN TWO DIMENSIONS

4.3. REACTIONS AT SUPPORTS AND CONNECTIONS FOR A TWO-DIMENSIONAL STRUCTURE

In the first part of this chapter, the equilibrium of a two-dimensional structure is considered; that is, it is assumed that the structure being analyzed and the forces applied to it are contained in the same plane. Clearly, the reactions needed to maintain the structure in the same position will also be contained in this plane.

The reactions exerted on a two-dimensional structure can be divided into three groups corresponding to three types of *supports*, or *connections:*

Photo 4.3 As the link of the awning window opening mechanism is extended, the force it exerts on the slider results in a normal force being applied to the rod, which causes the window to open.

1. *Reactions Equivalent to a Force with Known Line of Action.* Supports and connections causing reactions of this type include *rollers, rockers, frictionless surfaces, short links and cables, collars on frictionless rods,* and *frictionless pins in slots.* Each of these supports and connections can prevent motion in one direction only. They are shown in Fig. 4.1, together with the reactions they produce. Each of these reactions involves *one unknown,* namely, the magnitude of the reaction; this magnitude should be denoted by an appropriate letter. The line of action of the reaction is known and should be indicated clearly in the free-body diagram. The sense of the reaction must be as shown in Fig. 4.1 for the cases of a frictionless surface (toward the free body) or a cable (away from the free body). The reaction can be directed either way in the case of double-track rollers, links, collars on rods, and pins in slots. Single-track rollers and rockers are generally assumed to be reversible, and thus the corresponding reactions can also be directed either way.

Photo 4.4 The abutment-mounted rocker bearing shown is used to support the roadway of a bridge.

2. *Reactions Equivalent to a Force of Unknown Direction and Magnitude.* Supports and connections causing reactions of this type include *frictionless pins in fitted holes, hinges,* and *rough surfaces.* They can prevent translation of the free body in all directions, but they cannot prevent the body from rotating about the connection. Reactions of this group involve *two unknowns* and are usually represented by their x and y components. In the case of a rough surface, the component normal to the surface must be directed away from the surface, and thus is directed toward the free body.

Photo 4.5 Shown is the rocker expansion bearing of a plate girder bridge. The convex surface of the rocker allows the support of the girder to move horizontally.

3. *Reactions Equivalent to a Force and a Couple.* These reactions are caused by *fixed supports,* which oppose any motion of the free body and thus constrain it completely. Fixed supports actually produce forces over the entire surface of contact; these forces, however, form a system which can be reduced to a force and a couple. Reactions of this group involve *three unknowns,* consisting usually of the two components of the force and the moment of the couple.

Support or Connection	Reaction	Number of Unknowns
Rollers Rocker Frictionless surface	Force with known line of action	1
Short cable Short link	Force with known line of action	1
Collar on frictionless rod Frictionless pin in slot	90° Force with known line of action	1
Frictionless pin or hinge Rough surface	or α Force of unknown direction	2
Fixed support	or α Force and couple	3

Fig. 4.1 Reactions at supports and connections.

When the sense of an unknown force or couple is not readily apparent, no attempt should be made to determine it. Instead, the sense of the force or couple should be arbitrarily assumed; the sign of the answer obtained will indicate whether the assumption is correct or not.

4.4. EQUILIBRIUM OF A RIGID BODY IN TWO DIMENSIONS

The conditions stated in Sec. 4.1 for the equilibrium of a rigid body become considerably simpler for the case of a two-dimensional structure. Choosing the x and y axes to be in the plane of the structure, we have

$$F_z = 0 \qquad M_x = M_y = 0 \qquad M_z = M_O$$

for each of the forces applied to the structure. Thus, the six equations of equilibrium derived in Sec. 4.1 reduce to

$$\Sigma F_x = 0 \qquad \Sigma F_y = 0 \qquad \Sigma M_O = 0 \qquad (4.4)$$

and to three trivial identities, $0 = 0$. Since $\Sigma M_O = 0$ must be satisfied regardless of the choice of the origin O, we can write the equations of equilibrium for a two-dimensional structure in the more general form

$$\Sigma F_x = 0 \qquad \Sigma F_y = 0 \qquad \Sigma M_A = 0 \qquad (4.5)$$

where A is any point in the plane of the structure. The three equations obtained can be solved for no more than *three unknowns*.

We saw in the preceding section that unknown forces include reactions and that the number of unknowns corresponding to a given reaction depends upon the type of support or connection causing that reaction. Referring to Sec. 4.3, we observe that the equilibrium equations (4.5) can be used to determine the reactions associated with two rollers and one cable, one fixed support, or one roller and one pin in a fitted hole, etc.

Consider Fig. 4.2a, in which the truss shown is subjected to the given forces **P**, **Q**, and **S**. The truss is held in place by a pin at A and a roller at B. The pin prevents point A from moving by exerting on the truss a force which can be resolved into the components $\mathbf{A}_x$ and $\mathbf{A}_y$; the roller keeps the truss from rotating about A by exerting the vertical force **B**. The free-body diagram of the truss is shown in Fig. 4.2b; it includes the reactions $\mathbf{A}_x$, $\mathbf{A}_y$, and **B** as well as the applied forces **P**, **Q**, **S** and the weight **W** of the truss. Expressing that the sum of the moments about A of all of the forces shown in Fig. 4.2b is zero, we write the equation $\Sigma M_A = 0$, which can be used to determine the magnitude B since it does not contain A_x or A_y. Next, expressing that the sum of the x components and the sum of the y components of the forces are zero, we write the equations $\Sigma F_x = 0$ and $\Sigma F_y = 0$, from which we can obtain the components A_x and A_y, respectively.

An additional equation could be obtained by expressing that the sum of the moments of the external forces about a point other than A is zero. We could write, for instance, $\Sigma M_B = 0$. Such a statement, however, does not contain any new information, since it has already been established that the system of the forces shown in Fig. 4.2b is equivalent to zero. The additional equation *is not independent* and cannot be used to determine a fourth unknown. It will be useful,

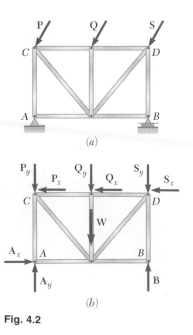

(a)

(b)

Fig. 4.2

however, for checking the solution obtained from the original three equations of equilibrium.

While the three equations of equilibrium cannot be *augmented* by additional equations, any of them can be *replaced* by another equation. Therefore, an alternative system of equations of equilibrium is

$$\Sigma F_x = 0 \qquad \Sigma M_A = 0 \qquad \Sigma M_B = 0 \qquad (4.6)$$

where the second point about which the moments are summed (in this case, point B) cannot lie on the line parallel to the y axis that passes through point A (Fig. 4.2b). These equations are sufficient conditions for the equilibrium of the truss. The first two equations indicate that the external forces must reduce to a single vertical force at A. Since the third equation requires that the moment of this force be zero about a point B which is not on its line of action, the force must be zero, and the rigid body is in equilibrium.

A third possible set of equations of equilibrium is

$$\Sigma M_A = 0 \qquad \Sigma M_B = 0 \qquad \Sigma M_C = 0 \qquad (4.7)$$

where the points A, B, and C do not lie in a straight line (Fig. 4.2b). The first equation requires that the external forces reduce to a single force at A; the second equation requires that this force pass through B; and the third equation requires that it pass through C. Since the points A, B, C do not lie in a straight line, the force must be zero, and the rigid body is in equilibrium.

The equation $\Sigma M_A = 0$, which expresses that the sum of the moments of the forces about pin A is zero, possesses a more definite physical meaning than either of the other two equations (4.7). These two equations express a similar idea of balance, but with respect to points about which the rigid body is not actually hinged. They are, however, as useful as the first equation, and our choice of equilibrium equations should not be unduly influenced by the physical meaning of these equations. Indeed, it will be desirable in practice to choose equations of equilibrium containing only one unknown, since this eliminates the necessity of solving simultaneous equations. Equations containing only one unknown can be obtained by summing moments about the point of intersection of the lines of action of two unknown forces or, if these forces are parallel, by summing components in a direction perpendicular to their common direction. For example, in Fig. 4.3, in which the truss shown is held by rollers at A and B and a short link at D, the reactions at A and B can be eliminated by summing x components. The reactions at A and D will be eliminated by summing moments about C, and the reactions at B and D by summing moments about D. The equations obtained are

(a)

(b)

Fig. 4.3

$$\Sigma F_x = 0 \qquad \Sigma M_C = 0 \qquad \Sigma M_D = 0$$

Each of these equations contains only one unknown.

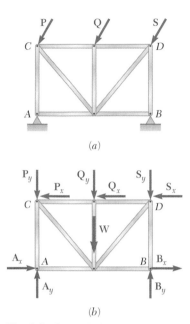

(a)

(b)

Fig. 4.4 Statically indeterminate reactions.

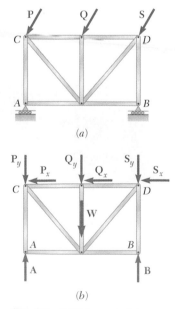

(a)

(b)

Fig. 4.5 Partial constraints.

4.5. STATICALLY INDETERMINATE REACTIONS. PARTIAL CONSTRAINTS

In the two examples considered in the preceding section (Figs. 4.2 and 4.3), the types of supports used were such that the rigid body could not possibly move under the given loads or under any other loading conditions. In such cases, the rigid body is said to be *completely constrained*. We also recall that the reactions corresponding to these supports involved *three unknowns* and could be determined by solving the three equations of equilibrium. When such a situation exists, the reactions are said to be *statically determinate*.

Consider Fig. 4.4a, in which the truss shown is held by pins at A and B. These supports provide more constraints than are necessary to keep the truss from moving under the given loads or under any other loading conditions. We also note from the free-body diagram of Fig. 4.4b that the corresponding reactions involve *four unknowns*. Since, as was pointed out in Sec. 4.4, only three independent equilibrium equations are available, there are *more unknowns than equations;* thus, all of the unknowns cannot be determined. While the equations $\Sigma M_A = 0$ and $\Sigma M_B = 0$ yield the vertical components B_y and A_y, respectively, the equation $\Sigma F_x = 0$ gives only the sum $A_x + B_x$ of the horizontal components of the reactions at A and B. The components A_x and B_x are said to be *statically indeterminate*. They could be determined by considering the deformations produced in the truss by the given loading, but this method is beyond the scope of statics and belongs to the study of mechanics of materials.

The supports used to hold the truss shown in Fig. 4.5a consist of rollers at A and B. Clearly, the constraints provided by these supports are not sufficient to keep the truss from moving. While any vertical motion is prevented, the truss is free to move horizontally. The truss is said to be *partially constrained.*† Turning our attention to Fig. 4.5b, we note that the reactions at A and B involve only *two unknowns*. Since three equations of equilibrium must still be satisfied, there are *fewer unknowns than equations*, and, in general, one of the equilibrium equations will not be satisfied. While the equations $\Sigma M_A = 0$ and $\Sigma M_B = 0$ can be satisfied by a proper choice of reactions at A and B, the equation $\Sigma F_x = 0$ will not be satisfied unless the sum of the horizontal components of the applied forces happens to be zero. We thus observe that the equilibrium of the truss of Fig. 4.5 cannot be maintained under general loading conditions.

It appears from the above that if a rigid body is to be completely constrained and if the reactions at its supports are to be statically determinate, *there must be as many unknowns as there are equations of equilibrium*. When this condition is *not* satisfied, we can be certain either that the rigid body is not completely constrained or that the reactions at its supports are not statically determinate; it is also possible that the rigid body is not completely constrained *and* that the reactions are statically indeterminate.

We should note, however, that, while *necessary*, the above condition is *not sufficient*. In other words, the fact that the number of

† Partially constrained bodies are often referred to as *unstable*. However, to avoid confusion between this type of instability, due to insufficient constraints, and the type of instability considered in Chap. 10, which relates to the behavior of a rigid body when its equilibrium is disturbed, we will restrict the use of the words *stable* and *unstable* to the latter case.

unknowns is equal to the number of equations is no guarantee that the body is completely constrained or that the reactions at its supports are statically determinate. Consider Fig. 4.6*a*, in which the truss shown is held by rollers at *A*, *B*, and *E*. While there are three unknown reactions, **A**, **B**, and **E** (Fig. 4.6*b*), the equation $\Sigma F_x = 0$ will not be satisfied unless the sum of the horizontal components of the applied forces happens to be zero. Although there are a sufficient number of constraints, these constraints are not properly arranged, and the truss is free to move horizontally. We say that the truss is *improperly constrained*. Since only two equilibrium equations are left for determining three unknowns, the reactions will be statically indeterminate. Thus, improper constraints also produce static indeterminacy.

Another example of improper constraints—and of static indeterminacy—is provided by the truss shown in Fig. 4.7. This truss is held by a pin at *A* and by rollers at *B* and *C*, which altogether involve four unknowns. Since only three independent equilibrium equations are available, the reactions at the supports are statically indeterminate. On the other hand, we note that the equation $\Sigma M_A = 0$ cannot be satisfied under general loading conditions, since the lines of action of the reactions **B** and **C** pass through *A*. We conclude that the truss can rotate about *A* and that it is improperly constrained.†

The examples of Figs. 4.6 and 4.7 lead us to conclude that *a rigid body is improperly constrained whenever the supports,* even though they may provide a sufficient number of reactions, *are arranged in such a way that the reactions must be either concurrent or parallel.*‡

In summary, to be sure that a two-dimensional rigid body is completely constrained and that the reactions at its supports are statically determinate, we should verify that the reactions involve three—and only three—unknowns and that the supports are arranged in such a way that they do not require the reactions to be either concurrent or parallel.

Supports involving statically indeterminate reactions should be used with care in the *design* of structures and only with a full knowledge of the problems they may cause. On the other hand, the *analysis* of structures possessing statically indeterminate reactions often can be partially carried out by the methods of statics. In the case of the truss of Fig. 4.4, for example, the vertical components of the reactions at *A* and *B* were obtained from the equilibrium equations.

For obvious reasons, supports producing partial or improper constraints should be avoided in the design of stationary structures. However, a partially or improperly constrained structure will not necessarily collapse; under particular loading conditions, equilibrium can be maintained. For example, the trusses of Figs. 4.5 and 4.6 will be in equilibrium if the applied forces **P**, **Q**, and **S** are vertical. Besides, structures which are designed to move *should* be only partially constrained. A railroad car, for instance, would be of little use if it were completely constrained by having its brakes applied permanently.

---

†Rotation of the truss about *A* requires some "play" in the supports at *B* and *C*. In practice such play will always exist. In addition, we note that if the play is kept small, the displacements of the rollers *B* and *C* and, thus, the distances from *A* to the lines of action of the reactions **B** and **C** will also be small. The equation $\Sigma M_A = 0$ then requires that **B** and **C** be very large, a situation which can result in the failure of the supports at *B* and *C*.

‡Because this situation arises from an inadequate arrangement or *geometry* of the supports, it is often referred to as *geometric instability*.

Fig. 4.6   Improper constraints.

Fig. 4.7   Improper constraints.

SAMPLE PROBLEM 4.1

A fixed crane has a mass of 1000 kg and is used to lift a 2400-kg crate. It is held in place by a pin at A and a rocker at B. The center of gravity of the crane is located at G. Determine the components of the reactions at A and B.

SOLUTION

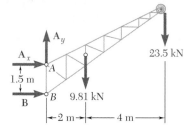

Free-Body Diagram.   A free-body diagram of the crane is drawn. By multiplying the masses of the crane and of the crate by $g = 9.81$ m/s^2, we obtain the corresponding weights, that is, 9810 N or 9.81 kN, and 23 500 N or 23.5 kN. The reaction at pin A is a force of unknown direction; it is represented by its components $\mathbf{A}_x$ and $\mathbf{A}_y$. The reaction at the rocker B is perpendicular to the rocker surface; thus, it is horizontal. We assume that $\mathbf{A}_x$, $\mathbf{A}_y$, and $\mathbf{B}$ act in the directions shown.

Determination of B.   We express that the sum of the moments of all external forces about point A is zero. The equation obtained will contain neither A_x nor A_y, since the moments of $\mathbf{A}_x$ and $\mathbf{A}_y$ about A are zero. Multiplying the magnitude of each force by its perpendicular distance from A, we write

$+\!\!\uparrow\!\Sigma M_A = 0$: $+B(1.5 \text{ m}) - (9.81 \text{ kN})(2 \text{ m}) - (23.5 \text{ kN})(6 \text{ m}) = 0$
$\qquad\qquad\qquad B = +107.1 \text{ kN}$ $\qquad\qquad$ $\mathbf{B} = 107.1 \text{ kN} \rightarrow$ ◄

Since the result is positive, the reaction is directed as assumed.

Determination of A_x.   The magnitude of $\mathbf{A}_x$ is determined by expressing that the sum of the horizontal components of all external forces is zero.

$\xrightarrow{+}\Sigma F_x = 0$: $A_x + B = 0$
$\qquad\qquad\qquad A_x + 107.1 \text{ kN} = 0$
$\qquad\qquad\qquad A_x = -107.1 \text{ kN}$ $\qquad\qquad$ $\mathbf{A}_x = 107.1 \text{ kN} \leftarrow$ ◄

Since the result is negative, the sense of $\mathbf{A}_x$ is opposite to that assumed originally.

Determination of A_y.   The sum of the vertical components must also equal zero.

$+\!\!\uparrow\!\Sigma F_y = 0$: $A_y - 9.81 \text{ kN} - 23.5 \text{ kN} = 0$
$\qquad\qquad\qquad A_y = +33.3 \text{ kN}$ $\qquad\qquad$ $\mathbf{A}_y = 33.3 \text{ kN} \uparrow$ ◄

Adding vectorially the components $\mathbf{A}_x$ and $\mathbf{A}_y$, we find that the reaction at A is 112.2 kN ⦨17.3°.

Check.   The values obtained for the reactions can be checked by recalling that the sum of the moments of all of the external forces about any point must be zero. For example, considering point B, we write

$+\!\!\uparrow\!\Sigma M_B = -(9.81 \text{ kN})(2 \text{ m}) - (23.5 \text{ kN})(6 \text{ m}) + (107.1 \text{ kN})(1.5 \text{ m}) = 0$

SAMPLE PROBLEM 4.2

Three loads are applied to a beam as shown. The beam is supported by a roller at A and by a pin at B. Neglecting the weight of the beam, determine the reactions at A and B when $P = 15$ kips.

SOLUTION

Free-Body Diagram. A free-body diagram of the beam is drawn. The reaction at A is vertical and is denoted by **A**. The reaction at B is represented by components $\mathbf{B}_x$ and $\mathbf{B}_y$. Each component is assumed to act in the direction shown.

Equilibrium Equations. We write the following three equilibrium equations and solve for the reactions indicated:

$$\xrightarrow{+}\Sigma F_x = 0: \qquad\qquad B_x = 0 \qquad\qquad \mathbf{B}_x = 0 \quad \blacktriangleleft$$

$$+\uparrow\Sigma M_A = 0:$$
$$-(15\text{ kips})(3\text{ ft}) + B_y(9\text{ ft}) - (6\text{ kips})(11\text{ ft}) - (6\text{ kips})(13\text{ ft}) = 0$$
$$B_y = +21.0\text{ kips} \qquad \mathbf{B}_y = 21.0\text{ kips} \uparrow \quad \blacktriangleleft$$

$$+\uparrow\Sigma M_B = 0:$$
$$-A(9\text{ ft}) + (15\text{ kips})(6\text{ ft}) - (6\text{ kips})(2\text{ ft}) - (6\text{ kips})(4\text{ ft}) = 0$$
$$A = +6.00\text{ kips} \qquad \mathbf{A} = 6.00\text{ kips} \uparrow \quad \blacktriangleleft$$

Check. The results are checked by adding the vertical components of all of the external forces:

$$+\uparrow\Sigma F_y = +6.00\text{ kips} - 15\text{ kips} + 21.0\text{ kips} - 6\text{ kips} - 6\text{ kips} = 0$$

Remark. In this problem the reactions at both A and B are vertical; however, these reactions are vertical for different reasons. At A, the beam is supported by a roller; hence the reaction cannot have a horizontal component. At B, the horizontal component of the reaction is zero because it must satisfy the equilibrium equation $\Sigma F_x = 0$ and because none of the other forces acting on the beam has a horizontal component.

We could have noticed at first glance that the reaction at B was vertical and dispensed with the horizontal component $\mathbf{B}_x$. This, however, is a bad practice. In following it, we would run the risk of forgetting the component $\mathbf{B}_x$ when the loading conditions require such a component (that is, when a horizontal load is included). Also, the component $\mathbf{B}_x$ was found to be zero by using and solving an equilibrium equation, $\Sigma F_x = 0$. By setting $\mathbf{B}_x$ equal to zero immediately, we might not realize that we actually made use of this equation and thus might lose track of the number of equations available for solving the problem.

SAMPLE PROBLEM 4.3

A loading car is at rest on a track forming an angle of 25° with the vertical. The gross weight of the car and its load is 5500 lb, and it is applied at a point 30 in. from the track, halfway between the two axles. The car is held by a cable attached 24 in. from the track. Determine the tension in the cable and the reaction at each pair of wheels.

SOLUTION

Free-Body Diagram. A free-body diagram of the car is drawn. The reaction at each wheel is perpendicular to the track, and the tension force **T** is parallel to the track. For convenience, we choose the x axis parallel to the track and the y axis perpendicular to the track. The 5500-lb weight is then resolved into x and y components.

$$W_x = +(5500 \text{ lb}) \cos 25° = +4980 \text{ lb}$$
$$W_y = -(5500 \text{ lb}) \sin 25° = -2320 \text{ lb}$$

Equilibrium Equations. We take moments about A to eliminate **T** and $\mathbf{R}_1$ from the computation.

$$+\curvearrowleft \Sigma M_A = 0: \quad -(2320 \text{ lb})(25 \text{ in.}) - (4980 \text{ lb})(6 \text{ in.}) + R_2(50 \text{ in.}) = 0$$
$$R_2 = +1758 \text{ lb} \qquad\qquad \mathbf{R}_2 = 1758 \text{ lb} \nearrow \quad \blacktriangleleft$$

Now, taking moments about B to eliminate **T** and $\mathbf{R}_2$ from the computation, we write

$$+\curvearrowleft \Sigma M_B = 0: \quad (2320 \text{ lb})(25 \text{ in.}) - (4980 \text{ lb})(6 \text{ in.}) - R_1(50 \text{ in.}) = 0$$
$$R_1 - +562 \text{ lb} \qquad\qquad \mathbf{R}_1 = 562 \text{ lb} \nearrow \quad \blacktriangleleft$$

The value of T is found by writing

$$\searrow + \Sigma F_x = 0: \quad +4980 \text{ lb} - T = 0$$
$$T = +4980 \text{ lb} \qquad\qquad \mathbf{T} = 4980 \text{ lb} \nwarrow \quad \blacktriangleleft$$

The computed values of the reactions are shown in the adjacent sketch.

Check. The computations are verified by writing

$$\nearrow + \Sigma F_y = +562 \text{ lb} + 1758 \text{ lb} - 2320 \text{ lb} = 0$$

The solution could also have been checked by computing moments about any point other than A or B.

SAMPLE PROBLEM 4.4

The frame shown supports part of the roof of a small building. Knowing that the tension in the cable is 150 kN, determine the reaction at the fixed end E.

SOLUTION

Free-Body Diagram. A free-body diagram of the frame and of the cable *BDF* is drawn. The reaction at the fixed end *E* is represented by the force components $\mathbf{E}_x$ and $\mathbf{E}_y$ and the couple $\mathbf{M}_E$. The other forces acting on the free body are the four 20-kN loads and the 150-kN force exerted at end *F* of the cable.

Equilibrium Equations. Noting that $DF = \sqrt{(4.5\text{ m})^2 + (6\text{ m})^2} = 7.5$ m, we write

$\xrightarrow{+} \Sigma F_x = 0:$ $E_x + \dfrac{4.5}{7.5}(150\text{ kN}) = 0$

$E_x = -90.0\text{ kN}$ $\mathbf{E}_x = 90.0\text{ kN} \leftarrow$ ◄

$+\uparrow \Sigma F_y = 0:$ $E_y - 4(20\text{ kN}) - \dfrac{6}{7.5}(150\text{ kN}) = 0$

$E_y = +200\text{ kN}$ $\mathbf{E}_y = 200\text{ kN} \uparrow$ ◄

$+\gamma \Sigma M_E = 0:$ $-(20\text{ kN})(7.2\text{ m}) + (20\text{ kN})(5.4\text{ m}) + (20\text{ kN})(3.6\text{ m})$

$+ (20\text{ kN})(1.8\text{ m}) + \dfrac{6}{7.5}(150\text{ kN})(4.5\text{ m}) + M_E = 0$

$M_E = +180.0\text{ kN}\cdot\text{m}$ $\mathbf{M}_E = 180.0\text{ kN}\cdot\text{m} \gamma$ ◄

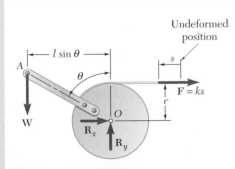

SAMPLE PROBLEM 4.5

A 400-lb weight is attached at *A* to the lever shown. The constant of the spring *BC* is $k = 250$ lb/in., and the spring is unstretched when $\theta = 0$. Determine the position of equilibrium.

SOLUTION

Free-Body Diagram. We draw a free-body diagram of the lever and cylinder. Denoting by *s* the deflection of the spring from its undeformed position, and noting that $s = r\theta$, we have $F = ks = kr\theta$.

Equilibrium Equation. Summing the moments of **W** and **F** about *O*, we write

$+\gamma \Sigma M_O = 0:$ $Wl \sin\theta - r(kr\theta) = 0$ $\sin\theta = \dfrac{kr^2}{Wl}\theta$

Substituting the given data, we obtain

$$\sin\theta = \frac{(250\text{ lb/in.})(3\text{ in.})^2}{(400\text{ lb})(8\text{ in.})}\theta \qquad \sin\theta = 0.703\theta$$

Solving numerically, we find $\theta = 0 \qquad \theta = 80.3°$ ◄

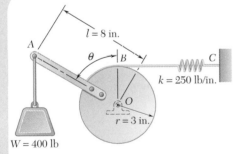

169

SOLVING PROBLEMS
ON YOUR OWN

You saw that the external forces acting on a rigid body in equilibrium form a system equivalent to zero. To solve an equilibrium problem your first task is to draw a neat, reasonably large *free-body diagram* on which you will show all external forces and relevant dimensions. Both known and unknown forces must be included.

For a two-dimensional rigid body, the reactions at the supports can involve one, two, or three unknowns depending on the type of support (Fig. 4.1). For the successful solution of a problem, a correct free-body diagram is essential. Never proceed with the solution of a problem until you are sure that your free-body diagram includes all loads, all reactions, and the weight of the body (if appropriate).

As you construct your free-body diagrams, it will be necessary to assign directions to the unknown reactions. We suggest you always assume these forces act in a positive direction, so that positive answers always imply forces acting in a positive direction, while negative answers always imply forces acting in a negative direction. Similarly, we recommend you always assume the unknown force in a rod or cable is tensile, so that a positive result always means a tensile reaction. While a negative or comprehensive reaction is possible for a rod, a negative answer for a cable is impossible and, therefore, implies that there is an error in your solution.

1. You can write three equilibrium equations and solve them for *three unknowns.* The three equations might be

$$\Sigma F_x = 0 \qquad \Sigma F_y = 0 \qquad \Sigma M_O = 0$$

However, there are usually several sets of equations that you can write, such as

$$\Sigma F_x = 0 \qquad \Sigma M_A = 0 \qquad \Sigma M_B = 0$$

where point B is chosen in such a way that the line AB is not parallel to the y axis, or

$$\Sigma M_A = 0 \qquad \Sigma M_B = 0 \qquad \Sigma M_C = 0$$

where the points A, B, and C do not lie in a straight line.

2. To simplify your solution, it may be helpful to use one of the following solution techniques if applicable.

   a. By summing moments about the point of intersection of the lines of action of two unknown forces, you will obtain an equation in a single unknown.

b. By summing components in a direction perpendicular to two unknown parallel forces, you will obtain an equation in a single unknown.

In some of the following problems you will be asked to determine the *allowable range of values* of the applied load for a given set of constraints, such as the maximum reaction at a support or the maximum force in one or more cables or rods. For problems of this type, you first assume a *maximum* loading situation (for example, the maximum allowed force in a rod), and then apply the equations of equilibrium to determine the corresponding unknown reactions and applied load. If the reactions satisfy the constraints, then the applied load is either the maximum or minimum value of the allowable range. However, if the solution violates a constraint (for example, the force in a cable is compressive), the initial assumption is wrong and another loading condition must be assumed (for the previous example, you would assume the force in the cable is zero, the *minimum* allowed reaction). The solution process is then repeated for another possible *maximum* loading to complete the determination of the allowable range of values of the applied load.

As in Chap. 2, we strongly recommend you always write the equations of equilibrium in the same form that we have used in the preceding sample problems. That is, both the known and unknown quantities are placed on the left side of the equation, and their sum is set equal to zero.

4.1 The boom on a 4300-kg truck is used to unload a pallet of shingles of mass 1600 kg. Determine the reaction at each of the two (*a*) rear wheels *B*, (*b*) front wheels *C*.

Fig. P4.1

4.2 Two children are standing on a diving board of mass 65 kg. Knowing that the masses of the children at *C* and *D* are 28 kg and 40 kg, respectively, determine (*a*) the reaction at *A*, (*b*) the reaction at *B*.

Fig. P4.2

4.3 Two crates, each weighing 250 lb, are placed as shown in the bed of a 3000-lb pickup truck. Determine the reactions at each of the two (*a*) rear wheels *A*, (*b*) front wheels *B*.

Fig. P4.3

4.4 Solve Prob. 4.3 assuming that crate *D* is removed and that the position of crate *C* is unchanged.

4.5 A T-shaped bracket supports the four loads shown. Determine the reactions at A and B if (a) $a = 100$ mm, (b) $a = 70$ mm.

4.6 For the bracket and loading of Prob. 4.5, determine the smallest distance a if the bracket is not to move.

4.7 A hand truck is used to move two barrels, each weighing 80 lb. Neglecting the weight of the hand truck, determine (a) the vertical force **P** which should be applied to the handle to maintain equilibrium when $\alpha = 35°$, (b) the corresponding reaction at each of the two wheels.

Fig. P4.5

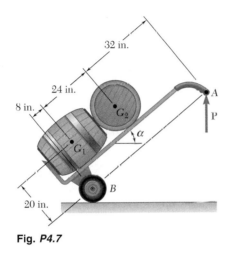

Fig. P4.7

4.8 Solve Prob. 4.7 when $\alpha = 40°$.

4.9 Four boxes are placed on a uniform 14-kg wooden plank which rests on two sawhorses. Knowing that the masses of boxes B and D are 4.5 kg and 45 kg, respectively, determine the range of values of the mass of box A so that the plank remains in equilibrium when box C is removed.

Fig. P4.9

4.10 A control rod is attached to a crank at A and cords are attached at B and C. For the given force in the rod, determine the range of values of the tension in the cord at C knowing that the cords must remain taut and that the maximum allowed tension in a cord is 180 N.

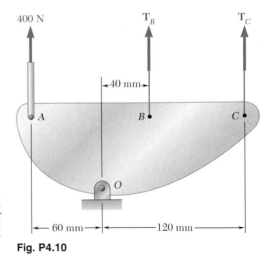

Fig. P4.10

4.11 The maximum allowable value of each of the reactions is 360 N. Neglecting the weight of the beam, determine the range of values of the distance d for which the beam is safe.

Fig. *P4.11*

4.12 Solve Prob. 4.11 assuming that the 100-N load is replaced by a 160-N load.

4.13 For the beam of Sample Prob. 4.2, determine the range of values of P for which the beam will be safe knowing that the maximum allowable value of each of the reactions is 45 kips and that the reaction at A must be directed upward.

4.14 For the beam and loading shown, determine the range of values of the distance a for which the reaction at B does not exceed 50 lb downward or 100 lb upward.

Fig. P4.14

4.15 A follower $ABCD$ is held against a circular cam by a stretched spring, which exerts a force of 21 N for the position shown. Knowing that the tension in rod BE is 14 N, determine (*a*) the force exerted on the roller at A, (*b*) the reaction at bearing C.

Fig. P4.15

4.16 A 6-m-long pole *AB* is placed in a hole and is guyed by three cables. Knowing that the tensions in cables *BD* and *BE* are 442 N and 322 N, respectively, determine (*a*) the tension in cable *CD*, (*b*) the reaction at *A*.

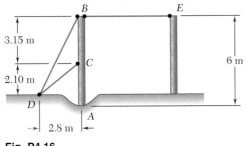

Fig. P4.16

4.17 Determine the reactions at *A* and *C* when (*a*) α = 0, (*b*) α = 30°.

Fig. P4.17

4.18 Determine the reactions at *A* and *B* when (*a*) *h* = 0, (*b*) *h* = 8 in.

Fig. P4.18

4.19 The lever *BCD* is hinged at *C* and is attached to a control rod at *B*. If *P* = 200 N, determine (*a*) the tension in rod *AB*, (*b*) the reaction at *C*.

4.20 The lever *BCD* is hinged at *C* and is attached to a control rod at *B*. Determine the maximum force **P** which can be safely applied at *D* if the maximum allowable value of the reaction at *C* is 500 N.

4.21 The required tension in cable *AB* is 800 N. Determine (*a*) the vertical force **P** which must be applied to the pedal, (*b*) the corresponding reaction at *C*.

4.22 Determine the maximum tension which can be developed in cable *AB* if the maximum allowable value of the reaction at *C* is 1000 N.

Fig. *P4.19* and *P4.20*

Fig. P4.21 and P4.22

4.23 and 4.24 A steel rod is bent to form a mounting bracket. For each of the mounting brackets and loadings shown, determine the reactions at *A* and *B*.

(a)　　　　　　　　　　　　(b)

Fig. P4.23

(a)　　　　　　　　　　　　(b)

Fig. P4.24

Fig. P4.25 and P4.26

4.25 A sign is hung by two chains from mast *AB*. The mast is hinged at *A* and is supported by cable *BC*. Knowing that the tensions in chains *DE* and *FH* are 50 lb and 30 lb, respectively, and that *d* = 1.3 ft, determine (*a*) the tension in cable *BC*, (*b*) the reaction at *A*.

4.26 A sign is hung by two chains from mast *AB*. The mast is hinged at *A* and is supported by cable *BC*. Knowing that the tensions in chains *DE* and *FH* are 30 lb and 20 lb, respectively, and that *d* = 1.54 ft, determine (*a*) the tension in cable *BC*, (*b*) the reaction at *A*.

4.27 For the frame and loading shown, determine the reactions at *A* and *E* when (*a*) α = 30°, (*b*) α = 45°.

Fig. P4.27

4.28 A lever AB is hinged at C and is attached to a control cable at A. If the lever is subjected to a 300-N vertical force at B, determine (*a*) the tension in the cable, (*b*) the reaction at C.

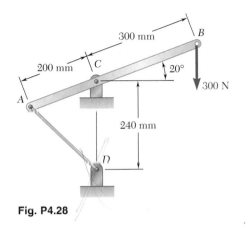

Fig. P4.28

4.29 Neglecting friction and the radius of the pulley, determine the tension in cable BCD and the reaction at support A when $d = 80$ mm.

4.30 Neglecting friction and the radius of the pulley, determine the tension in cable BCD and the reaction at support A when $d = 144$ mm.

4.31 Neglecting friction, determine the tension in cable ABD and the reaction at support C.

Fig. P4.29 and P4.30

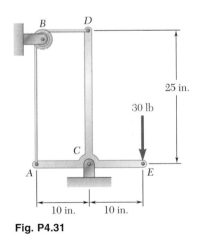

Fig. P4.31

4.32 Rod ABC is bent in the shape of a circular arc of radius R. Knowing that $\theta = 35°$, determine the reaction (*a*) at B, (*b*) at C.

4.33 Rod ABC is bent in the shape of a circular arc of radius R. Knowing that $\theta = 50°$, determine the reaction (*a*) at B, (*b*) at C.

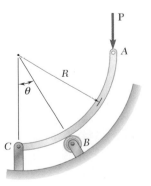

Fig. *P4.32* and P4.33

4.34 Neglecting friction and the radius of the pulley, determine (*a*) the tension in cable *ADB*, (*b*) the reaction at *C*.

Fig. P4.34

4.35 Neglecting friction, determine the tension in cable *ABD* and the reaction at *C* when $\theta = 60°$.

4.36 Neglecting friction, determine the tension in cable *ABD* and the reaction at *C* when $\theta = 30°$.

4.37 Determine the tension in each cable and the reaction at *D*.

Fig. P4.35 and *P4.36*

Fig. P4.37

4.38 Rod *ABCD* is bent in the shape of a circular arc of radius 80 mm and rests against frictionless surfaces at *A* and *D*. Knowing that the collar at *B* can move freely on the rod and that $\theta = 45°$, determine (*a*) the tension in cord *OB*, (*b*) the reactions at *A* and *D*.

4.39 Rod *ABCD* is bent in the shape of a circular arc of radius 80 mm and rests against frictionless surfaces at *A* and *D*. Knowing that the collar at *B* can move freely on the rod, determine (*a*) the value of θ for which the tension in cord *OB* is as small as possible, (*b*) the corresponding value of the tension, (*c*) the reactions at *A* and *D*.

Fig. P4.38 and P4.39

4.40 Bar *AC* supports two 100-lb loads as shown. Rollers *A* and *C* rest against frictionless surfaces and a cable *BD* is attached at *B*. Determine (*a*) the tension in cable *BD*, (*b*) the reaction at *A*, (*c*) the reaction at *C*.

4 in. — 12 in. — 4 in.

100 lb

C

100 lb

B

10 in.

D

A

6 in.

20 in.

Fig. *P4.40*

4.41 A parabolic slot has been cut in plate *AD*, and the plate has been placed so that the slot fits two fixed, frictionless pins *B* and *C*. The equation of the slot is $y = x^2/100$, where x and y are expressed in mm. Knowing that the input force $P = 4$ N, determine (*a*) the force each pin exerts on the plate, (*b*) the output force **Q**.

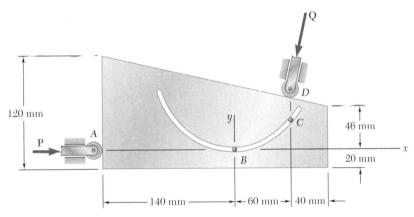

Q

120 mm

P

A

y

D

C

46 mm

B

20 mm

x

140 mm — 60 mm — 40 mm

Fig. *P4.41* and *P4.42*

4.42 A parabolic slot has been cut in plate *AD*, and the plate has been placed so that the slot fits two fixed, frictionless pins *B* and *C*. The equation of the slot is $y = x^2/100$, where x and y are expressed in mm. Knowing that the maximum allowable force exerted on the roller at *D* is 8.5 N, determine (*a*) the corresponding magnitude of the input force **P**, (*b*) the force each pin exerts on the plate.

4.43 A movable bracket is held at rest by a cable attached at *E* and by frictionless rollers. Knowing that the width of post *FG* is slightly less than the distance between the rollers, determine the force exerted on the post by each roller when $\alpha = 20°$.

4.44 Solve Prob. 4.43 when $\alpha = 30°$.

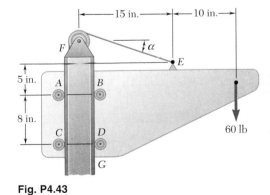

15 in. — 10 in.

F

α

E

5 in.

A

B

8 in.

C

D

60 lb

G

Fig. *P4.43*

4.45 A 20-lb weight can be supported in the three different ways shown. Knowing that the pulleys have a 4-in. radius, determine the reaction at A in each case.

(a) (b) (c)

Fig. P4.45

4.46 A belt passes over two 50-mm-diameter pulleys which are mounted on a bracket as shown. Knowing that $M = 0$ and $T_i = T_o = 24$ N, determine the reaction at C.

4.47 A belt passes over two 50-mm-diameter pulleys which are mounted on a bracket as shown. Knowing that $M = 0.40$ N $\cdot$ m and that T_i and T_o are equal to 32 N and 16 N, respectively, determine the reaction at C.

4.48 A 350-lb utility pole is used to support at C the end of an electric wire. The tension in the wire is 120 lb, and the wire forms an angle of 15° with the horizontal at C. Determine the largest and smallest allowable tensions in the guy cable BD if the magnitude of the couple at A may not exceed 200 lb $\cdot$ ft.

Fig. P4.46 and P4.47

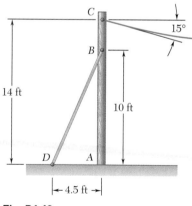

Fig. P4.48

4.49 In a laboratory experiment, students hang the masses shown from a beam of negligible mass. (a) Determine the reaction at the fixed support A knowing that end D of the beam does not touch support E. (b) Determine the reaction at the fixed support A knowing that the adjustable support E exerts an upward force of 6 N on the beam.

4.50 In a laboratory experiment, students hang the masses shown from a beam of negligible mass. Determine the range of values of the force exerted on the beam by the adjustable support E for which the magnitude of the couple at A does not exceed 2.5 N $\cdot$ m.

Fig. P4.49 and P4.50

4.51 Knowing that the tension in wire *BD* is 300 lb, determine the reaction at fixed support *C* for the frame shown.

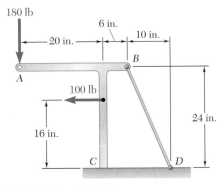

Fig. P4.51 and P4.52

4.52 Determine the range of allowable values of the tension in wire *BD* if the magnitude of the couple at the fixed support *C* is not to exceed 75 lb · ft.

4.53 Uniform rod *AB* of length *l* and weight *W* lies in a vertical plane and is acted upon by a couple **M**. The ends of the rod are connected to small rollers which rest against frictionless surfaces. (*a*) Express the angle θ corresponding to equilibrium in terms of *M*, *W*, and *l*. (*b*) Determine the value of θ corresponding to equilibrium when *M* = 1.5 lb · ft, *W* = 4 lb, and *l* = 2 ft.

Fig. P4.53

4.54 A slender rod *AB*, of weight *W*, is attached to blocks *A* and *B*, which move freely in the guides shown. The blocks are connected by an elastic cord which passes over a pulley at *C*. (*a*) Express the tension in the cord in terms of *W* and θ. (*b*) Determine the value of θ for which the tension in the cord is equal to 3*W*.

Fig. P4.54

4.55 A thin, uniform ring of mass *m* and radius *R* is attached by a frictionless pin to a collar at *A* and rests against a small roller at *B*. The ring lies in a vertical plane, and the collar can move freely on a horizontal rod and is acted upon by a horizontal force **P**. (*a*) Express the angle θ corresponding to equilibrium in terms of *m* and *P*. (*b*) Determine the value of θ corresponding to equilibrium when *m* = 500 g and *P* = 5 N.

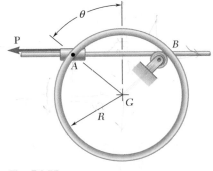

Fig. P4.55

4.56 Rod AB is acted upon by a couple $\mathbf{M}$ and two forces, each of magnitude P. (a) Derive an equation in θ, P, M, and l which must be satisfied when the rod is in equilibrium. (b) Determine the value of θ corresponding to equilibrium when $M = 150$ lb · in., $P = 20$ lb, and $l = 6$ in.

Fig. P4.56 **Fig. P4.57**

4.57 A vertical load $\mathbf{P}$ is applied at end B of rod BC. The constant of the spring is k, and the spring is unstretched when $\theta = 90°$. (a) Neglecting the weight of the rod, express the angle θ corresponding to equilibrium in terms of P, k, and l. (b) Determine the value of θ corresponding to equilibrium when $P = \frac{1}{4}kl$.

4.58 Solve Sample Prob. 4.5 assuming that the spring is unstretched when $\theta = 90°$.

4.59 A collar B of weight W can move freely along the vertical rod shown. The constant of the spring is k, and the spring is unstretched when $\theta = 0$. (a) Derive an equation in θ, W, k, and l which must be satisfied when the collar is in equilibrium. (b) Knowing that $W = 3$ lb, $l = 6$ in., and $k = 8$ lb/ft, determine the value of θ corresponding to equilibrium.

4.60 A slender rod AB, of mass m, is attached to blocks A and B which move freely in the guides shown. The constant of the spring is k, and the spring is unstretched when $\theta = 0$. (a) Neglecting the mass of the blocks, derive an equation in m, g, k, l, and θ which must be satisfied when the rod is in equilibrium. (b) Determine the value of θ when $m = 2$ kg, $l = 750$ mm, and $k = 30$ N/m.

Fig. P4.59

Fig. P4.60

4.61 The bracket *ABC* can be supported in the eight different ways shown. All connections consist of smooth pins, rollers, or short links. In each case, determine whether (*a*) the plate is completely, partially, or improperly constrained, (*b*) the reactions are statically determinate or indeterminate, (*c*) the equilibrium of the plate is maintained in the position shown. Also, wherever possible, compute the reactions assuming that the magnitude of the force **P** is 100 N.

4.62 Eight identical 20 × 30-in. rectangular plates, each weighing 50 lb, are held in a vertical plane as shown. All connections consist of frictionless pins, rollers, or short links. For each case, answer the questions listed in Prob. 4.61, and, wherever possible, compute the reactions.

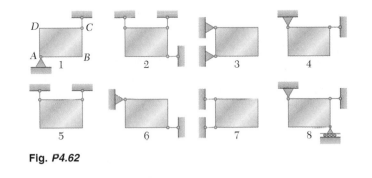

Fig. P4.62

Fig. P4.61

4.6. EQUILIBRIUM OF A TWO-FORCE BODY

A particular case of equilibrium which is of considerable interest is that of a rigid body subjected to two forces. Such a body is commonly called a *two-force body.* It will be shown that *if a two-force body is in equilibrium, the two forces must have the same magnitude, the same line of action, and opposite sense.*

Consider a corner plate subjected to two forces **F₁** and **F₂** acting at *A* and *B*, respectively (Fig. 4.8*a*). If the plate is to be in equilibrium, the sum of the moments of **F₁** and **F₂** about any point must be zero. First, we sum moments about *A*. Since the moment of **F₁** is obviously zero, the moment of **F₂** must also be zero and the line of action of **F₂** must pass through *A* (Fig. 4.8*b*). Summing moments about *B*, we prove similarly that the line of action of **F₁** must pass through *B* (Fig. 4.8*c*). Therefore, both forces have the same line of action (line *AB*). From either of the equations $\Sigma F_x = 0$ and $\Sigma F_y = 0$ it is seen that they must also have the same magnitude but opposite sense.

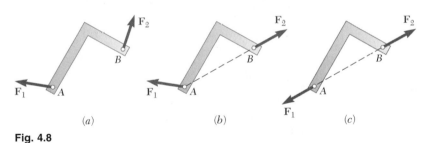

(*a*)   (*b*)   (*c*)

Fig. 4.8

Fig. 4.8   *(repeated)*

If several forces act at two points *A* and *B*, the forces acting at *A* can be replaced by their resultant **F**$_1$ and those acting at *B* can be replaced by their resultant **F**$_2$. Thus a two-force body can be more generally defined as *a rigid body subjected to forces acting at only two points.* The resultants **F**$_1$ and **F**$_2$ then must have the same line of action, the same magnitude, and opposite sense (Fig. 4.8).

In the study of structures, frames, and machines, you will see how the recognition of two-force bodies simplifies the solution of certain problems.

4.7. EQUILIBRIUM OF A THREE-FORCE BODY

Another case of equilibrium that is of great interest is that of a *three-force body,* that is, a rigid body subjected to three forces or, more generally, *a rigid body subjected to forces acting at only three points.* Consider a rigid body subjected to a system of forces which can be reduced to three forces **F**$_1$, **F**$_2$, and **F**$_3$ acting at *A*, *B*, and *C*, respectively (Fig. 4.9*a*). It will be shown that if the body is in equilibrium, *the lines of action of the three forces must be either concurrent or parallel.*

Since the rigid body is in equilibrium, the sum of the moments of **F**$_1$, **F**$_2$, and **F**$_3$ about any point must be zero. Assuming that the lines of action of **F**$_1$ and **F**$_2$ intersect and denoting their point of intersection by *D*, we sum moments about *D* (Fig. 4.9*b*). Since the moments of **F**$_1$ and **F**$_2$ about *D* are zero, the moment of **F**$_3$ about *D* must also be zero, and the line of action of **F**$_3$ must pass through *D* (Fig. 4.9*c*). Therefore, the three lines of action are concurrent. The only exception occurs when none of the lines intersect; the lines of action are then parallel.

Although problems concerning three-force bodies can be solved by the general methods of Secs. 4.3 to 4.5, the property just established can be used to solve them either graphically or mathematically from simple trigonometric or geometric relations.

Fig. 4.9

SAMPLE PROBLEM 4.6

A man raises a 10-kg joist, of length 4 m, by pulling on a rope. Find the tension T in the rope and the reaction at A.

SOLUTION

Free-Body Diagram. The joist is a three-force body, since it is acted upon by three forces: its weight $\mathbf{W}$, the force $\mathbf{T}$ exerted by the rope, and the reaction $\mathbf{R}$ of the ground at A. We note that

$$W = mg = (10 \text{ kg})(9.81 \text{ m/s}^2) = 98.1 \text{ N}$$

Three-Force Body. Since the joist is a three-force body, the forces acting on it must be concurrent. The reaction $\mathbf{R}$, therefore, will pass through the point of intersection C of the lines of action of the weight $\mathbf{W}$ and the tension force $\mathbf{T}$. This fact will be used to determine the angle α that $\mathbf{R}$ forms with the horizontal.

Drawing the vertical BF through B and the horizontal CD through C, we note that

$$AF = BF = (AB) \cos 45° = (4 \text{ m}) \cos 45° = 2.828 \text{ m}$$
$$CD = EF = AE = \tfrac{1}{2}(AF) = 1.414 \text{ m}$$
$$BD = (CD) \cot (45° + 25°) = (1.414 \text{ m}) \tan 20° = 0.515 \text{ m}$$
$$CE = DF = BF - BD = 2.828 \text{ m} - 0.515 \text{ m} = 2.313 \text{ m}$$

We write

$$\tan \alpha = \frac{CE}{AE} = \frac{2.313 \text{ m}}{1.414 \text{ m}} = 1.636$$

$$\alpha = 58.6° \quad \blacktriangleleft$$

We now know the direction of all the forces acting on the joist.

Force Triangle. A force triangle is drawn as shown, and its interior angles are computed from the known directions of the forces. Using the law of sines, we write

$$\frac{T}{\sin 31.4°} = \frac{R}{\sin 110°} = \frac{98.1 \text{ N}}{\sin 38.6°}$$

$$T = 81.9 \text{ N} \quad \blacktriangleleft$$
$$\mathbf{R} = 147.8 \text{ N} \; \angle 58.6° \quad \blacktriangleleft$$

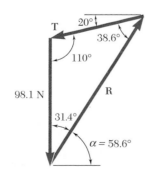

The preceding sections covered two particular cases of equilibrium of a rigid body.

1. A *two-force body* is a body subjected to forces at only two points.   The resultants of the forces acting at each of these points must have the *same magnitude, the same line of action, and opposite sense.* This property will allow you to simplify the solutions of some problems by replacing the two unknown components of a reaction by a single force of unknown magnitude but of *known direction.*

2. A *three-force body* is subjected to forces at only three points.   The resultants of the forces acting at each of these points must be *concurrent or parallel.* To solve a problem involving a three-force body with concurrent forces, draw your free-body diagram showing that the lines of action of these three forces pass through the same point. The use of geometry will then allow you to complete the solution using a force triangle [Sample Prob. 4.6].

Although the principle noted above for the solution of problems involving three-force bodies is easily understood, it can be difficult to determine the needed geometric constructions. If you encounter difficulty, first draw a reasonably large free-body diagram and then seek a relation between known or easily calculated lengths and a dimension that involves an unknown. This was done in Sample Prob. 4.6, where the easily calculated dimensions *AE* and *CE* were used to determine the angle α.

Problems

4.63 Horizontal and vertical links are hinged to a wheel, and forces are applied to the links as shown. Knowing that $a = 3.0$ in., determine the value of P and the reaction at A.

4.64 Horizontal and vertical links are hinged to a wheel, and forces are applied to the links as shown. Determine the range of values of the distance a for which the magnitude of the reaction at A does not exceed 12 lb.

4.65 Using the method of Sec. 4.7, solve Prob. 4.21.

4.66 Using the method of Sec. 4.7, solve Prob. 4.22.

4.67 To remove a nail, a small block of wood is placed under a crowbar, and a horizontal force **P** is applied as shown. Knowing that $l = 3.5$ in. and $P = 30$ lb, determine the vertical force exerted on the nail and the reaction at B.

Fig. P4.63 and P4.64

Fig. *P4.67* and *P4.68*

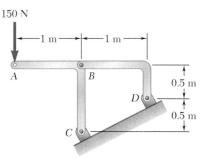

Fig. P4.69

4.68 To remove a nail, a small block of wood is placed under a crowbar, and a horizontal force **P** is applied as shown. Knowing that the maximum vertical force needed to extract the nail is 600 lb and that the horizontal force **P** is not to exceed 65 lb, determine the largest acceptable value of distance l.

4.69 For the frame and loading shown, determine the reactions at C and D.

4.70 For the frame and loading shown, determine the reactions at A and C.

Fig. P4.70

187

4.71 To remove the lid from a 5-gallon pail, the tool shown is used to apply an upward and radially outward force to the bottom inside rim of the lid. Assuming that the rim rests against the tool at *A* and that a 100-N force is applied as indicated to the handle, determine the force acting on the rim.

Fig. P4.71

4.72 To remove the lid from a 5-gallon pail, the tool shown is used to apply an upward and radially outward force to the bottom inside rim of the lid. Assuming that the top and the rim of the lid rest against the tool at *A* and *B*, respectively, and that a 60-N force is applied as indicated to the handle, determine the force acting on the rim.

Fig. P4.72

4.73 A 200-lb crate is attached to the trolley-beam system shown. Knowing that *a* = 1.5 ft, determine (*a*) the tension in cable *CD*, (*b*) the reaction at *B*.

4.74 Solve Prob. 4.73 assuming that *a* = 3 ft.

4.75 A 20-kg roller, of diameter 200 mm, which is to be used on a tile floor, is resting directly on the subflooring as shown. Knowing that the thickness of each tile is 8 mm, determine the force **P** required to move the roller onto the tiles if the roller is pushed to the left.

4.76 A 20-kg roller, of diameter 200 mm, which is to be used on a tile floor, is resting directly on the subflooring as shown. Knowing that the thickness of each tile is 8 mm, determine the force **P** required to move the roller onto the tiles if the roller is pulled to the right.

4.77 A small hoist is mounted on the back of a pickup truck and is used to lift a 120-kg crate. Determine (*a*) the force exerted on the hoist by the hydraulic cylinder *BC*, (*b*) the reaction at *A*.

Fig. P4.73

Fig. *P4.75* and P4.76

Fig. P4.77

4.78 The clamp shown is used to hold the rough workpiece C. Knowing that the maximum allowable compressive force on the workpiece is 200 N and neglecting the effect of friction at A, determine the corresponding (a) reaction at B, (b) reaction at A, (c) tension in the bolt.

4.79 A modified peavey is used to lift a 0.2-m-diameter log of mass 36 kg. Knowing that $\theta = 45°$ and that the force exerted at C by the worker is perpendicular to the handle of the peavey, determine (a) the force exerted at C, (b) the reaction at A.

Fig. P4.78

Fig. P4.81

Fig. P4.79 and P4.80

*4.80** A modified peavey is used to lift a 0.2-m-diameter log of mass 36 kg. Knowing that $\theta = 60°$ and that the force exerted at C by the worker is perpendicular to the handle of the peavey, determine (a) the force exerted at C, (b) the reaction at A.

4.81 Member ABC is supported by a pin and bracket at B and by an inextensible cord attached at A and C and passing over a frictionless pulley at D. The tension may be assumed to be the same in portion AD and CD of the cord. For the loading shown and neglecting the size of the pulley, determine the tension in the cord and the reaction at B.

4.82 Member ABCD is supported by a pin and bracket at C and by an inextensible cord attached at A and D and passing over frictionless pulleys at B and E. Neglecting the size of the pulleys, determine the tension in the cord and the reaction at C.

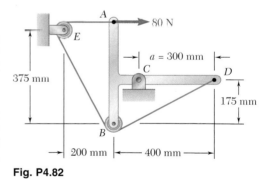

Fig. P4.82

4.83 Using the method of Sec. 4.7, solve Prob. 4.18.

4.84 Using the method of Sec. 4.7, solve Prob. 4.28.

4.85 Knowing that $\theta = 35°$, determine the reaction (a) at B, (b) at C.

4.86 Knowing that $\theta = 50°$, determine the reaction (a) at B, (b) at C.

Fig. P4.85 and P4.86

Fig. P4.87 and P4.88

Fig. *P4.89* and P4.90

Fig. *P4.92*

Fig. P4.94

4.87 A slender rod of length L and weight W is held in equilibrium as shown, with one end against a frictionless wall and the other end attached to a cord of length S. Derive an expression for the distance h in terms of L and S. Show that this position of equilibrium does not exist if $S > 2L$.

4.88 A slender rod of length $L = 200$ mm is held in equilibrium as shown, with one end against a frictionless wall and the other end attached to a cord of length $S = 300$ mm. Knowing that the mass of the rod is 1.5 kg, determine (*a*) the distance h, (*b*) the tension in the cord, (*c*) the reaction at B.

4.89 A slender rod of length L and weight W is attached to collars which can slide freely along the guides shown. Knowing that the rod is in equilibrium, derive an expression for the angle θ in terms of the angle β.

4.90 A 10-kg slender rod of length L is attached to collars which can slide freely along the guides shown. Knowing that the rod is in equilibrium and that $\beta = 25°$, determine (*a*) the angle θ that the rod forms with the vertical, (*b*) the reactions at A and B.

4.91 A uniform slender rod of mass 5 g and length 250 mm is balanced on a glass of inner diameter 70 mm. Neglecting friction, determine the angle θ corresponding to equilibrium.

Fig. P4.91

4.92 Rod AB is bent into the shape of a circular arc and is lodged between two pegs D and E. It supports a load **P** at end B. Neglecting friction and the weight of the rod, determine the distance c corresponding to equilibrium when $a = 1$ in. and $R = 5$ in.

4.93 A uniform rod AB of weight W and length $2R$ rests inside a hemispherical bowl of radius R as shown. Neglecting friction, determine the angle θ corresponding to equilibrium.

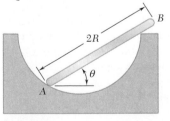

Fig. *P4.93*

4.94 A uniform slender rod of mass m and length $4r$ rests on the surface shown and is held in the given equilibrium position by the force **P**. Neglecting the effect of friction at A and C, (*a*) determine the angle θ, (*b*) derive an expression for P in terms of m.

4.95 A uniform slender rod of length $2L$ and mass m rests against a
roller at D and is held in the equilibrium position shown by a cord of length
a. Knowing that $L = 200$ mm, determine (a) the angle θ, (b) the length a.

Fig. P4.95

EQUILIBRIUM IN THREE DIMENSIONS

4.8. EQUILIBRIUM OF A RIGID BODY IN THREE DIMENSIONS

We saw in Sec. 4.1 that six scalar equations are required to express
the conditions for the equilibrium of a rigid body in the general three-
dimensional case:

$$\Sigma F_x = 0 \qquad \Sigma F_y = 0 \qquad \Sigma F_z = 0 \qquad (4.2)$$
$$\Sigma M_x = 0 \qquad \Sigma M_y = 0 \qquad \Sigma M_z = 0 \qquad (4.3)$$

These equations can be solved for no more than *six unknowns*, which
generally will represent reactions at supports or connections.

In most problems the scalar equations (4.2) and (4.3) will be more
conveniently obtained if we first express in vector form the conditions
for the equilibrium of the rigid body considered. We write

$$\Sigma \mathbf{F} = 0 \qquad \Sigma \mathbf{M}_O = \Sigma(\mathbf{r} \times \mathbf{F}) = 0 \qquad (4.1)$$

and express the forces $\mathbf{F}$ and position vectors $\mathbf{r}$ in terms of scalar components and unit vectors. Next, we compute all vector products, either
by direct calculation or by means of determinants (see Sec. 3.8). We
observe that as many as three unknown reaction components may be
eliminated from these computations through a judicious choice of the
point O. By equating to zero the coefficients of the unit vectors in each
of the two relations (4.1), we obtain the desired scalar equations.†

4.9. REACTIONS AT SUPPORTS AND CONNECTIONS FOR A THREE-DIMENSIONAL STRUCTURE

The reactions on a three-dimensional structure range from the single
force of known direction exerted by a frictionless surface to the force-
couple system exerted by a fixed support. Consequently, in problems
involving the equilibrium of a three-dimensional structure, there can
be between one and six unknowns associated with the reaction at each

---

†In some problems, it will be found convenient to eliminate the reactions at two points
A and B from the solution by writing the equilibrium equation $\Sigma M_{AB} = 0$, which involves
the determination of the moments of the forces about the axis AB joining points A and B
(see Sample Prob. 4.10).

support or connection. Various types of supports and connections are shown in Fig. 4.10 with their corresponding reactions. A simple way of determining the type of reaction corresponding to a given support or connection and the number of unknowns involved is to find which of the six fundamental motions (translation in the x, y, and z directions, rotation about the x, y, and z axes) are allowed and which motions are prevented.

Ball supports, frictionless surfaces, and cables, for example, prevent translation in one direction only and thus exert a single force whose line of action is known; each of these supports involves one unknown, namely, the magnitude of the reaction. Rollers on rough surfaces and wheels on rails prevent translation in two directions; the corresponding reactions consist of two unknown force components. Rough surfaces in direct contact and ball-and-socket supports prevent translation in three directions; these supports involve three unknown force components.

Some supports and connections can prevent rotation as well as translation; the corresponding reactions include couples as well as forces. For example, the reaction at a fixed support, which prevents any motion (rotation as well as translation), consists of three unknown forces and three unknown couples. A universal joint, which is designed to allow rotation about two axes, will exert a reaction consisting of three unknown force components and one unknown couple.

Other supports and connections are primarily intended to prevent translation; their design, however, is such that they also prevent some rotations. The corresponding reactions consist essentially of force components but *may* also include couples. One group of supports of this type includes hinges and bearings designed to support radial loads only (for example, journal bearings, roller bearings). The corresponding reactions consist of two force components but may also include two couples. Another group includes pin-and-bracket supports, hinges, and bearings designed to support an axial thrust as well as a radial load (for example, ball bearings). The corresponding reactions consist of three force components but may include two couples. However, these supports will not exert any appreciable couples under normal conditions of use. Therefore, *only* force components should be included in their analysis *unless* it is found that couples are necessary to maintain the equilibrium of the rigid body, or unless the support is known to have been specifically designed to exert a couple (see Probs. 4.128 through 4.131).

If the reactions involve more than six unknowns, there are more unknowns than equations, and some of the reactions are *statically indeterminate.* If the reactions involve fewer than six unknowns, there are more equations than unknowns, and some of the equations of equilibrium cannot be satisfied under general loading conditions; the rigid body is only *partially constrained.* Under the particular loading conditions corresponding to a given problem, however, the extra equations often reduce to trivial identities, such as $0 = 0$, and can be disregarded; although only partially constrained, the rigid body remains in equilibrium (see Sample Probs. 4.7 and 4.8). Even with six or more unknowns, it is possible that some equations of equilibrium will not be satisfied. This can occur when the reactions associated with the given supports either are parallel or intersect the same line; the rigid body is then *improperly constrained.*

Photo 4.6 Universal joints, easily seen on the drive shafts of rear-wheel-drive cars and trucks, allow rotational motion to be transferred between two non-collinear shafts.

Photo 4.7 The pillow block bearing shown supports the shaft of a fan used to ventilate a foundry.

Force with known
line of action
(one unknown)

Ball   Frictionless surface

Cable

Force with known
line of action
(one unknown)

$\mathbf{F}_y$
$\mathbf{F}_z$

Two force components

Roller on
rough surface

Wheel on rail

$\mathbf{F}_y$
$\mathbf{F}_x$
$\mathbf{F}_z$

Three force components

Rough surface   Ball and socket

$\mathbf{F}_y$
$\mathbf{F}_x$
$\mathbf{F}_z$
$\mathbf{M}_x$

Universal
joint

Three force components
and one couple

Fixed support

$\mathbf{M}_y$
$\mathbf{F}_y$
$\mathbf{M}_x$
$\mathbf{M}_z$
$\mathbf{F}_z$
$\mathbf{F}_x$

Three force components
and three couples

Hinge and bearing supporting radial load only

$(\mathbf{M}_y)$
$\mathbf{F}_y$
$(\mathbf{M}_z)$
$\mathbf{F}_z$

Two force components
(and two couples)

Pin and bracket

Hinge and bearing supporting
axial thrust and radial load

$(\mathbf{M}_y)$
$\mathbf{F}_y$
$(\mathbf{M}_z)$
$\mathbf{F}_z$
$\mathbf{F}_x$

Three force components
(and two couples)

Fig. 4.10   Reactions at supports and connections.

193

SAMPLE PROBLEM 4.7

A 20-kg ladder used to reach high shelves in a storeroom is supported by two flanged wheels A and B mounted on a rail and by an unflanged wheel C resting against a rail fixed to the wall. An 80-kg man stands on the ladder and leans to the right. The line of action of the combined weight $\mathbf{W}$ of the man and ladder intersects the floor at point D. Determine the reactions at A, B, and C.

SOLUTION

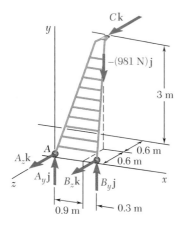

Free-Body Diagram. A free-body diagram of the ladder is drawn. The forces involved are the combined weight of the man and ladder.

$$\mathbf{W} = -mg\mathbf{j} = -(80 \text{ kg} + 20 \text{ kg})(9.81 \text{ m/s}^2)\mathbf{j} = -(981 \text{ N})\mathbf{j}$$

and five unknown reaction components, two at each flanged wheel and one at the unflanged wheel. The ladder is thus only partially constrained; it is free to roll along the rails. It is, however, in equilibrium under the given load since the equation $\Sigma F_x = 0$ is satisfied.

Equilibrium Equations. We express that the forces acting on the ladder form a system equivalent to zero:

$$\Sigma \mathbf{F} = 0: \qquad A_y\mathbf{j} + A_z\mathbf{k} + B_y\mathbf{j} + B_z\mathbf{k} - (981 \text{ N})\mathbf{j} + C\mathbf{k} = 0$$
$$(A_y + B_y - 981 \text{ N})\mathbf{j} + (A_z + B_z + C)\mathbf{k} = 0 \qquad (1)$$

$$\Sigma \mathbf{M}_A = \Sigma(\mathbf{r} \times \mathbf{F}) = 0: \qquad 1.2\mathbf{i} \times (B_y\mathbf{j} + B_z\mathbf{k}) + (0.9\mathbf{i} - 0.6\mathbf{k}) \times (-981\mathbf{j})$$
$$+ (0.6\mathbf{i} + 3\mathbf{j} - 1.2\mathbf{k}) \times C\mathbf{k} = 0$$

Computing the vector product, we have†

$$1.2B_y\mathbf{k} - 1.2B_z\mathbf{j} - 882.9\mathbf{k} - 588.6\mathbf{i} - 0.6C\mathbf{j} + 3C\mathbf{i} = 0$$
$$(3C - 588.6)\mathbf{i} - (1.2B_z + 0.6C)\mathbf{j} + (1.2B_y - 882.9)\mathbf{k} = 0 \qquad (2)$$

Setting the coefficients of $\mathbf{i}$, $\mathbf{j}$, $\mathbf{k}$ equal to zero in Eq. (2), we obtain the following three scalar equations, which express that the sum of the moments about each coordinate axis must be zero:

$$3C - 588.6 = 0 \qquad C - +196.2 \text{ N}$$
$$1.2B_z + 0.6C = 0 \qquad B_z = -98.1 \text{ N}$$
$$1.2B_y - 882.9 = 0 \qquad B_y = +736 \text{ N}$$

The reactions at B and C are therefore

$$\mathbf{B} = +(736 \text{ N})\mathbf{j} - (98.1 \text{ N})\mathbf{k} \qquad \mathbf{C} = +(196.2 \text{ N})\mathbf{k} \quad \blacktriangleleft$$

Setting the coefficients of $\mathbf{j}$ and $\mathbf{k}$ equal to zero in Eq. (1), we obtain two scalar equations expressing that the sums of the components in the y and z directions are zero. Substituting for B_y, B_z, and C the values obtained above, we write

$$A_y + B_y - 981 = 0 \qquad A_y + 736 - 981 = 0 \qquad A_y = +245 \text{ N}$$
$$A_z + B_z + C = 0 \qquad A_z - 98.1 + 196.2 = 0 \qquad A_z = -98.1 \text{ N}$$

We conclude that the reaction at A is $\qquad \mathbf{A} = +(245 \text{ N})\mathbf{j} - (98.1 \text{ N})\mathbf{k} \quad \blacktriangleleft$

†The moments in this sample problem and in Sample Probs. 4.8 and 4.9 can also be expressed in the form of determinants (see Sample Prob. 3.10).

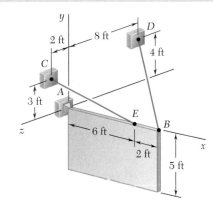

SAMPLE PROBLEM 4.8

A 5 × 8-ft sign of uniform density weighs 270 lb and is supported by a ball-and-socket joint at A and by two cables. Determine the tension in each cable and the reaction at A.

SOLUTION

Free-Body Diagram. A free-body diagram of the sign is drawn. The forces acting on the free body are the weight $\mathbf{W} = -(270 \text{ lb})\mathbf{j}$ and the reactions at A, B, and E. The reaction at A is a force of unknown direction and is represented by three unknown components. Since the directions of the forces exerted by the cables are known, these forces involve only one unknown each, namely, the magnitudes T_{BD} and T_{EC}. Since there are only five unknowns, the sign is partially constrained. It can rotate freely about the x axis; it is, however, in equilibrium under the given loading, since the equation $\Sigma M_x = 0$ is satisfied.

The components of the forces $\mathbf{T}_{BD}$ and $\mathbf{T}_{EC}$ can be expressed in terms of the unknown magnitudes T_{BD} and T_{EC} by writing

$$\overrightarrow{BD} = -(8 \text{ ft})\mathbf{i} + (4 \text{ ft})\mathbf{j} - (8 \text{ ft})\mathbf{k} \qquad BD = 12 \text{ ft}$$
$$\overrightarrow{EC} = -(6 \text{ ft})\mathbf{i} + (3 \text{ ft})\mathbf{j} + (2 \text{ ft})\mathbf{k} \qquad EC = 7 \text{ ft}$$
$$\mathbf{T}_{BD} = T_{BD}\left(\frac{\overrightarrow{BD}}{BD}\right) = T_{BD}(-\tfrac{2}{3}\mathbf{i} + \tfrac{1}{3}\mathbf{j} - \tfrac{2}{3}\mathbf{k})$$
$$\mathbf{T}_{EC} = T_{EC}\left(\frac{\overrightarrow{EC}}{EC}\right) = T_{EC}(-\tfrac{6}{7}\mathbf{i} + \tfrac{3}{7}\mathbf{j} + \tfrac{2}{7}\mathbf{k})$$

Equilibrium Equations. We express that the forces acting on the sign form a system equivalent to zero:

$\Sigma\mathbf{F} = 0$: $\qquad A_x\mathbf{i} + A_y\mathbf{j} + A_z\mathbf{k} + \mathbf{T}_{BD} + \mathbf{T}_{EC} - (270 \text{ lb})\mathbf{j} = 0$
$(A_x - \tfrac{2}{3}T_{BD} - \tfrac{6}{7}T_{EC})\mathbf{i} + (A_y + \tfrac{1}{3}T_{BD} + \tfrac{3}{7}T_{EC} - 270 \text{ lb})\mathbf{j}$
$$+ (A_z - \tfrac{2}{3}T_{BD} + \tfrac{2}{7}T_{EC})\mathbf{k} = 0 \qquad (1)$$

$\Sigma\mathbf{M}_A = \Sigma(\mathbf{r} \times \mathbf{F}) = 0$:
$(8 \text{ ft})\mathbf{i} \times T_{BD}(-\tfrac{2}{3}\mathbf{i} + \tfrac{1}{3}\mathbf{j} - \tfrac{2}{3}\mathbf{k}) + (6 \text{ ft})\mathbf{i} \times T_{EC}(-\tfrac{6}{7}\mathbf{i} + \tfrac{3}{7}\mathbf{j} + \tfrac{2}{7}\mathbf{k})$
$$+ (4 \text{ ft})\mathbf{i} \times (-270 \text{ lb})\mathbf{j} = 0$$
$$(2.667T_{BD} + 2.571T_{EC} - 1080 \text{ lb})\mathbf{k} + (5.333T_{BD} - 1.714T_{EC})\mathbf{j} = 0 \qquad (2)$$

Setting the coefficients of $\mathbf{j}$ and $\mathbf{k}$ equal to zero in Eq. (2), we obtain two scalar equations which can be solved for T_{BD} and T_{EC}:

$$T_{BD} = 101.3 \text{ lb} \qquad T_{EC} = 315 \text{ lb} \quad \blacktriangleleft$$

Setting the coefficients of $\mathbf{i}$, $\mathbf{j}$, and $\mathbf{k}$ equal to zero in Eq. (1), we obtain three more equations, which yield the components of $\mathbf{A}$. We have

$$\mathbf{A} = +(338 \text{ lb})\mathbf{i} + (101.2 \text{ lb})\mathbf{j} - (22.5 \text{ lb})\mathbf{k} \quad \blacktriangleleft$$

SAMPLE PROBLEM 4.9

A uniform pipe cover of radius $r = 240$ mm and mass 30 kg is held in a horizontal position by the cable CD. Assuming that the bearing at B does not exert any axial thrust, determine the tension in the cable and the reactions at A and B.

SOLUTION

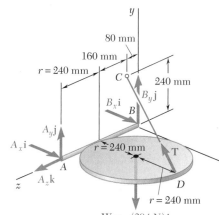

Free-Body Diagram. A free-body diagram is drawn with the coordinate axes shown. The forces acting on the free body are the weight of the cover,

$$\mathbf{W} = -mg\mathbf{j} = -(30 \text{ kg})(9.81 \text{ m/s}^2)\mathbf{j} = -(294 \text{ N})\mathbf{j}$$

and reactions involving six unknowns, namely, the magnitude of the force $\mathbf{T}$ exerted by the cable, three force components at hinge A, and two at hinge B. The components of $\mathbf{T}$ are expressed in terms of the unknown magnitude T by resolving the vector $\overrightarrow{DC}$ into rectangular components and writing

$$\overrightarrow{DC} = -(480 \text{ mm})\mathbf{i} + (240 \text{ mm})\mathbf{j} - (160 \text{ mm})\mathbf{k} \qquad DC = 560 \text{ mm}$$

$$\mathbf{T} = T\frac{\overrightarrow{DC}}{DC} = -\tfrac{6}{7}T\mathbf{i} + \tfrac{3}{7}T\mathbf{j} - \tfrac{2}{7}T\mathbf{k}$$

Equilibrium Equations. We express that the forces acting on the pipe cover form a system equivalent to zero:

$$\Sigma\mathbf{F} = 0: \qquad A_x\mathbf{i} + A_y\mathbf{j} + A_z\mathbf{k} + B_x\mathbf{i} + B_y\mathbf{j} + \mathbf{T} - (294 \text{ N})\mathbf{j} = 0$$

$$(A_x + B_x - \tfrac{6}{7}T)\mathbf{i} + (A_y + B_y + \tfrac{3}{7}T - 294 \text{ N})\mathbf{j} + (A_z - \tfrac{2}{7}T)\mathbf{k} = 0 \quad (1)$$

$$\Sigma\mathbf{M}_B = \Sigma(\mathbf{r} \times \mathbf{F}) = 0:$$
$$2r\mathbf{k} \times (A_x\mathbf{i} + A_y\mathbf{j} + A_z\mathbf{k})$$
$$+ (2r\mathbf{i} + r\mathbf{k}) \times (-\tfrac{6}{7}T\mathbf{i} + \tfrac{3}{7}T\mathbf{j} - \tfrac{2}{7}T\mathbf{k})$$
$$+ (r\mathbf{i} + r\mathbf{k}) \times (-294 \text{ N})\mathbf{j} = 0$$

$$(-2A_y - \tfrac{3}{7}T + 294 \text{ N})r\mathbf{i} + (2A_x - \tfrac{2}{7}T)r\mathbf{j} + (\tfrac{6}{7}T - 294 \text{ N})r\mathbf{k} = 0 \quad (2)$$

Setting the coefficients of the unit vectors equal to zero in Eq. (2), we write three scalar equations, which yield

$$A_x = +49.0 \text{ N} \qquad A_y = +73.5 \text{ N} \qquad T = 343 \text{ N} \quad \blacktriangleleft$$

Setting the coefficients of the unit vectors equal to zero in Eq. (1), we obtain three more scalar equations. After substituting the values of T, A_x, and A_y into these equations, we obtain

$$A_z = +98.0 \text{ N} \qquad B_x = +245 \text{ N} \qquad B_y = +73.5 \text{ N}$$

The reactions at A and B are therefore

$$\mathbf{A} = +(49.0 \text{ N})\mathbf{i} + (73.5 \text{ N})\mathbf{j} + (98.0 \text{ N})\mathbf{k} \quad \blacktriangleleft$$
$$\mathbf{B} = +(245 \text{ N})\mathbf{i} + (73.5 \text{ N})\mathbf{j} \quad \blacktriangleleft$$

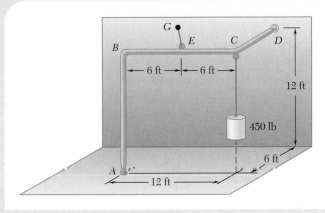

SAMPLE PROBLEM 4.10

A 450-lb load hangs from the corner C of a rigid piece of pipe $ABCD$ which has been bent as shown. The pipe is supported by the ball-and-socket joints A and D, which are fastened, respectively, to the floor and to a vertical wall, and by a cable attached at the midpoint E of the portion BC of the pipe and at a point G on the wall. Determine (a) where G should be located if the tension in the cable is to be minimum, (b) the corresponding minimum value of the tension.

SOLUTION

Free-Body Diagram. The free-body diagram of the pipe includes the load $\mathbf{W} = (-450 \text{ lb})\mathbf{j}$, the reactions at A and D, and the force $\mathbf{T}$ exerted by the cable. To eliminate the reactions at A and D from the computations, we express that the sum of the moments of the forces about AD is zero. Denoting by $\boldsymbol{\lambda}$ the unit vector along AD, we write

$$\Sigma M_{AD} = 0: \quad \boldsymbol{\lambda} \cdot (\overrightarrow{AE} \times \mathbf{T}) + \boldsymbol{\lambda} \cdot (\overrightarrow{AC} \times \mathbf{W}) = 0 \tag{1}$$

The second term in Eq. (1) can be computed as follows:

$$\overrightarrow{AC} \times \mathbf{W} = (12\mathbf{i} + 12\mathbf{j}) \times (-450\mathbf{j}) = -5400\mathbf{k}$$

$$\boldsymbol{\lambda} = \frac{\overrightarrow{AD}}{AD} = \frac{12\mathbf{i} + 12\mathbf{j} - 6\mathbf{k}}{18} = \tfrac{2}{3}\mathbf{i} + \tfrac{2}{3}\mathbf{j} - \tfrac{1}{3}\mathbf{k}$$

$$\boldsymbol{\lambda} \cdot (\overrightarrow{AC} \times \mathbf{W}) = (\tfrac{2}{3}\mathbf{i} + \tfrac{2}{3}\mathbf{j} - \tfrac{1}{3}\mathbf{k}) \cdot (-5400\mathbf{k}) = +1800$$

Substituting the value obtained into Eq. (1), we write

$$\boldsymbol{\lambda} \cdot (\overrightarrow{AE} \times \mathbf{T}) = -1800 \text{ lb} \cdot \text{ft} \tag{2}$$

Minimum Value of Tension. Recalling the commutative property for mixed triple products, we rewrite Eq. (2) in the form

$$\mathbf{T} \cdot (\boldsymbol{\lambda} \times \overrightarrow{AE}) = -1800 \text{ lb} \cdot \text{ft} \tag{3}$$

which shows that the projection of $\mathbf{T}$ on the vector $\boldsymbol{\lambda} \times \overrightarrow{AE}$ is a constant. It follows that $\mathbf{T}$ is minimum when parallel to the vector

$$\boldsymbol{\lambda} \times \overrightarrow{AE} = (\tfrac{2}{3}\mathbf{i} + \tfrac{2}{3}\mathbf{j} - \tfrac{1}{3}\mathbf{k}) \times (6\mathbf{i} + 12\mathbf{j}) = 4\mathbf{i} - 2\mathbf{j} + 4\mathbf{k}$$

Since the corresponding unit vector is $\tfrac{2}{3}\mathbf{i} - \tfrac{1}{3}\mathbf{j} + \tfrac{2}{3}\mathbf{k}$, we write

$$\mathbf{T}_{\min} = T(\tfrac{2}{3}\mathbf{i} - \tfrac{1}{3}\mathbf{j} + \tfrac{2}{3}\mathbf{k}) \tag{4}$$

Substituting for $\mathbf{T}$ and $\boldsymbol{\lambda} \times \overrightarrow{AE}$ in Eq. (3) and computing the dot products, we obtain $6T = -1800$ and, thus, $T = -300$. Carrying this value into (4), we obtain

$$\mathbf{T}_{\min} = -200\mathbf{i} + 100\mathbf{j} - 200\mathbf{k} \qquad T_{\min} = 300 \text{ lb} \blacktriangleleft$$

Location of G. Since the vector $\overrightarrow{EG}$ and the force $\mathbf{T}_{\min}$ have the same direction, their components must be proportional. Denoting the coordinates of G by $x, y, 0$, we write

$$\frac{x - 6}{-200} = \frac{y - 12}{+100} = \frac{0 - 6}{-200} \qquad x = 0 \qquad y = 15 \text{ ft} \blacktriangleleft$$

197

The equilibrium of a *three-dimensional body* was considered in the sections you just completed. It is again most important that you draw a complete *free-body diagram* as the first step of your solution.

1. As you draw the free-body diagram, pay particular attention to the reactions at the supports. The number of unknowns at a support can range from one to six (Fig. 4.10). To decide whether an unknown reaction or reaction component exists at a support, ask yourself whether the support prevents motion of the body in a certain direction or about a certain axis.

   a. If motion is prevented in a certain direction, include in your free-body diagram an unknown reaction or *reaction component* that acts in the *same direction.*

   b. If a support prevents rotation about a certain axis, include in your free-body diagram a *couple* of unknown magnitude that acts about the *same* axis.

2. The external forces acting on a three-dimensional body form a system equivalent to zero. Writing $\Sigma\mathbf{F} = 0$ and $\Sigma\mathbf{M}_A = 0$ about an appropriate point A, and setting the coefficients of $\mathbf{i}$, $\mathbf{j}$, $\mathbf{k}$ in both equations equal to zero will provide you with six scalar equations. In general, these equations will contain six unknowns and can be solved for these unknowns.

3. After completing your free-body diagram, you may want to seek equations involving as few unknowns as possible. The following strategies may help you.

   a. By summing moments about a ball-and-socket support or a hinge, you will obtain equations from which three unknown reaction components have been eliminated [Sample Probs. 4.8 and 4.9].

   b. If you can draw an axis through the points of application of all but one of the unknown reactions, summing moments about that axis will yield an equation in a single unknown [Sample Prob. 4.10].

4. After drawing your free-body diagram, we encourage you to compare the number of unknowns to the number of nontrivial, scalar equations of equilibrium for the given problem. Doing so will tell you if the body is properly or partially constrained and whether the problem is statically determinate or indeterminate. Further, as we consider more complex problems in later chapters, keeping track of the numbers of unknowns and equations will help you to develop correct solutions.

4.96 Gears A and B are attached to a shaft supported by bearings at C and D. The diameters of gears A and B are 150 mm and 75 mm, respectively, and the tangential and radial forces acting on the gears are as shown. Knowing that the system rotates at a constant rate, determine the reactions at C and D. Assume that the bearing at C does not exert any axial force, and neglect the weights of the gears and the shaft.

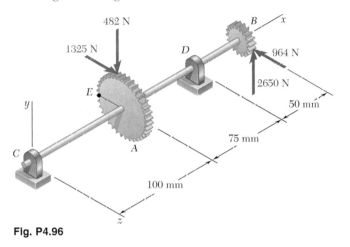

Fig. P4.96

4.97 Solve Prob. 4.96 assuming that for gear A the tangential and radial forces are acting at E, so that $\mathbf{F}_A = (1325 \text{ N})\mathbf{j} + (482 \text{ N})\mathbf{k}$.

4.98 Two transmission belts pass over sheaves welded to an axle supported by bearings at B and D. The sheave at A has a radius of 50 mm, and the sheave at C has a radius of 40 mm. Knowing that the system rotates with a constant rate, determine (a) the tension T, (b) the reactions at B and D. Assume that the bearing at D does not exert any axial thrust and neglect the weights of the sheaves and the axle.

Fig. *P4.98*

4.99 For the portion of a machine shown, the 4-in.-diameter pulley *A* and wheel *B* are fixed to a shaft supported by bearings at *C* and *D*. The spring of constant 2 lb/in. is unstretched when $\theta = 0$, and the bearing at *C* does not exert any axial force. Knowing that $\theta = 180°$ and that the machine is at rest and in equilibrium, determine (*a*) the tension *T*, (*b*) the reactions at *C* and *D*. Neglect the weights of the shaft, pulley, and wheel.

Fig. P4.99

4.100 Solve Prob. 4.99 for $\theta = 90°$.

4.101 A 1.2 × 2.4-m sheet of plywood having a mass of 17 kg has been temporarily placed among three pipe supports. The lower edge of the sheet rests on small collars *A* and *B* and its upper edge leans against pipe *C*. Neglecting friction at all surfaces, determine the reactions at *A*, *B*, and *C*.

4.102 The 200 × 200-mm square plate shown has a mass of 25 kg and is supported by three vertical wires. Determine the tension in each wire.

4.103 The 200 × 200-mm square plate shown has a mass of 25 kg and is supported by three vertical wires. Determine the mass and location of the lightest block which should be placed on the plate if the tensions in the three cables are to be equal.

Fig. *P4.101*

Fig. P4.102 and P4.103

4.104   A camera of mass 240 g is mounted on a small tripod of mass 200 g. Assuming that the mass of the camera is uniformly distributed and that the line of action of the weight of the tripod passes through D, determine (a) the vertical components of the reactions at A, B, and C when $\theta = 0$, (b) the maximum value of θ if the tripod is not to tip over.

120 mm

24 mm

Fig. P4.104

45 mm

35 mm

38 mm

38 mm

4.105   Two steel pipes AB and BC, each having a weight per unit length of 5 lb/ft, are welded together at B and are supported by three wires. Knowing that $a = 1.25$ ft, determine the tension in each wire.

Fig. P4.105

4 ft

2 ft

4.106   For the pipe assembly of Prob. 4.105, determine (a) the largest permissible value of a if the assembly is not to tip, (b) the corresponding tension in each wire.

4.107   A uniform aluminum rod of weight W is bent into a circular ring of radius R and is supported by three wires as shown. Determine the tension in each wire.

4.108   A uniform aluminum rod of weight W is bent into a circular ring of radius R and is supported by three wires as shown. A small collar of weight W' is then placed on the ring and positioned so that the tensions in the three wires are equal. Determine (a) the position of the collar, (b) the value of W', (c) the tension in the wires.

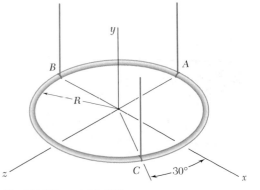

Fig. P4.107 and *P4.108*

R

30°

4.109 An opening in a floor is covered by a 3 × 4-ft sheet of plywood weighing 12 lb. The sheet is hinged at A and B and is maintained in a position slightly above the floor by a small block C. Determine the vertical component of the reaction (*a*) at A, (*b*) at B, (*c*) at C.

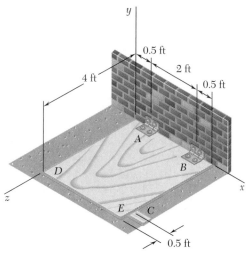

Fig. P4.109

4.110 Solve Prob. 4.109 assuming that the small block C is moved and placed under edge DE at a point 0.5 ft from corner E.

4.111 The 10-kg square plate shown is supported by three vertical wires. Determine (*a*) the tension in each wire when $a = 100$ mm, (*b*) the value of a for which tensions in the three wires are equal.

Fig. P4.111

4.112 The 3-m flagpole AC forms an angle of 30° with the z axis. It is held by a ball-and-socket joint at C and by two thin braces BD and BE. Knowing that the distance BC is 0.9 m, determine the tension in each brace and the reaction at C.

4.113 A 3-m boom is acted upon by the 4-kN force shown. Determine the tension in each cable and the reaction at the ball-and-socket joint at A.

Fig. P4.112

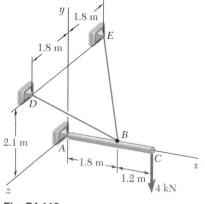

Fig. P4.113

4.114 An 8-ft-long boom is held by a ball-and-socket joint at C and by two cables AD and BE. Determine the tension in each cable and the reaction at C.

4.115 Solve Prob. 4.114 assuming that the given 198-lb load is replaced with two 99-lb loads applied at A and B.

4.116 The 18-ft pole ABC is acted upon by a 210-lb force as shown. The pole is held by a ball-and-socket joint at A and by two cables BD and BE. For $a = 9$ ft, determine the tension in each cable and the reaction at A.

4.117 Solve Prob. 4.116 for $a = 4.5$ ft.

4.118 Two steel pipes $ABCD$ and EBF are welded together at B to form the boom shown. The boom is held by a ball-and-socket joint at D and by two cables EG and $ICFH$; cable $ICFH$ passes around frictionless pulleys at C and F. For the loading shown, determine the tension in each cable and the reaction at D.

Fig. P4.114

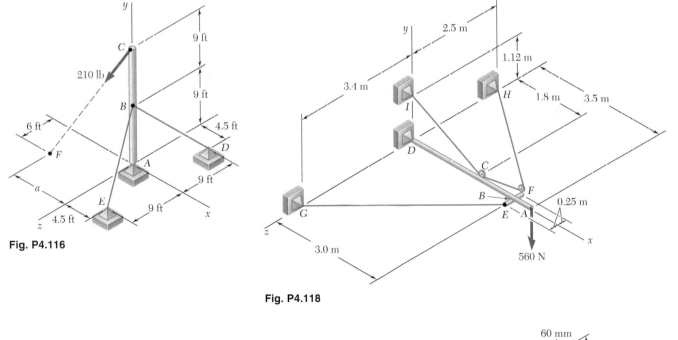

Fig. P4.116

Fig. P4.118

4.119 Solve Prob. 4.118 assuming that the 560-N load is applied at B.

4.120 The lever AB is welded to the bent rod BCD which is supported by bearings at E and F and by cable DG. Knowing that the bearing at E does not exert any axial thrust, determine (a) the tension in cable DG, (b) the reactions at E and F.

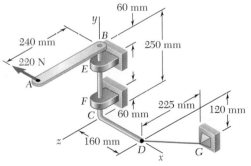

Fig. P4.120

4.121 A 30-kg cover for a roof opening is hinged at corners A and B. The roof forms an angle of 30° with the horizontal, and the cover is maintained in a horizontal position by the brace CE. Determine (a) the magnitude of the force exerted by the brace, (b) the reactions at the hinges. Assume that the hinge at A does not exert any axial thrust.

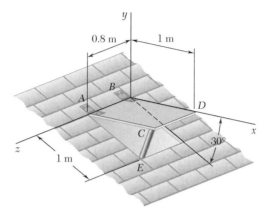

Fig. P4.121

4.122 The rectangular plate shown has a mass of 15 kg and is held in the position shown by hinges A and B and cable EF. Assuming that the hinge at B does not exert any axial thrust, determine (a) the tension in the cable, (b) the reactions at A and B.

4.123 Solve Prob. 4.122 assuming that cable EF is replaced by a cable attached at points E and H.

4.124 A small door weighing 16 lb is attached by hinges A and B to a wall and is held in the horizontal position shown by rope EFH. The rope passes around a small, frictionless pulley at F and is tied to a fixed cleat at H. Assuming that the hinge at A does not exert any axial thrust, determine (a) the tension in the rope, (b) the reactions at A and B.

Fig. P4.122

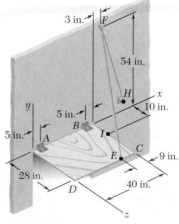

Fig. P4.124

4.125 Solve Prob. 4.124 assuming that the rope is attached to the door at I.

4.126 A 285-lb uniform rectangular plate is supported in the position shown by hinges A and B and by cable DCE, which passes over a frictionless hook at C. Assuming that the tension is the same in both parts of the cable, determine (*a*) the tension in the cable, (*b*) the reactions at A and B. Assume that the hinge at B does not exert any axial thrust.

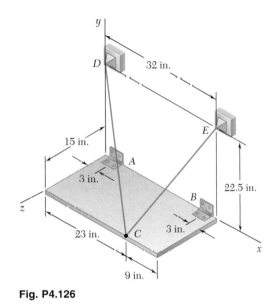

Fig. P4.126

4.127 Solve Prob. 4.126 assuming that cable DCE is replaced by a cable attached to point E and hook C.

4.128 The tensioning mechanism of a belt drive consists of frictionless pulley A, mounting plate B, and spring C. Attached below the mounting plate is slider block D which is free to move in the frictionless slot of bracket E. Knowing that the pulley and the belt lie in a horizontal plane, with portion F of the belt parallel to the x axis and portion G forming an angle of 30° with the x axis, determine (*a*) the force in the spring, (*b*) the reaction at D.

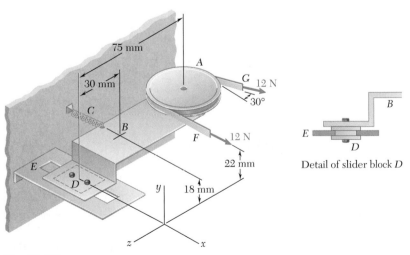

Detail of slider block D

Fig. P4.128

Fig. P4.129

4.129 The assembly shown is welded to collar *A* which fits on the vertical pin shown. The pin can exert couples about the *x* and *z* axes but does not prevent motion about or along the *y* axis. For the loading shown, determine the tension in each cable and the reaction at *A*.

4.130 The lever *AB* is welded to the bent rod *BCD* which is supported by bearing *E* and by cable *DG*. Assuming that the bearing can exert an axial thrust and couples about axes parallel to the *x* and *z* axes, determine (*a*) the tension in cable *DG*, (*b*) the reaction at *E*.

Fig. P4.130

4.131 Solve Prob. 4.124 assuming that the hinge at *A* is removed and that the hinge at *B* can exert couples about the *y* and *z* axes.

4.132 The frame shown is supported by three cables and a ball-and-socket joint at *A*. For **P** = 0, determine the tension in each cable and the reaction at *A*.

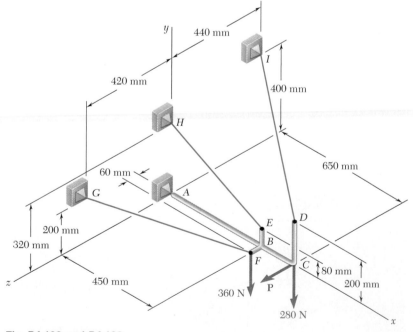

Fig. P4.132 and P4.133

4.133 The frame shown is supported by three cables and a ball-and-socket joint at *A*. For *P* = 50 N, determine the tension in each cable and the reaction at *A*.

4.134 The rigid L-shaped member *ABF* is supported by a ball-and-socket joint at *A* and by three cables. For the loading shown, determine the tension in each cable and the reaction at *A*.

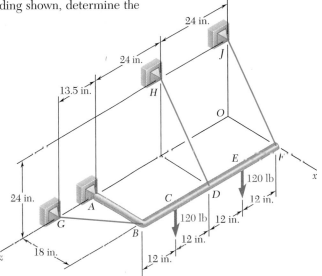

Fig. *P4.134*

4.135 Solve Prob. 4.134 assuming that the load at *C* has been removed.

4.136 In order to clean the clogged drainpipe *AE*, a plumber has disconnected both ends of the pipe and inserted a power snake through the opening at *A*. The cutting head of the snake is connected by a heavy cable to an electric motor which rotates at a constant speed as the plumber forces the cable into the pipe. The forces exerted by the plumber and the motor on the end of the cable can be represented by the wrench $\mathbf{F} = -(60 \text{ N})\mathbf{k}$, $\mathbf{M} = -(108 \text{ N} \cdot \text{m})\mathbf{k}$. Determine the additional reactions at *B*, *C*, and *D* caused by the cleaning operation. Assume that the reaction at each support consists of two force components perpendicular to the pipe.

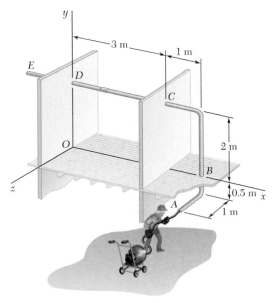

Fig. **P4.136**

4.137 Solve Prob. 4.136 assuming that the plumber exerts a force $\mathbf{F} = -(60 \text{ N})\mathbf{k}$ and that the motor is turned off ($\mathbf{M} = 0$).

4.138   Three rods are welded together to form a "corner" which is supported by three eyebolts. Neglecting friction, determine the reactions at A, B, and C when $P = 240$ N, $a = 120$ mm, $b = 80$ mm, and $c = 100$ mm.

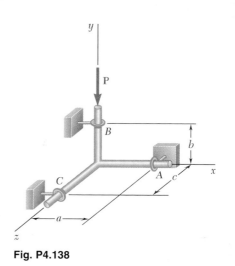

Fig. P4.138

4.139   Solve Prob. 4.138 assuming that the force **P** is removed and is replaced by a couple $\mathbf{M} = +(6 \text{ N} \cdot \text{m})\mathbf{j}$ acting at B.

4.140   The uniform 10-lb rod AB is supported by a ball-and-socket joint at A and leans against both the rod CD and the vertical wall. Neglecting the effects of friction, determine (*a*) the force which rod CD exerts on AB, (*b*) the reactions at A and B. (*Hint:* The force exerted by CD on AB must be perpendicular to both rods.)

Fig. P4.140

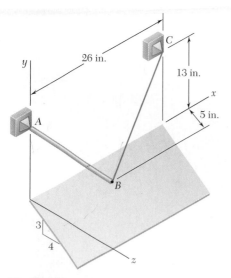

Fig. P4.141

4.141   A 21-in.-long uniform rod AB weighs 6.4 lb and is attached to a ball-and-socket joint at A. The rod rests against an inclined frictionless surface and is held in the position shown by cord BC. Knowing that the cord is 21 in. long, determine (*a*) the tension in the cord, (*b*) the reactions at A and B.

4.142 While being installed, the 56-lb chute *ABCD* is attached to a wall with brackets *E* and *F* and is braced with props *GH* and *IJ*. Assuming that the weight of the chute is uniformly distributed, determine the magnitude of the force exerted on the chute by prop *GH* if prop *IJ* is removed.

4.143 While being installed, the 56-lb chute *ABCD* is attached to a wall with brackets *E* and *F* and is braced with props *GH* and *IJ*. Assuming that the weight of the chute is uniformly distributed, determine the magnitude of the force exerted on the chute by prop *IJ* if prop *GH* is removed.

Fig. P4.142 and *P4.143*

4.144 To water seedlings, a gardener joins three lengths of pipe, *AB*, *BC*, and *CD*, fitted with spray nozzles and suspends the assembly using hinged supports at *A* and *D* and cable *EF*. Knowing that the pipe weighs 0.85 lb/ft, determine the tension in the cable.

4.145 Solve Prob. 4.144 assuming that cable *EF* is replaced by a cable connecting *E* and *C*.

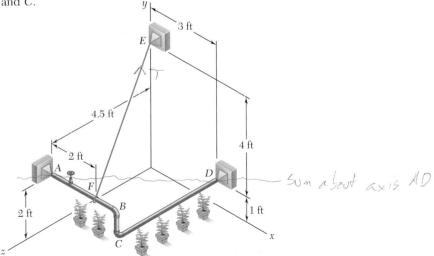

Fig. P4.144

4.146 The bent rod *ABDE* is supported by ball-and-socket joints at *A* and *E* and by the cable *DF*. If a 600-N load is applied at *C* as shown, determine the tension in the cable.

4.147 Solve Prob. 4.146 assuming that cable *DF* is replaced by a cable connecting *B* and *F*.

Fig. P4.146

4.148 Two rectangular plates are welded together to form the assembly shown. The assembly is supported by ball-and-socket joints at *B* and *D* and by a ball on a horizontal surface at *C*. For the loading shown, determine the reaction at *C*.

4.149 Two 1 × 2-m plywood panels, each of mass 15 kg, are nailed together as shown. The panels are supported by ball-and-socket joints at *A* and *F* and by the wire *BH*. Determine (*a*) the location of *H* in the *xy* plane if the tension in the wire is to be minimum, (*b*) the corresponding minimum tension.

4.150 Solve Prob. 4.149 subject to the restriction that *H* must lie on the *y* axis.

4.151 A uniform 20 × 30-in. steel plate *ABCD* weighs 85 lb and is attached to ball-and-socket joints at *A* and *B*. Knowing that the plate leans against a frictionless vertical wall at *D*, determine (*a*) the location of *D*, (*b*) the reaction at *D*.

Fig. P4.148

Fig. P4.149

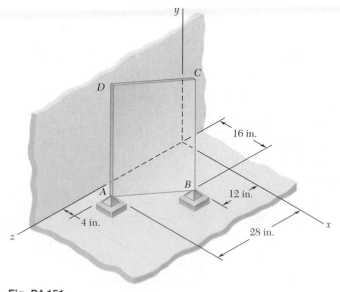

Fig. P4.151

This chapter was devoted to the study of the *equilibrium of rigid bodies,* that is, to the situation when the external forces acting on a rigid body *form a system equivalent to zero* [Sec. 4.1]. We then have

$$\Sigma \mathbf{F} = 0 \qquad \Sigma \mathbf{M}_O = \Sigma(\mathbf{r} \times \mathbf{F}) = 0 \qquad (4.1)$$

Resolving each force and each moment into its rectangular components, we can express the necessary and sufficient conditions for the equilibrium of a rigid body with the following six scalar equations:

$$\Sigma F_x = 0 \qquad \Sigma F_y = 0 \qquad \Sigma F_z = 0 \qquad (4.2)$$
$$\Sigma M_x = 0 \qquad \Sigma M_y = 0 \qquad \Sigma M_z = 0 \qquad (4.3)$$

These equations can be used to determine unknown forces applied to the rigid body or unknown reactions exerted by its supports.

When solving a problem involving the equilibrium of a rigid body, it is essential to consider *all* of the forces acting on the body. Therefore, the first step in the solution of the problem should be to draw a *free-body diagram* showing the body under consideration and all of the unknown as well as known forces acting on it [Sec. 4.2].

In the first part of the chapter, we considered the *equilibrium of a two-dimensional structure;* that is, we assumed that the structure considered and the forces applied to it were contained in the same plane. We saw that each of the reactions exerted on the structure by its supports could involve one, two, or three unknowns, depending upon the type of support [Sec. 4.3].

In the case of a two-dimensional structure, Eqs. (4.1), or Eqs. (4.2) and (4.3), reduce to *three equilibrium equations,* namely

$$\Sigma F_x = 0 \qquad \Sigma F_y = 0 \qquad \Sigma M_A = 0 \qquad (4.5)$$

where A is an arbitrary point in the plane of the structure [Sec. 4.4]. These equations can be used to solve for three unknowns. While the three equilibrium equations (4.5) cannot be *augmented* with additional equations, any of them can be *replaced* by another equation. Therefore, we can write alternative sets of equilibrium equations, such as

$$\Sigma F_x = 0 \qquad \Sigma M_A = 0 \qquad \Sigma M_B = 0 \qquad (4.6)$$

where point B is chosen in such a way that the line AB is not parallel to the y axis, or

$$\Sigma M_A = 0 \qquad \Sigma M_B = 0 \qquad \Sigma M_C = 0 \qquad (4.7)$$

where the points A, B, and C do not lie in a straight line.

Statical indeterminacy

Partial constraints

Improper constraints

Two-force body

Three-force body

Since any set of equilibrium equations can be solved for only three unknowns, the reactions at the supports of a rigid two-dimensional structure cannot be completely determined if they involve *more than three unknowns*; they are said to be *statically indeterminate* [Sec. 4.5]. On the other hand, if the reactions involve *fewer than three unknowns*, equilibrium will not be maintained under general loading conditions; the structure is said to be *partially constrained*. The fact that the reactions involve exactly three unknowns is no guarantee that the equilibrium equations can be solved for all three unknowns. If the supports are arranged in such a way that the reactions are *either concurrent or parallel*, the reactions are statically indeterminate, and the structure is said to be *improperly constrained*.

Two particular cases of equilibrium of a rigid body were given special attention. In Sec. 4.6, a *two-force body* was defined as a rigid body subjected to forces at only two points, and it was shown that the resultants $\mathbf{F}_1$ and $\mathbf{F}_2$ of these forces must have the *same magnitude, the same line of action, and opposite sense* (Fig. 4.11), a property which will simplify the solution of certain problems in later chapters. In Sec. 4.7, a *three-force body* was defined as a rigid body subjected to forces at only three points, and it was shown that the resultants $\mathbf{F}_1$, $\mathbf{F}_2$, and $\mathbf{F}_3$ of these forces must be *either concurrent* (Fig. 4.12) *or parallel*. This properly provides us with an alternative approach to the solution of problems involving a three-force body [Sample Prob. 4.6].

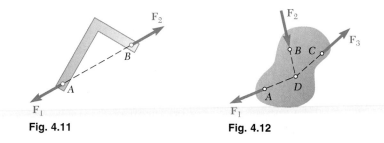

Fig. 4.11   **Fig. 4.12**

Equilibrium of a three-dimensional body

In the second part of the chapter, we considered the *equilibrium of a three-dimensional body* and saw that each of the reactions exerted on the body by its supports could involve between one and six unknowns, depending upon the type of support [Sec. 4.8].

In the general case of the equilibrium of a three-dimensional body, all of the six scalar equilibrium equations (4.2) and (4.3) listed at the beginning of this review should be used and solved for *six unknowns* [Sec. 4.9]. In most problems, however, these equations will be more conveniently obtained if we first write

$$\Sigma \mathbf{F} = 0 \qquad \Sigma \mathbf{M}_O = \Sigma(\mathbf{r} \times \mathbf{F}) = 0 \qquad (4.1)$$

and express the forces $\mathbf{F}$ and position vectors $\mathbf{r}$ in terms of scalar components and unit vectors. The vector products can then be

computed either directly or by means of determinants, and the desired scalar equations obtained by equating to zero the coefficients
of the unit vectors [Sample Probs. 4.7 through 4.9].

We noted that as many as three unknown reaction components
may be eliminated from the computation of $\Sigma \mathbf{M}_O$ in the second of
the relations (4.1) through a judicious choice of point O. Also, the
reactions at two points A and B can be eliminated from the solution of some problems by writing the equation $\Sigma M_{AB} = 0$, which
involves the computation of the moments of the forces about an
axis AB joining points A and B [Sample Prob. 4.10].

If the reactions involve more than six unknowns, some of the
reactions are *statically indeterminate;* if they involve fewer than six
unknowns, the rigid body is only *partially constrained.* Even with
six or more unknowns, the rigid body will be *improperly constrained* if the reactions associated with the given supports either
are parallel or intersect the same line.

Review Problems

4.152   Beam AD carries the two 40-lb loads shown. The beam is held by
a fixed support at D and by the cable BE which is attached to the counterweight W. Determine the reaction at D when (a) $W = 100$ lb, (b) $W = 90$ lb.

Fig. P4.152 and *P4.153*

4.153   For the beam and loading shown, determine the range of values
of W for which the magnitude of the couple at D does not exceed 40 lb · ft.

4.154   Determine the reactions at A and D when $\beta = 30°$.

4.155   Determine the reactions at A and D when $\beta = 60°$.

Fig. P4.154 and *P4.155*

900 lb

A

20 in.   40 in.   50 in.

B

Fig. P4.156

y

H

A B C

30°

250 mm

z 50 mm   300 mm

D

E

F 250 mm x

50 mm

400 N

Fig. P4.159

15 lb   20 lb   35 lb   20 lb   15 lb

A

B

C

6 in.   8 in.   8 in.   6 in.

Fig. _P4.160_

4.156 A 2100-lb tractor is used to lift 900 lb of gravel. Determine the reaction at each of the two (*a*) rear wheels *A*, (*b*) front wheels *B*.

3 in.   3 in.

C

1.8 in.

A B 5 lb

5 lb

Fig. P4.157

4.157 A tension of 5 lb is maintained in a tape as it passes the support system shown. Knowing that the radius of each pulley is 0.4 in., determine the reaction at *C*.

4.158 Solve Prob. 4.157 assuming that 0.6-in.-radius pulleys are used.

4.159 The bent rod *ABEF* is supported by bearings at *C* and *D* and by wire *AH*. Knowing that portion *AB* of the rod is 250 mm long, determine (*a*) the tension in wire *AH*, (*b*) the reactions at *C* and *D*. Assume that the bearing at *D* does not exert any axial thrust.

_4.160_ For the beam and loading shown, determine (*a*) the reaction at *A*, (*b*) the tension in cable *BC*.

4.161 Frame *ABCD* is supported by a ball-and-socket joint at *A* and by three cables. For *a* = 150 mm, determine the tension in each cable and the reaction at *A*.

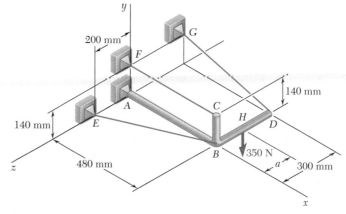

y

200 mm G

F

140 mm

A C

H

140 mm E

D

B 350 N

z 480 mm a 300 mm

x

Fig. P4.161 and _P4.162_

_4.162_ Frame *ABCD* is supported by a ball-and-socket joint at *A* and by three cables. Knowing that the 350-N load is applied at *D* (*a* = 300 mm), determine the tension in each cable and the reaction at *A*.

*4.163 In the problems listed below, the rigid bodies considered were completely constrained and the reactions were statically determinate. For each of these rigid bodies it is possible to create an improper set of constraints by changing a dimension of the body. In each of the following problems determine the value of *a* which results in improper constraints. (*a*) Prob. 4.81, (*b*) Prob. 4.82.

Computer Problems

4.C1 A slender rod AB of weight W is attached to blocks at A and B which can move freely in the guides shown. The constant of the spring is k, and the spring is unstretched when the rod is horizontal. Neglecting the weight of the blocks, derive an equation in terms of θ, W, l, and k which must be satisfied when the rod is in equilibrium. Knowing that $W = 10$ lb and $l = 40$ in., (a) calculate and plot the value of the spring constant k as a function of the angle θ for $15° \leq \theta \leq 40°$, (b) determine the two values of the angle θ corresponding to equilibrium when $k = 0.7$ lb/in.

Fig. P4.C1

4.C2 The position of the L-shaped rod shown is controlled by a cable attached at point B. Knowing that the rod supports a load of magnitude $P = 200$ N, use computational software to calculate and plot the tension T in the cable as a function of θ for values of θ from 0 to 120°. Determine the maximum tension T_{max} and the corresponding value of θ.

4.C3 The position of the 20-lb rod AB is controlled by the block shown, which is slowly moved to the left by the force $\mathbf{P}$. Neglecting the effect of friction, use computational software to calculate and plot the magnitude P of the force as a function of x for values of x decreasing from 30 in. to 0. Determine the maximum value of P and the corresponding value of x.

Fig. P4.C2

Fig. P4.C3

600 N

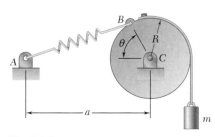

Fig. P4.C4

*4.C4 Member ABC is supported by a pin and bracket at C and by an inextensible cable of length 3.5 m that is attached at A and B and passes over a frictionless pulley at D. Neglecting the mass of ABC and the radius of the pulley, (*a*) plot the tension in the cable as a function of a for $0 \le a \le 2.4$ m, (*b*) determine the largest value of a for which equilibrium can be maintained.

4.C5 and 4.C6 The constant of spring AB is k, and the spring is unstretched when $\theta = 0$. Knowing that $R = 200$ mm, $a = 400$ mm, and $k = 1$ kN/m, use computational software to calculate and plot the mass m corresponding to equilibrium as a function of θ for values of θ from 0 to 90°. Determine the value of θ corresponding to equilibrium when $m = 2$ kg.

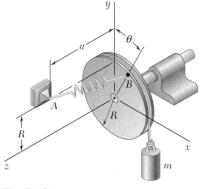

Fig. P4.C5

Fig. P4.C6

4.C7 An 8 × 10-in. panel of weight $W = 40$ lb is supported by hinges along edge AB. Cable CDE is attached to the panel at point C, passes over a small pulley at D, and supports a cylinder of weight W. Neglecting the effect of friction, use computational software to calculate and plot the weight of the cylinder corresponding to equilibrium as a function of θ for values of θ from 0 to 90°. Determine the value of θ corresponding to equilibrium when $W = 20$ lb.

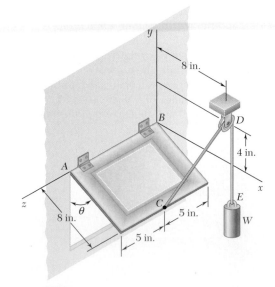

Fig. P4.C7

4.C8 A uniform circular plate of radius 300 mm and mass 26 kg is supported by three vertical wires that are equally spaced around its edge. A small 3-kg block E is placed on the plate at D and is then slowly moved along diameter CD until it reaches C. (a) Plot the tension in wires A and C as functions of a, where a is the distance of the block from D. (b) Determine the value of a for which the tension in wires A and C is minimum.

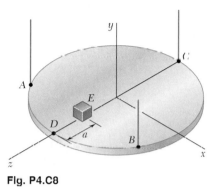

Fig. P4.C8

4.C9 The derrick shown supports a 4000-lb crate. It is held by a ball-and-socket joint at point A and by two cables attached at points D and E. Knowing that the derrick lies in a vertical plane forming an angle ϕ with the xy plane, use computational software to calculate and plot the tension in each cable as a function of ϕ for values of ϕ from 0 to 40°. Determine the value of ϕ for which the tension in cable BE is maximum.

4.C10 The 140-lb uniform steel plate $ABCD$ is welded to shaft EF and is maintained in the position shown by the couple **M**. Knowing that collars prevent the shaft from sliding in the bearings and that the shaft lies in the yz plane, plot the magnitude M of the couple as a function of θ for $0 \leq \theta \leq 90°$.

Fig. P4.C9

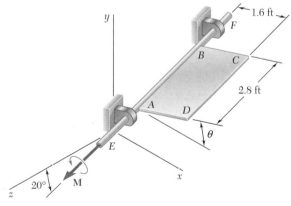

Fig. P4.C10

Distributed Forces: Centroids and Centers of Gravity

5

A precast section of roadway for a new interchange on Interstate 93 is shown being lowered from a gantry crane. In this chapter we will introduce the concept of the centroid of an area; in subsequent courses the relation between the location of the centroid and the behavior of the roadway under loading will be established.

DISTRIBUTED FORCES: CENTROIDS AND CENTERS OF GRAVITY

5.1   Introduction
      Areas and Lines
5.2   Center of Gravity of a
      Two-Dimensional Body
5.3   Centroids of Areas and Lines
5.4   First Moments of Areas and Lines
5.5   Composite Plates and Wires
5.6   Determination of Centroids by
      Integration
5.7   Theorems of Pappus-Guldinus
5.8   Distributed Loads on Beams
5.9   Forces on Submerged Surfaces
      Volumes
5.10 Center of Gravity of a Three-
      Dimensional Body. Centroid of a
      Volume
5.11 Composite Bodies
5.12 Determination of Centroids of
      Volumes by Integration

Photo 5.1 The precise balancing of the components of a mobile requires an understanding of centers of gravity and centroids, the main topics of this chapter.

5.1. INTRODUCTION

We have assumed so far that the attraction exerted by the earth on a rigid body could be represented by a single force **W**. This force, called the force of gravity or the weight of the body, was to be applied at the *center of gravity* of the body (Sec. 3.2). Actually, the earth exerts a force on each of the particles forming the body. The action of the earth on a rigid body should thus be represented by a large number of small forces distributed over the entire body. You will learn in this chapter, however, that all of these small forces can be replaced by a single equivalent force **W**. You will also learn how to determine the center of gravity, that is, the point of application of the resultant **W**, for bodies of various shapes.

In the first part of the chapter, two-dimensional bodies, such as flat plates and wires contained in a given plane, are considered. Two concepts closely associated with the determination of the center of gravity of a plate or a wire are introduced: the concept of the *centroid* of an area or a line and the concept of the *first moment* of an area or a line with respect to a given axis.

You will also learn that the computation of the area of a surface of revolution or of the volume of a body of revolution is directly related to the determination of the centroid of the line or area used to generate that surface or body of revolution (Theorems of Pappus-Guldinus). And, as is shown in Secs. 5.8 and 5.9, the determination of the centroid of an area simplifies the analysis of beams subjected to distributed loads and the computation of the forces exerted on submerged rectangular surfaces, such as hydraulic gates and portions of dams.

In the last part of the chapter, you will learn how to determine the center of gravity of a three-dimensional body as well as the centroid of a volume and the first moments of that volume with respect to the coordinate planes.

AREAS AND LINES

5.2. CENTER OF GRAVITY OF A TWO-DIMENSIONAL BODY

Let us first consider a flat horizontal plate (Fig. 5.1). We can divide the plate into n small elements. The coordinates of the first element

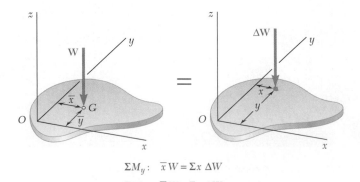

$$\Sigma M_y: \quad \overline{x}W = \Sigma x\,\Delta W$$
$$\Sigma M_x: \quad \overline{y}W = \Sigma y\,\Delta W$$

Fig. 5.1 Center of gravity of a plate.

are denoted by x_1 and y_1, those of the second element by x_2 and y_2, etc. The forces exerted by the earth on the elements of plate will be denoted, respectively, by $\Delta \mathbf{W}_1$, $\Delta \mathbf{W}_2$, . . . , $\Delta \mathbf{W}_n$. These forces or weights are directed toward the center of the earth; however, for all practical purposes they can be assumed to be parallel. Their resultant is therefore a single force in the same direction. The magnitude W of this force is obtained by adding the magnitudes of the elemental weights:

$$\Sigma F_z: \qquad W = \Delta W_1 + \Delta W_2 + \cdots + \Delta W_n$$

To obtain the coordinates $\bar{x}$ and $\bar{y}$ of the point G where the resultant $\mathbf{W}$ should be applied, we write that the moments of $\mathbf{W}$ about the y and x axes are equal to the sum of the corresponding moments of the elemental weights,

$$\begin{aligned} \Sigma M_y: &\quad \bar{x}W = x_1 \, \Delta W_1 + x_2 \, \Delta W_2 + \cdots + x_n \, \Delta W_n \\ \Sigma M_x: &\quad \bar{y}W = y_1 \, \Delta W_1 + y_2 \, \Delta W_2 + \cdots + y_n \, \Delta W_n \end{aligned} \qquad (5.1)$$

If we now increase the number of elements into which the plate is divided and simultaneously decrease the size of each element, we obtain in the limit the following expressions:

$$W = \int dW \qquad \bar{x}W = \int x \, dW \qquad \bar{y}W = \int y \, dW \qquad (5.2)$$

These equations define the weight $\mathbf{W}$ and the coordinates $\bar{x}$ and $\bar{y}$ of the center of gravity G of a flat plate. The same equations can be derived for a wire lying in the xy plane (Fig. 5.2). We note that the center of gravity G of a wire is usually not located on the wire.

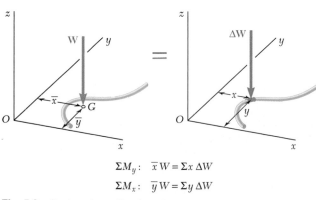

$$\Sigma M_y: \quad \bar{x}W = \Sigma x \, \Delta W$$
$$\Sigma M_x: \quad \bar{y}W = \Sigma y \, \Delta W$$

Fig. 5.2 Center of gravity of a wire.

5.3. CENTROIDS OF AREAS AND LINES

In the case of a flat homogeneous plate of uniform thickness, the magnitude ΔW of the weight of an element of the plate can be expressed as

$$\Delta W = \gamma t\, \Delta A$$

where γ = specific weight (weight per unit volume) of the material
t = thickness of the plate
ΔA = area of the element

Similarly, we can express the magnitude W of the weight of the entire plate as

$$W = \gamma t A$$

where A is the total area of the plate.

If U.S. customary units are used, the specific weight γ should be expressed in lb/ft^3, the thickness t in feet, and the areas ΔA and A in square feet. We observe that ΔW and W will then be expressed in pounds. If SI units are used, γ should be expressed in N/m^3, t in meters, and the areas ΔA and A in square meters; the weights ΔW and W will then be expressed in newtons.†

Substituting for ΔW and W in the moment equations (5.1) and dividing throughout by γt, we obtain

$$\Sigma M_y: \qquad \bar{x}A = x_1\,\Delta A_1 + x_2\,\Delta A_2 + \cdots + x_n\,\Delta A_n$$
$$\Sigma M_x: \qquad \bar{y}A = y_1\,\Delta A_1 + y_2\,\Delta A_2 + \cdots + y_n\,\Delta A_n$$

If we increase the number of elements into which the area A is divided and simultaneously decrease the size of each element, we obtain in the limit

$$\bar{x}A = \int x\, dA \qquad \bar{y}A = \int y\, dA \tag{5.3}$$

These equations define the coordinates $\bar{x}$ and $\bar{y}$ of the center of gravity of a homogeneous plate. The point whose coordinates are $\bar{x}$ and $\bar{y}$ is also known as the *centroid C of the area A* of the plate (Fig. 5.3). If the plate is not homogeneous, these equations cannot be used to determine the center of gravity of the plate; they still define, however, the centroid of the area.

In the case of a homogeneous wire of uniform cross section, the magnitude ΔW of the weight of an element of wire can be expressed as

$$\Delta W = \gamma a\, \Delta L$$

where γ = specific weight of the material
a = cross-sectional area of the wire
ΔL = length of the element

---

†It should be noted that in the SI system of units a given material is generally characterized by its density ρ (mass per unit volume) rather than by its specific weight γ. The specific weight of the material can then be obtained from the relation

$$\gamma = \rho g$$

where $g = 9.81\ \text{m/s}^2$. Since ρ is expressed in kg/m^3, we observe that γ will be expressed in $(\text{kg/m}^3)(\text{m/s}^2)$, that is, in N/m^3.

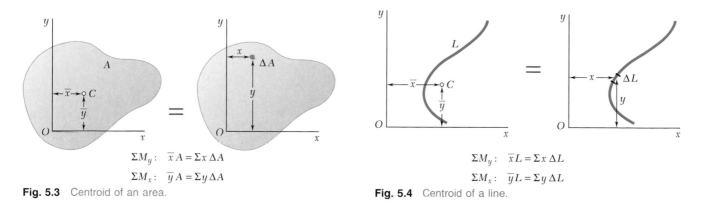

$$\Sigma M_y: \quad \bar{x}A = \Sigma x\, \Delta A$$
$$\Sigma M_x: \quad \bar{y}A = \Sigma y\, \Delta A$$

Fig. 5.3 Centroid of an area.

$$\Sigma M_y: \quad \bar{x}L = \Sigma x\, \Delta L$$
$$\Sigma M_x: \quad \bar{y}L = \Sigma y\, \Delta L$$

Fig. 5.4 Centroid of a line.

The center of gravity of the wire then coincides with the *centroid C of the line L* defining the shape of the wire (Fig. 5.4). The coordinates $\bar{x}$ and $\bar{y}$ of the centroid of the line L are obtained from the equations

$$\bar{x}L = \int x\, dL \qquad \bar{y}L = \int y\, dL \tag{5.4}$$

5.4. FIRST MOMENTS OF AREAS AND LINES

The integral $\int x\, dA$ in Eqs. (5.3) of the preceding section is known as the *first moment of the area A with respect to the y axis* and is denoted by Q_y. Similarly, the integral $\int y\, dA$ defines the *first moment of A with respect to the x axis* and is denoted by Q_x. We write

$$Q_y = \int x\, dA \qquad Q_x = \int y\, dA \tag{5.5}$$

Comparing Eqs. (5.3) with Eqs. (5.5), we note that the first moments of the area A can be expressed as the products of the area and the coordinates of its centroid:

$$Q_y = \bar{x}A \qquad Q_x = \bar{y}A \tag{5.6}$$

It follows from Eqs. (5.6) that the coordinates of the centroid of an area can be obtained by dividing the first moments of that area by the area itself. The first moments of the area are also useful in mechanics of materials for determining the shearing stresses in beams under transverse loadings. Finally, we observe from Eqs. (5.6) that if the centroid of an area is located on a coordinate axis, the first moment of the area with respect to that axis is zero. Conversely, if the first moment of an area with respect to a coordinate axis is zero, then the centroid of the area is located on that axis.

Relations similar to Eqs. (5.5) and (5.6) can be used to define the first moments of a line with respect to the coordinate axes and to express these moments as the products of the length L of the line and the coordinates $\bar{x}$ and $\bar{y}$ of its centroid.

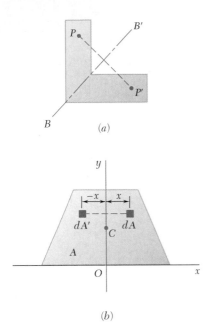

Fig. 5.5

An area A is said to be *symmetric with respect to an axis BB'* if for every point P of the area there exists a point P' of the same area such that the line PP' is perpendicular to BB' and is divided into two equal parts by that axis (Fig. 5.5a). A line L is said to be symmetric with respect to an axis BB' if it satisfies similar conditions. When an area A or a line L possesses an axis of symmetry BB', its first moment with respect to BB' is zero, and its centroid is located on that axis. For example, in the case of the area A of Fig. 5.5b, which is symmetric with respect to the y axis, we observe that for every element of area dA of abscissa x there exists an element dA' of equal area and with abscissa $-x$. It follows that the integral in the first of Eqs. (5.5) is zero and, thus, that $Q_y = 0$. It also follows from the first of the relations (5.3) that $\bar{x} = 0$. Thus, if an area A or a line L possesses an axis of symmetry, its centroid C is located on that axis.

We further note that if an area or line possesses two axes of symmetry, its centroid C must be located at the intersection of the two axes (Fig. 5.6). This property enables us to determine immediately the centroid of areas such as circles, ellipses, squares, rectangles, equilateral triangles, or other symmetric figures as well as the centroid of lines in the shape of the circumference of a circle, the perimeter of a square, etc.

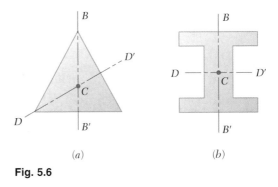

Fig. 5.6

An area A is said to be *symmetric with respect to a center O* if for every element of area dA of coordinates x and y there exists an element dA' of equal area with coordinates $-x$ and $-y$ (Fig. 5.7). It then follows that the integrals in Eqs. (5.5) are both zero and that $Q_x = Q_y = 0$. It also follows from Eqs. (5.3) that $\bar{x} = \bar{y} = 0$, that is, that the centroid of the area coincides with its center of symmetry O. Similarly, if a line possesses a center of symmetry O, the centroid of the line will coincide with the center O.

It should be noted that a figure possessing a center of symmetry does not necessarily possess an axis of symmetry (Fig. 5.7), while a figure possessing two axes of symmetry does not necessarily possess a center of symmetry (Fig. 5.6a). However, if a figure possesses two axes of symmetry at a right angle to each other, the point of intersection of these axes is a center of symmetry (Fig. 5.6b).

Determining the centroids of unsymmetrical areas and lines and of areas and lines possessing only one axis of symmetry will be discussed in Secs. 5.6 and 5.7. Centroids of common shapes of areas and lines are shown in Fig. 5.8A and B.

Fig. 5.7

Shape		$\bar{x}$	$\bar{y}$	Area
Triangular area			$\dfrac{h}{3}$	$\dfrac{bh}{2}$
Quarter-circular area		$\dfrac{4r}{3\pi}$	$\dfrac{4r}{3\pi}$	$\dfrac{\pi r^2}{4}$
Semicircular area		0	$\dfrac{4r}{3\pi}$	$\dfrac{\pi r^2}{2}$
Quarter-elliptical area		$\dfrac{4a}{3\pi}$	$\dfrac{4b}{3\pi}$	$\dfrac{\pi ab}{4}$
Semielliptical area		0	$\dfrac{4b}{3\pi}$	$\dfrac{\pi ab}{2}$
Semiparabolic area		$\dfrac{3a}{8}$	$\dfrac{3h}{5}$	$\dfrac{2ah}{3}$
Parabolic area		0	$\dfrac{3h}{5}$	$\dfrac{4ah}{3}$
Parabolic spandrel		$\dfrac{3a}{4}$	$\dfrac{3h}{10}$	$\dfrac{ah}{3}$
General spandrel		$\dfrac{n+1}{n+2}a$	$\dfrac{n+1}{4n+2}h$	$\dfrac{ah}{n+1}$
Circular sector		$\dfrac{2r\sin\alpha}{3\alpha}$	0	αr^2

Fig. 5.8A Centroids of common shapes of areas.

Shape		$\bar{x}$	$\bar{y}$	Length
Quarter-circular arc	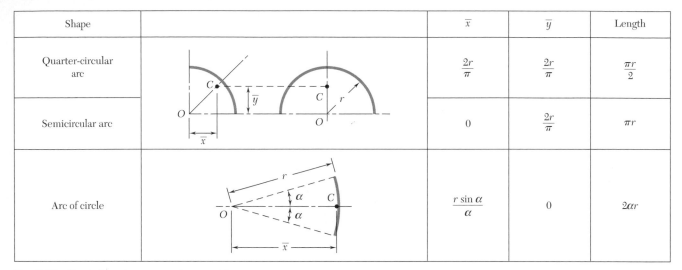	$\dfrac{2r}{\pi}$	$\dfrac{2r}{\pi}$	$\dfrac{\pi r}{2}$
Semicircular arc		0	$\dfrac{2r}{\pi}$	πr
Arc of circle		$\dfrac{r \sin \alpha}{\alpha}$	0	$2\alpha r$

Fig. 5.8B Centroids of common shapes of lines.

5.5. COMPOSITE PLATES AND WIRES

In many instances, a flat plate can be divided into rectangles, triangles, or the other common shapes shown in Fig. 5.8A. The abscissa $\overline{X}$ of its center of gravity G can be determined from the abscissas $\bar{x}_1$, $\bar{x}_2, \ldots, \bar{x}_n$ of the centers of gravity of the various parts by expressing that the moment of the weight of the whole plate about the y axis is equal to the sum of the moments of the weights of the various parts about the same axis (Fig. 5.9). The ordinate $\overline{Y}$ of the center of gravity of the plate is found in a similar way by equating moments about the x axis. We write

$$\Sigma M_y: \quad \overline{X}(W_1 + W_2 + \cdots + W_n) = \bar{x}_1 W_1 + \bar{x}_2 W_2 + \cdots + \bar{x}_n W_n$$

$$\Sigma M_x: \quad \overline{Y}(W_1 + W_2 + \cdots + W_n) = \bar{y}_1 W_1 + \bar{y}_2 W_2 + \cdots + \bar{y}_n W_n$$

$$\Sigma M_y: \quad \overline{X}\,\Sigma W = \Sigma\,\bar{x}\,W$$

$$\Sigma M_x: \quad \overline{Y}\,\Sigma W = \Sigma\,\bar{y}\,W$$

Fig. 5.9 Center of gravity of a composite plate.

or, for short,

$$\overline{X}\Sigma W = \Sigma \overline{x}W \qquad \overline{Y}\Sigma W = \Sigma \overline{y}W \qquad (5.7)$$

These equations can be solved for the coordinates $\overline{X}$ and $\overline{Y}$ of the center of gravity of the plate.

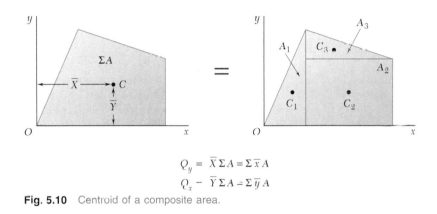

$$Q_y = \overline{X}\Sigma A = \Sigma \overline{x} A$$
$$Q_x = \overline{Y}\Sigma A = \Sigma \overline{y} A$$

Fig. 5.10 Centroid of a composite area.

If the plate is homogeneous and of uniform thickness, the center of gravity coincides with the centroid C of its area. The abscissa $\overline{X}$ of the centroid of the area can be determined by noting that the first moment Q_y of the composite area with respect to the y axis can be expressed both as the product of $\overline{X}$ and the total area and as the sum of the first moments of the elementary areas with respect to the y axis (Fig. 5.10). The ordinate $\overline{Y}$ of the centroid is found in a similar way by considering the first moment Q_x of the composite area. We have

$$Q_y = \overline{X}(A_1 + A_2 + \cdots + A_n) = \overline{x}_1 A_1 + \overline{x}_2 A_2 + \cdots + \overline{x}_n A_n$$
$$Q_x = \overline{Y}(A_1 + A_2 + \cdots + A_n) = \overline{y}_1 A_1 + \overline{y}_2 A_2 + \cdots + \overline{y}_n A_n$$

or, for short,

$$Q_y = \overline{X}\Sigma A = \Sigma \overline{x}A \qquad Q_x = \overline{Y}\Sigma A = \Sigma \overline{y}A \qquad (5.8)$$

These equations yield the first moments of the composite area, or they can be used to obtain the coordinates $\overline{X}$ and $\overline{Y}$ of its centroid.

Care should be taken to assign the appropriate sign to the moment of each area. First moments of areas, like moments of forces, can be positive or negative. For example, an area whose centroid is located to the left of the y axis will have a negative first moment with respect to that axis. Also, the area of a hole should be assigned a negative sign (Fig. 5.11).

Similarly, it is possible in many cases to determine the center of gravity of a composite wire or the centroid of a composite line by dividing the wire or line into simpler elements (see Sample Prob. 5.2).

	$\overline{x}$	A	$\overline{x}A$
A_1 Semicircle	−	+	−
A_2 Full rectangle	+	+	+
A_3 Circular hole	+	−	−

Fig. 5.11

SAMPLE PROBLEM 5.1

For the plane area shown, determine (a) the first moments with respect to the x and y axes, (b) the location of the centroid.

SOLUTION

Components of Area. The area is obtained by adding a rectangle, a triangle, and a semicircle and by then subtracting a circle. Using the coordinate axes shown, the area and the coordinates of the centroid of each of the component areas are determined and entered in the table below. The area of the circle is indicated as negative, since it is to be subtracted from the other areas. We note that the coordinate $\overline{y}$ of the centroid of the triangle is negative for the axes shown. The first moments of the component areas with respect to the coordinate axes are computed and entered in the table.

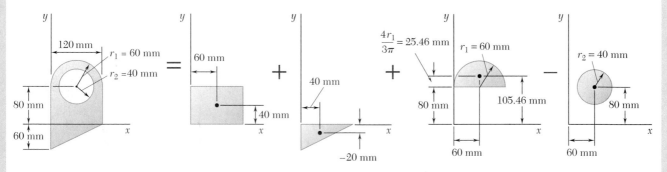

Component	A, mm²	$\overline{x}$, mm	$\overline{y}$, mm	$\overline{x}A$, mm³	$\overline{y}A$, mm³
Rectangle	$(120)(80) = 9.6 \times 10^3$	60	40	$+576 \times 10^3$	$+384 \times 10^3$
Triangle	$\frac{1}{2}(120)(60) = 3.6 \times 10^3$	40	-20	$+144 \times 10^3$	-72×10^3
Semicircle	$\frac{1}{2}\pi(60)^2 = 5.655 \times 10^3$	60	105.46	$+339.3 \times 10^3$	$+596.4 \times 10^3$
Circle	$-\pi(40)^2 = -5.027 \times 10^3$	60	80	-301.6×10^3	-402.2×10^3
	$\Sigma A = 13.828 \times 10^3$			$\Sigma \overline{x}A = +757.7 \times 10^3$	$\Sigma \overline{y}A = +506.2 \times 10^3$

a. **First Moments of the Area.** Using Eqs. (5.8), we write

$$Q_x = \Sigma \overline{y}A = 506.2 \times 10^3 \text{ mm}^3 \qquad Q_x = 506 \times 10^3 \text{ mm}^3 \blacktriangleleft$$
$$Q_y = \Sigma \overline{x}A = 757.7 \times 10^3 \text{ mm}^3 \qquad Q_y = 758 \times 10^3 \text{ mm}^3 \blacktriangleleft$$

b. **Location of Centroid.** Substituting the values given in the table into the equations defining the centroid of a composite area, we obtain

$$\overline{X}\Sigma A = \Sigma \overline{x}A: \qquad \overline{X}(13.828 \times 10^3 \text{ mm}^2) = 757.7 \times 10^3 \text{ mm}^3$$
$$\overline{X} = 54.8 \text{ mm} \blacktriangleleft$$

$$\overline{Y}\Sigma A = \Sigma \overline{y}A: \qquad \overline{Y}(13.828 \times 10^3 \text{ mm}^2) = 506.2 \times 10^3 \text{ mm}^3$$
$$\overline{Y} = 36.6 \text{ mm} \blacktriangleleft$$

SAMPLE PROBLEM 5.2

The figure shown is made from a piece of thin, homogeneous wire. Determine the location of its center of gravity.

SOLUTION

Since the figure is formed of homogeneous wire, its center of gravity coincides with the centroid of the corresponding line. Therefore, that centroid will be determined. Choosing the coordinate axes shown, with origin at A, we determine the coordinates of the centroid of each line segment and compute the first moments with respect to the coordinate axes.

Segment	L, in.	$\bar{x}$, in.	$\bar{y}$, in.	$\bar{x}L$, in^2	$\bar{y}L$, in^2
AB	24	12	0	288	0
BC	26	12	5	312	130
CA	10	0	5	0	50
	$\Sigma L = 60$			$\Sigma\bar{x}L = 600$	$\Sigma\bar{y}L = 180$

Substituting the values obtained from the table into the equations defining the centroid of a composite line, we obtain

$\bar{X}\Sigma L = \Sigma\bar{x}L:$ $\bar{X}(60 \text{ in.}) = 600 \text{ in}^2$ $\bar{X} = 10 \text{ in.}$ ◀

$\bar{Y}\Sigma L = \Sigma\bar{y}L:$ $\bar{Y}(60 \text{ in.}) = 180 \text{ in}^2$ $\bar{Y} = 3 \text{ in.}$ ◀

SAMPLE PROBLEM 5.3

A uniform semicircular rod of weight W and radius r is attached to a pin at A and rests against a frictionless surface at B. Determine the reactions at A and B.

SOLUTION

Free-Body Diagram.   A free-body diagram of the rod is drawn. The forces acting on the rod are its weight $\mathbf{W}$, which is applied at the center of gravity G (whose position is obtained from Fig. 5.8B); a reaction at A, represented by its components $\mathbf{A}_x$ and $\mathbf{A}_y$; and a horizontal reaction at B.

Equilibrium Equations

$$+\uparrow \Sigma M_A = 0: \qquad B(2r) - W\left(\frac{2r}{\pi}\right) = 0$$

$$B = +\frac{W}{\pi} \qquad\qquad \mathbf{B} = \frac{W}{\pi} \rightarrow \quad \blacktriangleleft$$

$$\xrightarrow{+} \Sigma F_x = 0: \qquad A_x + B = 0$$

$$A_x = -B = -\frac{W}{\pi} \qquad \mathbf{A}_x = \frac{W}{\pi} \leftarrow$$

$$+\uparrow \Sigma F_y = 0: \qquad A_y - W = 0 \qquad\qquad \mathbf{A}_y = W\uparrow$$

Adding the two components of the reaction at A:

$$A = \left[W^2 + \left(\frac{W}{\pi}\right)^2\right]^{1/2} \qquad\qquad A = W\left(1 + \frac{1}{\pi^2}\right)^{1/2} \quad \blacktriangleleft$$

$$\tan \alpha = \frac{W}{W/\pi} = \pi \qquad\qquad \alpha = \tan^{-1} \pi \quad \blacktriangleleft$$

The answers can also be expressed as follows:

$$\mathbf{A} = 1.049W \; \angle 72.3° \qquad \mathbf{B} = 0.318W \rightarrow \quad \blacktriangleleft$$

In this lesson we developed the general equations for locating the centers of gravity of two-dimensional bodies and wires [Eqs. (5.2)] and the centroids of plane areas [Eqs. (5.3)] and lines [Eqs. (5.4)]. In the following problems, you will have to locate the centroids of composite areas and lines or determine the first moments of the area for composite plates [Eqs. (5.8)].

1. *Locating the centroids of composite areas and lines.* Sample Problems 5.1 and 5.2 illustrate the procedure you should follow when solving problems of this type. There are, however, several points that should be emphasized.

   a. The first step in your solution should be to decide how to construct the given area or line from the common shapes of Fig. 5.8. You should recognize that for plane areas it is often possible to construct a particular shape in more than one way. Also, showing the different components (as is done in Sample Prob. 5.1) will help you to correctly establish their centroids and areas or lengths. Do not forget that you can subtract areas as well as add them to obtain a desired shape.

   b. We strongly recommend that for each problem you construct a table containing the areas or lengths and the respective coordinates of the centroids. It is essential for you to remember that areas which are "removed" (for example, holes) are treated as negative. Also, the sign of negative coordinates must be included. Therefore, you should always carefully note the location of the origin of the coordinate axes.

   c. When possible, use symmetry [Sec. 5.4] to help you determine the location of a centroid.

   d. In the formulas for the circular sector and for the arc of a circle in Fig. 5.8, the angle α must always be expressed in radians.

2. *Calculating the first moments of an area.* The procedures for locating the centroid of an area and for determining the first moments of an area are similar; however, for the latter it is not necessary to compute the total area. Also, as noted in Sec. 5.4, you should recognize that the first moment of an area relative to a centroidal axis is zero.

3. *Solving problems involving the center of gravity.* The bodies considered in the following problems are homogeneous; thus, their centers of gravity and centroids coincide. In addition, when a body that is suspended from a single pin is in equilibrium, the pin and the body's center of gravity must lie on the same vertical line.

It may appear that many of the problems in this lesson have little to do with the study of mechanics. However, being able to locate the centroid of composite shapes will be essential in several topics that you will soon encounter.

Problems

5.1 through 5.8   Locate the centroid of the plane area shown.

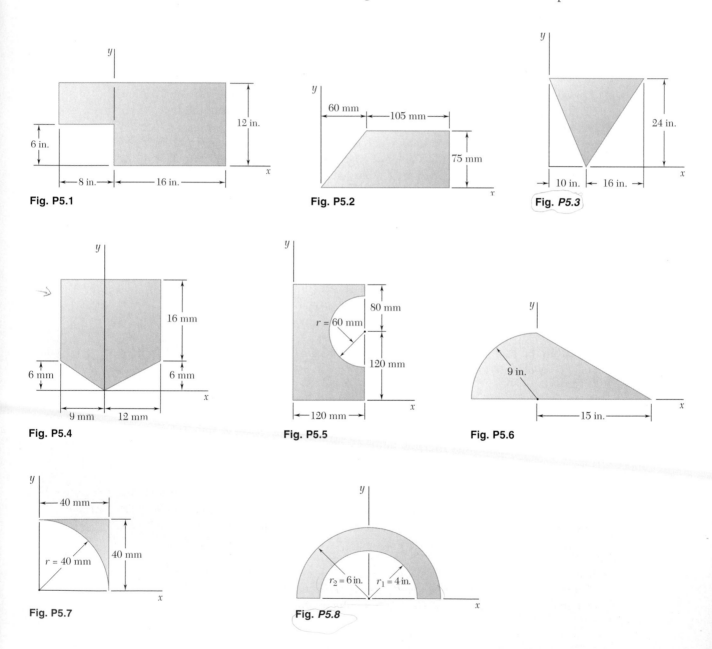

Fig. P5.1

Fig. P5.2

Fig. P5.3

Fig. P5.4

Fig. P5.5

Fig. P5.6

Fig. P5.7

Fig. P5.8

5.9   For the area of Prob. 5.8, determine the ratio r_2/r_1 so that $\overline{y} = 3r_1/4$.

5.10 Show that as r_1 approaches r_2, the location of the centroid approaches that of a circular arc of radius $(r_1 + r_2)/2$.

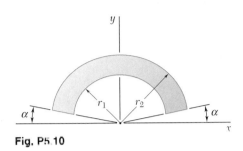

Fig. P5.10

5.11 through 5.16 Locate the centroid of the plane area shown.

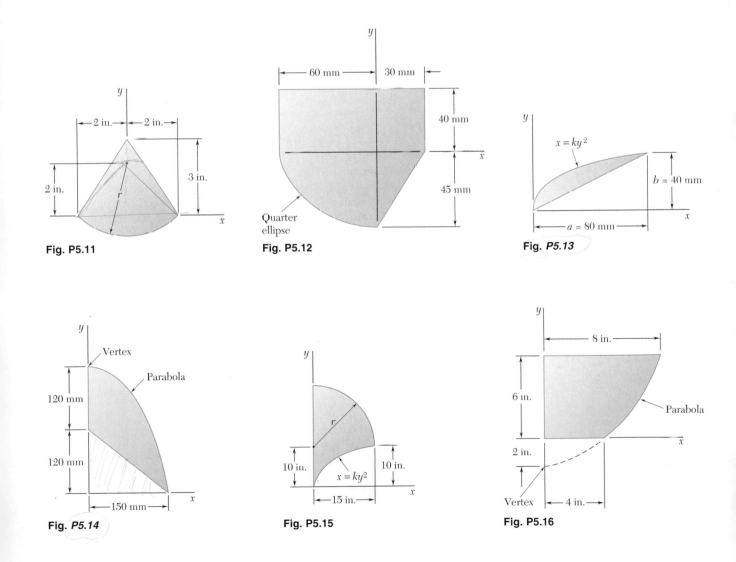

Fig. P5.11

Quarter ellipse
Fig. P5.12

$x = ky^2$
$b = 40$ mm
$a = 80$ mm
Fig. *P5.13*

Vertex
Parabola
120 mm
120 mm
150 mm
Fig. *P5.14*

r
10 in.
10 in.
$x = ky^2$
15 in.
Fig. P5.15

8 in.
6 in.
Parabola
2 in.
Vertex
4 in.
Fig. P5.16

5.17 and 5.18 The horizontal x axis is drawn through the centroid C of the area shown and divides the area into two component areas A_1 and A_2. Determine the first moment of each component area with respect to the x axis, and explain the results obtained.

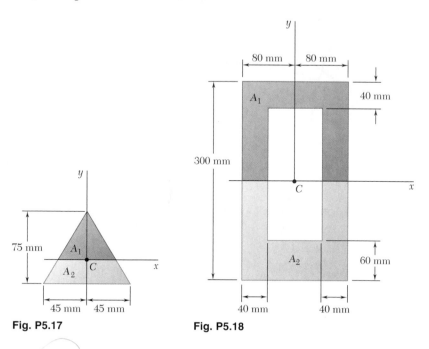

Fig. P5.17

Fig. P5.18

5.19 The first moment of the shaded area with respect to the x axis is denoted by Q_x. (a) Express Q_x in terms of r and θ. (b) For what value of θ is Q_x maximum, and what is the maximum value?

5.20 A composite beam is constructed by bolting four plates to four $2 \times 2 \times 3/8$-in. angles as shown. The bolts are equally spaced along the beam, and the beam supports a vertical load. As proved in mechanics of materials, the shearing forces exerted on the bolts at A and B are proportional to the first moments with respect to the centroidal x axis of the red shaded areas shown, respectively, in parts a and b of the figure. Knowing that the force exerted on the bolt at A is 70 lb, determine the force exerted on the bolt at B.

Fig. P5.19

Fig. P5.20

5.21 through 5.24 A thin, homogeneous wire is bent to form the perimeter of the figure indicated. Locate the center of gravity of the wire figure thus formed.

    5.21 Fig. P5.1.
    5.22 Fig. P5.2.
    5.23 Fig. P5.4.
    5.24 Fig. P5.8.

5.25 A 750-g uniform steel rod is bent into a circular arc of radius 500 mm as shown. The rod is supported by a pin at A and the cord BC. Determine (*a*) the tension in the cord, (*b*) the reaction at A.

5.26 The homogeneous wire $ABCD$ is bent as shown and is supported by a pin at B. Knowing that $l = 8$ in., determine the angle θ for which portion BC of the wire is horizontal.

Fig. P5.25

Fig. P5.26 and *P5.27*

5.27 The homogeneous wire $ABCD$ is bent as shown and is supported by a pin at B. Knowing that $\theta = 30°$, determine the length l for which portion CD of the wire is horizontal.

5.28 The homogeneous wire $ABCD$ is bent as shown and is attached to a hinge at C. Determine the length L for which the portion BCD of the wire is horizontal.

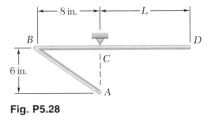

Fig. P5.28

5.29 Determine the distance h so that the centroid of the shaded area is as close to line BB' as possible when (*a*) $k = 0.2$, (*b*) $k = 0.6$.

5.30 Show when the distance h is selected to minimize the distance $\bar{y}$ from line BB' to the centroid of the shaded area that $\bar{y} = h$.

Fig. P5.29 and P5.30

5.6. DETERMINATION OF CENTROIDS BY INTEGRATION

The centroid of an area bounded by analytical curves (that is, curves defined by algebraic equations) is usually determined by evaluating the integrals in Eqs. (5.3) of Sec. 5.3:

$$\bar{x}A = \int x \, dA \qquad \bar{y}A = \int y \, dA \qquad (5.3)$$

If the element of area dA is a small rectangle of sides dx and dy, the evaluation of each of these integrals requires a *double integration* with respect to x and y. A double integration is also necessary if polar coordinates are used for which dA is a small element of sides dr and $r \, d\theta$.

In most cases, however, it is possible to determine the coordinates of the centroid of an area by performing a single integration. This is achieved by choosing dA to be a thin rectangle or strip or a thin sector or pie-shaped element (Fig. 5.12A); the centroid of the thin rectangle is located at its center, and the centroid of the thin sector is located at a distance $\frac{2}{3}r$ from its vertex (as it is for a triangle). The coordinates of the centroid of the area under consideration are then obtained by expressing that the first moment of the entire area with respect to each of the coordinate axes is equal to the sum (or integral) of the corresponding moments of the elements of area. Denoting by $\bar{x}_{el}$ and $\bar{y}_{el}$ the coordinates of the centroid of the element dA, we write

$$Q_y = \bar{x}A = \int \bar{x}_{el} \, dA$$
$$Q_x = \bar{y}A = \int \bar{y}_{el} \, dA \qquad (5.9)$$

If the area A is not already known, it can also be computed from these elements.

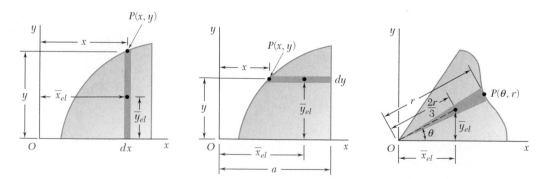

Fig. 5.12A   Centroids and areas of differential elements.

The coordinates $\bar{x}_{el}$ and $\bar{y}_{el}$ of the centroid of the element of area dA should be expressed in terms of the coordinates of a point located on the curve bounding the area under consideration. Also, the area of the element dA should be expressed in terms of the coordinates of that point and the appropriate differentials. This has been done in Fig. 5.12B for three common types of elements; the pie-shaped element of part c should be used when the equation of the curve bounding the area is given in polar coordinates. The appropriate expressions

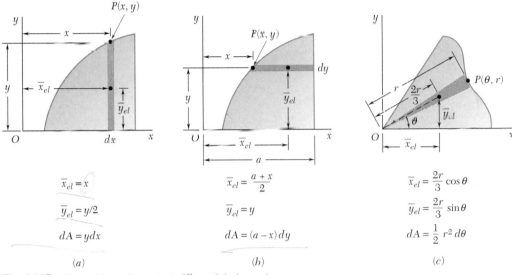

Fig. 5.12B   Centroids and areas of differential elements.

should be substituted into formulas (5.9), and the equation of the bounding curve should be used to express one of the coordinates in terms of the other. The integration is thus reduced to a single integration. Once the area has been determined and the integrals in Eqs. (5.9) have been evaluated, these equations can be solved for the coordinates $\bar{x}$ and $\bar{y}$ of the centroid of the area.

When a line is defined by an algebraic equation, its centroid can be determined by evaluating the integrals in Eqs. (5.4) of Sec. 5.3:

$$\bar{x}L = \int x \, dL \qquad \bar{y}L = \int y \, dL \qquad (5.4)$$

The differential length dL should be replaced by one of the following expressions, depending upon which coordinate, x, y, or θ, is chosen as the independent variable in the equation used to define the line (these expressions can be derived using the Pythagorean theorem):

$$dL = \sqrt{1 + \left(\frac{dy}{dx}\right)^2} \, dx \qquad dL = \sqrt{1 + \left(\frac{dx}{dy}\right)^2} \, dy$$

$$dL = \sqrt{r^2 + \left(\frac{dr}{d\theta}\right)^2} \, d\theta$$

After the equation of the line has been used to express one of the co-ordinates in terms of the other, the integration can be performed, and Eqs. (5.4) can be solved for the coordinates $\bar{x}$ and $\bar{y}$ of the centroid of the line.

5.7. THEOREMS OF PAPPUS-GULDINUS

These theorems, which were first formulated by the Greek geometer Pappus during the third century A.D. and later restated by the Swiss mathematician Guldinus, or Guldin, (1577–1643) deal with surfaces and bodies of revolution.

A *surface of revolution* is a surface which can be generated by rotating a plane curve about a fixed axis. For example (Fig. 5.13), the

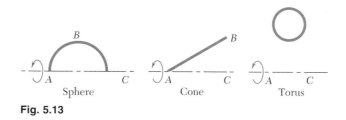

Fig. 5.13

surface of a sphere can be obtained by rotating a semicircular arc *ABC* about the diameter *AC*, the surface of a cone can be produced by rotating a straight line *AB* about an axis *AC*, and the surface of a torus or ring can be generated by rotating the circumference of a circle about a nonintersecting axis. A *body of revolution* is a body which can be generated by rotating a plane area about a fixed axis. As shown in Fig. 5.14, a sphere, a cone, and a torus can each be generated by rotating the appropriate shape about the indicated axis.

Photo 5.2 The storage tanks shown are all bodies of revolution. Thus, their surface areas and volumes can be determined using the theorems of Pappus-Guldinus.

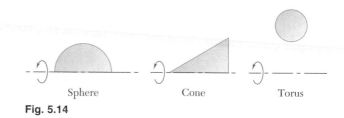

Fig. 5.14

THEOREM I. *The area of a surface of revolution is equal to the length of the generating curve times the distance traveled by the centroid of the curve while the surface is being generated.*

Proof. Consider an element dL of the line L (Fig. 5.15), which is revolved about the x axis. The area dA generated by the element dL is equal to $2\pi y\, dL$. Thus, the entire area generated by L is $A = \int 2\pi y\, dL$. Recalling that we found in Sec. 5.3 that the integral $\int y\, dL$

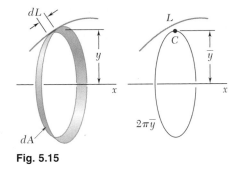

Fig. 5.15

is equal to $\bar{y}L$, we therefore have

$$A = 2\pi\bar{y}L \qquad (5.10)$$

where $2\pi\bar{y}$ is the distance traveled by the centroid of L (Fig. 5.15). It should be noted that the generating curve must not cross the axis about which it is rotated; if it did, the two sections on either side of the axis would generate areas having opposite signs, and the theorem would not apply.

THEOREM II. *The volume of a body of revolution is equal to the generating area times the distance traveled by the centroid of the area while the body is being generated.*

Proof. Consider an element dA of the area A which is revolved about the x axis (Fig. 5.16). The volume dV generated by the element

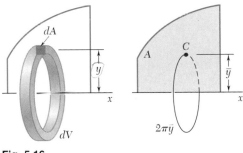

Fig. 5.16

dA is equal to $2\pi y \, dA$. Thus, the entire volume generated by A is $V = \int 2\pi y \, dA$, and since the integral $\int y \, dA$ is equal to $\bar{y}A$ (Sec. 5.3), we have

$$V = 2\pi\bar{y}A \qquad (5.11)$$

where $2\pi\bar{y}$ is the distance traveled by the centroid of A. Again, it should be noted that the theorem does not apply if the axis of rotation intersects the generating area.

The theorems of Pappus-Guldinus offer a simple way to compute the areas of surfaces of revolution and the volumes of bodies of revolution. Conversely, they can also be used to determine the centroid of a plane curve when the area of the surface generated by the curve is known or to determine the centroid of a plane area when the volume of the body generated by the area is known (see Sample Prob. 5.8).

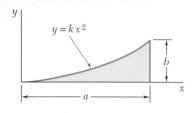

SAMPLE PROBLEM 5.4

Determine by direct integration the location of the centroid of a parabolic spandrel.

SOLUTION

Determination of the Constant k. The value of k is determined by substituting $x = a$ and $y = b$ into the given equation. We have $b = ka^2$ or $k = b/a^2$. The equation of the curve is thus

$$y = \frac{b}{a^2}x^2 \qquad \text{or} \qquad x = \frac{a}{b^{1/2}}y^{1/2}$$

Vertical Differential Element. We choose the differential element shown and find the total area of the figure.

$$A = \int dA = \int y\, dx = \int_0^a \frac{b}{a^2}x^2\, dx = \left[\frac{b}{a^2}\frac{x^3}{3}\right]_0^a = \frac{ab}{3}$$

The first moment of the differential element with respect to the y axis is $\bar{x}_{el}\, dA$; hence, the first moment of the entire area with respect to this axis is

$$Q_y = \int \bar{x}_{el}\, dA = \int xy\, dx = \int_0^a x\left(\frac{b}{a^2}x^2\right) dx = \left[\frac{b}{a^2}\frac{x^4}{4}\right]_0^a = \frac{a^2 b}{4}$$

Since $Q_y = \bar{x}A$, we have

$$\bar{x}A = \int \bar{x}_{el}\, dA \qquad \bar{x}\frac{ab}{3} = \frac{a^2 b}{4} \qquad \bar{x} = \tfrac{3}{4}a \quad \blacktriangleleft$$

Likewise, the first moment of the differential element with respect to the x axis is $\bar{y}_{el}\, dA$, and the first moment of the entire area is

$$Q_x = \int \bar{y}_{el}\, dA = \int \frac{y}{2} y\, dx = \int_0^a \frac{1}{2}\left(\frac{b}{a^2}x^2\right)^2 dx = \left[\frac{b^2}{2a^4}\frac{x^5}{5}\right]_0^a = \frac{ab^2}{10}$$

Since $Q_x = \bar{y}A$, we have

$$\bar{y}A = \int \bar{y}_{el}\, dA \qquad \bar{y}\frac{ab}{3} = \frac{ab^2}{10} \qquad \bar{y} = \tfrac{3}{10}b \quad \blacktriangleleft$$

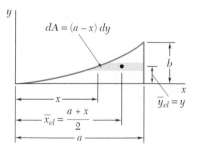

Horizontal Differential Element. The same results can be obtained by considering a horizontal element. The first moments of the area are

$$Q_y = \int \bar{x}_{el}\, dA = \int \frac{a + x}{2}(a - x)\, dy = \int_0^b \frac{a^2 - x^2}{2}\, dy$$

$$= \frac{1}{2}\int_0^b \left(a^2 - \frac{a^2}{b}y\right) dy = \frac{a^2 b}{4}$$

$$Q_x = \int \bar{y}_{el}\, dA = \int y(a - x)\, dy = \int y\left(a - \frac{a}{b^{1/2}}y^{1/2}\right) dy$$

$$= \int_0^b \left(ay - \frac{a}{b^{1/2}}y^{3/2}\right) dy = \frac{ab^2}{10}$$

To determine $\bar{x}$ and $\bar{y}$, the expressions obtained are again substituted into the equations defining the centroid of the area.

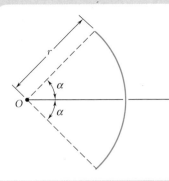

SAMPLE PROBLEM 5.5

Determine the location of the centroid of the circular arc shown.

SOLUTION

Since the arc is symmetrical with respect to the x axis, $\bar{y} = 0$. A differential element is chosen as shown, and the length of the arc is determined by integration.

$$L = \int dL = \int_{-\alpha}^{\alpha} r \, d\theta = r \int_{-\alpha}^{\alpha} d\theta = 2r\alpha$$

The first moment of the arc with respect to the y axis is

$$Q_y = \int x \, dL = \int_{-\alpha}^{\alpha} (r \cos \theta)(r \, d\theta) = r^2 \int_{-\alpha}^{\alpha} \cos \theta \, d\theta$$

$$= r^2[\sin \theta]_{-\alpha}^{\alpha} = 2r^2 \sin \alpha$$

Since $Q_y = \bar{x}L$, we write

$$\bar{x}(2r\alpha) = 2r^2 \sin \alpha \qquad \bar{x} = \frac{r \sin \alpha}{\alpha} \quad \blacktriangleleft$$

SAMPLE PROBLEM 5.6

Determine the area of the surface of revolution shown, which is obtained by rotating a quarter-circular arc about a vertical axis.

SOLUTION

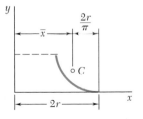

According to Theorem I of Pappus-Guldinus, the area generated is equal to the product of the length of the arc and the distance traveled by its centroid. Referring to Fig. 5.8B, we have

$$\bar{x} = 2r - \frac{2r}{\pi} = 2r\left(1 - \frac{1}{\pi}\right)$$

$$A = 2\pi \bar{x}L = 2\pi\left[2r\left(1 - \frac{1}{\pi}\right)\right]\left(\frac{\pi r}{2}\right)$$

$$A = 2\pi r^2(\pi - 1) \quad \blacktriangleleft$$

SAMPLE PROBLEM 5.7

The outside diameter of a pulley is 0.8 m, and the cross section of its rim is as shown. Knowing that the pulley is made of steel and that the density of steel is $\rho = 7.85 \times 10^3$ kg/m³, determine the mass and the weight of the rim.

SOLUTION

The volume of the rim can be found by applying Theorem II of Pappus-Guldinus, which states that the volume equals the product of the given cross-sectional area and the distance traveled by its centroid in one complete revolution. However, the volume can be more easily determined if we observe that the cross section can be formed from rectangle I, whose area is positive, and rectangle II, whose area is negative.

	Area, mm²	$\bar{y}$, mm	Distance Traveled by C, mm	Volume, mm³
I	+5000	375	$2\pi(375) = 2356$	$(5000)(2356) = 11.78 \times 10^6$
II	−1800	365	$2\pi(365) = 2293$	$(-1800)(2293) = -4.13 \times 10^6$
				Volume of rim $= 7.65 \times 10^6$

Since 1 mm = 10^{-3} m, we have 1 mm³ = $(10^{-3}$ m$)^3 = 10^{-9}$ m³, and we obtain $V = 7.65 \times 10^6$ mm³ = $(7.65 \times 10^6)(10^{-9}$ m³$) = 7.65 \times 10^{-3}$ m³.

$$m = \rho V = (7.85 \times 10^3 \text{ kg/m}^3)(7.65 \times 10^{-3} \text{ m}^3) \qquad m = 60.0 \text{ kg} \blacktriangleleft$$
$$W = mg = (60.0 \text{ kg})(9.81 \text{ m/s}^2) = 589 \text{ kg} \cdot \text{m/s}^2 \qquad W = 589 \text{ N} \blacktriangleleft$$

SAMPLE PROBLEM 5.8

Using the theorems of Pappus-Guldinus, determine (a) the centroid of a semicircular area, (b) the centroid of a semicircular arc. We recall that the volume and the surface area of a sphere are $\frac{4}{3}\pi r^3$ and $4\pi r^2$, respectively.

SOLUTION

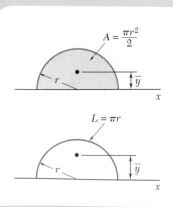

The volume of a sphere is equal to the product of the area of a semicircle and the distance traveled by the centroid of the semicircle in one revolution about the x axis.

$$V = 2\pi \bar{y} A \qquad \frac{4}{3}\pi r^3 = 2\pi \bar{y}(\frac{1}{2}\pi r^2) \qquad \bar{y} = \frac{4r}{3\pi} \blacktriangleleft$$

Likewise, the area of a sphere is equal to the product of the length of the generating semicircle and the distance traveled by its centroid in one revolution.

$$A = 2\pi \bar{y} L \qquad 4\pi r^2 = 2\pi \bar{y}(\pi r) \qquad \bar{y} = \frac{2r}{\pi} \blacktriangleleft$$

In the problems for this lesson, you will use the equations

$$\bar{x}A = \int x\, dA \qquad \bar{y}A = \int y\, dA \qquad (5.3)$$

$$\bar{x}L = \int x\, dL \qquad \bar{y}L = \int y\, dL \qquad (5.4)$$

to locate the centroids of plane areas and lines, respectively. You will also apply the theorems of Pappus-Guldinus (Sec. 5.7) to determine the areas of surfaces of revolution and the volumes of bodies of revolution.

1. *Determining by direct integration the centroids of areas and lines.* When solving problems of this type, you should follow the method of solution shown in Sample Probs. 5.4 and 5.5: compute A or L, determine the first moments of the area or the line, and solve Eqs. (5.3) or (5.4) for the coordinates of the centroid. In addition, you should pay particular attention to the following points.

a. Begin your solution by carefully defining or determining each term in the applicable integral formulas. We strongly encourage you to show on your sketch of the given area or line your choice for dA or dL and the distances to its centroid.

b. As explained in Sec. 5.6, the x and the y in the above equations represent the *coordinates of the centroid* of the differential elements dA and dL. It is important to recognize that the coordinates of the centroid of dA are not equal to the coordinates of a point located on the curve bounding the area under consideration. You should carefully study Fig. 5.12 until you fully understand this important point.

c. To possibly simplify or minimize your computations, always examine the shape of the given area or line before defining the differential element that you will use. For example, sometimes it may be preferable to use horizontal rectangular elements instead of vertical ones. Also, it will usually be advantageous to use polar coordinates when a line or an area has circular symmetry.

d. Although most of the integrations in this lesson are straightforward, at times it may be necessary to use more advanced techniques, such as trigonometric substitution or integration by parts. Of course, using a table of integrals is the fastest method to evaluate difficult integrals.

2. *Applying the theorems of Pappus-Guldinus.* As shown in Sample Probs. 5.6 through 5.8, these simple, yet very useful theorems allow you to apply your knowledge of centroids to the computation of areas and volumes. Although the theorems refer to the distance traveled by the centroid and to the length of the generating curve or to the generating area, the resulting equations [Eqs. (5.10) and (5.11)] contain the products of these quantities, which are simply the first moments of a line ($\bar{y}L$) and an area ($\bar{y}A$), respectively. Thus, for those problems for which the generating line or area consists of more than one common shape, you need only determine $\bar{y}L$ or $\bar{y}A$; you do not have to calculate the length of the generating curve or the generating area.

Problems

5.31 through 5.33 Determine by direct integration the centroid of the area shown. Express your answer in terms of a and h.

Fig. P5.31

Fig. P5.32

Fig. P5.33

5.34 through 5.36 Determine by direct integration the centroid of the area shown.

Fig. *P5.34*

Fig. P5.35

Fig. *P5.36*

5.37 and 5.38 Determine by direct integration the centroid of the area shown. Express your answer in terms of a and b.

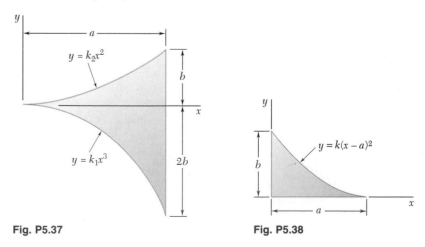

Fig. P5.37

Fig. P5.38

244

5.39 Determine by direct integration the centroid of the area shown.

5.40 and 5.41 Determine by direct integration the centroid of the area shown. Express your answer in terms of a and b.

Fig. P5.39

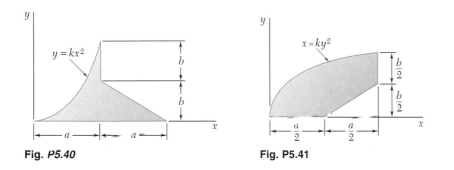

Fig. *P5.40*

Fig. P5.41

*5.42 A homogeneous wire is bent into the shape shown. Determine by direct integration the x coordinate of its centroid. Express your answer in terms of a.

5.43 and 5.44 A homogeneous wire is bent into the shape shown. Determine by direct integration the x coordinate of its centroid.

Fig. *P5.42*

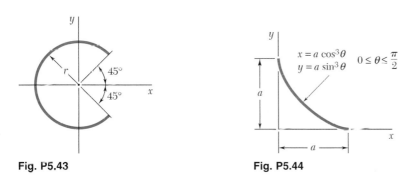

Fig. P5.43

Fig. P5.44

*5.45 and *5.46 Determine by direct integration the centroid of the area shown.

Fig. P5.45

Fig. P5.46

Fig. P5.50

Fig. P5.51

Fig. P5.53

5.47 Determine the volume and the surface area of the solid obtained by rotating the area of Prob. 5.2 about (*a*) the *x* axis, (*b*) the line $x = 165$ mm.

5.48 Determine the volume and the surface area of the solid obtained by rotating the area of Prob. 5.4 about (*a*) the line $y = 22$ mm, (*b*) the line $x = 12$ mm.

5.49 Determine the volume and the surface area of the solid obtained by rotating the area of Prob. 5.1 about (*a*) the *x* axis, (*b*) the line $x = 16$ in.

5.50 Determine the volume of the solid generated by rotating the semi-elliptical area shown about (*a*) the axis *AA'*, (*b*) the axis *BB'*, (*c*) the *y* axis.

5.51 Determine the volume and the surface area of the chain link shown, which is made from a 2-in.-diameter bar, if $R = 3$ in. and $L = 10$ in.

5.52 Verify that the expressions for the volumes of the first four shapes in Fig. 5.21 on page 261 are correct.

5.53 A 15-mm-diameter hole is drilled in a piece of 20-mm-thick steel; the hole is then countersunk as shown. Determine the volume of steel removed during the countersinking process.

5.54 Three different drive belt profiles are to be studied. If at any given time each belt makes contact with one-half of the circumference of its pulley, determine the *contact area* between the belt and the pulley for each design.

(*a*)   (*b*)   (*c*)

Fig. P5.54

5.55 Determine the capacity, in gallons, of the punch bowl shown if $R = 12$ in.

Fig. *P5.55*

5.56 The aluminum shade for a small high-intensity lamp has a uniform thickness of 3/32 in. Knowing that the specific weight of aluminum is 0.101 lb/in^3, determine the weight of the shade.

Fig. P5.56

5.57 The top of a round wooden table has the edge profile shown. Knowing that the diameter of the top is 1100 mm before shaping and that the density of the wood is 690 kg/m^3, determine the weight of the waste wood resulting from the production of 5000 tops.

Fig. P5.57 and *P5.58*

5.58 The top of a round wooden table has the shape shown. Determine how many liters of lacquer are required to finish 5000 tops knowing that each top is given three coats of lacquer and that 1 liter of lacquer covers 12 m^2.

5.59 The escutcheon (a decorative plate placed on a pipe where the pipe exits from a wall) shown is cast from yellow brass. Knowing that the specific weight of yellow brass is 0.306 lb/in^3, determine the weight of the escutcheon.

Fig. P5.59

***5.60** The reflector of a small flashlight has the parabolic shape shown. Determine the surface area of the inside of the reflector.

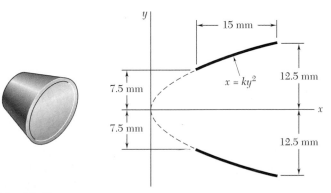

Fig. P5.60

*5.8. DISTRIBUTED LOADS ON BEAMS

The concept of the centroid of an area can be used to solve other problems besides those dealing with the weights of flat plates. Consider, for example, a beam supporting a *distributed load;* this load may consist of the weight of materials supported directly or indirectly by the beam, or it may be caused by wind or hydrostatic pressure. The distributed load can be represented by plotting the load w supported per unit length (Fig. 5.17); this load is expressed in N/m or in lb/ft. The magnitude of the force exerted on an element of beam of length dx is $dW = w\,dx$, and the total load supported by the beam is

$$W = \int_0^L w\,dx$$

We observe that the product $w\,dx$ is equal in magnitude to the element of area dA shown in Fig. 5.17a. The load W is thus equal in magnitude to the total area A under the load curve:

$$W = \int dA = A$$

We now determine where a *single concentrated load* **W**, of the same magnitude W as the total distributed load, should be applied on the beam if it is to produce the same reactions at the supports (Fig. 5.17b). However, this concentrated load **W**, which represents the resultant of the given distributed loading, is equivalent to the loading only when considering the free-body diagram of the entire beam. The point of application P of the equivalent concentrated load **W** is obtained by expressing that the moment of **W** about point O is equal to the sum of the moments of the elemental loads d**W** about O:

$$(OP)W = \int x\,dW$$

or, since $dW = w\,dx = dA$ and $W = A$,

$$(OP)A = \int_0^L x\,dA \qquad (5.12)$$

Since the integral represents the first moment with respect to the w axis of the area under the load curve, it can be replaced by the product $\bar{x}A$. We therefore have $OP = \bar{x}$, where $\bar{x}$ is the distance from the w axis to the centroid C of the area A (this is *not* the centroid of the beam).

A distributed load on a beam can thus be replaced by a concentrated load; the magnitude of this single load is equal to the area under the load curve, and its line of action passes through the centroid of that area. It should be noted, however, that the concentrated load is equivalent to the given loading only as far as external forces are concerned. It can be used to determine reactions but should not be used to compute internal forces and deflections.

(a)

(b)

Fig. 5.17

Photo 5.3 The roofs of the buildings shown must be able to support not only the total weight of the snow but also the nonsymmetric distributed loads resulting from drifting of the snow.

*5.9. FORCES ON SUBMERGED SURFACES

The approach used in the preceding section can be used to determine the resultant of the hydrostatic pressure forces exerted on a *rectangular surface* submerged in a liquid. Consider the rectangular plate shown in Fig. 5.18, which is of length L and width b, where b is measured perpendicular to the plane of the figure. As noted in Sec. 5.8, the load exerted on an element of the plate of length dx is $w\,dx$, where w is the load per unit length. However, this load can also be expressed as $p\,dA = pb\,dx$, where p is the gage pressure in the liquid† and b is the width of the plate; thus, $w = bp$. Since the gage pressure in a liquid is $p = \gamma h$, where γ is the specific weight of the liquid and h is the vertical distance from the free surface, it follows that

$$w = bp = b\gamma h \qquad (5.13)$$

which shows that the load per unit length w is proportional to h and, thus, varies linearly with x.

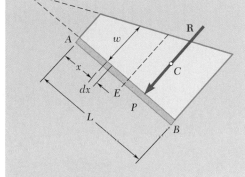

Fig. 5.18

Recalling the results of Sec. 5.8, we observe that the resultant **R** of the hydrostatic forces exerted on one side of the plate is equal in magnitude to the trapezoidal area under the load curve and that its line of action passes through the centroid C of that area. The point P of the plate where **R** is applied is known as the *center of pressure*.‡

Next, we consider the forces exerted by a liquid on a curved surface of constant width (Fig. 5.19*a*). Since the determination of the resultant **R** of these forces by direct integration would not be easy, we consider the free body obtained by detaching the volume of liquid ABD bounded by the curved surface AB and by the two plane surfaces AD and DB shown in Fig. 5.19*b*. The forces acting on the free body ABD are the weight **W** of the detached volume of liquid, the resultant $\mathbf{R}_1$ of the forces exerted on AD, the resultant $\mathbf{R}_2$ of the forces exerted on BD, and the resultant $-\mathbf{R}$ of the forces exerted *by the curved surface on the liquid*. The resultant $-\mathbf{R}$ is equal and opposite to, and has the same line of action as, the resultant **R** of the forces exerted *by the liquid on the curved surface*. The forces **W**, $\mathbf{R}_1$, and $\mathbf{R}_2$ can be determined by standard methods; after their values have been found, the force $-\mathbf{R}$ is obtained by solving the equations of equilibrium for the free body of Fig. 5.19*b*. The resultant **R** of the hydrostatic forces exerted on the curved surface is then obtained by reversing the sense of $-\mathbf{R}$.

The methods outlined in this section can be used to determine the resultant of the hydrostatic forces exerted on the surfaces of dams and rectangular gates and vanes. The resultants of forces on submerged surfaces of variable width will be determined in Chap. 9.

(a)

(b)

Fig. 5.19

Photo 5.4 As discussed in this section, the Grand Coulee Dam supports three different kinds of distributed forces: the weights of its constitutive elements, the pressure forces exerted by the water on its submerged face, and the pressure forces exerted by the ground on its base.

---

†The pressure p, which represents a load per unit area, is expressed in N/m² or in lb/ft². The derived SI unit N/m² is called a *pascal* (Pa).

‡Noting that the area under the load curve is equal to $w_E L$, where w_E is the load per unit length at the center E of the plate, and recalling Eq. (5.13), we can write

$$R = w_E L = (bp_E)L = p_E(bL) = p_E A$$

where A denotes the area of the *plate*. Thus, the magnitude of **R** can be obtained by multiplying the area of the plate by the pressure at its center E. The resultant **R**, however, *should be applied at P, not at E.*

$w_B = 4500$ N/m

$w_A = 1500$ N/m

A B

$L = 6$ m

SAMPLE PROBLEM 5.9

A beam supports a distributed load as shown. (*a*) Determine the equivalent concentrated load. (*b*) Determine the reactions at the supports.

SOLUTION

$\bar{x} = 4$ m

II   4.5 kN/m

1.5 kN/m   I

x

$\bar{x} = 2$ m

6 m

a. Equivalent Concentrated Load.
The magnitude of the resultant of the load is equal to the area under the load curve, and the line of action of the resultant passes through the centroid of the same area. We divide the area under the load curve into two triangles and construct the table below. To simplify the computations and tabulation, the given loads per unit length have been converted into kN/m.

Component	A, kN	$\bar{x}$, m	$\bar{x}A$, kN · m
Triangle I	4.5	2	9
Triangle II	13.5	4	54
	$\Sigma A = 18.0$		$\Sigma \bar{x}A = 63$

18 kN

$\bar{X} = 3.5$ m

A B

Thus, $\bar{X}\Sigma A = \Sigma\bar{x}A$: $\bar{X}(18$ kN$) = 63$ kN · m $\bar{X} = 3.5$ m

The equivalent concentrated load is

$$\mathbf{W} = 18 \text{ kN} \downarrow \quad \blacktriangleleft$$

and its line of action is located at a distance

$$\bar{X} = 3.5 \text{ m to the right of } A \quad \blacktriangleleft$$

b. Reactions.
The reaction at A is vertical and is denoted by $\mathbf{A}$; the reaction at B is represented by its components $\mathbf{B}_x$ and $\mathbf{B}_y$. The given load can be considered to be the sum of two triangular loads as shown. The resultant of each triangular load is equal to the area of the triangle and acts at its centroid. We write the following equilibrium equations for the free body shown:

4.5 kN   13.5 kN

B_x

A B_y

2 m

4 m

6 m

$\xrightarrow{+} \Sigma F_x = 0$: $\mathbf{B}_x = 0$ $\blacktriangleleft$

$+\uparrow\Sigma M_A = 0$: $-(4.5$ kN$)(2$ m$) - (13.5$ kN$)(4$ m$) + B_y(6$ m$) = 0$

$$\mathbf{B}_y = 10.5 \text{ kN} \uparrow \quad \blacktriangleleft$$

$+\uparrow\Sigma M_B = 0$: $+(4.5$ kN$)(4$ m$) + (13.5$ kN$)(2$ m$) - A(6$ m$) = 0$

$$\mathbf{A} = 7.5 \text{ kN} \uparrow \quad \blacktriangleleft$$

Alternative Solution. The given distributed load can be replaced by its resultant, which was found in part *a*. The reactions can be determined by writing the equilibrium equations $\Sigma F_x = 0$, $\Sigma M_A = 0$, and $\Sigma M_B = 0$. We again obtain

$$\mathbf{B}_x = 0 \quad \mathbf{B}_y = 10.5 \text{ kN} \uparrow \quad \mathbf{A} = 7.5 \text{ kN} \uparrow \quad \blacktriangleleft$$

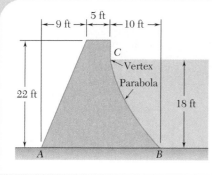

SAMPLE PROBLEM 5.10

The cross section of a concrete dam is as shown. Consider a 1-ft-thick section of the dam, and determine (a) the resultant of the reaction forces exerted by the ground on the base AB of the dam, (b) the resultant of the pressure forces exerted by the water on the face BC of the dam. The specific weights of concrete and water are 150 lb/ft³ and 62.4 lb/ft³, respectively.

SOLUTION

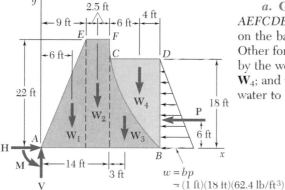

a. Ground Reaction. We choose as a free body the 1-ft-thick section AEFCDB of the dam and water. The reaction forces exerted by the ground on the base AB are represented by an equivalent force-couple system at A. Other forces acting on the free body are the weight of the dam, represented by the weights of its components W_1, W_2, and W_3; the weight of the water W_4; and the resultant P of the pressure forces exerted on section BD by the water to the right of section BD. We have

$$W_1 = \tfrac{1}{2}(9 \text{ ft})(22 \text{ ft})(1 \text{ ft})(150 \text{ lb/ft}^3) = 14,850 \text{ lb}$$
$$W_2 = (5 \text{ ft})(22 \text{ ft})(1 \text{ ft})(150 \text{ lb/ft}^3) = 16,500 \text{ lb}$$
$$W_3 = \tfrac{1}{3}(10 \text{ ft})(18 \text{ ft})(1 \text{ ft})(150 \text{ lb/ft}^3) = 9000 \text{ lb}$$
$$W_4 = \tfrac{2}{3}(10 \text{ ft})(18 \text{ ft})(1 \text{ ft})(62.4 \text{ lb/ft}^3) = 7488 \text{ lb}$$
$$P = \tfrac{1}{2}(18 \text{ ft})(1 \text{ ft})(18 \text{ ft})(62.4 \text{ lb/ft}^3) = 10,109 \text{ lb}$$

Equilibrium Equations

$$\xrightarrow{+}\Sigma F_x = 0: \qquad H - 10,109 \text{ lb} = 0 \qquad\qquad \mathbf{H} = 10,110 \text{ lb} \rightarrow \quad \blacktriangleleft$$

$$+\uparrow\Sigma F_y = 0: \qquad V - 14,850 \text{ lb} - 16,500 \text{ lb} - 9000 \text{ lb} - 7488 \text{ lb} = 0$$
$$\mathbf{V} = 47,840 \text{ lb} \uparrow \quad \blacktriangleleft$$

$$+\!\!\curvearrowleft\Sigma M_A = 0: \qquad -(14,850 \text{ lb})(6 \text{ ft}) - (16,500 \text{ lb})(11.5 \text{ ft}) - (9000 \text{ lb})(17 \text{ ft})$$
$$- (7488 \text{ lb})(20 \text{ ft}) + (10,109 \text{ lb})(6 \text{ ft}) + M = 0$$
$$\mathbf{M} = 520,960 \text{ lb} \cdot \text{ft} \curvearrowleft \quad \blacktriangleleft$$

We can replace the force-couple system obtained by a single force acting at a distance d to the right of A, where

$$d = \frac{520,960 \text{ lb} \cdot \text{ft}}{47,840 \text{ lb}} = 10.89 \text{ ft}$$

b. Resultant R of Water Forces. The parabolic section of water BCD is chosen as a free body. The forces involved are the resultant $-R$ of the forces exerted by the dam on the water, the weight W_4, and the force P. Since these forces must be concurrent, $-R$ passes through the point of intersection G of W_4 and P. A force triangle is drawn from which the magnitude and direction of $-R$ are determined. The resultant R of the forces exerted by the water on the face BC is equal and opposite:

$$\mathbf{R} = 12,580 \text{ lb} \, \measuredangle 36.5° \quad \blacktriangleleft$$

The problems in this lesson involve two common and very important types of loading: distributed loads on beams and forces on submerged surfaces of constant width. As we discussed in Secs. 5.8 and 5.9 and illustrated in Sample Probs. 5.9 and 5.10, determining the single equivalent force for each of these loadings requires a knowledge of centroids.

1. Analyzing beams subjected to distributed loads. In Sec. 5.8, we showed that a distributed load on a beam can be replaced by a single equivalent force. The magnitude of this force is equal to the area under the distributed load curve and its line of action passes through the centroid of that area. Thus, you should begin your solution by replacing the various distributed loads on a given beam by their respective single equivalent forces. The reactions at the supports of the beam can then be determined by using the methods of Chap. 4.

When possible, complex distributed loads should be divided into the common-shape areas shown in Fig. 5.8A [Sample Prob. 5.9]. Each of these areas can then be replaced by a single equivalent force. If required, the system of equivalent forces can be reduced further to a single equivalent force. As you study Sample Prob. 5.9, note how we have used the analogy between force and area and the techniques for locating the centroid of a composite area to analyze a beam subjected to a distributed load.

2. Solving problems involving forces on submerged bodies. The following points and techniques should be remembered when solving problems of this type.

    a. The pressure p at a depth h below the free surface of a liquid is equal to γh or $\rho g h$, where γ and ρ are the specific weight and the density of the liquid, respectively. The load per unit length w acting on a submerged surface of constant width b is then

$$w = bp = b\gamma h = b\rho gh$$

    b. The line of action of the resultant force **R** acting on a submerged plane surface is perpendicular to the surface.

    c. For a vertical or inclined plane rectangular surface of width b, the loading on the surface can be represented by a linearly distributed load which is trapezoidal in shape (Fig. 5.18). Further, the magnitude of **R** is given by

$$R = \gamma h_E A$$

where h_E is the vertical distance to the center of the surface and A is the area of the surface.

d. The load curve will be triangular (rather than trapezoidal) when the top edge of a plane rectangular surface coincides with the free surface of the liquid, since the pressure of the liquid at the free surface is zero. For this case, the line of action of **R** is easily determined, for it passes through the centroid of a *triangular* distributed load.

e. For the general case, rather than analyzing a trapezoid, we suggest that you use the method indicated in part *b* of Sample Prob. 5.9. First divide the trapezoidal distributed load into two triangles, and then compute the magnitude of the resultant of each triangular load. (The magnitude is equal to the area of the triangle times the width of the plate.) Note that the line of action of each resultant force passes through the centroid of the corresponding triangle and that the sum of these forces is equivalent to **R**. Thus, rather than using **R**, you can use the two equivalent resultant forces, whose points of application are easily calculated. Of course, the equation given for *R* in paragraph *c* should be used when only the magnitude of **R** is needed.

f. When the submerged surface of constant width is curved, the resultant force acting on the surface is obtained by considering the equilibrium of the volume of liquid bounded by the curved surface and by horizontal and vertical planes (Fig. 5.19). Observe that the force **R₁** of Fig. 5.19 is equal to the weight of the liquid lying above the plane *AD*. The method of solution for problems involving curved surfaces is shown in part *b* of Sample Prob. 5.10.

In subsequent mechanics courses (in particular, mechanics of materials and fluid mechanics), you will have ample opportunity to use the ideas introduced in this lesson.

5.61 and 5.62 For the beam and loading shown, determine (*a*) the magnitude and location of the resultant of the distributed load, (*b*) the reactions at the beam supports.

40 lb/ft

160 lb/ft

A

B

18 ft

Fig. P5.61

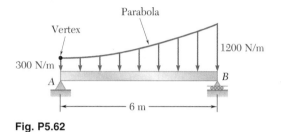

Parabola

Vertex

300 N/m

1200 N/m

A

B

6 m

Fig. P5.62

5.63 through 5.68 Determine the reactions at the beam supports for the given loading.

200 lb/ft

100 lb/ft

A

B

4 ft — 6 ft — 4 ft

Fig. P5.63

200 lb/ft

A

B

C

9 ft

3 ft

Fig. *P5.64*

200 N/m

A

B

18 N

120 mm — 200 mm

60 mm

Fig. P5.65

2 kN/m

A

B

2.5 kN/m

1.2 m — 1.8 m

Fig. P5.66

2 kN/m

Parabolas

1 kN/m

A

B

4 m — 2 m

Fig. P5.67

Vertex

Parabola

500 lb/ft

A

B

100 lb/ft

1 ft — 8 ft — 2 ft

Fig. *P5.68*

5.69 Determine (*a*) the distance *a* so that the vertical reactions at supports *A* and *B* are equal, (*b*) the corresponding reactions at the supports.

Fig. P5.69 and *P5.70*

5.70 Determine (*a*) the distance *a* so that the vertical reaction at support *B* is minimum, (*h*) the corresponding reactions at the supports.

5.71 Determine the reactions at the beam supports for the given loading when $w_0 = 1.5$ kN/m.

Fig. P5.71 and P5.72

5.72 Determine (*a*) the distributed load w_0 at the end *D* of the beam *ABCD* for which the reaction at *B* is zero, (*b*) the corresponding reactions at *C*.

5.73 A grade beam *AB* supports three concentrated loads and rests on soil and the top of a large rock. The soil exerts an upward distributed load, and the rock exerts a concentrated load $\mathbf{R}_R$ as shown. Knowing that $P = 4$ kN and $w_B = \frac{1}{2} w_A$, determine the values of w_A and R_R corresponding to equilibrium.

Fig. P5.73 and *P5.74*

5.74 A grade beam *AB* supports three concentrated loads and rests on soil and the top of a large rock. The soil exerts an upward distributed load, and the rock exerts a concentrated load $\mathbf{R}_R$ as shown. Knowing that $w_B = 0.4w_A$, determine (*a*) the largest value of $\mathbf{P}$ for which the beam is in equilibrium, (*b*) the corresponding value of w_A.

In the following problems, use $\gamma = 62.4$ lb/ft^3 for the specific weight of fresh water and $\gamma = 150$ lb/ft^3 for the specific weight of concrete if U.S. customary units are used. With SI units, use $\rho = 10^3$ kg/m^3 for the density of fresh water and $\rho_c = 2.40 \times 10^3$ kg/m^3 for the density of concrete. (See the footnote on page 222 for how to determine the specific weight of a material given its density.)

5.75 and 5.76 The cross section of a concrete dam is as shown. For a dam section of unit width, determine (a) the reaction forces exerted by the ground on the base AB of the dam, (b) the point of application of the resultant of the reaction forces of part a, (c) the resultant of the pressure forces exerted by the water on the face BC of the dam.

Fig. P5.75

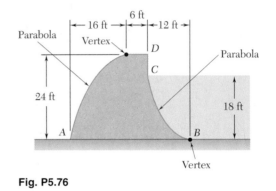

Fig. P5.76

5.77 The 9×12-ft side AB of a tank is hinged at its bottom A and is held in place by a thin rod BC. The maximum tensile force the rod can withstand without breaking is 40 kips, and the design specifications require the force in the rod not to exceed 20 percent of this value. If the tank is slowly filled with water, determine the maximum allowable depth of water d in the tank.

Fig. P5.77 and P5.78

5.78 The 9×12-ft side of an open tank is hinged at its bottom A and is held in place by a thin rod. The tank is filled with glycerine, whose specific weight is 80 lb/ft^3. Determine the force $\mathbf{T}$ in the rod and the reactions at the hinge after the tank is filled to a depth of 8 ft.

5.79 The friction force between a 2×2-m square sluice gate AB and its guides is equal to 10 percent of the resultant of the pressure forces exerted by the water on the face of the gate. Determine the initial force needed to lift the gate knowing that its mass is 500 kg.

Fig. P5.79

5.80 The dam for a lake is designed to withstand the additional force caused by silt which has settled on the lake bottom. Assuming that silt is equivalent to a liquid of density $\rho_s = 1.76 \times 10^3$ kg/m^3 and considering a 1-m-wide section of dam, determine the percentage increase in the force acting on the dam face for a silt accumulation of depth 1.5 m.

Fig. P5.80 and *P5.81*

5.81 The base of a dam for a lake is designed to resist up to 150 percent of the horizontal force of the water. After construction, it is found that silt (which is equivalent to a liquid of density $\rho_s = 1.76 \times 10^3$ kg/m^3) is settling on the lake bottom at a rate of 20 mm/yr. Considering a 1-m-wide section of dam, determine the number of years until the dam becomes unsafe.

5.82 The square gate AB is held in the position shown by hinges along its top edge A and by a shear pin at B. For a depth of water $d = 3.5$ m, determine the force exerted on the gate by the shear pin.

Fig. P5.82

5.83 A temporary dam is constructed in a 5-ft-wide fresh water channel by nailing two boards to pilings located at the sides of the channel and propping a third board AB against the pilings and the floor of the channel. Neglecting friction, determine the reactions at A and B when rope BC is slack.

Fig. P5.83 and *P5.84*

5.84 A temporary dam is constructed in a 5-ft-wide fresh water channel by nailing two boards to pilings located at the sides of the channel and propping a third board AB against the pilings and the floor of the channel. Neglecting friction, determine the magnitude and direction of the minimum tension required in rope BC to move board AB.

Fig. P5.85

5.85 A 2 × 3-m gate is hinged at A and is held in position by rod CD. End D rests against a spring whose constant is 12 kN/m. The spring is undeformed when the gate is vertical. Assuming that the force exerted by rod CD on the gate remains horizontal, determine the minimum depth of water d for which the bottom B of the gate will move to the end of the cylindrical portion of the floor.

5.86 Solve Prob. 5.85 if the mass of the gate is 500 kg.

5.87 The gate at the end of a 3-ft-wide fresh water channel is fabricated from three 240-lb, rectangular steel plates. The gate is hinged at A and rests against a frictionless support at D. Knowing that $d = 2.5$ ft, determine the reactions at A and D.

Fig. P5.87 and *P5.88*

5.88 The gate at the end of a 3-ft-wide fresh water channel is fabricated from three 240-lb, rectangular steel plates. The gate is hinged at A and rests against a frictionless support at D. Determine the depth of water d for which the gate will open.

5.89 A rain gutter is supported from the roof of a house by hangers that are spaced 0.6 m apart. After leaves clog the gutter's drain, the gutter slowly fills with rainwater. When the gutter is completely filled with water, determine (a) the resultant of the pressure force exerted by the water on the 0.6-m section of the curved surface of the gutter, (b) the force-couple system exerted on a hanger where it is attached to the gutter.

Fig. P5.89

VOLUMES

5.10. Center of Gravity of a Three-
Dimensional Body. Centroid of a Volume **259**

5.10. CENTER OF GRAVITY OF A THREE-DIMENSIONAL BODY. CENTROID OF A VOLUME

The *center of gravity* G of a three-dimensional body is obtained by dividing the body into small elements and by then expressing that the weight **W** of the body acting at G is equivalent to the system of distributed forces Δ**W** representing the weights of the small elements. Choosing the y axis to be vertical with positive sense upward (Fig. 5.20) and denoting by $\bar{\mathbf{r}}$ the position vector of G, we write that

Photo 5.5 To predict the flight characteristics of the modified Boeing 747 when used to transport a *Space Shuttle Orbiter*, the center of gravity of each craft had to be determined.

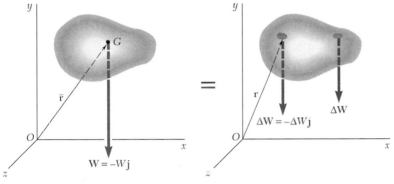

Fig. 5.20

W is equal to the sum of the elemental weights Δ**W** and that its moment about O is equal to the sum of the moments about O of the elemental weights:

$$\Sigma\mathbf{F}: \qquad -W\mathbf{j} = \Sigma(-\Delta W\mathbf{j})$$
$$\Sigma\mathbf{M}_O: \qquad \bar{\mathbf{r}} \times (-W\mathbf{j}) = \Sigma[\mathbf{r} \times (-\Delta W\mathbf{j})] \qquad (5.13)$$

Rewriting the last equation in the form

$$\bar{\mathbf{r}}W \times (-\mathbf{j}) = (\Sigma\mathbf{r}\,\Delta W) \times (-\mathbf{j}) \qquad (5.14)$$

we observe that the weight **W** of the body is equivalent to the system of the elemental weights Δ**W** if the following conditions are satisfied:

$$W = \Sigma\,\Delta W \qquad \bar{\mathbf{r}}W = \Sigma\mathbf{r}\,\Delta W$$

Increasing the number of elements and simultaneously decreasing the size of each element, we obtain in the limit

$$W = \int dW \qquad \bar{\mathbf{r}}W = \int \mathbf{r}\,dW \qquad (5.15)$$

We note that the relations obtained are independent of the orientation of the body. For example, if the body and the coordinate axes were rotated so that the z axis pointed upward, the unit vector $-\mathbf{j}$ would be replaced by $-\mathbf{k}$ in Eqs. (5.13) and (5.14), but the relations (5.15) would remain unchanged. Resolving the vectors $\bar{\mathbf{r}}$ and $\mathbf{r}$ into rectangular components, we note that the second of the relations (5.15) is equivalent to the three scalar equations

$$\bar{x}W = \int x\,dW \qquad \bar{y}W = \int y\,dW \qquad \bar{z}W = \int z\,dW \qquad (5.16)$$

If the body is made of a homogeneous material of specific weight γ, the magnitude dW of the weight of an infinitesimal element can be expressed in terms of the volume dV of the element, and the magnitude W of the total weight can be expressed in terms of the total volume V. We write

$$dW = \gamma \, dV \quad W = \gamma V$$

Substituting for dW and W in the second of the relations (5.15), we write

$$\overline{\mathbf{r}}V = \int \mathbf{r} \, dV \tag{5.17}$$

or, in scalar form,

$$\overline{x}V = \int x \, dV \qquad \overline{y}V = \int y \, dV \qquad \overline{z}V = \int z \, dV \tag{5.18}$$

The point whose coordinates are $\overline{x}, \overline{y}, \overline{z}$ is also known as the *centroid C of the volume V* of the body. If the body is not homogeneous, Eqs. (5.18) cannot be used to determine the center of gravity of the body; however, Eqs. (5.18) still define the centroid of the volume.

The integral $\int x \, dV$ is known as the *first moment of the volume with respect to the yz plane.* Similarly, the integrals $\int y \, dV$ and $\int z \, dV$ define the first moments of the volume with respect to the zx plane and the xy plane, respectively. It is seen from Eqs. (5.18) that if the centroid of a volume is located in a coordinate plane, the first moment of the volume with respect to that plane is zero.

A volume is said to be symmetrical with respect to a given plane if for every point P of the volume there exists a point P' of the same volume, such that the line PP' is perpendicular to the given plane and is bisected by that plane. The plane is said to be a *plane of symmetry* for the given volume. When a volume V possesses a plane of symmetry, the first moment of V with respect to that plane is zero, and the centroid of the volume is located in the plane of symmetry. When a volume possesses two planes of symmetry, the centroid of the volume is located on the line of intersection of the two planes. Finally, when a volume possesses three planes of symmetry which intersect at a well-defined point (that is, not along a common line), the point of intersection of the three planes coincides with the centroid of the volume. This property enables us to determine immediately the locations of the centroids of spheres, ellipsoids, cubes, rectangular parallelepipeds, etc.

The centroids of unsymmetrical volumes or of volumes possessing only one or two planes of symmetry should be determined by integration (Sec. 5.12). The centroids of several common volumes are shown in Fig. 5.21. It should be observed that in general the centroid of a volume of revolution *does not coincide* with the centroid of its cross section. Thus, the centroid of a hemisphere is different from that of a semicircular area, and the centroid of a cone is different from that of a triangle.

Shape		$\bar{x}$	Volume
Hemisphere		$\dfrac{3a}{8}$	$\dfrac{2}{3}\pi a^3$
Semiellipsoid of revolution		$\dfrac{3h}{8}$	$\dfrac{2}{3}\pi a^2 h$
Paraboloid of revolution		$\dfrac{h}{3}$	$\dfrac{1}{2}\pi a^2 h$
Cone		$\dfrac{h}{4}$	$\dfrac{1}{3}\pi a^2 h$
Pyramid		$\dfrac{h}{4}$	$\dfrac{1}{3}abh$

Fig. 5.21 Centroids of common shapes and volumes.

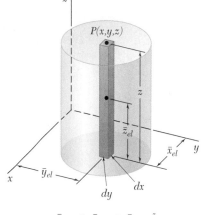

$$\bar{x}_{el} = x, \ \bar{y}_{el} = y, \ \bar{z}_{el} = \frac{z}{2}$$
$$dV = z \, dx \, dy$$

Fig. 5.22 Determination of the centroid of a volume by double integration.

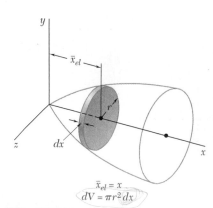

$$\bar{x}_{el} = x$$
$$dV = \pi r^2 \, dx$$

Fig. 5.23 Determination of the centroid of a body of revolution.

5.11. COMPOSITE BODIES

If a body can be divided into several of the common shapes shown in Fig. 5.21, its center of gravity G can be determined by expressing that the moment about O of its total weight is equal to the sum of the moments about O of the weights of the various component parts. Proceeding as in Sec. 5.10, we obtain the following equations defining the coordinates $\overline{X}, \overline{Y}, \overline{Z}$ of the center of gravity G:

$$\overline{X} \Sigma W = \Sigma \bar{x} W \qquad \overline{Y} \Sigma W = \Sigma \bar{y} W \qquad \overline{Z} \Sigma W = \Sigma \bar{z} W \quad (5.19)$$

If the body is made of a homogeneous material, its center of gravity coincides with the centroid of its volume, and we obtain

$$\overline{X} \Sigma V = \Sigma \bar{x} V \qquad \overline{Y} \Sigma V = \Sigma \bar{y} V \qquad \overline{Z} \Sigma V = \Sigma \bar{z} V \quad (5.20)$$

5.12. DETERMINATION OF CENTROIDS OF VOLUMES BY INTEGRATION

The centroid of a volume bounded by analytical surfaces can be determined by evaluating the integrals given in Sec. 5.10:

$$\bar{x} V = \int x \, dV \qquad \bar{y} V = \int y \, dV \qquad \bar{z} V = \int z \, dV \quad (5.21)$$

If the element of volume dV is chosen to be equal to a small cube of sides dx, dy, and dz, the evaluation of each of these integrals requires a *triple integration*. However, it is possible to determine the coordinates of the centroid of most volumes by *double integration* if dV is chosen to be equal to the volume of a thin filament (Fig. 5.22). The coordinates of the centroid of the volume are then obtained by rewriting Eqs. (5.21) as

$$\bar{x} V = \int \bar{x}_{el} \, dV \qquad \bar{y} V = \int \bar{y}_{el} \, dV \qquad \bar{z} V = \int \bar{z}_{el} \, dV \quad (5.22)$$

and by then substituting the expressions given in Fig. 5.22 for the volume dV and the coordinates $\bar{x}_{el}$, $\bar{y}_{el}$, $\bar{z}_{el}$. By using the equation of the surface to express z in terms of x and y, the integration is reduced to a double integration in x and y.

If the volume under consideration possesses *two planes of symmetry*, its centroid must be located on the line of intersection of the two planes. Choosing the x axis to lie along this line, we have

$$\bar{y} = \bar{z} = 0$$

and the only coordinate to determine is $\bar{x}$. This can be done with a *single integration* by dividing the given volume into thin slabs parallel to the yz plane and expressing dV in terms of x and dx in the equation

$$\bar{x} V = \int \bar{x}_{el} \, dV \quad (5.23)$$

For a body of revolution, the slabs are circular and their volume is given in Fig. 5.23.

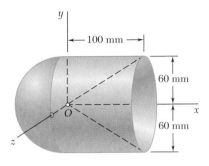

SAMPLE PROBLEM 5.11

Determine the location of the center of gravity of the homogeneous body of revolution shown, which was obtained by joining a hemisphere and a cylinder and carving out a cone.

SOLUTION

Because of symmetry, the center of gravity lies on the x axis. As shown in the figure below, the body can be obtained by adding a hemisphere to a cylinder and then subtracting a cone. The volume and the abscissa of the centroid of each of these components are obtained from Fig. 5.21 and are entered in the table below. The total volume of the body and the first moment of its volume with respect to the yz plane are then determined.

$\frac{3}{8}(60 \text{ mm}) = 22.5 \text{ mm}$ 50 mm $\frac{3}{4}(100 \text{ mm}) = 75 \text{ mm}$

Component	Volume, mm³		$\bar{x}$, mm	$\bar{x}V$, mm⁴
Hemisphere	$\frac{1}{2}\frac{4\pi}{3}(60)^3 =$	0.4524×10^6	-22.5	-10.18×10^6
Cylinder	$\pi(60)^2(100) =$	1.1310×10^6	$+50$	$+56.55 \times 10^6$
Cone	$-\frac{\pi}{3}(60)^2(100) =$	-0.3770×10^6	$+75$	-28.28×10^6
	$\Sigma V =$	1.206×10^6		$\Sigma\bar{x}V = +18.09 \times 10^6$

Thus,

$$\bar{X}\Sigma V = \Sigma\bar{x}V: \quad \bar{X}(1.206 \times 10^6 \text{ mm}^3) = 18.09 \times 10^6 \text{ mm}^4$$

$$\bar{X} = 15 \text{ mm} \quad \blacktriangleleft$$

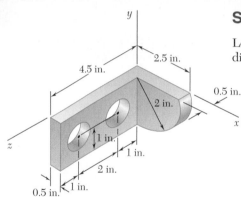

SAMPLE PROBLEM 5.12

Locate the center of gravity of the steel machine element shown. The diameter of each hole is 1 in.

SOLUTION

The machine element can be obtained by adding a rectangular parallelepiped (I) to a quarter cylinder (II) and then subtracting two 1-in.-diameter cylinders (III and IV). The volume and the coordinates of the centroid of each component are determined and are entered in the table below. Using the data in the table, we then determine the total volume and the moments of the volume with respect to each of the coordinate planes.

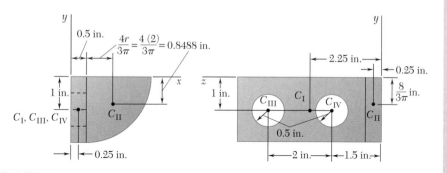

	V, in³	$\bar{x}$, in.	$\bar{y}$, in.	$\bar{z}$, in.	$\bar{x}V$, in⁴	$\bar{y}V$, in⁴	$\bar{z}V$, in⁴
I	$(4.5)(2)(0.5) = 4.5$	0.25	-1	2.25	1.125	-4.5	10.125
II	$\frac{1}{4}\pi(2)^2(0.5) = 1.571$	1.3488	-0.8488	0.25	2.119	-1.333	0.393
III	$-\pi(0.5)^2(0.5) = -0.3927$	0.25	-1	3.5	-0.098	0.393	-1.374
IV	$-\pi(0.5)^2(0.5) = -0.3927$	0.25	-1	1.5	-0.098	0.393	-0.589
	$\Sigma V = 5.286$				$\Sigma \bar{x}V = 3.048$	$\Sigma \bar{y}V = -5.047$	$\Sigma \bar{z}V = 8.555$

Thus,

$$\overline{X}\Sigma V = \Sigma \bar{x}V: \quad \overline{X}(5.286 \text{ in}^3) = \quad 3.048 \text{ in}^4 \qquad \overline{X} = \quad 0.577 \text{ in.} \blacktriangleleft$$
$$\overline{Y}\Sigma V = \Sigma \bar{y}V: \quad \overline{Y}(5.286 \text{ in}^3) = -5.047 \text{ in}^4 \qquad \overline{Y} = -0.955 \text{ in.} \blacktriangleleft$$
$$\overline{Z}\Sigma V = \Sigma \bar{z}V: \quad \overline{Z}(5.286 \text{ in}^3) = \quad 8.555 \text{ in}^4 \qquad \overline{Z} = \quad 1.618 \text{ in.} \blacktriangleleft$$

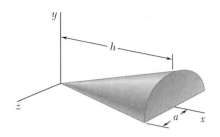

SAMPLE PROBLEM 5.13

Determine the location of the centroid of the half right circular cone shown.

SOLUTION

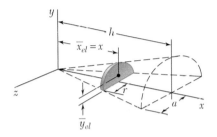

Since the xy plane is a plane of symmetry, the centroid lies in this plane and $\bar{z} = 0$. A slab of thickness dx is chosen as a differential element. The volume of this element is

$$dV = \tfrac{1}{2}\pi r^2\, dx$$

The coordinates $\bar{x}_{el}$ and $\bar{y}_{el}$ of the centroid of the element are

$$\bar{x}_{el} = x \qquad \bar{y}_{el} = \frac{4r}{3\pi}$$

where $\bar{y}_{el}$ is obtained from Fig. 5.8 (semicircular area).

We observe that r is proportional to x and write

$$\frac{r}{x} = \frac{a}{h} \qquad r = \frac{a}{h}x$$

The volume of the body is

$$V = \int dV = \int_0^h \tfrac{1}{2}\pi r^2\, dx = \int_0^h \tfrac{1}{2}\pi \left(\frac{a}{h}x\right)^2 dx = \frac{\pi a^2 h}{6}$$

The moment of the differential element with respect to the yz plane is $\bar{x}_{el}\, dV$; the total moment of the body with respect to this plane is

$$\int \bar{x}_{el}\, dV = \int_0^h x(\tfrac{1}{2}\pi r^2)\, dx = \int_0^h x(\tfrac{1}{2}\pi)\left(\frac{a}{h}x\right)^2 dx = \frac{\pi a^2 h^2}{8}$$

Thus,

$$\bar{x}V = \int \bar{x}_{el}\, dV \qquad \bar{x}\frac{\pi a^2 h}{6} = \frac{\pi a^2 h^2}{8} \qquad \bar{x} = \tfrac{3}{4}h \quad \blacktriangleleft$$

Likewise, the moment of the differential element with respect to the zx plane is $\bar{y}_{el}\, dV$; the total moment is

$$\int \bar{y}_{el}\, dV = \int_0^h \frac{4r}{3\pi}(\tfrac{1}{2}\pi r^2)\, dx = \frac{2}{3}\int_0^h \left(\frac{a}{h}x\right)^3 dx = \frac{a^3 h}{6}$$

Thus,

$$\bar{y}V = \int \bar{y}_{el}\, dV \qquad \bar{y}\frac{\pi a^2 h}{6} = \frac{a^3 h}{6} \qquad \bar{y} = \frac{a}{\pi} \quad \blacktriangleleft$$

SOLVING PROBLEMS
ON YOUR OWN

In the problems for this lesson, you will be asked to locate the centers of gravity of three-dimensional bodies or the centroids of their volumes. All of the techniques we previously discussed for two-dimensional bodies—using symmetry, dividing the body into common shapes, choosing the most efficient differential element, etc.—may also be applied to the general three-dimensional case.

1. Locating the centers of gravity of composite bodies. In general, Eqs. (5.19) must be used:

$$\overline{X}\Sigma W = \Sigma \overline{x}W \qquad \overline{Y}\Sigma W = \Sigma \overline{y}W \qquad \overline{Z}\Sigma W = \Sigma \overline{z}W \qquad (5.19)$$

However, for the case of a *homogeneous body*, the center of gravity of the body coincides with the *centroid of its volume*. Therefore, for this special case, the center of gravity of the body can also be located using Eqs. (5.20):

$$\overline{X}\Sigma V = \Sigma \overline{x}V \qquad \overline{Y}\Sigma V = \Sigma \overline{y}V \qquad \overline{Z}\Sigma V = \Sigma \overline{z}V \qquad (5.20)$$

You should realize that these equations are simply an extension of the equations used for the two-dimensional problems considered earlier in the chapter. As the solutions of Sample Probs. 5.11 and 5.12 illustrate, the methods of solution for two- and three-dimensional problems are identical. Thus, we once again strongly encourage you to construct appropriate diagrams and tables when analyzing composite bodies. Also, as you study Sample Prob. 5.12, observe how the x and y coordinates of the centroid of the quarter cylinder were obtained using the equations for the centroid of a quarter circle.

We note that *two special cases* of interest occur when the given body consists of either uniform wires or uniform plates made of the same material.

    a. For a body made of *several wire elements* of the *same uniform cross section*, the cross-sectional area A of the wire elements will factor out of Eqs. (5.20) when V is replaced with the product AL, where L is the length of a given element. Equations (5.20) thus reduce in this case to

$$\overline{X}\Sigma L = \Sigma \overline{x}L \qquad \overline{Y}\Sigma L = \Sigma \overline{y}L \qquad \overline{Z}\Sigma L = \Sigma \overline{z}L$$

    b. For a body made of *several plates* of the *same uniform thickness*, the thickness t of the plates will factor out of Eqs. (5.20) when V is replaced with the product tA, where A is the area of a given plate. Equations (5.20) thus reduce in this case to

$$\overline{X}\Sigma A = \Sigma \overline{x}A \qquad \overline{Y}\Sigma A = \Sigma \overline{y}A \qquad \overline{Z}\Sigma A = \Sigma \overline{z}A$$

2. Locating the centroids of volumes by direct integration. As explained in Sec. 5.11, evaluating the integrals of Eqs. (5.21) can be simplified by choosing either a thin filament (Fig. 5.22) or a thin slab (Fig. 5.23) for the element of volume dV. Thus, you should begin your solution by identifying, if possible, the dV which produces the single or double integrals that are the easiest to compute. For bodies of revolution, this may be a thin slab (as in Sample Prob. 5.13) or a thin cylindrical shell. However, it is important to remember that the relationship that you establish among the variables (like the relationship between r and x in Sample Prob. 5.13) will directly affect the complexity of the integrals you will have to compute. Finally, we again remind you that $\overline{x}_{el}$, $\overline{y}_{el}$, and $\overline{z}_{el}$ in Eqs. (5.22) are the coordinates of the centroid of dV.

Problems

5.90 The composite body shown is formed by removing a hemisphere of radius r from a cylinder of radius R and height $2R$. Determine (*a*) the y coordinate of the centroid when $r = 3R/4$, (*b*) the ratio r/R for which $\bar{y} = -1.2R$.

Fig. P5.90

5.91 Determine the y coordinate of the centroid of the body shown.

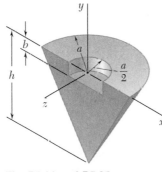

Fig. P5.91 and P5.92

5.92 Determine the z coordinate of the centroid of the body shown. (*Hint:* Use the result of Sample Prob. 5.13.)

5.93 Consider the composite body shown. Determine (*a*) the value of $\bar{x}$ when $h = L/2$, (*b*) the ratio h/L for which $\bar{x} = L$.

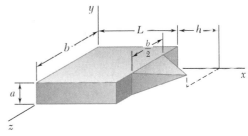

Fig. P5.93

5.94 For the machine element shown, determine the x coordinate of the center of gravity.

Fig. P5.94 and P5.95

5.95 For the machine element shown, determine the y coordinate of the center of gravity.

5.96 For the machine element shown, locate the x coordinate of the center of gravity.

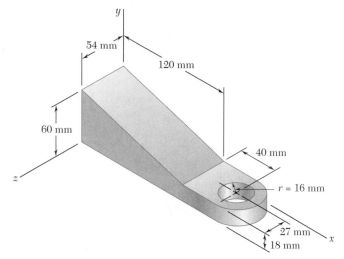

Fig. P5.96 and P5.97

5.97 For the machine element shown, locate the y coordinate of the center of gravity.

5.98 For the stop bracket shown, locate the x coordinate of the center of gravity.

5.99 For the stop bracket shown, locate the z coordinate of the center of gravity.

5.100 Locate the center of gravity of the sheet-metal form shown.

Fig. *P5.98* and *P5.99*

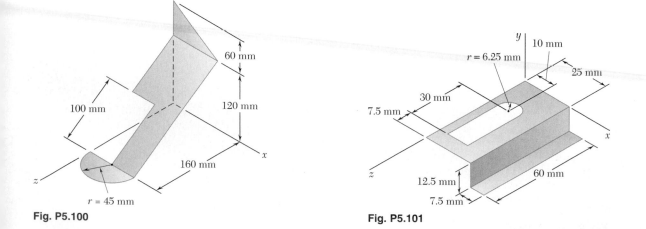

Fig. P5.100

Fig. P5.101

5.101 A mounting bracket for electronic components is formed from sheet metal of uniform thickness. Locate the center of gravity of the bracket.

5.102 Locate the center of gravity of the sheet-metal form shown.

Problems **269**

Fig. **P5.102**

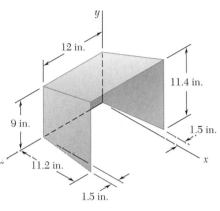

Fig. P5.103

*5.103 An enclosure for an electronic device is formed from sheet metal of uniform thickness. Locate the center of gravity of the enclosure.

Fig. P5.104

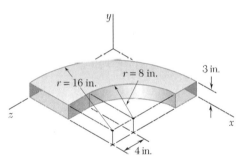

Fig. **P5.105**

5.104 A 200-mm-diameter cylindrical duct and a 100 × 200-mm rectangular duct are to be joined as indicated. Knowing that the ducts are fabricated from the same sheet metal, which is of uniform thickness, locate the center of gravity of the assembly.

5.105 An elbow for the duct of a ventilating system is made of sheet metal of uniform thickness. Locate the center of gravity of the elbow.

5.106 A window awning is fabricated from sheet metal of uniform thickness. Locate the center of gravity of the awning.

Fig. P5.106

5.107 The thin, plastic front cover of a wall clock is of uniform thickness. Locate the center of gravity of the cover.

Fig. P5.107

5.108 and 5.109 A thin steel wire of uniform cross section is bent into the shape shown, where arc BC is a quarter circle of radius R. Locate its center of gravity.

Fig. P5.108

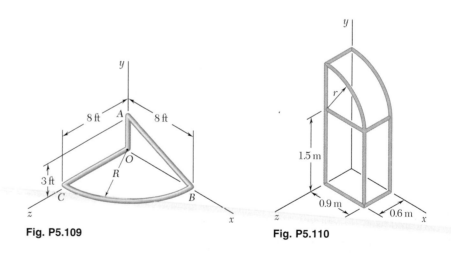

Fig. P5.109

Fig. P5.110

5.110 The frame of a greenhouse is constructed from uniform aluminum channels. Locate the center of gravity of the portion of the frame shown.

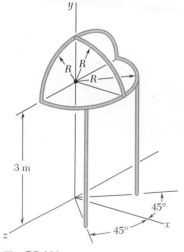

Fig. P5.111

5.111 The decorative metalwork at the entrance of a store is fabricated from uniform steel structural tubing. Knowing that $R = 1.2$ m, locate the center of gravity of the metalwork.

5.112 A scratch awl has a plastic handle and a steel blade and shank. Knowing that the specific weight of plastic is 0.0374 lb/in³ and of steel is 0.284 lb/in³, locate the center of gravity of the awl.

Fig. P5.112

Fig. P5.113

5.113 A bronze bushing is mounted inside a steel sleeve. Knowing that the density of bronze is 8800 kg/m³ and of steel is 7860 kg/m³, determine the center of gravity of the assembly.

5.114 A marker for a garden path consists of a truncated regular pyramid carved from stone of specific weight 160 lb/ft³. The pyramid is mounted on a steel base of thickness h. Knowing that the specific weight of steel is 490 lb/ft³ and that steel plate is available in ¼-in. increments, specify the minimum thickness h for which the center of gravity of the marker is approximately 12 in. above the top of the base.

*5.115 The ends of the park bench shown are made of concrete, while the seat and back are wooden boards. Each piece of wood is 36 × 120 × 1180 mm. Knowing that the density of concrete is 2320 kg/m³ and of wood is 470 kg/m³, determine the x and y coordinates of the center of gravity of the bench.

Fig. P5.114

Fig. P5.115

5.116 through 5.118 Determine by direct integration the values of $\bar{x}$ for the two volumes obtained by passing a vertical cutting plane through the given shape of Fig. 5.21. The cutting plane is parallel to the base of the given shape and divides the shape into two volumes of equal height.

  5.116 A hemisphere.
  5.117 A semiellipsoid of revolution.
  5.118 A paraboloid of revolution.

5.119 and 5.120 Locate the centroid of the volume obtained by rotating the shaded area about the x axis.

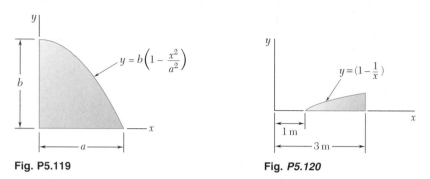

Fig. P5.119

Fig. P5.120

5.121 Locate the centroid of the volume obtained by rotating the shaded area about the line $x = h$.

*5.122 Locate the centroid of the volume generated by revolving the portion of the sine curve shown about the x axis.

Fig. P5.121

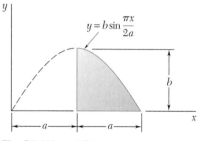

Fig. P5.122 and P5.123

*5.123 Locate the centroid of the volume generated by revolving the portion of the sine curve shown about the y axis. (*Hint:* Use a thin cylindrical shell of radius r and thickness dr as the element of volume.)

*5.124 Show that for a regular pyramid of height h and n sides ($n = 3$, 4, . . .) the centroid of the volume of the pyramid is located at a distance $h/4$ above the base.

5.125 Determine by direct integration the location of the centroid of one-half of a thin, uniform hemispherical shell of radius R.

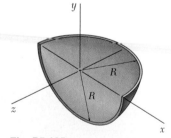

Fig. P5.125

5.126 The sides and the base of a punch bowl are of uniform thickness t. If $t \ll R$ and $R = 350$ mm, determine the location of the center of gravity of (*a*) the bowl, (*b*) the punch.

Fig. P5.126

5.127 After grading a lot, a builder places four stakes to designate the corners of the slab for a house. To provide a firm, level base for the slab, the builder places a minimum of 60 mm of gravel beneath the slab. Determine the volume of gravel needed and the x coordinate of the centroid of the volume of the gravel. (*Hint:* The bottom of the gravel is an oblique plane, which can be represented by the equation $y = a + bx + cz$.)

Fig. P5.127

5.128 Determine by direct integration the location of the centroid of the volume between the xz plane and the portion shown of the surface $y = 16h(ax - x^2)(bz - z^2)/a^2b^2$.

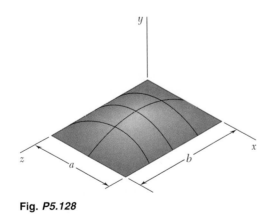

Fig. *P5.128*

5.129 Locate the centroid of the section shown, which was cut from a circular cylinder by an inclined plane.

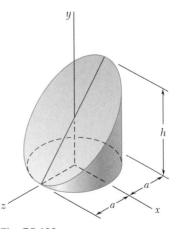

Fig. P5.129

This chapter was devoted chiefly to the determination of the *center of gravity* of a rigid body, that is, to the determination of the point G where a single force $\mathbf{W}$, called the *weight* of the body, can be applied to represent the effect of the earth's attraction on the body.

Center of gravity of a two-dimensional body

In the first part of the chapter, we considered *two-dimensional bodies*, such as flat plates and wires contained in the xy plane. By adding force components in the vertical z direction and moments about the horizontal y and x axes [Sec. 5.2], we derived the relations

$$W = \int dW \qquad \bar{x}W = \int x \, dW \qquad \bar{y}W = \int y \, dW \qquad (5.2)$$

which define the weight of the body and the coordinates $\bar{x}$ and $\bar{y}$ of its center of gravity.

Centroid of an area or line

In the case of a *homogeneous flat plate of uniform thickness* [Sec. 5.3], the center of gravity G of the plate coincides with the *centroid C of the area A* of the plate, the coordinates of which are defined by the relations

$$\bar{x}A = \int x \, dA \qquad \bar{y}A = \int y \, dA \qquad (5.3)$$

Similarly, the determination of the center of gravity of a *homogeneous wire of uniform cross section* contained in a plane reduces to the determination of the *centroid C of the line L* representing the wire; we have

$$\bar{x}L = \int x \, dL \qquad \bar{y}L = \int y \, dL \qquad (5.4)$$

First moments

The integrals in Eqs. (5.3) are referred to as the *first moments* of the area A with respect to the y and x axes and are denoted by Q_y and Q_x, respectively [Sec. 5.4]. We have

$$Q_y = \bar{x}A \qquad Q_x = \bar{y}A \qquad (5.6)$$

The first moments of a line can be defined in a similar way.

Properties of symmetry

The determination of the centroid C of an area or line is simplified when the area or line possesses certain *properties of symmetry*. If the area or line is symmetric with respect to an axis, its centroid C lies on that axis; if it is symmetric with respect to two axes, C is located at the intersection of the two axes; if it is symmetric with respect to a center O, C coincides with O.

The *areas and the centroids of various common shapes* are tabulated in Fig. 5.8. When a flat plate can be divided into several of these shapes, the coordinates $\overline{X}$ and $\overline{Y}$ of its center of gravity G can be determined from the coordinates $\overline{x}_1, \overline{x}_2, \ldots$ and $\overline{y}_1, \overline{y}_2, \ldots$ of the centers of gravity $G_1, G_2, \ldots$ of the various parts [Sec. 5.5]. Equating moments about the y and x axes, respectively (Fig. 5.24), we have

$$\overline{X}\Sigma W = \Sigma\overline{x}W \qquad \overline{Y}\Sigma W = \Sigma\overline{y}W \qquad (5.7)$$

Center of gravity of a composite body

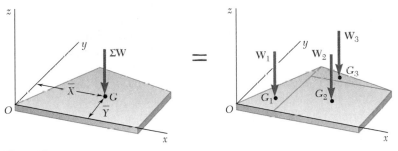

Fig. 5.24

If the plate is homogeneous and of uniform thickness, its center of gravity coincides with the centroid C of the area of the plate, and Eqs. (5.7) reduce to

$$Q_y = \overline{X}\Sigma A = \Sigma\overline{x}A \qquad Q_x = \overline{Y}\Sigma A = \Sigma\overline{y}A \qquad (5.8)$$

These equations yield the first moments of the composite area, or they can be solved for the coordinates $\overline{X}$ and $\overline{Y}$ of its centroid [Sample Prob. 5.1]. The determination of the center of gravity of a composite wire is carried out in a similar fashion [Sample Prob. 5.2].

When an area is bounded by analytical curves, the coordinates of its centroid can be determined by *integration* [Sec. 5.6]. This can be done by evaluating either the double integrals in Eqs. (5.3) or a *single integral* which uses one of the thin rectangular or pie-shaped elements of area shown in Fig. 5.12. Denoting by $\overline{x}_{el}$ and $\overline{y}_{el}$ the coordinates of the centroid of the element dA, we have

Determination of centroid by integration

$$Q_y = \overline{x}A = \int \overline{x}_{el}\, dA \qquad Q_x = \overline{y}A = \int \overline{y}_{el}\, dA \qquad (5.9)$$

It is advantageous to use the same element of area to compute both of the first moments Q_y and Q_x; the same element can also be used to determine the area A [Sample Prob. 5.4].

Theorems of Pappus-Guldinus

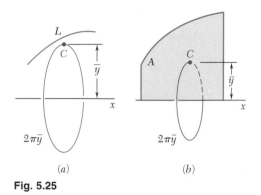

Fig. 5.25

The *theorems of Pappus-Guldinus* relate the determination of the area of a surface of revolution or the volume of a body of revolution to the determination of the centroid of the generating curve or area [Sec. 5.7]. The area A of the surface generated by rotating a curve of length L about a fixed axis (Fig. 5.25a) is

$$A = 2\pi\bar{y}L \qquad (5.10)$$

where $\bar{y}$ represents the distance from the centroid C of the curve to the fixed axis. Similarly, the volume V of the body generated by rotating an area A about a fixed axis (Fig. 5.25b) is

$$V = 2\pi\bar{y}A \qquad (5.11)$$

where $\bar{y}$ represents the distance from the centroid C of the area to the fixed axis.

Distributed loads

The concept of centroid of an area can also be used to solve problems other than those dealing with the weight of flat plates. For example, to determine the reactions at the supports of a beam [Sec. 5.8], we can replace a *distributed load w* by a concentrated load $\mathbf{W}$ equal in magnitude to the area A under the load curve and passing through the centroid C of that area (Fig. 5.26). The same approach can be used to determine the resultant of the hydrostatic forces exerted on a *rectangular plate submerged in a liquid* [Sec. 5.9].

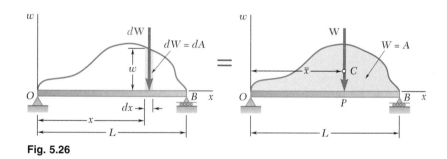

Fig. 5.26

Center of gravity of a three-dimensional body

The last part of the chapter was devoted to the determination of the *center of gravity G of a three-dimensional body*. The coordinates $\bar{x}$, $\bar{y}$, $\bar{z}$ of G were defined by the relations

$$\bar{x}W = \int x\,dW \qquad \bar{y}W = \int y\,dW \qquad \bar{z}W = \int z\,dW \qquad (5.16)$$

In the case of a *homogeneous body*, the center of gravity G coincides with the *centroid C of the volume V* of the body; the coordinates of C are defined by the relations

Centroid of a volume

$$\bar{x}V = \int x\,dV \qquad \bar{y}V = \int y\,dV \qquad \bar{z}V = \int z\,dV \qquad (5.18)$$

If the volume possesses a *plane of symmetry,* its centroid C will lie in that plane; if it possesses two planes of symmetry, C will be located on the line of intersection of the two planes; if it possesses three planes of symmetry which intersect at only one point, C will coincide with that point [Sec. 5.10].

The *volumes and centroids of various common three-dimensional shapes* are tabulated in Fig. 5.21. When a body can be divided into several of these shapes, the coordinates $\overline{X}, \overline{Y}, \overline{Z}$ of its center of gravity G can be determined from the corresponding coordinates of the centers of gravity of its various parts [Sec. 5.11]. We have

$$\overline{X}\Sigma W = \Sigma \overline{x}\,W \qquad \overline{Y}\Sigma W = \Sigma \overline{y}\,W \qquad \overline{Z}\Sigma W = \Sigma \overline{z}\,W \quad (5.19)$$

If the body is made of a homogeneous material, its center of gravity coincides with the centroid C of its volume, and we write [Sample Probs. 5.11 and 5.12]

$$\overline{X}\Sigma V = \Sigma \overline{x}\,V \qquad \overline{Y}\Sigma V = \Sigma \overline{y}\,V \qquad \overline{Z}\Sigma V = \Sigma \overline{z}\,V \qquad (5.20)$$

When a volume is bounded by analytical surfaces, the coordinates of its centroid can be determined by *integration* [Sec. 5.12]. To avoid the computation of the triple integrals in Eqs. (5.18), we

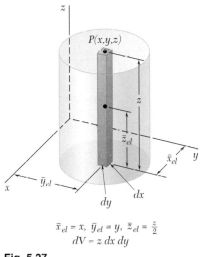

$\overline{x}_{el} = x, \ \overline{y}_{el} = y, \ \overline{z}_{el} = \frac{z}{2}$
$dV = z\,dx\,dy$

Fig. 5.27

can use elements of volume in the shape of thin filaments, as shown in Fig. 5.27. Denoting by $\overline{x}_{el}$, $\overline{y}_{el}$, and $\overline{z}_{el}$ the coordinates of the centroid of the element dV, we rewrite Eqs. (5.18) as

$$\overline{x}V = \int \overline{x}_{el}\,dV \qquad \overline{y}V = \int \overline{y}_{el}\,dV \qquad \overline{z}V = \int \overline{z}_{el}\,dV \qquad (5.22)$$

which involve only double integrals. If the volume possesses *two planes of symmetry*, its centroid C is located on their line of intersection. Choosing the x axis to lie along that line and dividing the volume into thin slabs parallel to the yz plane, we can determine C from the relation

$$\overline{x}V = \int \overline{x}_{el}\,dV \qquad (5.23)$$

with a *single integration* [Sample Prob. 5.13]. For a body of revolution, these slabs are circular and their volume is given in Fig. 5.28.

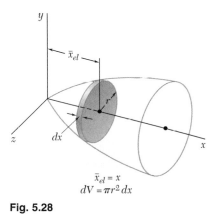

$\overline{x}_{el} = x$
$dV = \pi r^2\,dx$

Fig. 5.28

Review Problems

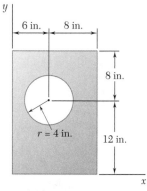

Fig. P5.130

5.130 Locate the centroid of the plane area shown.

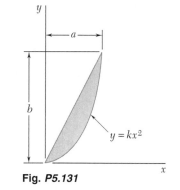

Fig. *P5.131*

5.131 For the area shown, determine the ratio a/b for which $\bar{x} = \bar{y}$.

5.132 Locate the centroid of the plane area shown.

Fig. P5.132

5.133 Determine by direct integration the centroid of the area shown. Express your answer in terms of a and h.

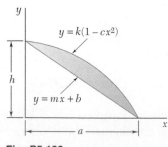

Fig. *P5.133*

5.134   Member *ABCDE* is a component of a mobile and is formed from a single piece of aluminum tubing. Knowing that the member is supported at *C* and that *l* = 2 m, determine the distance *d* so that portion *BCD* of the member is horizontal.

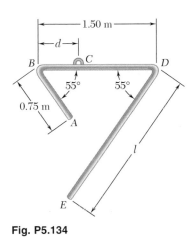

Fig. P5.134

5.135   A cylindrical hole is drilled through the center of a steel ball bearing shown here in cross section. Denoting the length of the *hole* by *L*, show that the volume of the steel remaining is equal to the volume of a sphere of diameter *L*.

Fig. P5.135

5.136   For the beam and loading shown, determine (*a*) the magnitude and location of the resultant of the distributed load, (*b*) the reactions at the beam supports.

Fig. *P5.136*

5.137   Determine the reactions at the beam supports for the given loading.

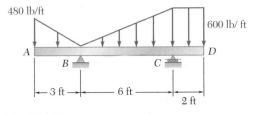

Fig. P5.137

5.138 Locate the center of gravity of the sheet-metal form shown.

$r = 1.8$ m

1.2 m

0.8 m

1.5 m

Fig. P5.138

5.139 The composite body shown is formed by removing a semiellipsoid of revolution of semimajor axis h and semiminor axis $a/2$ from a hemisphere of radius a. Determine (a) the y coordinate of the centroid when $h = a/2$, (b) the ratio h/a for which $\bar{y} = -0.4a$.

5.140 A thin steel wire of uniform cross section is bent into the shape shown. Locate its center of gravity.

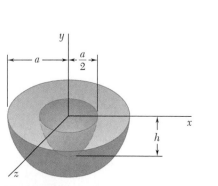

a $\dfrac{a}{2}$

h

Fig. P5.139

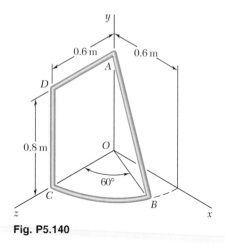

0.6 m 0.6 m

A

D

0.8 m

O

$60°$

C

B

Fig. P5.140

5.141 Locate the centroid of the volume obtained by rotating the shaded area about the x axis.

$y = k(x - h)^2$

a

h

Fig. P5.141

Computer Problems

5.C1 For the trapezoid shown, determine the x and y coordinates of the centroid knowing that $h_1 = a/n^2$ and $h_2 = a/n$. Plot the values of $\bar{x}$ and $\bar{y}$ for $1 \leq n \leq 4$ and $a = 6$ in.

Fig. P5.C1

5.C2 Determine the distance h for which the centroid of the shaded area is as high above BB' as possible. Plot the ratio h/b as a function of k for $0.125 \leq k \leq 0.875$.

Fig. P5.C2

5.C3 Approximate the general spandrel shown using a series of n rectangles, each of width Δa and of the form $bcc'b'$, and then use computational software to calculate the coordinates of the centroid of the area. Locate the centroid when (a) $m = 2$, $a = 4$ in., $h = 4$ in.; (b) $m = 2$, $a = 4$ in., $h = 25$ in.; (c) $m = 5$, $a = 4$ in., $h = 4$ in.; (d) $m = 5$, $a = 4$ in., $h = 25$ in. In each case compare the answers obtained to the exact values of $\bar{x}$ and $\bar{y}$ computed from the formulas given in Fig. 5.8A and determine the percentage error.

Fig. P5.C3

5.C4 Determine the volume and the surface area of the solid obtained by rotating the area shown about the y axis when $a = 80$ mm and (a) $n = 1$, (b) $n = 2$, (c) $n = 3$.

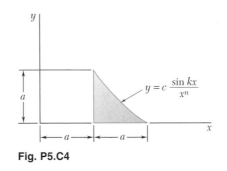

$$y = c \frac{\sin kx}{x^n}$$

Fig. P5.C4

5.C5 Gate AB of uniform width w is held in the position shown by a hinge along its top edge A and by a single shear pin located at the center of its lower edge B. Determine the force on shear pin B as a function of the water depth d. Plot the force on pin B as a function of water depth for 0.54 m $\leq d \leq 1.5$ m when (a) $w = 0.25$ m, (b) $w = 0.50$ m, (c) $w = 0.75$ m, (d) $w = 1$ m. The density of water is $\rho = 10^3$ kg/m³.

Fig. P5.C5

5.C6 An open tank is to be slowly filled with water. Determine the resultant and direction of the pressure force exerted by the water on a 4-ft-wide section of side ABC of the tank as a function of the depth d. Using computational software, plot the pressure force as a function of d for $0 \leq d \leq 10$ ft. The specific weight of water is 62.4 lb/ft³.

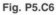

Fig. P5.C6

5.C7 The cross section of a concrete dam is as shown. For a dam section of 1-m width, plot the magnitude of the resultant of the reaction forces exerted by the ground on the base AB of the dam and the resultant of the pressure forces exerted by the water on the face BC of the dam as functions of the depth d of the water for $0 \leq d \leq 16$ m. The densities of concrete and water are 2.40×10^3 kg/m^3 and 10^3 kg/m^3, respectively.

Fig. P5.C7

5.C8 A beam is subjected to the loading shown. Plot the magnitude of the vertical reactions at supports A and B as functions of distance a for $0 \leq a \leq 3$ m.

Fig. P5.C8

5.C9 The three-dimensional structure shown is fabricated from five thin steel rods of equal diameter. Using computational software, determine the coordinates of the center of gravity of the structure. Locate the coordinates of the center of gravity when (a) $h = 40$ ft, $R = 15$ ft, $\alpha = 90°$; (b) $h = 22$ in., $R = 30$ in., $\alpha = 30°$; (c) $h = 70$ ft, $R = 65$ ft, $\alpha = 135°$.

5.C10 Determine the y coordinate of the centroid of the body shown when $h = nb$ and $h = n^2b$. Plot $\bar{y}$ as a function of n for both cases using $1 \leq n \leq 10$ and (a) $b = 4$ in., (b) $b = 6$ in., (c) $b = 8$ in.

Fig. P5.C9

Fig. P5.C10

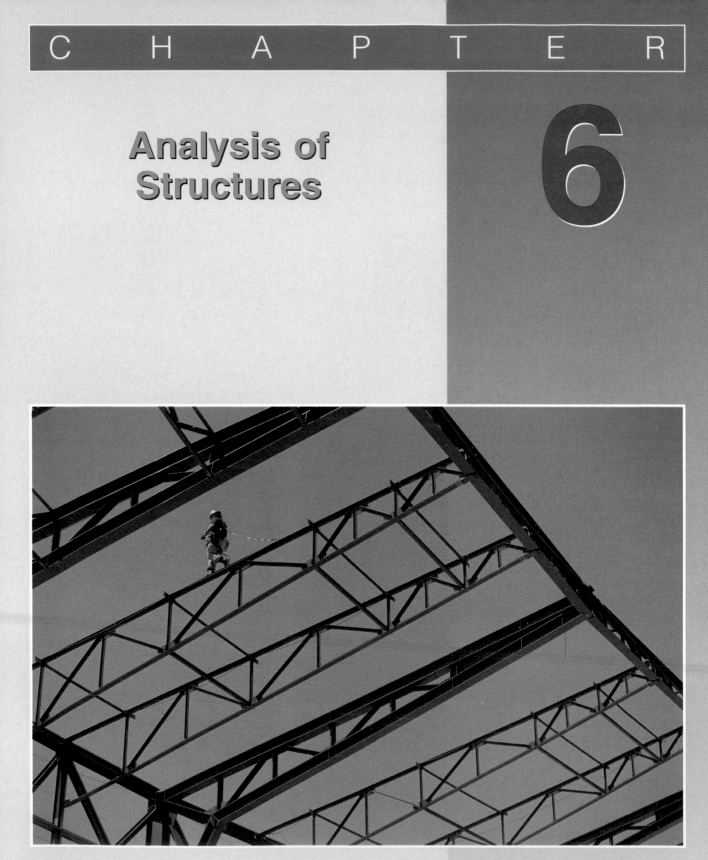

CHAPTER

Analysis of Structures

6

Trusses, such as those used for the roof of Safeco Field in Seattle, provide both a practical and an economical solution to many engineering problems.

6.1. INTRODUCTION

The problems considered in the preceding chapters concerned the equilibrium of a single rigid body, and all forces involved were external to the rigid body. We now consider problems dealing with the equilibrium of structures made of several connected parts. These problems call for the determination not only of the external forces acting on the structure but also of the forces which hold together the various parts of the structure. From the point of view of the structure as a whole, these forces are *internal forces.*

Consider, for example, the crane shown in Fig. 6.1*a*, which carries a load W. The crane consists of three beams AD, CF, and BE connected by frictionless pins; it is supported by a pin at A and by a cable DG. The free-body diagram of the crane has been drawn in Fig. 6.1*b*. The external forces, which are shown in the diagram, include the weight W, the two components A_x and A_y of the reaction at A, and the force T exerted by the cable at D. The internal forces holding the various parts of the crane together do not appear in the diagram. If, however, the crane is dismembered and if a free-body diagram is drawn for each of its component parts, the forces holding the three beams together will also be represented, since these forces are external forces from the point of view of each component part (Fig. 6.1*c*).

ANALYSIS OF STRUCTURES

6.1 Introduction
    Trusses
6.2 Definition of a Truss
6.3 Simple Trusses
6.4 Analysis of Trusses by the Method of Joints
6.5 Joints under Special Loading Conditions
6.6 Space Trusses
6.7 Analysis of Trusses by the Method of Sections
6.8 Trusses Made of Several Simple Trusses
    Frames and Machines
6.9 Structures Containing Multiforce Members
6.10 Analysis of a Frame
6.11 Frames Which Cease to Be Rigid When Detached from Their Supports
6.12 Machines

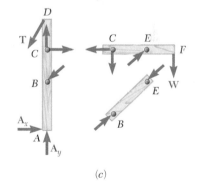

(a)　　　　　(b)　　　　　(c)

Fig. 6.1

It will be noted that the force exerted at B by member BE on member AD has been represented as equal and opposite to the force exerted at the same point by member AD on member BE; the force exerted at E by BE on CF is shown equal and opposite to the force exerted by CF on BE; and the components of the force exerted at C by CF on AD are shown equal and opposite to the components of the force exerted by AD on CF. This is in conformity with Newton's third law, which states that *the forces of action and reaction between bodies in contact have the same magnitude, same line of action, and opposite sense.* As pointed out in Chap. 1, this law, which is based on experimental evidence, is one of the six fundamental principles of elementary mechanics, and its application is essential to the solution of problems involving connected bodies.

285

Photo 6.1 Shown is a pin-jointed connection on the approach span to the San Francisco–Oakland Bay Bridge.

In this chapter, three broad categories of engineering structures will be considered:

1. *Trusses,* which are designed to support loads and are usually stationary, fully constrained structures. Trusses consist exclusively of straight members connected at joints located at the ends of each member. Members of a truss, therefore, are *two-force members,* that is, members each acted upon by two equal and opposite forces directed along the member.

2. *Frames,* which are also designed to support loads and are also usually stationary, fully constrained structures. However, like the crane of Fig. 6.1, frames always contain at least one *multiforce member,* that is, a member acted upon by three or more forces which, in general, are not directed along the member.

3. *Machines,* which are designed to transmit and modify forces and are structures containing moving parts. Machines, like frames, always contain at least one multiforce member.

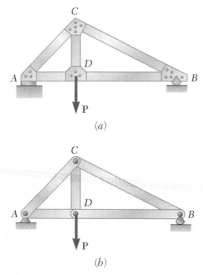

Fig. 6.2

TRUSSES

6.2. DEFINITION OF A TRUSS

The truss is one of the major types of engineering structures. It provides both a practical and an economical solution to many engineering situations, especially in the design of bridges and buildings. A typical truss is shown in Fig. 6.2*a*. A truss consists of straight members connected at joints. Truss members are connected at their extremities only; thus no member is continuous through a joint. In Fig. 6.2*a*, for example, there is no member *AB*; there are instead two distinct members *AD* and *DB*. Most actual structures are made of several trusses joined together to form a space framework. Each truss is designed to carry those loads which act in its plane and thus may be treated as a two-dimensional structure.

In general, the members of a truss are slender and can support little lateral load; all loads, therefore, must be applied to the various joints, and not to the members themselves. When a concentrated load is to be applied between two joints, or when a distributed load is to be supported by the truss, as in the case of a bridge truss, a floor system must be provided which, through the use of stringers and floor beams, transmits the load to the joints (Fig. 6.3).

The weights of the members of the truss are also assumed to be applied to the joints, half of the weight of each member being applied to each of the two joints the member connects. Although the members are actually joined together by means of bolted or welded connections, it is customary to assume that the members are pinned together; therefore, the forces acting at each end of a member reduce to a single force and no couple. Thus, the only forces assumed to be applied to a truss member are a single force at each end of the

Fig. 6.3

member. Each member can then be treated as a two-force member, and the entire truss can be considered as a group of pins and two-force members (Fig. 6.2*b*). An individual member can be acted upon as shown in either of the two sketches of Fig. 6.4. In Fig. 6.4*a*, the forces tend to pull the member apart, and the member is in tension; in Fig. 6.4*b*, the forces tend to compress the member, and the member is in compression. A number of typical trusses are shown in Fig. 6.5.

(*a*)　　　(*b*)

Fig. 6.4

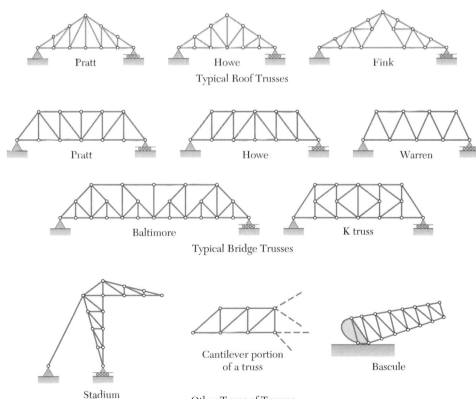

Pratt　　Howe　　Fink

Typical Roof Trusses

Pratt　　Howe　　Warren

Baltimore　　K truss

Typical Bridge Trusses

Stadium

Cantilever portion of a truss

Bascule

Other Types of Trusses

Fig. 6.5

6.3. SIMPLE TRUSSES

Consider the truss of Fig. 6.6*a*, which is made of four members connected by pins at *A*, *B*, *C*, and *D*. If a load is applied at *B*, the truss will greatly deform, completely losing its original shape. In contrast, the truss of Fig. 6.6*b*, which is made of three members connected by pins at *A*, *B*, and *C*, will deform only slightly under a load applied at *B*. The only possible deformation for this truss is one involving small changes in the length of its members. The truss of Fig. 6.6*b* is said to be a *rigid truss*, the term rigid being used here to indicate that the truss *will not collapse.*

Fig. 6.6

Photo 6.2 Two K-trusses were used as the main components of the movable bridge shown which moved above a large stockpile of ore. The bucket below the trusses picked up ore and redeposited it until the ore was thoroughly mixed. The ore was then sent to the mill for processing into steel.

As shown in Fig. 6.6*c*, a larger rigid truss can be obtained by adding two members *BD* and *CD* to the basic triangular truss of Fig. 6.6*b*. This procedure can be repeated as many times as desired, and the resulting truss will be rigid if each time two new members are added, they are attached to two existing joints and are connected at a new joint.† A truss which can be constructed in this manner is called a *simple truss.*

It should be noted that a simple truss is not necessarily made only of triangles. The truss of Fig. 6.6*d*, for example, is a simple truss which was constructed from triangle *ABC* by adding successively the joints *D*, *E*, *F*, and *G*. On the other hand, rigid trusses are not always simple trusses, even when they appear to be made of triangles. The Fink and Baltimore trusses shown in Fig. 6.5, for instance, are not simple trusses, since they cannot be constructed from a single triangle in the manner described above. All the other trusses shown in Fig. 6.5 are simple trusses, as may be easily checked. (For the K truss, start with one of the central triangles.)

Returning to Fig. 6.6, we note that the basic triangular truss of Fig. 6.6*b* has three members and three joints. The truss of Fig. 6.6*c* has two more members and one more joint, that is, five members and four joints altogether. Observing that every time two new members are added, the number of joints is increased by one, we find that in a simple truss the total number of members is $m = 2n - 3$, where n is the total number of joints.

---

†The three joints must not be in a straight line.

6.4. ANALYSIS OF TRUSSES BY THE METHOD OF JOINTS

We saw in Sec. 6.2 that a truss can be considered as a group of pins and two-force members. The truss of Fig. 6.2, whose free-body diagram is shown in Fig. 6.7a, can thus be dismembered, and a free-body diagram can be drawn for each pin and each member (Fig. 6.7b). Each member is acted upon by two forces, one at each end; these forces have the same magnitude, same line of action, and opposite sense (Sec. 4.6). Furthermore, Newton's third law indicates that the forces of action and reaction between a member and a pin are equal and opposite. Therefore, the forces exerted by a member on the two pins it connects must be directed along that member and be equal and opposite. The common magnitude of the forces exerted by a member on the two pins it connects is commonly referred to as the *force in the member* considered, even though this quantity is actually a scalar. Since the lines of action of all the internal forces in a truss are known, the analysis of a truss reduces to computing the forces in its various members and to determining whether each of its members is in tension or in compression.

Since the entire truss is in equilibrium, each pin must be in equilibrium. The fact that a pin is in equilibrium can be expressed by drawing its free-body diagram and writing two equilibrium equations (Sec. 2.9). If the truss contains n pins, there will, therefore, be $2n$ equations available, which can be solved for $2n$ unknowns. In the case of a simple truss, we have $m = 2n - 3$, that is, $2n = m + 3$, and the number of unknowns which can be determined from the free-body diagrams of the pins is thus $m + 3$. This means that the forces in all the members, the two components of the reaction $\mathbf{R}_A$, and the reaction $\mathbf{R}_B$ can be found by considering the free-body diagrams of the pins.

The fact that the entire truss is a rigid body in equilibrium can be used to write three more equations involving the forces shown in the free-body diagram of Fig. 6.7a. Since they do not contain any new information, these equations are not independent of the equations associated with the free-body diagrams of the pins. Nevertheless, they can be used to determine the components of the reactions at the supports. The arrangement of pins and members in a simple truss is such that it will then always be possible to find a joint involving only two unknown forces. These forces can be determined by the methods of Sec. 2.11 and their values transferred to the adjacent joints and treated as known quantities at these joints. This procedure can be repeated until all unknown forces have been determined.

As an example, the truss of Fig. 6.7 will be analyzed by considering the equilibrium of each pin successively, starting with a joint at which only two forces are unknown. In the truss considered, all pins are subjected to at least three unknown forces. Therefore, the reactions at the supports must first be determined by considering the entire truss as a free body and using the equations of equilibrium of a rigid body. We find in this way that $\mathbf{R}_A$ is vertical and determine the magnitudes of $\mathbf{R}_A$ and $\mathbf{R}_B$.

The number of unknown forces at joint A is thus reduced to two, and these forces can be determined by considering the equilibrium of pin A. The reaction $\mathbf{R}_A$ and the forces $\mathbf{F}_{AC}$ and $\mathbf{F}_{AD}$ exerted on pin

Fig. 6.7

Photo 6.3 Because roof trusses, such as those shown, require support only at their ends, it is possible to construct buildings with large, unobstructed floor areas.

A by members *AC* and *AD*, respectively, must form a force triangle. First we draw $\mathbf{R}_A$ (Fig. 6.8); noting that $\mathbf{F}_{AC}$ and $\mathbf{F}_{AD}$ are directed along *AC* and *AD*, respectively, we complete the triangle and determine the magnitude and sense of $\mathbf{F}_{AC}$ and $\mathbf{F}_{AD}$. The magnitudes F_{AC} and F_{AD} represent the forces in members *AC* and *AD*. Since $\mathbf{F}_{AC}$ is directed down and to the left, that is, *toward* joint *A*, member *AC* pushes on pin *A* and is in compression. Since $\mathbf{F}_{AD}$ is directed *away* from joint *A*, member *AD* pulls on pin *A* and is in tension.

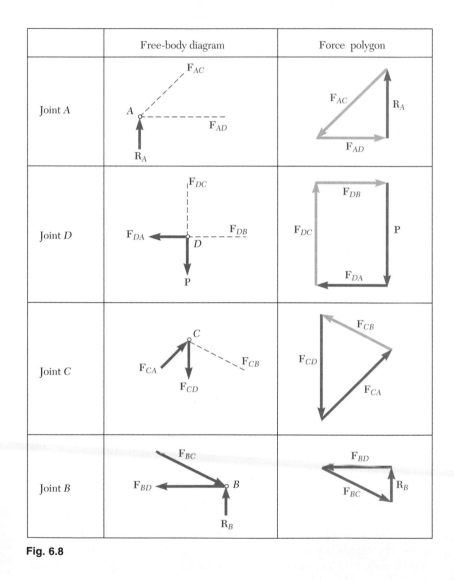

	Free-body diagram	Force polygon
Joint *A*		
Joint *D*		
Joint *C*		
Joint *B*		

Fig. 6.8

We can now proceed to joint *D*, where only two forces, $\mathbf{F}_{DC}$ and $\mathbf{F}_{DB}$, are still unknown. The other forces are the load $\mathbf{P}$, which is given, and the force $\mathbf{F}_{DA}$ exerted on the pin by member *AD*. As indicated above, this force is equal and opposite to the force $\mathbf{F}_{AD}$ exerted by the same member on pin *A*. We can draw the force polygon corresponding to joint *D*, as shown in Fig. 6.8, and determine the forces

$\mathbf{F}_{DC}$ and $\mathbf{F}_{DB}$ from that polygon. However, when more than three forces are involved, it is usually more convenient to solve the equations of equilibrium $\Sigma F_x = 0$ and $\Sigma F_y = 0$ for the two unknown forces. Since both of these forces are found to be directed away from joint D, members DC and DB pull on the pin and are in tension.

Next, joint C is considered; its free-body diagram is shown in Fig. 6.8. It is noted that both $\mathbf{F}_{CD}$ and $\mathbf{F}_{CA}$ are known from the analysis of the preceding joints and that only $\mathbf{F}_{CB}$ is unknown. Since the equilibrium of each pin provides sufficient information to determine two unknowns, a check of our analysis is obtained at this joint. The force triangle is drawn, and the magnitude and sense of $\mathbf{F}_{CB}$ are determined. Since $\mathbf{F}_{CB}$ is directed toward joint C, member CB pushes on pin C and is in compression. The check is obtained by verifying that the force $\mathbf{F}_{CB}$ and member CB are parallel.

At joint B, all of the forces are known. Since the corresponding pin is in equilibrium, the force triangle must close and an additional check of the analysis is obtained.

It should be noted that the force polygons shown in Fig. 6.8 are not unique. Each of them could be replaced by an alternative configuration. For example, the force triangle corresponding to joint A could be drawn as shown in Fig. 6.9. The triangle actually shown in Fig. 6.8 was obtained by drawing the three forces $\mathbf{R}_A$, $\mathbf{F}_{AC}$, and $\mathbf{F}_{AD}$ in tip-to-tail fashion in the order in which their lines of action are encountered when moving clockwise around joint A. The other force polygons in Fig. 6.8, having been drawn in the same way, can be made to fit into a single diagram, as shown in Fig. 6.10. Such a diagram, known as *Maxwell's diagram*, greatly facilitates the *graphical analysis* of truss problems.

Fig. 6.9

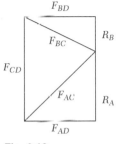

Fig. 6.10

*6.5. JOINTS UNDER SPECIAL LOADING CONDITIONS

Consider Fig. 6.11a, in which the joint shown connects four members lying in two intersecting straight lines. The free-body diagram of Fig. 6.11b shows that pin A is subjected to two pairs of directly opposite forces. The corresponding force polygon, therefore, must be a parallelogram (Fig. 6.11c), and *the forces in opposite members must be equal*.

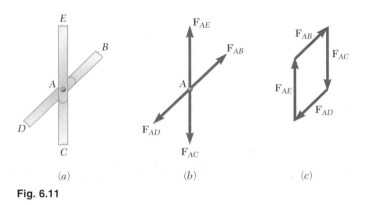

(a)   (b)   (c)

Fig. 6.11

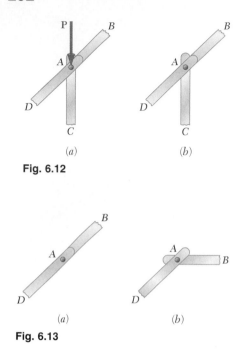

(a)   (b)

Fig. 6.12

(a)   (b)

Fig. 6.13

Consider next Fig. 6.12*a*, in which the joint shown connects three members and supports a load **P**. Two of the members lie in the same line, and the load **P** acts along the third member. The free-body diagram of pin *A* and the corresponding force polygon will be as shown in Fig. 6.11*b* and *c*, with $\mathbf{F}_{AE}$ replaced by the load **P**. Thus, *the forces in the two opposite members must be equal, and the force in the other member must equal P.* A particular case of special interest is shown in Fig. 6.12*b*. Since, in this case, no external load is applied to the joint, we have *P* = 0, and the force in member *AC* is zero. Member *AC* is said to be a *zero-force member.*

Consider now a joint connecting two members only. From Sec. 2.9, we know that a particle which is acted upon by two forces will be in equilibrium if the two forces have the same magnitude, same line of action, and opposite sense. In the case of the joint of Fig. 6.13*a*, which connects two members *AB* and *AD* lying in the same line, *the forces in the two members must be equal* for pin *A* to be in equilibrium. In the case of the joint of Fig. 6.13*b*, pin *A* cannot be in equilibrium unless the forces in both members are zero. Members connected as shown in Fig. 6.13*b*, therefore, must be *zero-force members.*

Spotting the joints which are under the special loading conditions listed above will expedite the analysis of a truss. Consider, for example, a Howe truss loaded as shown in Fig. 6.14. All of the members represented by green lines will be recognized as zero-force members. Joint *C* connects three members, two of which lie in the same line, and is not subjected to any external load; member *BC* is thus a zero-force member. Applying the same reasoning to joint *K*, we find that member *JK* is also a zero-force member. But joint *J* is now in the same situation as joints *C* and *K*, and member *IJ* must be a zero-force member. The examination of joints *C*, *J*, and *K* also shows that the forces in members *AC* and *CE* are equal, that the forces in members *HJ* and *JL* are equal, and that the forces in members *IK* and *KL* are equal. Turning our attention to joint *I*, where the 20-kN load and member *HI* are collinear, we note that the force in member *HI* is 20 kN (tension) and that the forces in members *GI* and *IK* are equal. Hence, the forces in members *GI*, *IK*, and *KL* are equal.

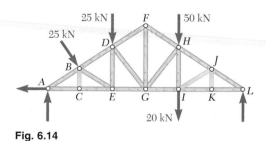

Fig. 6.14

Note that the conditions described above do not apply to joints *B* and *D* in Fig. 6.14, and it would be wrong to assume that the force in member *DE* is 25 kN or that the forces in members *AB* and *BD* are equal. The forces in these members and in all remaining members should be found by carrying out the analysis of joints *A*, *B*, *D*, *E*, *F*, *G*, *H*, and *L* in the usual manner. Thus, until you have become thoroughly familiar with the conditions under which the rules established in this section can be applied, you should draw the free-body diagrams of all pins and write the corresponding equilibrium equations (or draw the corresponding force polygons) whether or not the joints being considered are under one of the special loading conditions described above.

A final remark concerning zero-force members: These members are not useless. For example, although the zero-force members of Fig. 6.14 do not carry any loads under the loading conditions shown, the same members would probably carry loads if the loading conditions were changed. Besides, even in the case considered, these members are needed to support the weight of the truss and to maintain the truss in the desired shape.

*6.6. SPACE TRUSSES

When several straight members are joined together at their extremities to form a three-dimensional configuration, the structure obtained is called a *space truss*.

We recall from Sec. 6.3 that the most elementary two-dimensional rigid truss consisted of three members joined at their extremities to form the sides of a triangle; by adding two members at a time to this basic configuration, and connecting them at a new joint, it was possible to obtain a larger rigid structure which was defined as a simple truss. Similarly, the most elementary rigid space truss consists of six members joined at their extremities to form the edges of a tetrahedron *ABCD* (Fig. 6.15*a*). By adding three members at a time to this basic configuration, such as *AE*, *BE*, and *CE*, attaching them to three existing joints, and connecting them at a new joint,† we can obtain a larger rigid structure which is defined as a *simple space truss* (Fig. 6.15*b*). Observing that the basic tetrahedron has six members and four joints and that every time three members are added, the number of joints is increased by one, we conclude that in a simple space truss the total number of members is $m = 3n - 6$, where n is the total number of joints.

If a space truss is to be completely constrained and if the reactions at its supports are to be statically determinate, the supports should consist of a combination of balls, rollers, and balls and sockets which provides six unknown reactions (see Sec. 4.9). These unknown reactions may be readily determined by solving the six equations expressing that the three-dimensional truss is in equilibrium.

Although the members of a space truss are actually joined together by means of bolted or welded connections, it is assumed that each joint consists of a ball-and-socket connection. Thus, no couple will be applied to the members of the truss, and each member can be treated as a two-force member. The conditions of equilibrium for each joint will be expressed by the three equations $\Sigma F_x = 0$, $\Sigma F_y = 0$, and $\Sigma F_z = 0$. In the case of a simple space truss containing n joints, writing the conditions of equilibrium for each joint will thus yield $3n$ equations. Since $m = 3n - 6$, these equations suffice to determine all unknown forces (forces in m members and six reactions at the supports). However, to avoid the necessity of solving simultaneous equations, care should be taken to select joints in such an order that no selected joint will involve more than three unknown forces.

†The four joints must not lie in a plane.

(a)

(b)

Fig. 6.15

Photo 6.4 Three-dimensional or space trusses are used for broadcast and power transmission line towers, roof framing, and spacecraft applications, such as components of the *International Space Station*.

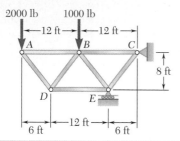

SAMPLE PROBLEM 6.1

Using the method of joints, determine the force in each member of the truss shown.

SOLUTION

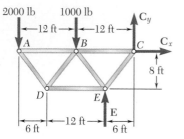

Free-Body: Entire Truss. A free-body diagram of the entire truss is drawn; external forces acting on this free body consist of the applied loads and the reactions at C and E. We write the following equilibrium equations.

$$+\uparrow\Sigma M_C = 0: \qquad (2000 \text{ lb})(24 \text{ ft}) + (1000 \text{ lb})(12 \text{ ft}) - E(6 \text{ ft}) = 0$$
$$E = +10{,}000 \text{ lb} \qquad\qquad \mathbf{E} = 10{,}000 \text{ lb}\uparrow$$

$$\xrightarrow{+}\Sigma F_x = 0: \qquad\qquad\qquad\qquad\qquad\qquad\qquad \mathbf{C}_x = 0$$

$$+\uparrow\Sigma F_y = 0: \qquad -2000 \text{ lb} - 1000 \text{ lb} + 10{,}000 \text{ lb} + C_y = 0$$
$$C_y = -7000 \text{ lb} \qquad\qquad \mathbf{C}_y = 7000 \text{ lb}\downarrow$$

Free-Body: Joint A. This joint is subjected to only two unknown forces, namely, the forces exerted by members AB and AD. A force triangle is used to determine $\mathbf{F}_{AB}$ and $\mathbf{F}_{AD}$. We note that member AB pulls on the joint and thus is in tension and that member AD pushes on the joint and thus is in compression. The magnitudes of the two forces are obtained from the proportion

$$\frac{2000 \text{ lb}}{4} = \frac{F_{AB}}{3} = \frac{F_{AD}}{5}$$

$$F_{AB} = 1500 \text{ lb } T \quad \blacktriangleleft$$
$$F_{AD} = 2500 \text{ lb } C \quad \blacktriangleleft$$

Free-Body: Joint D. Since the force exerted by member AD has been determined, only two unknown forces are now involved at this joint. Again, a force triangle is used to determine the unknown forces in members DB and DE.

$$F_{DB} = F_{DA} \qquad\qquad F_{DB} = 2500 \text{ lb } T \quad \blacktriangleleft$$
$$F_{DE} = 2(\tfrac{3}{5})F_{DA} \qquad F_{DE} = 3000 \text{ lb } C \quad \blacktriangleleft$$

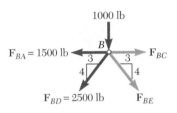

Free-Body: Joint B. Since more than three forces act at this joint, we determine the two unknown forces $\mathbf{F}_{BC}$ and $\mathbf{F}_{BE}$ by solving the equilibrium equations $\Sigma F_x = 0$ and $\Sigma F_y = 0$. We arbitrarily assume that both unknown forces act away from the joint, that is, that the members are in tension. The positive value obtained for F_{BC} indicates that our assumption was correct; member BC is in tension. The negative value of F_{BE} indicates that our assumption was wrong; member BE is in compression.

$$+\uparrow\Sigma F_y = 0: \qquad -1000 - \tfrac{4}{5}(2500) - \tfrac{4}{5}F_{BE} = 0$$
$$F_{BE} = -3750 \text{ lb} \qquad F_{BE} = 3750 \text{ lb } C \quad \blacktriangleleft$$

$$\xrightarrow{+}\Sigma F_x = 0: \qquad F_{BC} - 1500 - \tfrac{3}{5}(2500) - \tfrac{3}{5}(3750) = 0$$
$$F_{BC} = +5250 \text{ lb} \qquad F_{BC} = 5250 \text{ lb } T \quad \blacktriangleleft$$

Free-Body: Joint E. The unknown force $\mathbf{F}_{EC}$ is assumed to act away from the joint. Summing x components, we write

$$\xrightarrow{+}\Sigma F_x = 0: \qquad \tfrac{3}{5}F_{EC} + 3000 + \tfrac{3}{5}(3750) = 0$$
$$F_{EC} = -8750 \text{ lb} \qquad F_{EC} = 8750 \text{ lb } C \quad \blacktriangleleft$$

Summing y components, we obtain a check of our computations:

$$+\uparrow\Sigma F_y = 10{,}000 - \tfrac{4}{5}(3750) - \tfrac{4}{5}(8750)$$
$$= 10{,}000 - 3000 - 7000 = 0 \qquad \text{(checks)}$$

Free-Body: Joint C. Using the computed values of $\mathbf{F}_{CB}$ and $\mathbf{F}_{CE}$, we can determine the reactions $\mathbf{C}_x$ and $\mathbf{C}_y$ by considering the equilibrium of this joint. Since these reactions have already been determined from the equilibrium of the entire truss, we will obtain two checks of our computations. We can also simply use the computed values of all forces acting on the joint (forces in members and reactions) and check that the joint is in equilibrium:

$$\xrightarrow{+}\Sigma F_x = -5250 + \tfrac{3}{5}(8750) = -5250 + 5250 = 0 \qquad \text{(checks)}$$
$$+\uparrow\Sigma F_y = -7000 + \tfrac{4}{5}(8750) = -7000 + 7000 = 0 \qquad \text{(checks)}$$

SOLVING PROBLEMS
ON YOUR OWN

In this lesson you learned to use the *method of joints* to determine the forces in the members of a *simple truss*, that is, a truss that can be constructed from a basic triangular truss by adding to it two new members at a time and connecting them at a new joint. We note that the analysis of a *simple* truss can always be carried out by the method of joints.

Your solution will consist of the following steps:

1. Draw a free-body diagram of the entire truss, and use this diagram to determine the reactions at the supports.

2. Note that the choice of the first joint is not unique. Once you have determined the reactions at the supports of the truss, you can choose either of two joints as a starting point for your analysis. In Sample Prob. 6.1, we started at joint A and proceeded through joints D, B, E, and C, but we could also have started at joint C and proceeded through joints E, B, D, and A. On the other hand, having selected a first joint, you may in some cases reach a point in your analysis beyond which you cannot proceed. You must then start again from another joint to complete your solution. Because there is usually more than one way to analyze a truss, we encourage you to outline your solution *before* starting any computations.

3. Locate a joint connecting only two members, and draw the free-body diagram of its pin. Use this free-body diagram to determine the unknown force in each of the two members. If only three forces are involved (the two unknown forces and a known one), you will probably find it more convenient to draw and solve the corresponding force triangle. If more than three forces are involved, you should write and solve the equilibrium equations for the pin, $\Sigma F_x = 0$ and $\Sigma F_y = 0$, assuming that the members are in tension. A positive answer means that the member is in tension, a negative answer that the member is in compression. Once the forces have been found, enter their values on a sketch of the truss, with T for tension and C for compression.

In Sec. 6.4, we discussed how to use a force triangle to determine the unknown forces at a joint, and in Sample Prob. 6.1 we applied this technique to joints A and D. In each of these cases, the directions of the unknown forces at a joint were not specified in the associated free-body diagram—only their lines of action were shown. Thus, when you construct a force triangle, draw the known force first and then add the lines of action of the unknown forces, remembering that the forces must be arranged tip to tail to form a triangle. It is in this last step that the directions of the unknown forces are established.

4. Next, locate a joint where the forces in only two of the connected members are still unknown. Draw the free-body diagram of the pin and use it as indicated above to determine the two unknown forces.

5. Repeat this procedure until the forces in all the members of the truss have been found. Since you previously used the three equilibrium equations associated with the free-body diagram of the entire truss to determine the reactions at the supports, you will end up with three extra equations. These equations can be used to check your computations.

Problems

6.1 through 6.8 Using the method of joints, determine the force in each member of the truss shown. State whether each member is in tension or compression.

Fig. P6.1

Fig. P6.2

Fig. *P6.3*

Fig. P6.4

Fig. P6.5

Fig. P6.6

Fig. P6.7

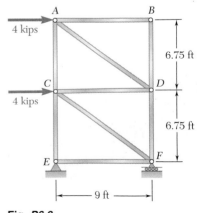

Fig. *P6.8*

6.9 Determine the force in each member of the Gambrel roof truss shown. State whether each member is in tension or compression.

Fig. P6.9

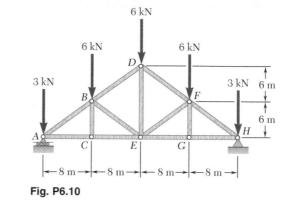

Fig. P6.10

6.10 Determine the force in each member of the Howe roof truss shown. State whether each member is in tension or compression.

6.11 Determine the force in each member of the Fink roof truss shown. State whether each member is in tension or compression.

Fig. P6.11

Fig. *P6.12*

6.12 Determine the force in each member of the fan roof truss shown. State whether each member is in tension or compression.

6.13 Determine the force in each member of the roof truss shown. State whether each member is in tension or compression.

Fig. P6.13

Fig. *P6.14*

6.14 Determine the force in each member of the Gambrel roof truss shown. State whether each member is in tension or compression.

6.15   Determine the force in each member of the Pratt bridge truss shown. State whether each member is in tension or compression.

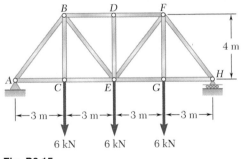

Fig. P6.15

6.16   Solve Prob. 6.15 assuming that the load at *G* has been removed.

6.17   Determine the force in member *DE* and in each of the members located to the left of *DE* for the inverted Howe roof truss shown. State whether each member is in tension or compression.

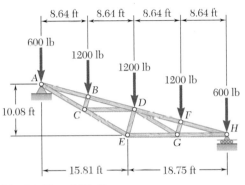

Fig. P6.17 and P6.18

6.18   Determine the force in each of the members located to the right of *DE* for the inverted Howe roof truss shown. State whether each member is in tension or compression.

6.19   Determine the force in each of the members located to the left of member *FG* for the scissor roof truss shown. State whether each member is in tension or compression.

Fig. P6.19 and P6.20

6.20   Determine the force in member *FG* and in each of the members located to the right of member *FG* for the scissor roof truss shown. State whether each member is in tension or compression.

6.21 The portion of truss shown represents the upper part of a power transmission line tower. For the given loading, determine the force in each of the members located above *HJ*. State whether each member is in tension or compression.

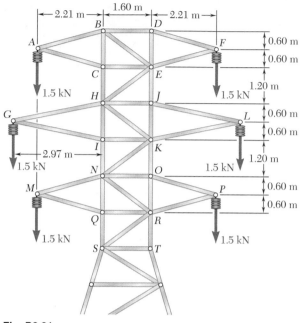

Fig. P6.21

6.22 For the tower and loading of Prob. 6.21 and knowing that $F_{CH} = F_{EJ} = 1.5$ kN C and $F_{EH} = 0$, determine the force in member *HJ* and in each of the members located between *HJ* and *NO*. State whether each member is in tension or compression.

6.23 Determine the force in each member of the truss shown. State whether each member is in tension or compression.

6.24 Determine the force in each member of the truss shown. State whether each member is in tension or compression.

Fig. P6.23

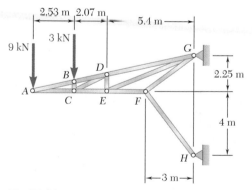

Fig. P6.24

6.25 For the roof truss shown in Fig. *P6.25* and *P6.26*, determine the force in each of the members located to the left of member *GH*. State whether each member is in tension or compression.

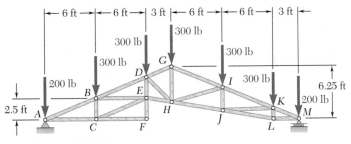

Fig. *P6.25* and *P6.26*

6.26 Determine the force in member *GH* and in each of the members located to the right of *GH* for the roof truss shown. State whether each member is in tension or compression.

6.27 Determine whether the trusses of Probs. 6.13, 6.14, and 6.25 are simple trusses.

6.28 Determine whether the trusses of Probs. 6.19, 6.21, and 6.23 are simple trusses.

6.29 and **6.30** For the given loading, determine the zero-force members in the truss shown.

Fig. *P6.29*

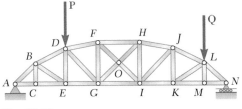

Fig. P6.30

6.31 and **6.32** For the given loading, determine the zero-force members in the truss shown.

Fig. *P6.31*

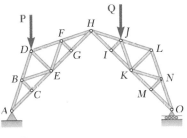

Fig. P6.32

6.33 and **6.34** For the given loading, determine the zero-force members in the truss shown.

Fig. P6.33

Fig. P6.34

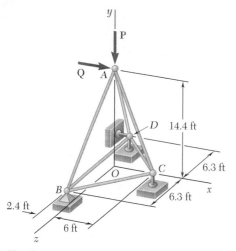

Fig. P6.36 and P6.37

6.35 Determine the zero-force members in the truss of (*a*) Prob. 6.9, (*b*) Prob. 6.19.

***6.36** The truss shown consists of six members and is supported by a ball and socket at *B*, a short link at *C*, and two short links at *D*. Determine the force in each of the members for **P** = (−5670 lb)**j** and **Q** = 0.

***6.37** The truss shown consists of six members and is supported by a ball and socket at *B*, a short link at *C*, and two short links at *D*. Determine the force in each of the members for **P** = 0 and **Q** = (7420 lb)**i**.

***6.38** The truss shown consists of six members and is supported by a short link at *A*, two short links at *B*, and a ball and socket at *D*. Determine the force in each of the members for the given loading.

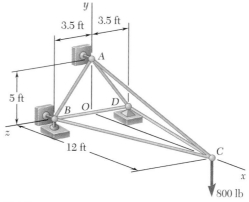

Fig. P6.38

***6.39** The portion of a power line transmission tower shown consists of nine members and is supported by a ball and socket at *B* and short links at *C*, *D*, and *E*. Determine the force in each of the members for the given loading.

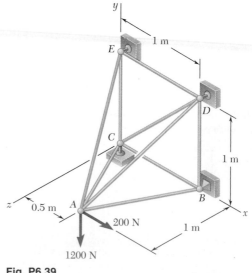

Fig. P6.39

*6.40 The truss shown consists of 18 members and is supported by a ball and socket at A, two short links at B, and one short link at G. (*a*) Check that this truss is a simple truss, that it is completely constrained, and that the reactions at its supports are statically determinate. (*b*) For the given loading, determine the force in each of the six members joined at E.

*6.41 The truss shown consists of 18 members and is supported by a ball and socket at A, two short links at B, and one short link at G. (*a*) Check that this truss is a simple truss, that it is completely constrained, and that the reactions at its supports are statically determinate. (*b*) For the given loading, determine the force in each of the six members joined at G.

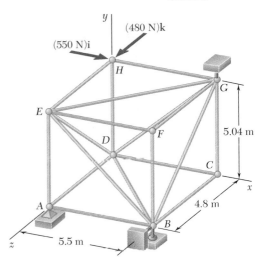

Fig. P6.40 and P6.41

6.7. ANALYSIS OF TRUSSES BY THE METHOD OF SECTIONS

The method of joints is most effective when the forces in all the members of a truss are to be determined. If, however, the force in only one member or the forces in a very few members are desired, another method, the method of sections, is more efficient.

Assume, for example, that we want to determine the force in member BD of the truss shown in Fig. 6.16*a*. To do this, we must determine the force with which member BD acts on either joint B or joint D. If we were to use the method of joints, we would choose either joint B or joint D as a free body. However, we can also choose as a free body a larger portion of the truss, composed of several joints and members, provided that the desired force is one of the external forces acting on that portion. If, in addition, the portion of the truss is chosen so that there is a total of only three unknown forces acting upon it, the desired force can be obtained by solving the equations of equilibrium for this portion of the truss. In practice, the portion of the truss to be utilized is obtained by *passing a section* through three members of the truss, one of which is the desired member, that is, by drawing a line which divides the truss into two completely separate parts but does not intersect more than three members. Either of the two portions of the truss obtained after the intersected members have been removed can then be used as a free body.†

In Fig. 6.16*a*, the section *nn* has been passed through members BD, BE, and CE, and the portion ABC of the truss is chosen as the free body (Fig. 6.16*b*). The forces acting on the free body are the loads $\mathbf{P}_1$ and $\mathbf{P}_2$ at points A and B and the three unknown forces $\mathbf{F}_{BD}$, $\mathbf{F}_{BE}$, and $\mathbf{F}_{CE}$. Since it is not known whether the members removed were in tension or compression, the three forces have been arbitrarily drawn away from the free body as if the members were in tension.

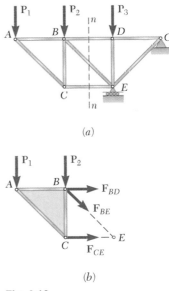

(*a*)

(*b*)

Fig. 6.16

---

†In the analysis of certain trusses, sections are passed which intersect more than three members; the forces in one, or possibly two, of the intersected members may be obtained if equilibrium equations can be found, each of which involves only one unknown (see Probs. 6.60 through 6.63).

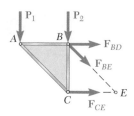

Fig. 6.16b (repeated)

The fact that the rigid body ABC is in equilibrium can be expressed by writing three equations which can be solved for the three unknown forces. If only the force $\mathbf{F}_{BD}$ is desired, we need write only one equation, provided that the equation does not contain the other unknowns. Thus the equation $\Sigma M_E = 0$ yields the value of the magnitude F_{BD} of the force $\mathbf{F}_{BD}$ (Fig. 6.16b). A positive sign in the answer will indicate that our original assumption regarding the sense of $\mathbf{F}_{BD}$ was correct and that member BD is in tension; a negative sign will indicate that our assumption was incorrect and that BD is in compression.

On the other hand, if only the force $\mathbf{F}_{CE}$ is desired, an equation which does not involve $\mathbf{F}_{BD}$ or $\mathbf{F}_{BE}$ should be written; the appropriate equation is $\Sigma M_B = 0$. Again a positive sign for the magnitude F_{CE} of the desired force indicates a correct assumption, that is, tension; and a negative sign indicates an incorrect assumption, that is, compression.

If only the force $\mathbf{F}_{BE}$ is desired, the appropriate equation is $\Sigma F_y = 0$. Whether the member is in tension or compression is again determined from the sign of the answer.

When the force in only one member is determined, no independent check of the computation is available. However, when all the unknown forces acting on the free body are determined, the computations can be checked by writing an additional equation. For instance, if $\mathbf{F}_{BD}$, $\mathbf{F}_{BE}$, and $\mathbf{F}_{CE}$ are determined as indicated above, the computation can be checked by verifying that $\Sigma F_x = 0$.

*6.8. TRUSSES MADE OF SEVERAL SIMPLE TRUSSES

Consider two simple trusses ABC and DEF. If they are connected by three bars BD, BE, and CE as shown in Fig. 6.17a, they will form together a rigid truss $ABDF$. The trusses ABC and DEF can also be combined into a single rigid truss by joining joints B and D into a single joint B and by connecting joints C and E by a bar CE (Fig. 6.17b). The truss thus obtained is known as a *Fink truss*. It should be noted that the trusses of Fig. 6.17a and b are *not* simple trusses; they cannot be constructed from a triangular truss by adding successive pairs of members as prescribed in Sec. 6.3. They are rigid trusses, however, as we can check by comparing the systems of connections used to hold the simple trusses ABC and DEF together (three bars in Fig. 6.17a, one pin and one bar in Fig. 6.17b) with the systems of supports discussed in Secs. 4.4 and 4.5. Trusses made of several simple trusses rigidly connected are known as *compound trusses*.

(a)   (b)

Fig. 6.17

In a compound truss the number of members m and the number of joints n are still related by the formula $m = 2n - 3$. This can be verified by observing that, if a compound truss is supported by a frictionless pin and a roller (involving three unknown reactions), the total number of unknowns is $m + 3$, and this number must be equal to the number $2n$ of equations obtained by expressing that the n pins are in equilibrium; it follows that $m = 2n - 3$. Compound trusses supported by a pin and a roller, or by an equivalent system of supports, are *statically determinate, rigid,* and *completely constrained.* This means that all of the unknown reactions and the forces in all the members can be determined by the methods of statics, and that the truss will neither collapse nor move. The forces in the members, however, cannot all be determined by the method of joints, except by solving a large number of simultaneous equations. In the case of the compound truss of Fig. 6.17a, for example, it is more efficient to pass a section through members BD, BE, and CE to determine the forces in these members.

Suppose, now, that the simple trusses ABC and DEF are connected by *four* bars BD, BE, CD, and CE (Fig. 6.18). The number of members m is now larger than $2n - 3$; the truss obtained is *overrigid,* and one of the four members BD, BE, CD, or CE is said to be *redundant.* If the truss is supported by a pin at A and a roller at F, the total number of unknowns is $m + 3$. Since $m > 2n - 3$, the number $m + 3$ of unknowns is now larger than the number $2n$ of available independent equations; the truss is *statically indeterminate.*

Finally, let us assume that the two simple trusses ABC and DEF are joined by a pin as shown in Fig. 6.19a. The number of members m is smaller than $2n - 3$. If the truss is supported by a pin at A and a roller at F, the total number of unknowns is $m + 3$. Since $m < 2n - 3$, the number $m + 3$ of unknowns is now smaller than the number $2n$ of equilibrium equations which should be satisfied; the truss is *nonrigid* and will collapse under its own weight. However, if two pins are used to support it, the truss becomes *rigid* and will not collapse (Fig. 6.19b). We note that the total number of unknowns is now $m + 4$ and is equal to the number $2n$ of equations. More generally, if the reactions at the supports involve r unknowns, the condition for a compound truss to be statically determinate, rigid, and completely constrained is $m + r = 2n$. However, while necessary, this condition is not sufficient for the equilibrium of a structure which ceases to be rigid when detached from its supports (see Sec. 6.11).

Fig. 6.18

(a)   (b)

Fig. 6.19

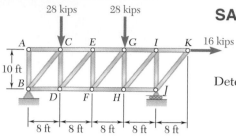

28 kips   28 kips

A   C   E   G   I   K   16 kips

10 ft

B

D   F   H   J

8 ft   8 ft   8 ft   8 ft   8 ft

SAMPLE PROBLEM 6.2

Determine the force in members EF and GI of the truss shown.

SOLUTION

Free-Body: Entire Truss. A free-body diagram of the entire truss is drawn; external forces acting on this free body consist of the applied loads and the reactions at B and J. We write the following equilibrium equations.

$+\curvearrowleft \Sigma M_B = 0$:

$$-(28 \text{ kips})(8 \text{ ft}) - (28 \text{ kips})(24 \text{ ft}) - (16 \text{ kips})(10 \text{ ft}) + J(32 \text{ ft}) = 0$$
$$J = +33 \text{ kips} \qquad \mathbf{J} = 33 \text{ kips} \uparrow$$

$\xrightarrow{+} \Sigma F_x = 0$: $\qquad B_x + 16 \text{ kips} = 0$
$$B_x = -16 \text{ kips} \qquad \mathbf{B}_x = 16 \text{ kips} \leftarrow$$

$+\curvearrowleft \Sigma M_J = 0$:
$$(28 \text{ kips})(24 \text{ ft}) + (28 \text{ kips})(8 \text{ ft}) - (16 \text{ kips})(10 \text{ ft}) - B_y(32 \text{ ft}) = 0$$
$$B_y = +23 \text{ kips} \qquad \mathbf{B}_y = 23 \text{ kips} \uparrow$$

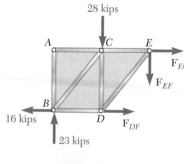

Force in Member EF. Section nn is passed through the truss so that it intersects member EF and only two additional members. After the intersected members have been removed, the left-hand portion of the truss is chosen as a free body. Three unknowns are involved; to eliminate the two horizontal forces, we write

$+\uparrow \Sigma F_y = 0$: $\qquad +23 \text{ kips} - 28 \text{ kips} - F_{EF} = 0$
$$F_{EF} = -5 \text{ kips}$$

The sense of $\mathbf{F}_{EF}$ was chosen assuming member EF to be in tension; the negative sign obtained indicates that the member is in compression.

$$F_{EF} = 5 \text{ kips } C \qquad \blacktriangleleft$$

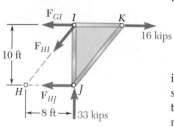

Force in Member GI. Section mm is passed through the truss so that it intersects member GI and only two additional members. After the intersected members have been removed, we choose the right-hand portion of the truss as a free body. Three unknown forces are again involved; to eliminate the two forces passing through point H, we write

$+\curvearrowleft \Sigma M_H = 0$: $\qquad (33 \text{ kips})(8 \text{ ft}) - (16 \text{ kips})(10 \text{ ft}) + F_{GI}(10 \text{ ft}) = 0$
$$F_{GI} = -10.4 \text{ kips} \qquad F_{GI} = 10.4 \text{ kips } C \qquad \blacktriangleleft$$

SAMPLE PROBLEM 6.3

Determine the force in members FH, GH, and GI of the roof truss shown.

SOLUTION

Free Body: Entire Truss. From the free-body diagram of the entire truss, we find the reactions at A and L:

$$\mathbf{A} = 12.50 \text{ kN} \uparrow \qquad \mathbf{L} = 7.50 \text{ kN} \uparrow$$

We note that

$$\tan \alpha = \frac{FG}{GL} = \frac{8 \text{ m}}{15 \text{ m}} = 0.5333 \qquad \alpha = 28.07°$$

Force in Member GI. Section nn is passed through the truss as shown. Using the portion HLI of the truss as a free body, the value of F_{GI} is obtained by writing

$$+\curvearrowleft \Sigma M_H = 0: \qquad (7.50 \text{ kN})(10 \text{ m}) - (1 \text{ kN})(5 \text{ m}) - F_{GI}(5.33 \text{ m}) = 0$$
$$F_{GI} = +13.13 \text{ kN} \qquad F_{GI} = 13.13 \text{ kN } T \quad \blacktriangleleft$$

Force in Member FH. The value of F_{FH} is obtained from the equation $\Sigma M_G = 0$. We move $\mathbf{F}_{FH}$ along its line of action until it acts at point F, where it is resolved into its x and y components. The moment of $\mathbf{F}_{FH}$ with respect to point G is now equal to $(F_{FH} \cos \alpha)(8 \text{ m})$.

$$+\curvearrowleft \Sigma M_G = 0:$$
$$(7.50 \text{ kN})(15 \text{ m}) - (1 \text{ kN})(10 \text{ m}) - (1 \text{ kN})(5 \text{ m}) + (F_{FH} \cos \alpha)(8 \text{ m}) = 0$$
$$F_{FH} = -13.81 \text{ kN} \qquad F_{FH} = 13.81 \text{ kN } C \quad \blacktriangleleft$$

Force in Member GH. We first note that

$$\tan \beta = \frac{GI}{HI} = \frac{5 \text{ m}}{\frac{2}{3}(8 \text{ m})} = 0.9375 \qquad \beta = 43.15°$$

The value of F_{GH} is then determined by resolving the force $\mathbf{F}_{GH}$ into x and y components at point G and solving the equation $\Sigma M_L = 0$.

$$+\curvearrowleft \Sigma M_L = 0: \qquad (1 \text{ kN})(10 \text{ m}) + (1 \text{ kN})(5 \text{ m}) + (F_{GH} \cos \beta)(15 \text{ m}) = 0$$
$$F_{GH} = -1.371 \text{ kN} \qquad F_{GH} = 1.371 \text{ kN } C \quad \blacktriangleleft$$

SOLVING PROBLEMS
ON YOUR OWN

The *method of joints* that you studied earlier is usually the best method to use when the forces *in all the members* of a simple truss are to be found. However, the method of sections, which was covered in this lesson, is more effective when the force *in only one member* or the forces *in a very few members* of a simple truss are desired. The method of sections must also be used when the truss *is not a simple truss.*

A. *To determine the force in a given truss member* by the method of sections, you should follow these steps.

1. Draw a free-body diagram of the entire truss, and use this diagram to determine the reactions at the supports.

2. Pass a section through three members of the truss, one of which is the desired member. After you have removed these members, you will obtain two separate portions of the truss.

3. Select one of the two portions of the truss you have obtained, and draw its free-body diagram. This diagram should include the external forces applied to the selected portion as well as the forces exerted on it by the intersected members before these members were removed.

4. You can now write three equilibrium equations which can be solved for the forces in the three intersected members.

5. An alternative approach is to write a single equation, which can be solved for the force in the desired member. To do so, first observe whether the forces exerted by the other two members on the free body are parallel or whether their lines of action intersect.
   a. If these forces are parallel, they can be eliminated by writing an equilibrium equation involving *components in a direction perpendicular* to these two forces.
   b. If their lines of action intersect at a point H, they can be eliminated by writing an equilibrium equation involving *moments about H.*

6. Keep in mind that the section you use must intersect three members only. This is because the equilibrium equations in step 4 can be solved for three unknowns only. However, you can pass a section through more than three members to find the force in one of those members if you can write an equilibrium equation containing only that force as an unknown. Examples of such special situations are found in Probs. 6.60 through 6.63. Also, although we have considered separately the method of joints and the method of sections, you should recognize that these two techniques can be used sequentially to determine the force in a given member of a truss when the force cannot be found using a single section.

B. In Probs. 6.64–6.67 you will be asked to determine the tensile forces in the active counters of trusses (the forces in the other counters are zero). To determine the forces in each pair of counters, begin by determining the reactions at the supports and then pass a section through the truss which passes through the pair of counters. After selecting the portion of the truss to be analyzed and drawing its free-body diagram, next *assume* which counter is acting and use an equilibrium equation to determine the force in that counter. If the force is tensile, your assumption was correct; if not, you must repeat the analysis using the other counter of the pair as the active counter. However, you can often deduce which is the active counter by examining the free-body diagram, as a mental summation of forces of the external forces (loads and reactions at the supports) will imply which counter must act for equilibrium to exist (remembering that the force in a counter can only be tensile).

C. About completely constrained and determinate trusses:

1. First note that any simple truss which is simply supported is a completely constrained and determinate truss.

2. To determine whether any other truss is or is not completely constrained and determinate, you first count the number m of its members, the number n of its joints, and the number r of the reaction components at its supports. You then compare the sum $m + r$ representing the number of unknowns and the product $2n$ representing the number of available independent equilibrium equations.
   a. **If $m + r < 2n$,** there are fewer unknowns than equations. Thus, some of the equations cannot be satisfied; the truss is only *partially constrained.*
   b. **If $m + r > 2n$,** there are more unknowns than equations. Thus, some of the unknowns cannot be determined; the truss is *indeterminate.*
   c. **If $m + r = 2n$,** there are as many unknowns as there are equations. This, however, does not mean that all the unknowns can be determined and that all the equations can be satisfied. To find out whether the truss is *completely* or *improperly constrained,* you should try to determine the reactions at its supports and the forces in its members. If all can be found, the truss is *completely constrained and determinate.*

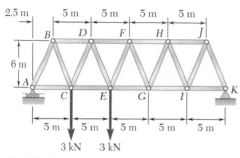

Fig. P6.42 and *P6.43*

6.42 A Warren bridge truss is loaded as shown. Determine the force in members *CE*, *DE*, and *DF*.

6.43 A Warren bridge truss is loaded as shown. Determine the force in members *EG*, *FG*, and *FH*.

6.44 A parallel chord Howe truss is loaded as shown. Determine the force in members *CE*, *DE*, and *DF*.

Fig. P6.44 and P6.45

Fig. P6.46 and *P6.47*

6.45 A parallel chord Howe truss is loaded as shown. Determine the force in members *GI*, *GJ*, and *HJ*.

6.46 A floor truss is loaded as shown. Determine the force in members *CF*, *EF*, and *EG*.

6.47 A floor truss is loaded as shown. Determine the force in members *FI*, *HI*, and *HJ*.

6.48 A pitched flat roof truss is loaded as shown. Determine the force in members *CE*, *DE*, and *DF*.

Fig. P6.48 and *P6.49*

6.49 A pitched flat roof truss is loaded as shown. Determine the force in members *EG*, *GH*, and *HJ*.

6.50 A Howe scissors roof truss is loaded as shown. Determine the force in members *DF*, *DG*, and *EG*.

Fig. P6.50 and P6.51

6.51 A Howe scissors roof truss is loaded as shown. Determine the force in members *GI*, *HI*, and *HJ*.

6.52 A Fink roof truss is loaded as shown. Determine the force in members *BD*, *CD*, and *CE*.

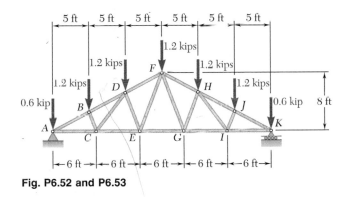

Fig. P6.52 and P6.53

6.53 A Fink roof truss is loaded as shown. Determine the force in members *FH*, *FG*, and *EG*.

6.54 A Fink roof truss is loaded as shown. Determine the force in members *DF*, *DG*, and *EG*. (*Hint:* First determine the force in member *EK*.)

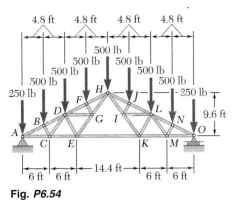

Fig. *P6.54*

6.55 A roof truss is loaded as shown. Determine the force in members *FH*, *GJ*, and *GI*.

Fig. P6.55

Fig. P6.56 and P6.57

Fig. P6.60 and *P6.61*

6.56 A stadium roof truss is loaded as shown. Determine the force in members *AB*, *AG*, and *FG*.

6.57 A stadium roof truss is loaded as shown. Determine the force in members *AE*, *EF*, and *FJ*.

6.58 A vaulted roof truss is loaded as shown. Determine the force in members *BE*, *CE*, and *DF*.

Fig. P6.58 and P6.59

6.59 A vaulted roof truss is loaded as shown. Determine the force in members *HJ*, *IJ*, and *GI*.

6.60 Determine the force in members *AF* and *EJ* of the truss shown when *P* = *Q* = 2 kips. (*Hint:* Use section *aa*.)

6.61 Determine the force in members *AF* and *EJ* of the truss shown when *P* = 2 kips and *Q* = 0. (*Hint:* Use section *aa*.)

6.62 Determine the force in members *DG* and *FH* of the truss shown. (*Hint:* Use section *aa*.)

Fig. P6.62 and P6.63

6.63 Determine the force in members *IL*, *GJ*, and *HK* of the truss shown. (*Hint:* Begin with pins *I* and *J* and then use section *bb*.)

6.64 and 6.65 The diagonal members in the center panels of the truss shown are very slender and can act only in tension; such members are known as *counters*. Determine the forces in the counters which are acting under the given loading.

Fig. P6.64

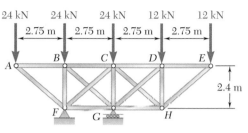

Fig. P6.65

6.66 The diagonal members in the center panel of the truss shown are very slender and can act only in tension; such members are known as *counters*. Determine the force in members *BD* and *CE* and in the counter which is acting when *P* = 3 kips.

6.67 Solve Prob. 6.66 when *P* = 15 kips.

6.68 The diagonal members *CF* and *DE* of the truss shown are very slender and can act only in tension; such members are known as *counters*. Determine the force in members *CE* and *DF* and in the counter which is acting.

Fig. P6.66

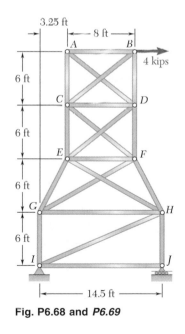

Fig. P6.68 and *P6.69*

6.69 The diagonal members *EH* and *FG* of the truss shown are very slender and can act only in tension; such members are known as *counters*. Determine the force in members *EG* and *FH* and in the counter which is acting.

6.70 through 6.75 Classify each of the structures shown as completely, partially, or improperly constrained; if completely constrained, further classify it as determinate or indeterminate. (All members can act both in tension and in compression.)

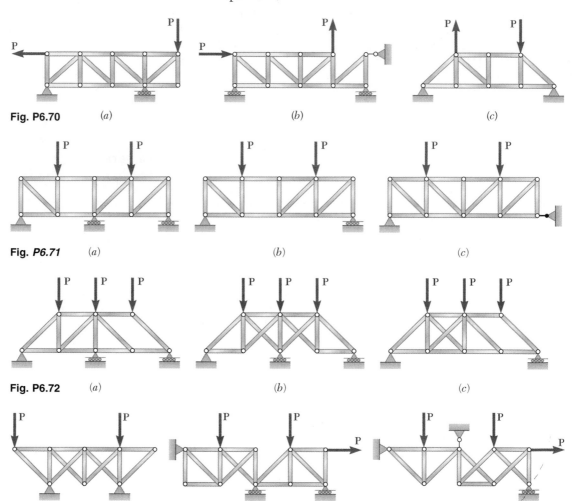

Fig. P6.70   (a)   (b)   (c)

Fig. *P6.71*   (a)   (b)   (c)

Fig. P6.72   (a)   (b)   (c)

Fig. *P6.73*   (a)   (b)   (c)

Fig. P6.74   (a)   (b)   (c)

Fig. P6.75   (a)   (b)   (c)

6.9. STRUCTURES CONTAINING MULTIFORCE MEMBERS

Under trusses, we have considered structures consisting entirely of pins and straight two-force members. The forces acting on the two-force members were known to be directed along the members themselves. We now consider structures in which at least one of the members is a *multiforce* member, that is, a member acted upon by three or more forces. These forces will generally not be directed along the members on which they act; their direction is unknown, and they should be represented therefore by two unknown components.

Frames and machines are structures containing multiforce members. *Frames* are designed to support loads and are usually stationary, fully constrained structures. *Machines* are designed to transmit and modify forces; they may or may not be stationary and will always contain moving parts.

6.10. ANALYSIS OF A FRAME

As a first example of analysis of a frame, the crane described in Sec. 6.1, which carries a given load W (Fig. 6.20*a*), will again be considered. The free-body diagram of the entire frame is shown in Fig. 6.20*b*. This diagram can be used to determine the external forces acting on the frame. Summing moments about A, we first determine the force **T** exerted by the cable; summing x and y components, we then determine the components $\mathbf{A}_x$ and $\mathbf{A}_y$ of the reaction at the pin A.

In order to determine the internal forces holding the various parts of a frame together, we must dismember the frame and draw a free-body diagram for each of its component parts (Fig. 6.20*c*). First, the two-force members should be considered. In this frame, member BE is the only two-force member. The forces acting at each end of this member must have the same magnitude, same line of action, and opposite sense (Sec. 4.6). They are therefore directed along BE and will be denoted, respectively, by $\mathbf{F}_{BE}$ and $-\mathbf{F}_{BE}$. Their sense will be arbitrarily assumed as shown in Fig. 6.20*c*; later the sign obtained for the common magnitude F_{BE} of the two forces will confirm or deny this assumption.

Next, we consider the multiforce members, that is, the members which are acted upon by three or more forces. According to Newton's third law, the force exerted at B by member BE on member AD must be equal and opposite to the force $\mathbf{F}_{BE}$ exerted by AD on BE. Similarly, the force exerted at E by member BE on member CF must be equal and opposite to the force $-\mathbf{F}_{BE}$ exerted by CF on BE. Thus the forces that the two-force member BE exerts on AD and CF are, respectively, equal to $-\mathbf{F}_{BE}$ and $\mathbf{F}_{BE}$; they have the same magnitude F_{BE} and opposite sense, and should be directed as shown in Fig. 6.20*c*.

At C two multiforce members are connected. Since neither the direction nor the magnitude of the forces acting at C is known, these forces will be represented by their x and y components. The components $\mathbf{C}_x$ and $\mathbf{C}_y$ of the force acting on member AD will be arbitrarily directed to the right and upward. Since, according to Newton's third law, the forces exerted by member CF on AD and by member AD on

Fig. 6.20

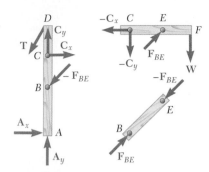

Fig. 6.20c (repeated)

CF are equal and opposite, the components of the force acting on member *CF must* be directed to the left and downward; they will be denoted, respectively, by $-\mathbf{C}_x$ and $-\mathbf{C}_y$. Whether the force $\mathbf{C}_x$ is actually directed to the right and the force $-\mathbf{C}_x$ is actually directed to the left will be determined later from the sign of their common magnitude C_x, a plus sign indicating that the assumption made was correct, and a minus sign that it was wrong. The free-body diagrams of the multiforce members are completed by showing the external forces acting at *A*, *D*, and *F*.†

The internal forces can now be determined by considering the free-body diagram of either of the two multiforce members. Choosing the free-body diagram of *CF*, for example, we write the equations $\Sigma M_C = 0$, $\Sigma M_E = 0$, and $\Sigma F_x = 0$, which yield the values of the magnitudes F_{BE}, C_y, and C_x, respectively. These values can be checked by verifying that member *AD* is also in equilibrium.

It should be noted that the pins in Fig. 6.20 were assumed to form an integral part of one of the two members they connected and so it was not necessary to show their free-body diagrams. This assumption can always be used to simplify the analysis of frames and machines. When a pin connects three or more members, however, or when a pin connects a support and two or more members, or when a load is applied to a pin, a clear decision must be made in choosing the member to which the pin will be assumed to belong. (If multiforce members are involved, the pin should be attached to one of these members.) The various forces exerted on the pin should then be clearly identified. This is illustrated in Sample Prob. 6.6.

6.11. FRAMES WHICH CEASE TO BE RIGID WHEN DETACHED FROM THEIR SUPPORTS

The crane analyzed in Sec. 6.10 was so constructed that it could keep the same shape without the help of its supports; it was therefore considered as a rigid body. Many frames, however, will collapse if detached from their supports; such frames cannot be considered as rigid bodies. Consider, for example, the frame shown in Fig. 6.21a, which consists of two members *AC* and *CB* carrying loads **P** and **Q** at their midpoints; the members are supported by pins at *A* and *B* and are connected by a pin at *C*. If detached from its supports, this frame will not maintain its shape; it should therefore be considered as made of *two distinct rigid parts AC and CB.*

---

†It is not strictly necessary to use a minus sign to distinguish the force exerted by one member on another from the equal and opposite force exerted by the second member on the first, since the two forces belong to different free-body diagrams and thus cannot easily be confused. In the Sample Problems, the same symbol is used to represent equal and opposite forces which are applied to different free bodies. It should be noted that, under these conditions, the sign obtained for a given force component will not directly relate the sense of that component to the sense of the corresponding coordinate axis. Rather, a positive sign will indicate that *the sense assumed for that component in the free-body diagram* is correct, and a negative sign will indicate that it is wrong.

The equations $\Sigma F_x = 0$, $\Sigma F_y = 0$, $\Sigma M = 0$ (about any given point) express the conditions for the *equilibrium of a rigid body* (Chap. 4); we should use them, therefore, in connection with the free-body diagrams of rigid bodies, namely, the free-body diagrams of members AC and CB (Fig. 6.21*b*). Since these members are multiforce members, and since pins are used at the supports and at the connection, the reactions at A and B and the forces at C will each be represented by two components. In accordance with Newton's third law, the components of the force exerted by CB on AC and the components of the force exerted by AC on CB will be represented by vectors of the same magnitude and opposite sense; thus, if the first pair of components consists of $\mathbf{C}_x$ and $\mathbf{C}_y$, the second pair will be represented by $-\mathbf{C}_x$ and $-\mathbf{C}_y$. We note that four unknown force components act on free body AC, while only three independent equations can be used to express that the body is in equilibrium; similarly, four unknowns, but only three equations, are associated with CB. However, only six different unknowns are involved in the analysis of the two members, and altogether six equations are available to express that the members are in equilibrium. Writing $\Sigma M_A = 0$ for free body AC and $\Sigma M_B = 0$ for CB, we obtain two simultaneous equations which may be solved for the common magnitude C_x of the components $\mathbf{C}_x$ and $-\mathbf{C}_x$ and for the common magnitude C_y of the components $\mathbf{C}_y$ and $-\mathbf{C}_y$. We then write $\Sigma F_x = 0$ and $\Sigma F_y = 0$ for each of the two free bodies, obtaining, successively, the magnitudes A_x, A_y, B_x, and B_y.

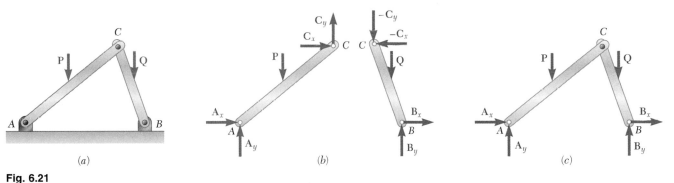

(*a*)   (*b*)   (*c*)

Fig. 6.21

It can now be observed that since the equations of equilibrium $\Sigma F_x = 0$, $\Sigma F_y = 0$, and $\Sigma M = 0$ (about any given point) are satisfied by the forces acting on free body AC, and since they are also satisfied by the forces acting on free body CB, they must be satisfied when the forces acting on the two free bodies are considered simultaneously. Since the internal forces at C cancel each other, we find that the equations of equilibrium must be satisfied by the external forces shown on the free-body diagram of the frame ACB itself (Fig. 6.21*c*), although the frame is not a rigid body. These equations can be used to determine some of the components of the reactions at A and B. We will also find, however, that *the reactions cannot be completely determined from the free-body diagram of the whole frame*. It is thus

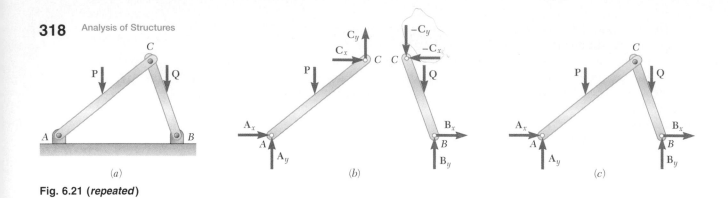

Fig. 6.21 (repeated)

necessary to dismember the frame and to consider the free-body diagrams of its component parts (Fig. 6.21*b*), even when we are interested in determining external reactions only. This is because the equilibrium equations obtained for free body *ACB are necessary conditions* for the equilibrium of a nonrigid structure, *but are not sufficient conditions.*

The method of solution outlined in the second paragraph of this section involved simultaneous equations. A more efficient method is now presented, which utilizes the free body *ACB* as well as the free bodies *AC* and *CB*. Writing $\Sigma M_A = 0$ and $\Sigma M_B = 0$ for free body *ACB*, we obtain B_y and A_y. Writing $\Sigma M_C = 0$, $\Sigma F_x = 0$, and $\Sigma F_y = 0$ for free body *AC*, we obtain, successively, A_x, C_x, and C_y. Finally, writing $\Sigma F_x = 0$ for *ACB*, we obtain B_x.

We noted above that the analysis of the frame of Fig. 6.21 involves six unknown force components and six independent equilibrium equations. (The equilibrium equations for the whole frame were obtained from the original six equations and, therefore, are not independent.) Moreover, we checked that all unknowns could be actually determined and that all equations could be satisfied. The frame considered is *statically determinate and rigid*.[†] In general, to determine whether a structure is statically determinate and rigid, we should draw a free-body diagram for each of its component parts and count the reactions and internal forces involved. We should also determine the number of independent equilibrium equations (excluding equations expressing the equilibrium of the whole structure or of groups of component parts already analyzed). If there are more unknowns than equations, the structure is *statically indeterminate*. If there are fewer unknowns than equations, the structure is *nonrigid*. If there are as many unknowns as equations, *and if all unknowns can be determined and all equations satisfied* under general loading conditions, the structure is *statically determinate and rigid*. If, however, due to an *improper arrangement* of members and supports, all unknowns cannot be determined and all equations cannot be satisfied, the structure is *statically indeterminate and nonrigid.*

[†]The word *rigid* is used here to indicate that the frame will maintain its shape as long as it remains attached to its supports.

SAMPLE PROBLEM 6.4

In the frame shown, members *ACE* and *BCD* are connected by a pin at *C* and by the link *DE*. For the loading shown, determine the force in link *DE* and the components of the force exerted at *C* on member *BCD*.

$$\alpha = \tan^{-1}\frac{80}{150} = 28.07°$$

SOLUTION

Free Body: Entire Frame. Since the external reactions involve only three unknowns, we compute the reactions by considering the free-body diagram of the entire frame.

$$+\uparrow\Sigma F_y = 0: \qquad A_y - 480 \text{ N} = 0 \qquad A_y = +480 \text{ N} \qquad \mathbf{A}_y = 480 \text{ N}\uparrow$$
$$+\gamma\Sigma M_A = 0: \qquad -(480 \text{ N})(100 \text{ mm}) + B(160 \text{ mm}) = 0$$
$$B = +300 \text{ N} \qquad \mathbf{B} = 300 \text{ N}\rightarrow$$
$$\xrightarrow{+}\Sigma F_x = 0: \qquad B + A_x = 0$$
$$300 \text{ N} + A_x = 0 \qquad A_x = -300 \text{ N} \qquad \mathbf{A}_x = 300 \text{ N}\leftarrow$$

Members. We now dismember the frame. Since only two members are connected at *C*, the components of the unknown forces acting on *ACE* and *BCD* are, respectively, equal and opposite and are assumed directed as shown. We assume that link *DE* is in tension and exerts equal and opposite forces at *D* and *E*, directed as shown.

Free Body: Member BCD. Using the free body *BCD*, we write

$$+\gamma\Sigma M_C = 0:$$
$$-(F_{DE}\sin\alpha)(250 \text{ mm}) - (300 \text{ N})(80 \text{ mm}) - (480 \text{ N})(100 \text{ mm}) = 0$$
$$F_{DE} = -561 \text{ N} \qquad\qquad F_{DE} = 561 \text{ N } C \blacktriangleleft$$
$$\xrightarrow{+}\Sigma F_x = 0: \qquad C_x - F_{DE}\cos\alpha + 300 \text{ N} = 0$$
$$C_x - (-561 \text{ N})\cos 28.07° + 300 \text{ N} = 0 \qquad C_x = -795 \text{ N}$$
$$+\uparrow\Sigma F_y = 0: \qquad C_y - F_{DE}\sin\alpha - 480 \text{ N} = 0$$
$$C_y - (-561 \text{ N})\sin 28.07° - 480 \text{ N} = 0 \qquad C_y = +216 \text{ N}$$

From the signs obtained for C_x and C_y we conclude that the force components $\mathbf{C}_x$ and $\mathbf{C}_y$ exerted on member *BCD* are directed, respectively, to the left and up. We have

$$\mathbf{C}_x = 795 \text{ N}\leftarrow, \quad \mathbf{C}_y = 216 \text{ N}\uparrow \blacktriangleleft$$

Free Body: Member ACE (Check). The computations are checked by considering the free body *ACE*. For example,

$$+\gamma\Sigma M_A = (F_{DE}\cos\alpha)(300 \text{ mm}) + (F_{DE}\sin\alpha)(100 \text{ mm}) - C_x(220 \text{ mm})$$
$$= (-561\cos\alpha)(300) + (-561\sin\alpha)(100) - (-795)(220) = 0$$

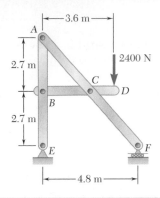

SAMPLE PROBLEM 6.5

Determine the components of the forces acting on each member of the frame shown.

SOLUTION

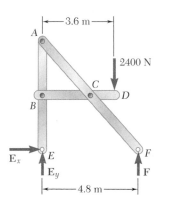

Free Body: Entire Frame. Since the external reactions involve only three unknowns, we compute the reactions by considering the free-body diagram of the entire frame.

$+\gamma\Sigma M_E = 0$: $-(2400 \text{ N})(3.6 \text{ m}) + F(4.8 \text{ m}) = 0$
                $F = +1800 \text{ N}$ $\mathbf{F} = 1800 \text{ N}\uparrow$ ◀
$+\uparrow\Sigma F_y = 0$: $-2400 \text{ N} + 1800 \text{ N} + E_y = 0$
                $E_y = +600 \text{ N}$ $\mathbf{E}_y = 600 \text{ N}\uparrow$ ◀
$\xrightarrow{+}\Sigma F_x = 0$: $E_x = 0$ $\mathbf{E}_x = 0$ ◀

Members. The frame is now dismembered; since only two members are connected at each joint, equal and opposite components are shown on each member at each joint.

Free Body: Member BCD

$+\gamma\Sigma M_B = 0$: $-(2400 \text{ N})(3.6 \text{ m}) + C_y(2.4 \text{ m}) = 0$ $C_y = +3600 \text{ N}$ ◀
$+\gamma\Sigma M_C = 0$: $-(2400 \text{ N})(1.2 \text{ m}) + B_y(2.4 \text{ m}) = 0$ $B_y = +1200 \text{ N}$ ◀
$\xrightarrow{+}\Sigma F_x = 0$: $-B_x + C_x = 0$

We note that neither B_x nor C_x can be obtained by considering only member *BCD*. The positive values obtained for B_y and C_y indicate that the force components $\mathbf{B}_y$ and $\mathbf{C}_y$ are directed as assumed.

Free Body: Member ABE

$+\gamma\Sigma M_A = 0$: $B_x(2.7 \text{ m}) = 0$ $B_x = 0$ ◀
$\xrightarrow{+}\Sigma F_x = 0$: $+B_x - A_x = 0$ $A_x = 0$ ◀
$+\uparrow\Sigma F_y = 0$: $-A_y + B_y + 600 \text{ N} = 0$
              $-A_y + 1200 \text{ N} + 600 \text{ N} = 0$ $A_y = +1800 \text{ N}$ ◀

Free Body: Member BCD. Returning now to member *BCD*, we write

$\xrightarrow{+}\Sigma F_x = 0$: $-B_x + C_x = 0$ $0 + C_x = 0$ $C_x = 0$ ◀

Free Body: Member ACF (Check). All unknown components have now been found; to check the results, we verify that member *ACF* is in equilibrium.

$+\gamma\Sigma M_C = (1800 \text{ N})(2.4 \text{ m}) - A_y(2.4 \text{ m}) - A_x(2.7 \text{ m})$
        $= (1800 \text{ N})(2.4 \text{ m}) - (1800 \text{ N})(2.4 \text{ m}) - 0 = 0$ (checks)

SAMPLE PROBLEM 6.6

A 600-lb horizontal force is applied to pin A of the frame shown. Determine the forces acting on the two vertical members of the frame.

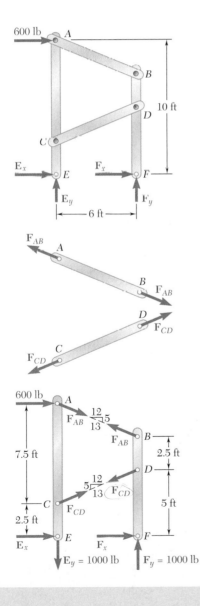

SOLUTION

Free Body: Entire Frame. The entire frame is chosen as a free body; although the reactions involve four unknowns, $\mathbf{E}_y$ and $\mathbf{F}_y$ may be determined by writing

$+\circlearrowleft\Sigma M_E = 0$: $-(600 \text{ lb})(10 \text{ ft}) + F_y(6 \text{ ft}) = 0$

           $F_y = +1000 \text{ lb}$ $\mathbf{F}_y = 1000 \text{ lb}\uparrow$ ◄

$+\uparrow\Sigma F_y = 0$: $E_y + F_y = 0$

           $E_y = -1000 \text{ lb}$ $\mathbf{E}_y = 1000 \text{ lb}\downarrow$ ◄

Members. The equations of equilibrium of the entire frame are not sufficient to determine $\mathbf{E}_x$ and $\mathbf{F}_x$. The free-body diagrams of the various members must now be considered in order to proceed with the solution. In dismembering the frame we will assume that pin A is attached to the multi-force member ACE and, thus, that the 600-lb force is applied to that member. We also note that AB and CD are two-force members.

Free Body: Member ACE

$+\uparrow\Sigma F_y = 0$: $-\frac{5}{13}F_{AB} + \frac{5}{13}F_{CD} - 1000 \text{ lb} = 0$

$+\circlearrowleft\Sigma M_E = 0$: $-(600 \text{ lb})(10 \text{ ft}) - (\frac{12}{13}F_{AB})(10 \text{ ft}) - (\frac{12}{13}F_{CD})(2.5 \text{ ft}) = 0$

Solving these equations simultaneously, we find

$$F_{AB} = -1040 \text{ lb} \qquad F_{CD} = +1560 \text{ lb} \quad ◄$$

The signs obtained indicate that the sense assumed for F_{CD} was correct and the sense for F_{AB} was incorrect. Summing now x components,

$\xrightarrow{+}\Sigma F_x = 0$: $600 \text{ lb} + \frac{12}{13}(-1040 \text{ lb}) + \frac{12}{13}(+1560 \text{ lb}) + E_x = 0$

           $E_x = -1080 \text{ lb}$ $\mathbf{E}_x = 1080 \text{ lb}\leftarrow$ ◄

Free Body: Entire Frame. Since $\mathbf{E}_x$ has been determined, we can return to the free-body diagram of the entire frame and write

$\xrightarrow{+}\Sigma F_x = 0$: $600 \text{ lb} - 1080 \text{ lb} + F_x = 0$

           $F_x = +480 \text{ lb}$ $\mathbf{F}_x = 480 \text{ lb}\rightarrow$ ◄

Free Body: Member BDF (Check). We can check our computations by verifying that the equation $\Sigma M_B = 0$ is satisfied by the forces acting on member BDF.

$+\circlearrowleft\Sigma M_B = -(\frac{12}{13}F_{CD})(2.5 \text{ ft}) + (F_x)(7.5 \text{ ft})$

           $= -\frac{12}{13}(1560 \text{ lb})(2.5 \text{ ft}) + (480 \text{ lb})(7.5 \text{ ft})$

           $= -3600 \text{ lb} \cdot \text{ft} + 3600 \text{ lb} \cdot \text{ft} = 0$ (checks)

In this lesson you learned to analyze *frames containing one or more multiforce members*. In the problems that follow you will be asked to determine the external reactions exerted on the frame and the internal forces that hold together the members of the frame.

In solving problems involving frames containing one or more multiforce members, follow these steps:

1. Draw a free-body diagram of the entire frame. Use this free-body diagram to calculate, to the extent possible, the reactions at the supports. (In Sample Prob. 6.6 only two of the four reaction components could be found from the free body of the entire frame.)

2. Dismember the frame, and draw a free-body diagram of each member.

3. Considering first the two-force members, apply equal and opposite forces to each two-force member at the points where it is connected to another member. If the two-force member is a straight member, these forces will be directed along the axis of the member. If you cannot tell at this point whether the member is in tension or compression, just *assume* that the member is in tension and *direct both of the forces away from the member*. Since these forces have the same unknown magnitude, give them both the *same name* and, to avoid any confusion later, *do not use a plus sign or a minus sign*.

4. Next, consider the multiforce members. For each of these members, show all the forces acting on the member, including *applied loads, reactions, and internal forces at connections*. The magnitude and direction of any reaction or reaction component found earlier from the free-body diagram of the entire frame should be clearly indicated.

   a. Where a multiforce member is connected to a two-force member, apply to the multiforce member a force *equal and opposite* to the force drawn on the free-body diagram of the two-force member, *giving it the same name*.

   b. Where a multiforce member is connected to another multiforce member, use *horizontal and vertical components* to represent the internal forces at that point, since neither the direction nor the magnitude of these forces is known. The direction you choose for each of the two force components exerted on the first multiforce member is arbitrary, but *you must apply equal and opposite force components of the same name* to the other multiforce member. Again, *do not use a plus sign or a minus sign*.

5. The internal forces may now be determined, as well as any *reactions* that you have not already found.

a. The free-body diagram of each of the multiforce members can provide you with *three equilibrium equations.*

b. To simplify your solution, you should seek a way to write an equation involving a single unknown. If you can locate *a point where all but one of the unknown force components intersect,* you will obtain an equation in a single unknown by summing moments about that point. *If all unknown forces except one are parallel,* you will obtain an equation in a single unknown by summing force components in a direction perpendicular to the parallel forces.

c. Since you arbitrarily chose the direction of each of the unknown forces, you cannot determine until the solution is completed whether your guess was correct. To do that, consider the *sign* of the value found for each of the unknowns: a *positive* sign means that the direction you selected was *correct;* a *negative* sign means that the direction is *opposite* to the direction you assumed.

6. To be more effective and efficient as you proceed through your solution, observe the following rules:

a. If an equation involving only one unknown can be found, write that equation and *solve it for that unknown.* Immediately *replace* that unknown wherever it appears on other free-body diagrams *by the value you have found.* Repeat this process by seeking equilibrium equations involving only one unknown until you have found all of the internal forces and unknown reactions.

b. If an equation involving only one unknown cannot be found, you may have to *solve a pair of simultaneous equations.* Before doing so, check that you have shown the values of all of the reactions that were obtained from the free-body diagram of the entire frame.

c. The total number of equations of equilibrium for the entire frame and for the individual members *will be larger than the number of unknown forces and reactions.* After you have found all the reactions and all the internal forces, you can use the remaining dependent equations to check the accuracy of your solution.

Problems

Fig. P6.76

Fig. P6.78

Fig. P6.82

6.76 For the frame and loading shown, determine the force acting on member *ABC* (*a*) at *B*, (*b*) at *C*.

6.77 Determine the force in member *AC* and the reaction at *B* when (*a*) θ = 30°, (*b*) θ = 60°.

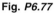

Fig. *P6.77*

6.78 A circular ring of radius 200 mm is pinned at *A* and is supported by rod *BC*, which is fitted with a collar at *C* that can be moved along the ring. For the position when θ = 35°, determine (*a*) the force in rod *BC*, (*b*) the reaction at *A*.

6.79 Solve Prob. 6.78 when θ = −20°.

6.80 For the frame and loading shown, determine the components of all forces acting on member *ABC*.

Fig. P6.80

6.81 Solve Prob. 6.80 assuming that the 80-kN load is replaced by a clockwise couple of magnitude 120 kN · m applied to member *EDC* at point *D*.

6.82 For the frame and loading shown, determine the components of all forces acting on member *ABC*.

6.83 Solve Prob. 6.82 assuming that the 4.5-kip load is replaced by a clockwise couple of magnitude 54 kip · ft applied to member *CDEF* at point *D*.

6.84 Determine the components of the reactions at A and E when a 160-lb force directed vertically downward is applied (*a*) at B, (*b*) at D.

6.85 Determine the components of the reactions at A and E when a 120-N force directed vertically downward is applied (*a*) at B, (*b*) at D.

6.86 Determine the components of the reactions at A and E when the frame is loaded by a clockwise couple of magnitude 360 lb · in. applied (*a*) at B, (*b*) at D.

Fig. P6.84 and P6.86

Fig. P6.85 and *P6.87*

6.87 Determine the components of the reactions at A and E when the frame is loaded by a counterclockwise couple of magnitude 24 N · m applied (*a*) at B, (*b*) at D.

6.88 Determine the components of the reactions at A and B when (*a*) the 240-N load is applied as shown, (*b*) the 240-N load is moved along its line of action and is applied at E.

6.89 The 192-N load can be moved along the line of action shown and applied at A, D, or E. Determine the components of the reactions at B and F when the 192-N load is applied (*a*) at A, (*b*) at D, (*c*) at E.

6.90 The 192-N load is removed and a 288 N · m clockwise couple is applied successively at A, D, and E. Determine the components of the reactions at B and F when the couple is applied (*a*) at A, (*b*) at B, (*c*) at E.

6.91 (*a*) Show that when a frame supports a pulley at A, an equivalent loading of the frame and of each of its component parts can be obtained by removing the pulley and applying at A two forces equal and parallel to the forces that the cable exerted on the pulley. (*b*) Show that if one end of the cable is attached to the frame at point B, a force of magnitude equal to the tension in the cable should also be applied at B.

Fig. P6.88

Fig. P6.89 and *P6.90*

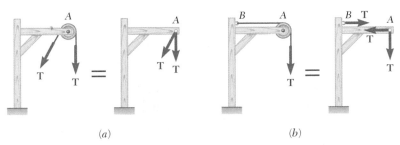

(*a*)   (*b*)

Fig. P6.91

6.92 Knowing that the pulley has a radius of 1.5 ft, determine the components of the reactions at A and E.

Fig. P6.92

6.93 Knowing that the pulley has a radius of 1.25 in., determine the components of the reactions at B and E.

Fig. P6.93

Fig. P6.94

6.94 Two 200-mm-diameter pipes (pipe *1* and pipe *2*) are supported every 3 m by a small frame like the one shown. Knowing that the combined mass per unit length of each pipe and its contents is 32 kg/m and assuming frictionless surfaces, determine the components of the reactions at A and E when $a = 0$.

6.95 Solve Prob. 6.94 when $a = 280$ mm.

6.96 The tractor and scraper units shown are connected by a vertical pin located 1.8 ft behind the tractor wheels. The distance from C to D is 2.25 ft. The center of gravity of the 20-kip tractor unit is located at G_t, while the centers of gravity of the 16-kip scraper unit and the 90-kip load are located at G_s and G_l, respectively. Knowing that the tractor is at rest with its brakes released, determine (*a*) the reactions at each of the four wheels, (*b*) the forces exerted on the tractor unit at C and D.

Fig. P6.96

6.97   Solve Prob. 6.96 assuming that the 90-kip load has been removed.

6.98 and 6.99   For the frame and loading shown, determine the components of all forces acting on member *ABE*.

Fig. P6.98

Fig. P6.99

6.100   For the frame and loading shown, determine the components of the forces acting on member *ABC* at *B* and *C*.

Fig. P6.100

6.101   For the frame and loading shown, determine the components of the forces acting on member *CDE* at *C* and *D*.

Fig. *P6.101*

6.102 Knowing that $P = 15$ N and $Q = 65$ N, determine the components of the forces exerted (a) on member $BCDF$ at C and D, (b) on member $ACEG$ at E.

Fig. P6.102 and *P6.103*

6.103 Knowing that $P = 25$ N and $Q = 55$ N, determine the components of the forces exerted (a) on member $BCDF$ at C and D, (b) on member $ACEG$ at E.

6.104 Knowing that $P = 822$ N and $Q = 0$, determine for the frame and loading shown (a) the reaction at D, (b) the force in member BF.

6.105 Knowing that $P = 0$ and $Q = 1096$ N, determine for the frame and loading shown (a) the reaction at D, (b) the force in member BF.

Fig. P6.104 and P6.105

6.106 The axis of the three-hinge arch ABC is a parabola with vertex at B. Knowing that $P = 28$ kips and $Q = 35$ kips, determine (a) the components of the reaction at A, (b) the components of the force exerted at B on segment AB.

Fig. P6.106 and *P6.107*

6.107 The axis of the three-hinge arch ABC is a parabola with vertex at B. Knowing that $P = 35$ kips and $Q = 28$ kips, determine (a) the components of the reaction at A, (b) the components of the force exerted at B on segment AB.

6.108 For the frame and loading shown, determine the reactions at *A*, *B*, *D*, and *E*. Assume that the surface at each support is frictionless.

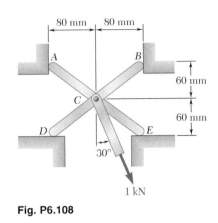

80 mm   80 mm

A   *B*

60 mm

C

60 mm

D   *E*

30°

1 kN

Fig. P6.108

6.109 through 6.111 Members *ABC* and *CDE* are pin-connected at *C* and are supported by four links. For the loading shown, determine the force in each link.

Fig. P6.109

Fig. *P6.110*

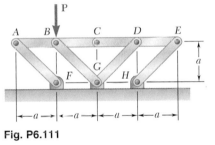

Fig. P6.111

6.112 Solve Prob. 6.109 assuming that the force **P** is replaced by a clockwise couple of moment **M**$_0$ applied to member *CDE* at *D*.

6.113 Four wooden beams, each of length 2*a*, are nailed together at their midpoints to form the support system shown. Assuming that only vertical forces are exerted at the connections, determine the vertical reactions at *A*, *D*, *E*, and *H*.

Fig. P6.113

6.114 Three wooden beams, each of length of 3*a*, are nailed together to form the support system shown. Assuming that only vertical forces are exerted at the connections, determine the vertical reactions at *A*, *D*, and *F*.

Fig. P6.114

6.115 through 6.117 Each of the frames shown consists of two L-shaped members connected by two rigid links. For each frame, determine the reactions at the supports and indicate whether the frame is rigid.

(*a*)

(*b*)

(*c*)

Fig. P6.115

(*a*)

(*b*)

(*c*)

Fig. P6.116

(*a*)

(*b*)

(*c*)

Fig. P6.117

6.12. MACHINES

Machines are structures designed to transmit and modify forces. Whether they are simple tools or include complicated mechanisms, their main purpose is to transform *input forces* into *output forces*. Consider, for example, a pair of cutting pliers used to cut a wire (Fig. 6.22*a*). If we apply two equal and opposite forces **P** and −**P** on their handles, they will exert two equal and opposite forces **Q** and −**Q** on the wire (Fig. 6.22*b*).

Fig. 6.22

Photo 6.5 The lamp shown can be placed in many positions. By considering various free bodies, the force in the springs and the internal forces at the joints can be determined.

To determine the magnitude Q of the output forces when the magnitude P of the input forces is known (or, conversely, to determine P when Q is known), we draw a free-body diagram of the pliers *alone*, showing the input forces **P** and −**P** and the *reactions* **Q** and **Q** that the wire exerts on the pliers (Fig. 6.23). However, since

Fig. 6.23

a pair of pliers forms a nonrigid structure, we must use one of the component parts as a free body in order to determine the unknown forces. Considering Fig. 6.24*a*, for example, and taking moments about A, we obtain the relation $Pa = Qb$, which defines the magnitude Q in terms of P or P in terms of Q. The same free-body diagram can be used to determine the components of the internal force at A; we find $A_x = 0$ and $A_y = P + Q$.

In the case of more complicated machines, it generally will be necessary to use several free-body diagrams and, possibly, to solve simultaneous equations involving various internal forces. The free bodies should be chosen to include the input forces and the reactions to the output forces, and the total number of unknown force components involved should not exceed the number of available independent equations. It is advisable, before attempting to solve a problem, to determine whether the structure considered is determinate. There is no point, however, in discussing the rigidity of a machine, since a machine includes moving parts and thus *must* be nonrigid.

Fig. 6.24

SAMPLE PROBLEM 6.7

A hydraulic-lift table is used to raise a 1000-kg crate. It consists of a platform and two identical linkages on which hydraulic cylinders exert equal forces. (Only one linkage and one cylinder are shown.) Members EDB and CG are each of length $2a$, and member AD is pinned to the midpoint of EDB. If the crate is placed on the table, so that half of its weight is supported by the system shown, determine the force exerted by each cylinder in raising the crate for $\theta = 60°$, $a = 0.70$ m, and $L = 3.20$ m. Show that the result obtained is independent of the distance d.

SOLUTION

The machine considered consists of the platform and the linkage. Its free-body diagram includes a force $\mathbf{F}_{DH}$ exerted by the cylinder, the weight $\frac{1}{2}\mathbf{W}$, and reactions at E and G that we assume to be directed as shown. Since more than three unknowns are involved, this diagram will not be used. The mechanism is dismembered and a free-body diagram is drawn for each of its component parts. We note that AD, BC, and CG are two-force members. We already assumed member CG to be in compression. We now assume that AD and BC are in tension; the forces exerted on them are then directed as shown. Equal and opposite vectors will be used to represent the forces exerted by the two-force members on the platform, on member BDE, and on roller C.

Free Body: Platform *ABC*.

$$\xrightarrow{+}\Sigma F_x = 0: \qquad F_{AD}\cos\theta = 0 \qquad F_{AD} = 0$$
$$+\uparrow\Sigma F_y = 0: \qquad B + C - \tfrac{1}{2}W = 0 \qquad B + C = \tfrac{1}{2}W \qquad (1)$$

Free Body: Roller *C*.
We draw a force triangle and obtain $F_{BC} = C\cot\theta$.

Free Body: Member *BDE*.
Recalling that $F_{AD} = 0$,

$$+\!\!\upharpoonleft\Sigma M_E = 0: \qquad F_{DH}\cos(\phi - 90^{\circ})a - B(2a\cos\theta) - F_{BC}(2a\sin\theta) = 0$$
$$F_{DH}a\sin\phi - B(2a\cos\theta) - (C\cot\theta)(2a\sin\theta) = 0$$
$$F_{DH}\sin\phi - 2(B + C)\cos\theta = 0$$

Recalling Eq. (1), we have

$$F_{DH} = W\frac{\cos\theta}{\sin\phi} \qquad (2)$$

and we observe that *the result obtained is independent of d.* ◀

Applying first the law of sines to triangle *EDH*, we write

$$\frac{\sin\phi}{EH} = \frac{\sin\theta}{DH} \qquad \sin\phi = \frac{EH}{DH}\sin\theta \qquad (3)$$

Using now the law of cosines, we have

$$(DH)^2 = a^2 + L^2 - 2aL\cos\theta$$
$$= (0.70)^2 + (3.20)^2 - 2(0.70)(3.20)\cos 60^{\circ}$$
$$(DH)^2 = 8.49 \qquad DH = 2.91\text{ m}$$

We also note that

$$W = mg = (1000\text{ kg})(9.81\text{ m/s}^2) = 9810\text{ N} = 9.81\text{ kN}$$

Substituting for $\sin\phi$ from (3) into (2) and using the numerical data, we write

$$F_{DH} = W\frac{DH}{EH}\cot\theta = (9.81\text{ kN})\frac{2.91\text{ m}}{3.20\text{ m}}\cot 60^{\circ}$$

$$F_{DH} = 5.15\text{ kN} \quad ◀$$

This lesson was devoted to the analysis of *machines*. Since machines are designed to transmit or modify forces, they always contain moving parts. However, the machines considered here will always be at rest, and you will be working with the set of *forces required to maintain the equilibrium of the machine.*

Known forces that act on a machine are called *input forces*. *A machine transforms the input forces into output forces,* such as the cutting forces applied by the pliers of Fig. 6.22. You will determine the output forces by finding the forces equal and opposite to the output forces that should be applied to the machine to maintain its equilibrium.

In the preceding lesson you analyzed frames; you will now use almost the same procedure to analyze machines:

1. *Draw a free-body diagram of the whole machine,* and use it to determine as many as possible of the unknown forces exerted on the machine.

2. *Dismember the machine, and draw a free-body diagram of each member.*

3. *Considering first the two-force members,* apply equal and opposite forces to each two-force member at the points where it is connected to another member. If you cannot tell at this point whether the member is in tension or in compression just *assume* that the member is in tension and *direct both of the forces away from the member.* Since these forces have the same unknown magnitude, *give them both the same name.*

4. *Next consider the multiforce members.* For each of these members, show all the forces acting on the member, including applied loads and forces, reactions, and internal forces at connections.
   a. Where a multiforce member is connected to a two-force member, apply to the multiforce member a force *equal and opposite* to the force drawn on the free-body diagram of the two-force member, *giving it the same name.*
   b. Where a multiforce member is connected to another multiforce member, use *horizontal and vertical components* to represent the internal forces at that point. The directions you choose for each of the two force components exerted on the first multiforce member are arbitrary, but *you must apply equal and opposite force components of the same name* to the other multiforce member.

5. *Equilibrium equations can be written* after you have completed the various free-body diagrams.
   a. To simplify your solution, you should, whenever possible, write and solve equilibrium equations involving single unknowns.
   b. Since you arbitrarily chose the direction of each of the unknown forces, you must determine at the end of the solution whether your guess was correct. To that effect, *consider the sign* of the value found for each of the unknowns. A *positive* sign indicates that your guess was correct, and a *negative* sign indicates that it was not.

6. *Finally, you should check your solution* by substituting the results obtained into an equilibrium equation that you have not previously used.

Problems

6.118 A 50-N force directed vertically downward is applied to the toggle vise at C. Knowing that link BD is 150 mm long and that a = 100 mm, determine the horizontal force exerted on block E.

6.119 A 50-N force directed vertically downward is applied to the toggle vise at C. Knowing that link BD is 150 mm long and that a = 200 mm, determine the horizontal force exerted on block E.

6.120 The press shown is used to emboss a small seal at E. Knowing that P = 60 lb, determine (a) the vertical component of the force exerted on the seal, (b) the reaction at A.

Fig. P6.118 and P6.119

Figs. P6.120 and P6.121

6.121 The press shown is used to emboss a small seal at E. Knowing that the vertical component of the force exerted on the seal must be 240 lb, determine (a) the required vertical force **P**, (b) the corresponding reaction at A.

6.122 The double toggle latching mechanism shown is used to hold member G against the support. Knowing that α = 60°, determine the force exerted on G.

6.123 The double toggle latching mechanism shown is used to hold member G against the support. Knowing that α = 75°, determine the force exerted on G.

6.124 For the system and loading shown, determine (a) the force **P** required for equilibrium, (b) the corresponding force in member BD, (c) the corresponding reaction at C.

Fig. P6.122 and P6.123

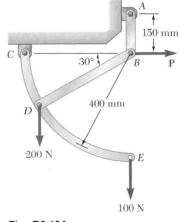

Fig. P6.124

6.125 A couple **M** of magnitude 15 kip · in. is applied to the crank of the engine system shown. For each of the two positions shown, determine the force **P** required to hold the system in equilibrium.

(a)   (b)

Fig. P6.125 and *P6.126*

6.126 A force **P** of magnitude 4 kips is applied to the piston of the engine system shown. For each of the two positions shown, determine the couple **M** required to hold the system in equilibrium.

6.127 Arm *BCD* is connected by pins to crank *AB* at *B* and to a collar at *C*. Neglecting the effect of friction, determine the couple **M** required to hold the system in equilibrium when $\theta = 0$.

6.128 Arm *BCD* is connected by pins to crank *AB* at *B* and to a collar at *C*. Neglecting the effect of friction, determine the couple **M** required to hold the system in equilibrium when $\theta = 45°$.

Fig. P6.127 and P6.128

6.129 The double toggle mechanism shown is used in a punching machine. Knowing that links *AB* and *BC* are each of length 150 mm, determine the couple **M** required to hold the system in equilibrium when $\phi = 20°$.

Fig. P6.129 and P6.130

6.130 The double toggle mechanism shown is used in a punching machine. Knowing that links *AB* and *BC* are each of length 150 mm and that $M = 75$ N · m, determine the angle ϕ if the system is in equilibrium.

6.131 and 6.132   Two rods are connected by a slider block as shown. Neglecting the effect of friction, determine the couple $\mathbf{M}_A$ required to hold the system in equilibrium.

Fig. P6.131

Fig. P6.132

6.133 and *6.134*   Rod CD is attached to the collar D and passes through a collar welded to end B of lever AB. Neglecting the effect of friction, determine the couple $\mathbf{M}$ required to hold the system in equilibrium when $\theta = 30°$.

Fig. P6.133

Fig. *P6.134*

6.135   The slab tongs shown are used to lift a 960-lb block of granite. Determine the forces exerted at D and F on tong BDF.

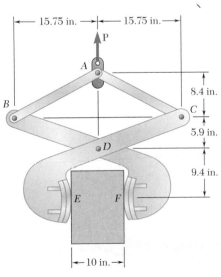

Fig. P6.135

6.136 The pallet puller shown is used to pull a loaded pallet to the rear of a truck. Knowing that $P = 420$ lb, determine the forces exerted at G and H on tong FGH.

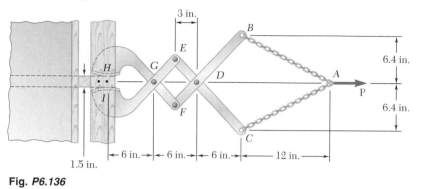

Fig. P6.136

6.137 The drum lifter shown is used to lift a steel drum. Knowing that the mass of the drum and its contents is 240 kg, determine the forces exerted at F and H on member DFH.

6.138 A small barrel having a mass of 72 kg is lifted by a pair of tongs as shown. Knowing that $a = 100$ mm, determine the forces exerted at B and D on tong ABD.

Fig. P6.137

Fig. P6.138

6.139 Determine the magnitude of the gripping forces exerted along line aa on the nut when two 240-N forces are applied to the handles as shown. Assume that pins A and D slide freely in slots cut in the jaws.

Fig. P6.139

6.140   In using the bolt cutter shown, a worker applies two 75-lb forces to the handles. Determine the magnitude of the forces exerted by the cutter on the bolt.

Fig. *P6.140*

6.141   Determine the magnitude of the gripping forces produced when two 50-lb forces are applied as shown.

Fig. P6.141

Fig. P6.142

6.142   The compound-lever pruning shears shown can be adjusted by placing pin *A* at various ratchet positions on blade *ACE*. Knowing that 1.5-kN vertical forces are required to complete the pruning of a small branch, determine the magnitude *P* of the forces that must be applied to the handles when the shears are adjusted as shown.

6.143   Determine the force **P** which must be applied to the toggle *BCD* to maintain equilibrium in the position shown.

Fig. P6.143

6.144 In the locked position shown, the toggle clamp exerts at A a vertical 1.2-kN force on the wooden block, and handle CF rests against the stop at G. Determine the force **P** required to release the clamp. (*Hint:* To release the clamp, the forces of contact at G must be zero.)

Fig. P6.144

6.145 The garden shears shown consist of two blades and two handles. The two handles are connected by pin C and the two blades are connected by pin D. The left blade and the right handle are connected by pin A; the right blade and the left handle are connected by pin B. Determine the magnitude of the forces exerted on the small branch E when two 20-lb forces are applied to the handles as shown.

Fig. P6.145

6.146 The bone rongeur shown is used in surgical procedures to cut small bones. Determine the magnitude of the forces exerted on the bone at E when two 25-lb forces are applied as shown.

Fig. P6.146

6.147 The telescoping arm *ABC* is used to provide an elevated platform for construction workers. The workers and the platform together have a mass of 240 kg and have a combined center of gravity located directly above *C*. For the position when $\theta = 24°$, determine (*a*) the force exerted at *B* by the single hydraulic cylinder *BD*, (*b*) the force exerted on the supporting carriage at *A*.

6.148 The telescoping arm *ABC* can be lowered until end *C* is close to the ground, so that workers can easily board the platform. For the position when $\theta = -18°$, determine (*a*) the force exerted at *B* by the single hydraulic cylinder *BD*, (*b*) the force exerted on the supporting carriage at *A*.

6.149 The bucket of the front-end loader shown carries a 12-kN load. The motion of the bucket is controlled by two identical mechanisms, only one of which is shown. Knowing that the mechanism shown supports one-half of the 12-kN load, determine the force exerted (*a*) by cylinder *CD*, (*b*) by cylinder *FH*.

Fig. P6.147 and P6.148

Dimensions in mm

Fig. P6.149

6.150 The motion of the bucket of the front-end loader shown is controlled by two arms and a linkage which are pin-connected at *D*. The arms are located symmetrically with respect to the central, vertical, and longitudinal plane of the loader; one arm *AFJ* and its control cylinder *EF* are shown. The single linkage *GHDB* and its control cylinder *BC* are located in the plane of symmetry. For the position shown, determine the force exerted (*a*) by cylinder *BC*, (*b*) by cylinder *EF*.

Dimensions in mm

Fig. P6.150

6.151 In the planetary gear system shown, the radius of the central gear A is $a = 20$ mm, the radius of the planetary gears is b, and the radius of the outer gear E is $(a + 2b)$. A clockwise couple of magnitude $M_A = 20$ N · m is applied to the central gear A, and a counterclockwise couple of magnitude $M_S = 100$ N · m is applied to the spider BCD. If the system is to be in equilibrium, determine (a) the required radius b of the planetary gears, (b) the couple $\mathbf{M}_E$ that must be applied to the outer gear E.

Fig. P6.151

6.152 Gears A and D are rigidly attached to horizontal shafts that are held by frictionless bearings. Determine (a) the couple $\mathbf{M}_0$ that must be applied to shaft DEF to maintain equilibrium, (b) the reactions at G and H.

*6.153 Two shafts AC and CF, which lie in the vertical xy plane, are connected by a universal joint at C. The bearings at B and D do not exert any axial force. A couple of magnitude 50 N · m (clockwise when viewed from the positive x axis) is applied to shaft CF at F. At a time when the arm of the crosspiece attached to shaft CF is horizontal, determine (a) the magnitude of the couple which must be applied to shaft AC at A to maintain equilibrium, (b) the reactions at B, D, and E. (*Hint:* The sum of the couples exerted on the crosspiece must be zero.)

Fig. P6.153

*6.154 Solve Prob. 6.153 assuming that the arm of the crosspiece attached to shaft CF is vertical.

*6.155 The large mechanical tongs shown are used to grab and lift a thick 1800-lb steel slab HJ. Knowing that slipping does not occur between the tong grips and the slab at H and J, determine the components of all forces acting on member EFH. (*Hint:* Consider the symmetry of the tongs to establish relationships between the components of the force acting at E on EFH and the components of the force acting at D on CDF.)

Fig. P6.155

In this chapter you learned to determine the *internal forces* holding together the various parts of a structure.

The first half of the chapter was devoted to the analysis of *trusses*, that is, to the analysis of structures consisting of *straight members connected at their extremities only*. The members being slender and unable to support lateral loads, all the loads must be applied at the joints; a truss may thus be assumed to consist of *pins and two-force members* [Sec. 6.2].

Analysis of trusses

A truss is said to be *rigid* if it is designed in such a way that it will not greatly deform or collapse under a small load. A triangular truss consisting of three members connected at three joints is clearly a rigid truss (Fig. 6.25*a*) and so will be the truss obtained by adding two new members to the first one and connecting them at a new joint (Fig. 6.25*b*). Trusses obtained by repeating this procedure are called *simple trusses*. We may check that in a simple truss the total number of members is $m - 2n - 3$, where n is the total number of joints [Sec. 6.3].

Simple trusses

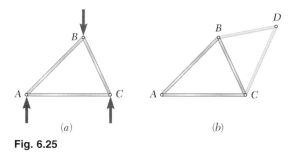

Fig. 6.25

The forces in the various members of a simple truss can be determined by the *method of joints* [Sec. 6.4]. First, the reactions at the supports can be obtained by considering the entire truss as a free body. The free-body diagram of each pin is then drawn, showing the forces exerted on the pin by the members or supports it connects. Since the members are straight two-force members, the force exerted by a member on the pin is directed along that member, and only the magnitude of the force is unknown. It is always possible in the case of a simple truss to draw the free-body diagrams of the pins in such an order that only two unknown forces are included in each diagram. These forces can be obtained from the corresponding two equilibrium equations or—if only three forces are involved—from the corresponding force triangle. If the force

Method of joints

exerted by a member on a pin is directed toward that pin, the member is in *compression;* if it is directed away from the pin, the member is in *tension* [Sample Prob. 6.1]. The analysis of a truss is sometimes expedited by first recognizing *joints under special loading conditions* [Sec. 6.5]. The method of joints can also be extended to the analysis of three-dimensional or *space trusses* [Sec. 6.6].

Method of sections

The *method of sections* is usually preferred to the method of joints when the force in only one member—or very few members—of a truss is desired [Sec. 6.7]. To determine the force in member *BD* of the truss of Fig. 6.26*a*, for example, we *pass a section* through members *BD, BE,* and *CE,* remove these members, and use the portion *ABC* of the truss as a free body (Fig. 6.26*b*). Writing $\Sigma M_E = 0$, we determine the magnitude of the force $\mathbf{F}_{BD}$, which represents the force in member *BD*. A positive sign indicates that the member is in *tension;* a negative sign indicates that it is in *compression* [Sample Probs. 6.2 and 6.3].

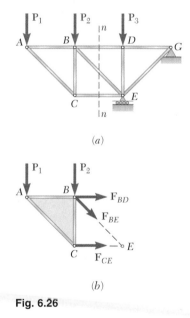

(a)

(b)

Fig. 6.26

Compound trusses

The method of sections is particularly useful in the analysis of *compound trusses,* that is, trusses which cannot be constructed from the basic triangular truss of Fig. 6.25*a* but which can be obtained by rigidly connecting several simple trusses [Sec. 6.8]. If the component trusses have been properly connected (for example, one pin and one link, or three nonconcurrent and nonparallel links) and if the resulting structure is properly supported (for example, one pin and one roller), the compound truss is *statically determinate, rigid, and completely constrained.* The following necessary—but not sufficient—condition is then satisfied: $m + r = 2n$, where m is the number of members, r is the number of unknowns representing the reactions at the supports, and n is the number of joints.

The second part of the chapter was devoted to the analysis of *frames and machines*. Frames and machines are structures which contain *multiforce members*, that is, members acted upon by three or more forces. Frames are designed to support loads and are usually stationary, fully constrained structures. Machines are designed to transmit or modify forces and always contain moving parts [Sec. 6.9].

Frames and machines

To *analyze a frame*, we first consider the *entire frame as a free body* and write three equilibrium equations [Sec. 6.10]. If the frame remains rigid when detached from its supports, the reactions involve only three unknowns and may be determined from these equations [Sample Probs. 6.4 and 6.5]. On the other hand, if the frame ceases to be rigid when detached from its supports, the reactions involve more than three unknowns and cannot be completely determined from the equilibrium equations of the frame [Sec. 6.11; Sample Prob. 6.6].

Analysis of a frame

We then *dismember the frame* and identify the various members as either two-force members or multiforce members; pins are assumed to form an integral part of one of the members they connect. We draw the free-body diagram of each of the multiforce members, noting that when two multiforce members are connected to the same two-force member, they are acted upon by that member with *equal and opposite forces of unknown magnitude but known direction*. When two multiforce members are connected by a pin, they exert on each other *equal and opposite forces of unknown direction*, which should be represented by *two unknown components*. The equilibrium equations obtained from the free-body diagrams of the multiforce members can then be solved for the various internal forces [Sample Probs. 6.4 and 6.5]. The equilibrium equations can also be used to complete the determination of the reactions at the supports [Sample Prob. 6.6]. Actually, if the frame is *statically determinate and rigid*, the free-body diagrams of the multiforce members could provide as many equations as there are unknown forces (including the reactions) [Sec. 6.11]. However, as suggested above, it is advisable to first consider the free-body diagram of the entire frame to minimize the number of equations that must be solved simultaneously.

Multiforce members

To *analyze a machine*, we dismember it and, following the same procedure as for a frame, draw the free-body diagram of each of the multiforce members. The corresponding equilibrium equations yield the *output forces* exerted by the machine in terms of the *input forces* applied to it, as well as the *internal forces* at the various connections [Sec. 6.12; Sample Prob. 6.7].

Analysis of a machine

Fig. P6.156

6.156 For the frame and loading shown, determine the force acting on member ABC (a) at B, (b) at C.

6.157 Determine the force in each member of the truss shown. State whether each member is in tension or compression.

Fig. P6.157

Fig. P6.158

6.158 The tongs shown are used to apply a total upward force of 45 kN on a pipe cap. Determine the forces exerted at D and F on tong ADF.

6.159 A stadium roof truss is loaded as shown. Determine the force in members BC, BH, and GH.

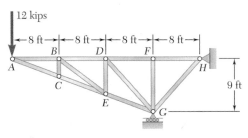

Fig. P6.159 and P6.160

6.160 A stadium roof truss is loaded as shown. Determine the force in members EJ, FJ, and EI.

6.161 For the frame and loading shown, determine the components of the forces acting on member *DABC* at *B* and at *D*.

Fig. P6.161

6.162 Using the method of joints, determine the force in each member of the truss shown. State whether each member is in tension or compression.

Fig. P6.162

Fig. *P6.163*

6.163 For the frame and loading shown, determine the components of the forces acting on member *CFE* at *C* and at *F*.

6.164 A Mansard roof truss is loaded as shown. Determine the force in members *DF*, *DG*, and *EG*.

Fig. P6.164 and P6.165

6.165 A Mansard roof truss is loaded as shown. Determine the force in members *GI*, *HI*, and *HJ*.

6.166 Rod CD is fitted with a collar at D that can be moved along rod AB, which is bent in the shape of a circular arc. For the position when $\theta = 30°$, determine (a) the force in rod CD, (b) the reaction at B.

Fig. P6.166

6.167 A log weighing 800 lb is lifted by a pair of tongs as shown. Determine the forces exerted at E and at F on tong DEF.

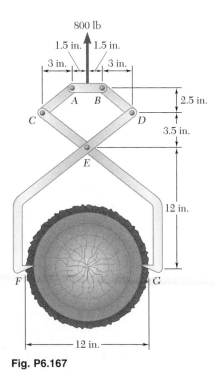

Fig. P6.167

Computer Problems

6.C1 For the loading shown, determine the force in each member of the truss as a function of the dimension a. Plot the force in each member for 40 in. $\leq a \leq$ 240 in., with tensile forces plotted as positive and compressive forces as negative.

Fig. P6.C1

6.C2 In the Fink truss shown, a single 3.75-kip load is to be applied to the top chord of the truss at one of the numbered joints. Calculate the force in member CD as the load is successively applied at joints 1, 2, . . . , 9.

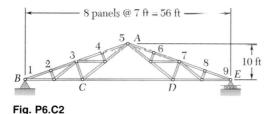

Fig. P6.C2

6.C3 The floor of a bridge will rest on stringers that will be simply supported by transverse floor beams, as in Fig 6.3. The ends of the beams will be connected to the upper joints of two trusses, one of which is shown in Fig. P6.C3. As part of the design of the bridge, it is desired to simulate the effect on this truss of driving a 3-kip truck over the bridge. Knowing that the distance between the truck's axles is $b = 7.5$ ft and assuming that the weight of the truck is equally distributed over its four wheels, use computational software to calculate and plot the forces created by the truck in members BH and GH as a function of x for $0 \leq x \leq 57.5$ ft. From the results obtained, determine (a) the maximum tensile force in BH, (b) the maximum compressive force in BH, (c) the maximum tensile force in GH. Indicate in each case the corresponding value of x.

Fig. P6.C3

Fig. P6.C4

6.C4 Knowing that the radius of the pulley is 300 mm, plot the magnitude of the reactions at A and G as functions of θ for 0 ≤ θ ≤ 90°.

6.C5 The design of a robotic system calls for the two-rod mechanism shown. Rods AC and BD are connected by a slider block D as shown. Neglecting the effect of friction, calculate and plot the couple $\mathbf{M}_A$ required to hold the rods in equilibrium for values of θ from 0 to 120°. For the same values of θ, calculate and plot the magnitude of the force $\mathbf{F}$ exerted by rod AC on the slider block.

Fig. P6.C5

6.C6 The magnitude of the force $\mathbf{P}$ applied to the piston of an engine system during one revolution of crank AB is shown in the figure. Plot the magnitude of the couple $\mathbf{M}$ required to hold the system in equilibrium as a function of θ for 0 ≤ θ ≤ 2π.

Fig. P6.C6

6.C7 The shelf ABC is held horizontally by a self-locking brace that consists of two parts BDE and EDF hinged at E and bearing against each other at D. Determine the force $\mathbf{P}$ required to release the brace as a function of the angle θ. Knowing that the mass of the shelf is 18 kg, plot the magnitude of $\mathbf{P}$ as a function of θ for 0° ≤ θ ≤ 90°.

Fig. P6.C7

6.C8 In the mechanism shown, the position of boom *AC* is controlled by arm *BD*. For the loading shown, calculate and plot the reaction at *A* and the couple **M** required to hold the system in equilibrium as functions of *θ* for −30° ≤ *θ* ≤ 90°. As part of the design process of the mechanism, determine (*a*) the value of *θ* for which *M* is maximum and the corresponding value of *M*, (*b*) the value of *θ* for which the reaction at *A* is maximum and the corresponding magnitude of this reaction.

Fig. P6.C8

6.C9 Rod *CD* is attached to collar *D* and passes through a collar welded to end *B* of lever *AB*. As an initial step in the design of lever *AB*, use computational software to calculate and plot the magnitude *M* of the couple required to hold the system in equilibrium as a function of *θ* for 15° ≤ *θ* ≤ 90°. Determine the value of *θ* for which *M* is minimum and the corresponding value of *M*.

6.C10 The tree trimmer shown is used to prune a 25-mm-diameter branch. Knowing that the tension in the rope is 70 N and neglecting the radii of the pulleys, plot the magnitude of the force exerted on the branch by blade *BC* as a function of *θ* for 0 ≤ *θ* ≤ 90°, where the cutting edge of the blade is vertical and just contacts the branch when *θ* = 0. (*Hint:* Assume that the line of action of the force acting on the branch passes through the center of the branch and is perpendicular to the blade and that portions *AD* and *AE* of the rope are parallel to a line drawn through points *A* and *E*.)

Fig. P6.C9

Fig. P6.C10

Forces in Beams
and Cables

The Gateshead Millennium Bridge for pedestrians in Newcastle-upon-Tyne, England is unique in both appearance and how it rotates to permit the passage of taller ships. The determination of the internal forces in a member, such as those in the curved footpath and supporting arch of the bridge, is the focus of this chappter.

FORCES IN BEAMS AND CABLES

7.1 Introduction
7.2 Internal Forces in Members
Beams
7.3 Various Types of Loading and Support
7.4 Shear and Bending Moment in a Beam
7.5 Shear and Bending-Moment Diagrams
7.6 Relations among Load, Shear, and Bending Moment
Cables
7.7 Cables with Concentrated Loads
7.8 Cables with Distributed Loads
7.9 Parabolic Cable
7.10 Catenary

*7.1. INTRODUCTION

In preceding chapters, two basic problems involving structures were considered: (1) determining the external forces acting on a structure (Chap. 4) and (2) determining the forces which hold together the various members forming a structure (Chap. 6). The problem of determining the internal forces which hold together the various parts of a given member will now be considered.

We will first analyze the internal forces in the members of a frame, such as the crane considered in Secs. 6.1 and 6.10, noting that whereas the internal forces in a straight two-force member can produce only *tension* or *compression* in that member, the internal forces in any other type of member usually produce *shear* and *bending* as well.

Most of this chapter will be devoted to the analysis of the internal forces in two important types of engineering structures, namely,

1. *Beams,* which are usually long, straight prismatic members designed to support loads applied at various points along the member.
2. *Cables,* which are flexible members capable of withstanding only tension and are designed to support either concentrated or distributed loads. Cables are used in many engineering applications, such as suspension bridges and transmission lines.

*7.2. INTERNAL FORCES IN MEMBERS

Let us first consider a *straight two-force member AB* (Fig. 7.1*a*). From Sec. 4.6, we know that the forces **F** and −**F** acting at *A* and *B*, respectively, must be directed along *AB* in opposite sense and have the same magnitude *F.* Now, let us cut the member at *C.* To maintain the equilibrium of the free bodies *AC* and *CB* thus obtained, we must apply to *AC* a force −**F** equal and opposite to **F**, and to *CB* a force **F** equal and opposite to −**F** (Fig. 7.1*b*). These new forces are directed along *AB* in opposite sense and have the same magnitude *F.* Since the two parts *AC* and *CB* were in equilibrium before the member was cut, *internal forces* equivalent to these new forces must have existed in the member itself. We conclude that in the case of a straight two-force member, the internal forces that the two portions of the member exert on each other are equivalent to *axial forces.* The common magnitude *F* of these forces does not depend upon the location of the section *C* and is referred to as the *force in member AB.* In the case considered, the member is in tension and will elongate under the action of the internal forces. In the case represented in Fig. 7.2, the member is in compression and will decrease in length under the action of the internal forces.

Next, let us consider a *multiforce member.* Take, for instance, member *AD* of the crane analyzed in Sec. 6.10. This crane is shown again in Fig. 7.3*a,* and the free-body diagram of member *AD* is drawn in Fig. 7.3*b.* We now cut member *AD* at *J* and draw a free-body diagram for each of the portions *JD* and *AJ* of the member (Fig. 7.3*c* and *d*). Considering the free body *JD,* we find that its equilibrium will

Fig. 7.1

(a)   (b)

Fig. 7.2

(a)   (b)

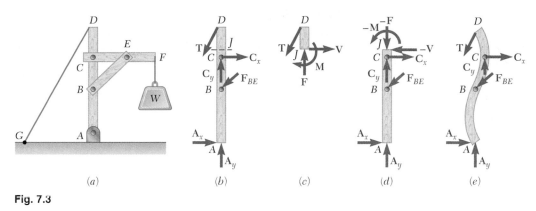

(a) (b) (c) (d) (e)

Fig. 7.3

be maintained if we apply at J a force $\mathbf{F}$ to balance the vertical component of $\mathbf{T}$, a force $\mathbf{V}$ to balance the horizontal component of $\mathbf{T}$, and a couple $\mathbf{M}$ to balance the moment of $\mathbf{T}$ about J. Again we conclude that internal forces must have existed at J before member AD was cut. The internal forces acting on the portion JD of member AD are equivalent to the force-couple system shown in Fig. 7.3c. According to Newton's third law, the internal forces acting on AJ must be equivalent to an equal and opposite force-couple system, as shown in Fig. 7.3d. It is clear that the action of the internal forces in member AD *is not limited to producing tension or compression* as in the case of straight two-force members; the internal forces *also produce shear and bending*. The force $\mathbf{F}$ is an *axial force;* the force $\mathbf{V}$ is called a *shearing force;* and the moment $\mathbf{M}$ of the couple is known as the *bending moment at J*. We note that when determining internal forces in a member, we should clearly indicate on which portion of the member the forces are supposed to act. The deformation which will occur in member AD is sketched in Fig. 7.3e. The actual analysis of such a deformation is part of the study of mechanics of materials.

It should be noted that in a *two-force member which is not straight*, the internal forces are also equivalent to a force-couple system. This is shown in Fig. 7.4, where the two-force member ABC has been cut at D.

Photo 7.1 The design of the shaft of a circular saw must account for the internal forces resulting from the forces applied to the teeth of the blade. At a given point in the shaft, these internal forces are equivalent to a force-couple system consisting of axial and shearing forces and a couple representing the bending and torsional moments.

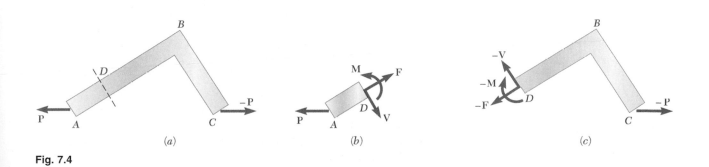

(a) (b) (c)

Fig. 7.4

SAMPLE PROBLEM 7.1

In the frame shown, determine the internal forces (*a*) in member *ACF* at point *J*, (*b*) in member *BCD* at point *K*. This frame has been previously considered in Sample Prob. 6.5.

SOLUTION

Reactions and Forces at Connections.   The reactions and the forces acting on each member of the frame are determined; this has been previously done in Sample Prob. 6.5, and the results are repeated here.

a. **Internal Forces at *J*.**   Member *ACF* is cut at point *J*, and the two parts shown are obtained. The internal forces at *J* are represented by an equivalent force-couple system and can be determined by considering the equilibrium of either part. Considering the *free body AJ*, we write

$+\curvearrowleft\Sigma M_J = 0$: $-(1800 \text{ N})(1.2 \text{ m}) + M = 0$
              $M = +2160 \text{ N} \cdot \text{m}$ $\mathbf{M} = 2160 \text{ N} \cdot \text{m} \, \curvearrowleft$ ◄

$+\searrow\Sigma F_x = 0$: $F - (1800 \text{ N}) \cos 41.7° = 0$
              $F = +1344 \text{ N}$ $\mathbf{F} = 1344 \text{ N} \searrow$ ◄

$+\nearrow\Sigma F_y = 0$: $-V + (1800 \text{ N}) \sin 41.7° = 0$
              $V = +1197 \text{ N}$ $\mathbf{V} = 1197 \text{ N} \swarrow$ ◄

The internal forces at *J* are therefore equivalent to a couple **M**, an axial force **F**, and a shearing force **V**. The internal force-couple system acting on part *JCF* is equal and opposite.

b. **Internal Forces at *K*.**   We cut member *BCD* at *K* and obtain the two parts shown. Considering the *free body BK*, we write

$+\curvearrowleft\Sigma M_K = 0$: $(1200 \text{ N})(1.5 \text{ m}) + M = 0$
              $M = -1800 \text{ N} \cdot \text{m}$ $\mathbf{M} = 1800 \text{ N} \cdot \text{m} \, \curvearrowright$ ◄

$\xrightarrow{+}\Sigma F_x = 0$: $F = 0$ $\mathbf{F} = 0$ ◄

$+\uparrow\Sigma F_y = 0$: $-1200 \text{ N} - V = 0$
              $V = -1200 \text{ N}$ $\mathbf{V} = 1200 \text{ N} \uparrow$ ◄

In this lesson you learned to determine the internal forces in the member of a frame. The internal forces at a given point in a *straight two-force member* reduce to an axial force, but in all other cases, they are equivalent to a *force-couple system* consisting of an *axial force* **F**, a *shearing force* **V**, and a couple **M** representing the *bending moment* at that point.

To determine the internal forces at a given point *J* of the member of a frame, you should take the following steps.

1. Draw a free-body diagram of the entire frame, and use it to determine as many of the reactions at the supports as you can.

2. Dismember the frame, and draw a free-body diagram of each of its members. Write as many equilibrium equations as are necessary to find all the forces acting on the member on which point *J* is located.

3. Cut the member at point J, and draw a free-body diagram of each of the two portions of the member that you have obtained, applying to each portion at point *J* the force components and couple representing the internal forces exerted by the other portion. Note that these force components and couples are equal in magnitude and opposite in sense.

4. Select one of the two free-body diagrams you have drawn, and use it to write three equilibrium equations for the corresponding portion of the member.
   a. Summing moments about J and equating them to zero will yield the bending moment at point *J*.
   b. Summing components in directions parallel and perpendicular to the member at *J* and equating them to zero will yield, respectively, the axial and shearing force.

5. When recording your answers, be sure to specify the portion of the member you have used, since the forces and couples acting on the two portions have opposite senses.

Since the solutions of the problems in this lesson require the determination of the forces exerted on each other by the various members of a frame, be sure to review the methods used in Chap. 6 to solve this type of problem. When frames involve pulleys and cables, for instance, remember that the forces exerted by a pulley on the member of the frame to which it is attached have the same magnitude and direction as the forces exerted by the cable on the pulley [Prob. 6.91].

Problems

Fig. P7.5 and *P7.6*

7.1 and 7.2 Determine the internal forces (axial force, shearing force, and bending moment) at point *J* of the structure indicated:
  7.1 Frame and loading of Prob. 6.77.
  7.2 Frame and loading of Prob. 6.76.

7.3 For the frame and loading of Prob. 6.80, determine the internal forces at a point *J* located halfway between points *A* and *B*.

7.4 For the frame and loading of Prob. 6.101, determine the internal forces at a point *J* located halfway between points *A* and *B*.

7.5 Determine the internal forces at point *J* of the structure shown.

7.6 Determine the internal forces at point *K* of the structure shown.

7.7 A semicircular rod is loaded as shown. Determine the internal forces at point *J*.

Fig. P7.7 and *P7.8*

7.8 A semicircular rod is loaded as shown. Determine the internal forces at point *K*.

7.9 An archer aiming at a target is pulling with a 210-N force on the bowstring. Assuming that the shape of the bow can be approximated by a parabola, determine the internal forces at point *J*.

7.10 For the bow of Prob. 7.9, determine the magnitude and location of the maximum (*a*) axial force, (*b*) shearing force, (*c*) bending moment.

Fig. P7.9

7.11 A semicircular rod is loaded as shown. Determine the internal forces at point J knowing that $\theta = 30°$.

7.12 A semicircular rod is loaded as shown. Determine the magnitude and location of the maximum bending moment in the rod.

Fig. P7.11 and *P7.12*

Fig. P7.13 and P7.14

7.13 Two members, each consisting of straight and 168-mm-radius quarter-circle portions, are connected as shown and support a 480-N load at D. Determine the internal forces at point J.

7.14 Two members, each consisting of straight and 168-mm-radius quarter-circle portions, are connected as shown and support a 480-N load at D. Determine the internal forces at point K.

7.15 Knowing that the radius of each pulley is 7.2 in. and neglecting friction, determine the internal forces at point J of the frame shown.

7.16 Knowing that the radius of each pulley is 7.2 in. and neglecting friction, determine the internal forces at point K of the frame shown.

Fig. P7.15 and P7.16

Fig. *P7.17* and *P7.18*

7.17 Knowing that the radius of each pulley is 7.2 in. and neglecting friction, determine the internal forces at point J of the frame shown.

7.18 Knowing that the radius of each pulley is 7.2 in. and neglecting friction, determine the internal forces at point K of the frame shown.

Fig. P7.19 and P7.20

7.19 A 140-mm-diameter pipe is supported every 3 m by a small frame consisting of two members as shown. Knowing that the combined mass per unit length of the pipe and its contents is 28 kg/m and neglecting the effect of friction, determine the internal forces at point *J*.

7.20 A 140-mm-diameter pipe is supported every 3 m by a small frame consisting of two members as shown. Knowing that the combined mass per unit length of the pipe and its contents is 28 kg/m and neglecting the effect of friction, determine the internal forces at point *K*.

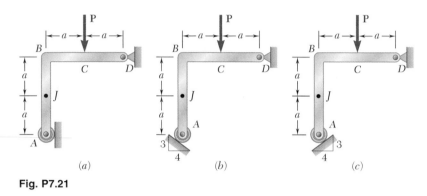

Fig. P7.21

7.21 and 7.22 A force **P** is applied to a bent rod which is supported by a roller and a pin and bracket. For each of the three cases shown, determine the internal forces at point *J*.

Fig. P7.22

Fig. P7.23 and P7.24

7.23 A rod of weight *W* and uniform cross section is bent into the circular arc of radius *r* shown. Determine the bending moment at point *J* when $\theta = 30°$.

7.24 A rod of weight *W* and uniform cross section is bent into the circular arc of radius *r* shown. Determine the bending moment at point *J* when $\theta = 120°$.

7.25 and *7.26* A quarter-circular rod of weight W and uniform cross section is supported as shown. Determine the bending moment at point J when $\theta = 30°$.

Fig. P7.25

Fig. *P7.26*

7.27 For the rod of Prob. 7.26, determine the magnitude and location of the maximum bending moment.

7.28 For the rod of Prob. 7.25, determine the magnitude and location of the maximum bending moment.

BEAMS

*7.3. VARIOUS TYPES OF LOADING AND SUPPORT

A structural member designed to support loads applied at various points along the member is known as a *beam*. In most cases, the loads are perpendicular to the axis of the beam and will cause only shear and bending in the beam. When the loads are not at a right angle to the beam, they will also produce axial forces in the beam.

Beams are usually long, straight prismatic bars. Designing a beam for the most effective support of the applied loads is a two-part process: (1) determining the shearing forces and bending moments produced by the loads and (2) selecting the cross section best suited to resist the shearing forces and bending moments determined in the first part. Here we are concerned with the first part of the problem of beam design. The second part belongs to the study of mechanics of materials.

A beam can be subjected to *concentrated loads* P_1, P_2, ..., expressed in newtons, pounds, or their multiples kilonewtons and kips (Fig. 7.5*a*), to a *distributed load w,* expressed in N/m, kN/m, lb/ft, or kips/ft (Fig. 7.5*b*), or to a combination of both. When the load w per unit length has a constant value over part of the beam (as between A and B in Fig. 7.5*b*), the load is said to be *uniformly distributed* over that part of the beam. The determination of the reactions at the supports is considerably simplified if distributed loads are replaced by equivalent concentrated loads, as explained in Sec. 5.8. This substitution, however, should not be performed, or at least should be performed with care, when internal forces are being computed (see Sample Prob. 7.3).

Beams are classified according to the way in which they are supported. Several types of beams frequently used are shown in

(*a*) Concentrated loads

(*b*) Distributed load

Fig. 7.5

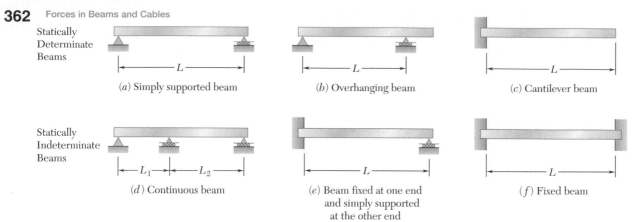

Statically
Determinate
Beams

(a) Simply supported beam

(b) Overhanging beam

(c) Cantilever beam

Statically
Indeterminate
Beams

(d) Continuous beam

(e) Beam fixed at one end
and simply supported
at the other end

(f) Fixed beam

Fig. 7.6

(a)

(b)

Fig. 7.7

Photo 7.2 The internal forces in the beams of
the overpass shown vary as the truck crosses the
overpass.

Fig. 7.6. The distance L between supports is called the *span*. It should
be noted that the reactions will be determinate if the supports involve
only three unknowns. If more unknowns are involved, the reactions
will be statically indeterminate and the methods of statics will not be
sufficient to determine the reactions; the properties of the beam with
regard to its resistance to bending must then be taken into consideration. Beams supported by two rollers are not shown here; they are
only partially constrained and will move under certain loadings.

Sometimes two or more beams are connected by hinges to form
a single continuous structure. Two examples of beams hinged at a
point H are shown in Fig. 7.7. It will be noted that the reactions at
the supports involve four unknowns and cannot be determined from
the free-body diagram of the two-beam system. They can be determined, however, by considering the free-body diagram of each beam
separately; six unknowns are involved (including two force components at the hinge), and six equations are available.

*7.4. SHEAR AND BENDING MOMENT IN A BEAM

Consider a beam AB subjected to various concentrated and distributed loads (Fig. 7.8a). We propose to determine the shearing force
and bending moment at any point of the beam. In the example considered here, the beam is simply supported, but the method used
could be applied to any type of statically determinate beam.

First we determine the reactions at A and B by choosing the entire beam as a free body (Fig. 7.8b); writing $\Sigma M_A = 0$ and $\Sigma M_B = 0$,
we obtain, respectively, $\mathbf{R}_B$ and $\mathbf{R}_A$.

To determine the internal forces at C, we cut the beam at C and
draw the free-body diagrams of the portions AC and CB of the beam
(Fig. 7.8c). Using the free-body diagram of AC, we can determine the
shearing force $\mathbf{V}$ at C by equating to zero the sum of the vertical components of all forces acting on AC. Similarly, the bending moment $\mathbf{M}$
at C can be found by equating to zero the sum of the moments about
C of all forces and couples acting on AC. Alternatively, we could use
the free-body diagram of CB† and determine the shearing force $\mathbf{V}'$
and the bending moment $\mathbf{M}'$ by equating to zero the sum of the

---

†The force and couple representing the internal forces acting on CB will now be de-
noted by $\mathbf{V}'$ and $\mathbf{M}'$, rather than by $-\mathbf{V}$ and $-\mathbf{M}$ as done earlier, in order to avoid confusion when applying the sign convention which we are about to introduce.

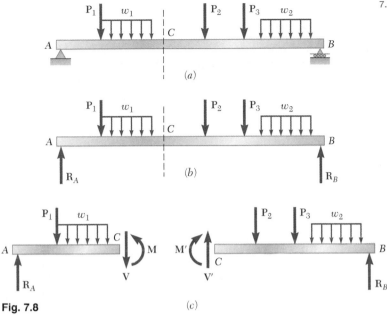

Fig. 7.8

vertical components and the sum of the moments about C of all forces and couples acting on CB. While this choice of free bodies may facilitate the computation of the numerical values of the shearing force and bending moment, it makes it necessary to indicate on which portion of the beam the internal forces considered are acting. If the shearing force and bending moment are to be computed at every point of the beam and efficiently recorded, we must find a way to avoid having to specify every time which portion of the beam is used as a free body. We shall adopt, therefore, the following conventions:

In determining the shearing force in a beam, *it will always be assumed* that the internal forces $\mathbf{V}$ and $\mathbf{V}'$ are directed as shown in Fig. 7.8c. A positive value obtained for their common magnitude V will indicate that this assumption was correct and that the shearing forces are actually directed as shown. A negative value obtained for V will indicate that the assumption was wrong and that the shearing forces are directed in the opposite way. Thus, only the magnitude V, together with a plus or minus sign, needs to be recorded to define completely the shearing forces at a given point of the beam. The scalar V is commonly referred to as the *shear* at the given point of the beam.

Similarly, *it will always be assumed* that the internal couples $\mathbf{M}$ and $\mathbf{M}'$ are directed as shown in Fig. 7.8c. A positive value obtained for their magnitude M, commonly referred to as the bending moment, will indicate that this assumption was correct, and a negative value will indicate that it was wrong. Summarizing the sign conventions we have presented, we state:

The shear V and the bending moment M at a given point of a beam are said to be positive when the internal forces and couples acting on each portion of the beam are directed as shown in Fig. 7.9a.

These conventions can be more easily remembered if we note that:

1. *The shear at C is positive when the **external** forces (loads and reactions) acting on the beam tend to shear off the beam at C as indicated in Fig. 7.9b.*

(a) Internal forces at section
(positive shear and positive bending moment)

(b) Effect of external forces
(positive shear)

(c) Effect of external forces
(positive bending moment)

Fig. 7.9

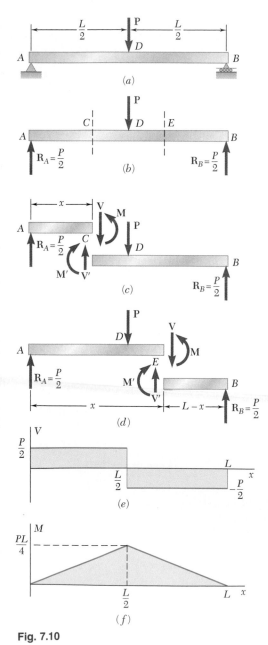

Fig. 7.10

2. *The bending moment at C is positive when the* **external** *forces acting on the beam tend to bend the beam at C as indicated in Fig. 7.9c.*

It may also help to note that the situation described in Fig. 7.9, in which the values of the shear and of the bending moment are positive, is precisely the situation which occurs in the left half of a simply supported beam carrying a single concentrated load at its midpoint. This particular example is fully discussed in the following section.

*7.5. SHEAR AND BENDING-MOMENT DIAGRAMS

Now that shear and bending moment have been clearly defined in sense as well as in magnitude, we can easily record their values at any point of a beam by plotting these values against the distance x measured from one end of the beam. The graphs obtained in this way are called, respectively, the *shear diagram* and the *bending-moment diagram*. As an example, consider a simply supported beam AB of span L subjected to a single concentrated load $\mathbf{P}$ applied at its midpoint D (Fig. 7.10a). We first determine the reactions at the supports from the free-body diagram of the entire beam (Fig. 7.10b); we find that the magnitude of each reaction is equal to $P/2$.

Next we cut the beam at a point C between A and D and draw the free-body diagrams of AC and CB (Fig. 7.10c). *Assuming that shear and bending moment are positive,* we direct the internal forces $\mathbf{V}$ and $\mathbf{V'}$ and the internal couples $\mathbf{M}$ and $\mathbf{M'}$ as indicated in Fig. 7.9a. Considering the free body AC and writing that the sum of the vertical components and the sum of the moments about C of the forces acting on the free body are zero, we find $V = +P/2$ and $M = +Px/2$. Both shear and bending moment are therefore positive; this can be checked by observing that the reaction at A tends to shear off and to bend the beam at C as indicated in Fig. 7.9b and c. We can plot V and M between A and D (Fig. 7.10e and f); the shear has a constant value $V = P/2$, while the bending moment increases linearly from $M = 0$ at $x = 0$ to $M = PL/4$ at $x = L/2$.

Cutting, now, the beam at a point E between D and B and considering the free body EB (Fig. 7.10d), we write that the sum of the vertical components and the sum of the moments about E of the forces acting on the free body are zero. We obtain $V = -P/2$ and $M = P(L - x)/2$. The shear is therefore negative and the bending moment positive; this can be checked by observing that the reaction at B bends the beam at E as indicated in Fig. 7.9c but tends to shear it off in a manner opposite to that shown in Fig. 7.9b. We can complete, now, the shear and bending-moment diagrams of Fig. 7.10e and f; the shear has a constant value $V = -P/2$ between D and B, while the bending moment decreases linearly from $M = PL/4$ at $x = L/2$ to $M = 0$ at $x = L$.

It should be noted that when a beam is subjected to concentrated loads only, the shear is of constant value between loads and the bending moment varies linearly between loads, but when a beam is subjected to distributed loads, the shear and bending moment vary quite differently (see Sample Prob. 7.3).

20 kN   40 kN

SAMPLE PROBLEM 7.2

Draw the shear and bending-moment diagrams for the beam and loading shown.

SOLUTION

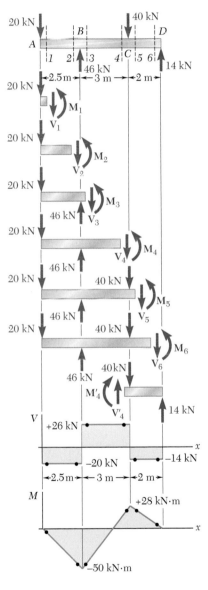

Free-Body: Entire Beam. From the free-body diagram of the entire beam, we find the reactions at B and D:

$$\mathbf{R}_B = 46 \text{ kN} \uparrow \qquad \mathbf{R}_D = 14 \text{ kN} \uparrow$$

Shear and Bending Moment. We first determine the internal forces just to the right of the 20-kN load at A. Considering the stub of the beam to the left of section 1 as a free body and assuming V and M to be positive (according to the standard convention), we write

$+\uparrow\Sigma F_y = 0:$ $-20 \text{ kN} - V_1 = 0$ $V_1 = -20 \text{ kN}$

$+\curvearrowleft\Sigma M_1 = 0:$ $(20 \text{ kN})(0 \text{ m}) + M_1 = 0$ $M_1 = 0$

We next consider as a free body the portion of the beam to the left of section 2 and write

$+\uparrow\Sigma F_y = 0:$ $-20 \text{ kN} - V_2 = 0$ $V_2 = -20 \text{ kN}$

$+\curvearrowleft\Sigma M_2 = 0:$ $(20 \text{ kN})(2.5 \text{ m}) + M_2 = 0$ $M_2 = -50 \text{ kN} \cdot \text{m}$

The shear and bending moment at sections $3, 4, 5,$ and 6 are determined in a similar way from the free-body diagrams shown. We obtain

$$\begin{aligned}
V_3 &= +26 \text{ kN} & M_3 &= -50 \text{ kN} \cdot \text{m} \\
V_4 &= +26 \text{ kN} & M_4 &= +28 \text{ kN} \cdot \text{m} \\
V_5 &= -14 \text{ kN} & M_5 &= +28 \text{ kN} \cdot \text{m} \\
V_6 &= -14 \text{ kN} & M_6 &= 0
\end{aligned}$$

For several of the latter sections, the results are more easily obtained by considering as a free body the portion of the beam to the right of the section. For example, considering the portion of the beam to the right of section 4, we write

$+\uparrow\Sigma F_y = 0:$ $V_4 - 40 \text{ kN} + 14 \text{ kN} = 0$ $V_4 = +26 \text{ kN}$

$+\curvearrowleft\Sigma M_4 = 0:$ $-M_4 + (14 \text{ kN})(2 \text{ m}) = 0$ $M_4 = +28 \text{ kN} \cdot \text{m}$

Shear and Bending-Moment Diagrams. We can now plot the six points shown on the shear and bending-moment diagrams. As indicated in Sec. 7.5, the shear is of constant value between concentrated loads, and the bending moment varies linearly; we therefore obtain the shear and bending-moment diagrams shown.

40 lb/in.

A *C* *D* *E* *B*

400 lb

|←— 12 in. —→|←6 in.→|4 in.|←—— 10 in. ——→|

|←—————————— 32 in. ——————————→|

SAMPLE PROBLEM 7.3

Draw the shear and bending-moment diagrams for the beam *AB*. The distributed load of 40 lb/in. extends over 12 in. of the beam, from *A* to *C*, and the 400-lb load is applied at *E*.

SOLUTION

Free-Body: Entire Beam. The reactions are determined by considering the entire beam as a free body.

$+\uparrow\Sigma M_A = 0$: $\quad B_y(32\text{ in.}) - (480\text{ lb})(6\text{ in.}) - (400\text{ lb})(22\text{ in.}) = 0$

$\quad\quad\quad\quad\quad\quad\quad B_y = +365\text{ lb}$ $\quad\quad\quad\quad\quad$ $\mathbf{B_y} = 365\text{ lb}\uparrow$

$+\uparrow\Sigma M_B = 0$: $\quad (480\text{ lb})(26\text{ in.}) + (400\text{ lb})(10\text{ in.}) - A(32\text{ in.}) = 0$

$\quad\quad\quad\quad\quad\quad\quad A = +515\text{ lb}$ $\quad\quad\quad\quad\quad\quad$ $\mathbf{A} = 515\text{ lb}\uparrow$

$\xrightarrow{+}\Sigma F_x = 0$: $\quad B_x = 0$ $\quad\quad\quad\quad\quad\quad\quad\quad\quad\quad$ $\mathbf{B_x} = 0$

The 400-lb load is now replaced by an equivalent force-couple system acting on the beam at point *D*.

Shear and Bending Moment. *From A to C.* We determine the internal forces at a distance *x* from point *A* by considering the portion of the beam to the left of section *1*. That part of the distributed load acting on the free body is replaced by its resultant, and we write

$+\uparrow\Sigma F_y = 0$: $\quad 515 - 40x - V = 0$ $\quad\quad\quad\quad$ $V = 515 - 40x$

$+\uparrow\Sigma M_1 = 0$: $\quad -515x + 40x(\frac{1}{2}x) + M = 0$ $\quad\quad$ $M = 515x - 20x^2$

Since the free-body diagram shown can be used for all values of *x* smaller than 12 in., the expressions obtained for *V* and *M* are valid throughout the region $0 < x < 12$ in.

From C to D. Considering the portion of the beam to the left of section *2* and again replacing the distributed load by its resultant, we obtain

$+\uparrow\Sigma F_y = 0$: $\quad 515 - 480 - V = 0$ $\quad\quad\quad$ $V = 35\text{ lb}$

$+\uparrow\Sigma M_2 = 0$: $\quad -515x + 480(x - 6) + M = 0$ $\quad M = (2880 + 35x)\text{ lb}\cdot\text{in.}$

These expressions are valid in the region 12 in. $< x <$ 18 in.

From D to B. Using the portion of the beam to the left of section *3*, we obtain for the region 18 in. $< x <$ 32 in.

$+\uparrow\Sigma F_y = 0$: $\quad 515 - 480 - 400 - V = 0$ $\quad\quad$ $V = -365\text{ lb}$

$+\uparrow\Sigma M_3 = 0$: $\quad -515x + 480(x - 6) - 1600 + 400(x - 18) + M = 0$

$\quad\quad\quad\quad\quad\quad\quad\quad\quad\quad\quad\quad$ $M = (11{,}680 - 365x)\text{ lb}\cdot\text{in.}$

Shear and Bending-Moment Diagrams. The shear and bending-moment diagrams for the entire beam can be plotted. We note that the couple of moment 1600 lb · in. applied at point *D* introduces a discontinuity into the bending-moment diagram.

SOLVING PROBLEMS
ON YOUR OWN

In this lesson you learned to determine the *shear V* and the *bending moment M* at any point in a beam. You also learned to draw the *shear diagram* and the *bending-moment diagram* for the beam by plotting, respectively, V and M against the distance x measured along the beam.

A. Determining the shear and bending moment in a beam. To determine the shear V and the bending moment M at a given point C of a beam, you should take the following steps.

1. Draw a free-body diagram of the entire beam, and use it to determine the reactions at the beam supports.

2. Cut the beam at point C, and, using the original loading, select one of the two portions of the beam you have obtained.

3. Draw the free-body diagram of the portion of the beam you have selected, showing:

    a. The loads and the reaction exerted on that portion of the beam, replacing each distributed load by an equivalent concentrated load as explained earlier in Sec. 5.8.

    b. The shearing force and the bending couple representing the internal forces at C. To facilitate recording the shear V and the bending moment M after they have been determined, follow the convention indicated in Figs. 7.8 and 7.9. Thus, if you are using the portion of the beam located to the *left of C*, apply at C a *shearing force* **V** *directed downward* and a *bending moment* **M** *directed counterclockwise.* If you are using the portion of the beam located to the *right of C*, apply at C a *shearing force* **V′** *directed upward* and a *bending moment* **M′** *directed clockwise* [Sample Prob. 7.2].

4. Write the equilibrium equations for the portion of the beam you have selected. Solve the equation $\Sigma F_y = 0$ for V and the equation $\Sigma M_C = 0$ for M.

5. Record the values of V and M with the sign obtained for each of them. A positive sign for V means that the shearing forces exerted at C on each of the two portions of the beam are directed as shown in Figs. 7.8 and 7.9; a negative sign means that they have the opposite sense. Similarly, a positive sign for M means that the bending couples at C are directed as shown in these figures, and a negative sign means that they have the opposite sense. In addition, a positive sign for M means that the concavity of the beam at C is directed upward, and a negative sign means that it is directed downward.

(continued)

B. Drawing the shear and bending-moment diagrams for a beam. These diagrams are obtained by plotting, respectively, V and M against the distance x measured along the beam. However, in most cases the values of V and M need to be
computed only at a few points.

1. For a beam supporting only concentrated loads, we note [Sample Prob.
7.2] that
     a. The shear diagram consists of segments of horizontal lines. Thus, to
draw the shear diagram of the beam you will need to compute V only just to the
left or just to the right of the points where the loads or the reactions are applied.
     *b. The bending-moment diagram consists of segments of oblique straight
lines.* Thus, to draw the bending-moment diagram of the beam you will need to
compute M only at the points where the loads or the reactions are applied.

2. For a beam supporting uniformly distributed loads, we note [Sample Prob.
7.3] that under each of the distributed loads:
     a. The shear diagram consists of a segment of an oblique straight line.
Thus, you will need to compute V only where the distributed load begins and where
it ends.
     b. The bending-moment diagram consists of a parabolic arc. In most cases
you will need to compute M only where the distributed load begins and where it
ends.

3. For a beam with a more complicated loading, it is necessary to consider
the free-body diagram of a portion of the beam of arbitrary length x and determine
V and M as functions of x. This procedure may have to be repeated several times,
since V and M are often represented by different functions in various parts of the
beam [Sample Prob. 7.3].

4. When a couple is applied to a beam, the shear has the same value on both
sides of the point of application of the couple, but the bending-moment diagram
will show a discontinuity at that point, rising or falling by an amount equal to the
magnitude of the couple. Note that a couple can either be applied directly to
the beam, or result from the application of a load to a member rigidly attached to the
beam [Sample Prob. 7.3].

Problems

7.29 through 7.32 For the beam and loading shown, (*a*) draw the shear and bending-moment diagrams, (*b*) determine the maximum absolute values of the shear and bending moment.

Fig. P7.29

Fig. P7.30

Fig. P7.31

Fig. P7.32

7.33 and 7.34 For the beam and loading shown, (*a*) draw the shear and bending-moment diagrams, (*b*) determine the maximum absolute values of the shear and bending moment.

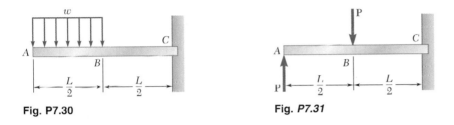

Fig. P7.33

Fig. P7.34

Fig. P7.35

7.35 and 7.36 For the beam and loading shown, (*a*) draw the shear and bending-moment diagrams, (*b*) determine the maximum absolute values of the shear and bending moment.

Fig. P7.36

Fig. P7.37

Fig. P7.38

7.37 and 7.38 For the beam and loading shown, (*a*) draw the shear and bending-moment diagrams, (*b*) determine the maximum absolute values of the shear and bending moment.

369

7.39 and 7.40 For the beam and loading shown, (*a*) draw the shear and bending-moment diagrams, (*b*) determine the maximum absolute values of the shear and bending moment.

Fig. P7.39

Fig. P7.42

Fig. P7.43

Fig. P7.45

Fig. P7.40

Fig. P7.41

7.41 and 7.42 Assuming the upward reaction of the ground on beam *AB* to be uniformly distributed, (*a*) draw the shear and bending-moment diagrams, (*b*) determine the maximum absolute values of the shear and bending moment.

7.43 Assuming the upward reaction of the ground on beam *AB* to be uniformly distributed and knowing that *a* = 0.9 ft, (*a*) draw the shear and bending-moment diagrams, (*b*) determine the maximum absolute values of the shear and bending moment.

7.44 Solve Prob. 7.43 assuming that *a* = 1.5 ft.

7.45 Two short angle sections *CE* and *DF* are bolted to the uniform beam *AB* of weight 3.33 kN, and the assembly is temporarily supported by the vertical cables *EG* and *FH* as shown. A second beam resting on beam *AB* at *I* exerts a downward force of 3 kN on *AB*. Knowing that *a* = 0.3 m and neglecting the weight of the angle sections, (*a*) draw the shear and bending-moment diagrams for beam *AB*, (*b*) determine the maximum absolute values of the shear and bending moment in the beam.

7.46 Solve Prob. 7.45 when *a* = 0.6 m.

Fig. P7.47

Fig. P7.48

7.47 Draw the shear and bending-moment diagrams for the beam *AB*, and determine the shear and bending moment (*a*) just to the left of *C*, (*b*) just to the right of *C*.

7.48 and 7.49 Draw the shear and bending-moment diagrams for the beam *AB*, and determine the maximum absolute values of the shear and bending moment.

Fig. P7.49

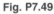

7.50 Neglecting the size of the pulley at *G*, (*a*) draw the shear and bending-moment diagrams for the beam *AB*, (*b*) determine the maximum absolute values of the shear and bending moment.

7.51 For the beam of Prob. 7.43, determine (*a*) the distance *a* for which the maximum absolute value of the bending moment in the beam is as small as possible, (*b*) the corresponding value of $|M|_{max}$. (*Hint:* Draw the bending-moment diagram and then equate the absolute values of the largest positive and negative bending moments obtained.)

7.52 For the assembly of Prob. 7.45, determine (*a*) the distance *a* for which the maximum absolute value of the bending moment in beam *AB* is as small as possible, (*b*) the corresponding value of $|M|_{max}$. (See hint for Prob. 7.51.)

7.53 For the beam shown, determine (*a*) the magnitude *P* of the two upward forces for which the maximum value of the bending moment is as small as possible, (*b*) the corresponding value of $|M|_{max}$. (See hint for Prob. 7.51.)

Fig. P7.50

Fig. P7.53

7.54 For the beam and loading shown, determine (*a*) the distance *a* for which the maximum absolute value of the bending moment in the beam is as small as possible, (*b*) the corresponding value of $|M|_{max}$. (See hint for Prob. 7.51.)

Fig. P7.54

7.55 Knowing that $P = Q = 375$ lb, determine (*a*) the distance *a* for which the maximum absolute value of the bending moment in beam *AB* is as small as possible, (*b*) the corresponding value of $|M|_{max}$. (See hint for Prob. 7.51.)

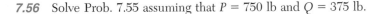

Fig. P7.55

7.56 Solve Prob. 7.55 assuming that $P = 750$ lb and $Q = 375$ lb.

*7.57 In order to reduce the bending moment in the cantilever beam *AB*, a cable and counterweight are permanently attached at end *B*. Determine the magnitude of the counterweight for which the maximum absolute value of the bending moment in the beam is as small as possible and the corresponding value of $|M|_{\max}$. Consider (*a*) the case when the distributed load is permanently applied to the beam, (*b*) the more general case when the distributed load may either be applied or removed.

Fig. P7.57

*7.6. RELATIONS AMONG LOAD, SHEAR, AND BENDING MOMENT

When a beam carries more than two or three concentrated loads, or when it carries distributed loads, the method outlined in Sec. 7.5 for plotting shear and bending moment is likely to be quite cumbersome. The construction of the shear diagram and, especially, of the bending-moment diagram will be greatly facilitated if certain relations existing among load, shear, and bending moment are taken into consideration.

Let us consider a simply supported beam *AB* carrying a distributed load *w* per unit length (Fig. 7.11*a*), and let *C* and *C'* be two points of the beam at a distance Δx from each other. The shear and bending moment at *C* will be denoted by *V* and *M*, respectively, and will be assumed positive; the shear and bending moment at *C'* will be denoted by $V + \Delta V$ and $M + \Delta M$.

Let us now detach the portion of beam *CC'* and draw its free-body diagram (Fig. 7.11*b*). The forces exerted on the free body include a load of magnitude $w\,\Delta x$ and internal forces and couples at *C* and *C'*. Since shear and bending moment have been assumed positive, the forces and couples will be directed as shown in the figure.

Relations between Load and Shear. We write that the sum of the vertical components of the forces acting on the free body *CC'* is zero:

$$V - (V + \Delta V) - w\,\Delta x = 0$$
$$\Delta V = -w\,\Delta x$$

Dividing both members of the equation by Δx and then letting Δx approach zero, we obtain

$$\frac{dV}{dx} = -w \tag{7.1}$$

Formula (7.1) indicates that for a beam loaded as shown in Fig. 7.11*a*, the slope dV/dx of the shear curve is negative; the absolute value of

(*a*)

(*b*)

Fig. 7.11

the slope at any point is equal to the load per unit length at that point. Integrating (7.1) between points C and D, we obtain

$$V_D - V_C = -\int_{x_C}^{x_D} w \, dx \qquad (7.2)$$

$$V_D - V_C = -(\text{area under load curve between } C \text{ and } D) \qquad (7.2')$$

Note that this result could also have been obtained by considering the equilibrium of the portion of beam CD, since the area under the load curve represents the total load applied between C and D.

It should be observed that formula (7.1) *is not valid* at a point where a concentrated load is applied; the shear curve is discontinuous at such a point, as seen in Sec. 7.5. Similarly, formulas (7.2) and (7.2') cease to be valid when concentrated loads are applied between C and D, since they do not take into account the sudden change in shear caused by a concentrated load. Formulas (7.2) and (7.2'), therefore, should be applied only between successive concentrated loads.

Relations between Shear and Bending Moment. Returning to the free-body diagram of Fig. 7.11b, and writing now that the sum of the moments about C' is zero, we obtain

$$(M + \Delta M) - M - V \, \Delta x + w \, \Delta x \frac{\Delta x}{2} = 0$$

$$\Delta M = V \, \Delta x - \tfrac{1}{2} w (\Delta x)^2$$

Dividing both members of the equation by Δx and then letting Δx approach zero, we obtain

$$\frac{dM}{dx} = V \qquad (7.3)$$

Formula (7.3) indicates that the slope dM/dx of the bending-moment curve is equal to the value of the shear. This is true at any point where the shear has a well-defined value, that is, at any point where no concentrated load is applied. Formula (7.3) also shows that the shear is zero at points where the bending moment is maximum. This property facilitates the determination of the points where the beam is likely to fail under bending.

Integrating (7.3) between points C and D, we obtain

$$M_D - M_C = \int_{x_C}^{x_D} V \, dx \qquad (7.4)$$

$$M_D - M_C = \text{area under shear curve between } C \text{ and } D \qquad (7.4')$$

Note that the area under the shear curve should be considered positive where the shear is positive and negative where the shear is negative. Formulas (7.4) and (7.4') are valid even when concentrated loads are applied between C and D, as long as the shear curve has been correctly drawn. The formulas cease to be valid, however, if a *couple* is applied at a point between C and D, since they do not take into account the sudden change in bending moment caused by a couple (see Sample Prob. 7.7).

Example.   Let us consider a simply supported beam *AB* of span *L* carrying a uniformly distributed load *w* (Fig. 7.12*a*). From the free-body diagram of the entire beam we determine the magnitude of the reactions at the supports: $R_A = R_B = wL/2$ (Fig. 7.12*b*). Next, we draw the shear diagram. Close to the end *A* of the beam, the shear is equal to R_A, that is, to $wL/2$, as we can check by considering a very small portion of the beam as a free body. Using formula (7.2), we can then determine the shear *V* at any distance *x* from *A*. We write

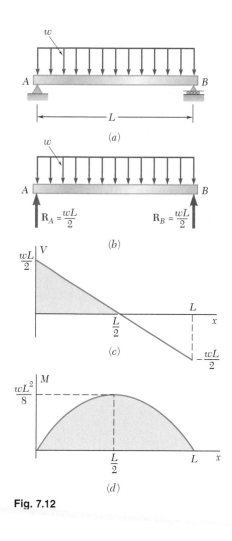

$$V - V_A = -\int_0^x w \, dx = -wx$$

$$V = V_A - wx = \frac{wL}{2} - wx = w\left(\frac{L}{2} - x\right)$$

The shear curve is thus an oblique straight line which crosses the *x* axis at $x = L/2$ (Fig. 7.12*c*). Considering, now, the bending moment, we first observe that $M_A = 0$. The value *M* of the bending moment at any distance *x* from *A* can then be obtained from formula (7.4); we have

$$M - M_A = \int_0^x V \, dx$$

$$M = \int_0^x w\left(\frac{L}{2} - x\right) dx = \frac{w}{2}(Lx - x^2)$$

The bending-moment curve is a parabola. The maximum value of the bending moment occurs when $x = L/2$, since *V* (and thus dM/dx) is zero for that value of *x*. Substituting $x = L/2$ in the last equation, we obtain $M_{max} = wL^2/8$.

In most engineering applications, the value of the bending moment needs to be known only at a few specific points. Once the shear diagram has been drawn, and after *M* has been determined at one of the ends of the beam, the value of the bending moment can then be obtained at any given point by computing the area under the shear curve and using formula (7.4'). For instance, since $M_A = 0$ for the beam of Fig. 7.12, the maximum value of the bending moment for that beam can be obtained simply by calculating the area of the shaded triangle in the shear diagram:

$$M_{max} = \frac{1}{2}\frac{L}{2}\frac{wL}{2} = \frac{wL^2}{8}$$

In this example, the load curve is a horizontal straight line, the shear curve is an oblique straight line, and the bending-moment curve is a parabola. If the load curve had been an oblique straight line (first degree), the shear curve would have been a parabola (second degree), and the bending-moment curve would have been a cubic (third degree). The shear and bending-moment curves will always be, respectively, one and two degrees higher than the load curve. Thus, once a few values of the shear and bending moment have been computed, we should be able to sketch the shear and bending-moment diagrams without actually determining the functions *V*(*x*) and *M*(*x*). The sketches obtained will be more accurate if we make use of the fact that at any point where the curves are continuous, the slope of the shear curve is equal to −*w* and the slope of the bending-moment curve is equal to *V*.

Fig. 7.12

SAMPLE PROBLEM 7.4

Draw the shear and bending-moment diagrams for the beam and loading shown.

SOLUTION

Free-Body: Entire Beam. Considering the entire beam as a free body, we determine the reactions:

$+\uparrow\Sigma M_A = 0$:

$D(24\text{ ft}) - (20\text{ kips})(6\text{ ft}) - (12\text{ kips})(14\text{ ft}) - (12\text{ kips})(28\text{ ft}) = 0$

$$D = +26\text{ kips} \qquad\qquad \mathbf{D} = 26\text{ kips} \uparrow$$

$+\uparrow\Sigma F_y = 0$: $\quad A_y - 20\text{ kips} - 12\text{ kips} + 26\text{ kips} - 12\text{ kips} = 0$

$$A_y = +18\text{ kips} \qquad\qquad \mathbf{A}_y = 18\text{ kips} \uparrow$$

$\xrightarrow{+}\Sigma F_x = 0$: $\quad A_x = 0 \qquad\qquad\qquad\qquad \mathbf{A}_x = 0$

We also note that at both A and E the bending moment is zero; thus two points (indicated by small circles) are obtained on the bending-moment diagram.

Shear Diagram. Since $dV/dx = -w$, we find that between concentrated loads and reactions the slope of the shear diagram is zero (that is, the shear is constant). The shear at any point is determined by dividing the beam into two parts and considering either part as a free body. For example, using the portion of beam to the left of section *1*, we obtain the shear between B and C:

$+\uparrow\Sigma F_y = 0$: $\quad +18\text{ kips} - 20\text{ kips} - V = 0 \qquad\qquad V = -2\text{ kips}$

We also find that the shear is $+12$ kips just to the right of D and zero at end E. Since the slope $dV/dx = -w$ is constant between D and E, the shear diagram between these two points is a straight line.

Bending-Moment Diagram. We recall that the area under the shear curve between two points is equal to the change in bending moment between the same two points. For convenience, the area of each portion of the shear diagram is computed and is indicated on the diagram. Since the bending moment M_A at the left end is known to be zero, we write

$$\begin{array}{ll} M_B - M_A = +108 & M_B = +108\text{ kip}\cdot\text{ft} \\ M_C - M_B = -\ 16 & M_C = +\ 92\text{ kip}\cdot\text{ft} \\ M_D - M_C = -140 & M_D = -\ 48\text{ kip}\cdot\text{ft} \\ M_E - M_D = +\ 48 & M_E = 0 \end{array}$$

Since M_E is known to be zero, a check of the computations is obtained.

Between the concentrated loads and reactions the shear is constant; thus the slope dM/dx is constant, and the bending-moment diagram is drawn by connecting the known points with straight lines. Between D and E, where the shear diagram is an oblique straight line, the bending-moment diagram is a parabola.

From the V and M diagrams we note that $V_{\max} = 18$ kips and $M_{\max} = 108$ kip $\cdot$ ft.

20 kN/m

A   C

B

|← 6 m →|← 3 m →|

SAMPLE PROBLEM 7.5

Draw the shear and bending-moment diagrams for the beam and loading shown, and determine the location and magnitude of the maximum bending moment.

SOLUTION

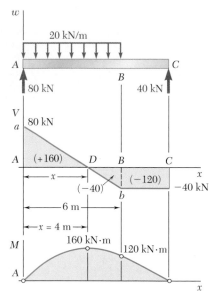

Free-Body: Entire Beam. Considering the entire beam as a free body, we obtain the reactions

$$\mathbf{R}_A = 80 \text{ kN} \uparrow \qquad \mathbf{R}_C = 40 \text{ kN} \uparrow$$

Shear Diagram. The shear just to the right of A is $V_A = +80$ kN. Since the change in shear between two points is equal to *minus* the area under the load curve between the same two points, we obtain V_B by writing

$$V_B - V_A = -(20 \text{ kN/m})(6 \text{ m}) = -120 \text{ kN}$$
$$V_B = -120 + V_A = -120 + 80 = -40 \text{ kN}$$

Since the slope $dV/dx = -w$ is constant between A and B, the shear diagram between these two points is represented by a straight line. Between B and C, the area under the load curve is zero; therefore,

$$V_C - V_B = 0 \qquad V_C = V_B = -40 \text{ kN}$$

and the shear is constant between B and C.

Bending-Moment Diagram. We note that the bending moment at each end of the beam is zero. In order to determine the maximum bending moment, we locate the section D of the beam where $V = 0$. We write

$$V_D - V_A = -wx$$
$$0 - 80 \text{ kN} = -(20 \text{ kN/m})x$$

and, solving for x: $\qquad\qquad\qquad\qquad\qquad\qquad x = 4 \text{ m}$ ◄

The maximum bending moment occurs at point D, where we have $dM/dx = V = 0$. The areas of the various portions of the shear diagram are computed and are given (in parentheses) on the diagram. Since the area of the shear diagram between two points is equal to the change in bending moment between the same two points, we write

$$M_D - M_A = +160 \text{ kN} \cdot \text{m} \qquad M_D = +160 \text{ kN} \cdot \text{m}$$
$$M_B - M_D = - 40 \text{ kN} \cdot \text{m} \qquad M_B = +120 \text{ kN} \cdot \text{m}$$
$$M_C - M_B = -120 \text{ kN} \cdot \text{m} \qquad M_C = 0$$

The bending-moment diagram consists of a parabolic arc followed by a segment of a straight line; the slope of the parabola at A is equal to the value of V at that point.

The maximum bending moment is

$$M_{\text{max}} = M_D = +160 \text{ kN} \cdot \text{m} \quad ◄$$

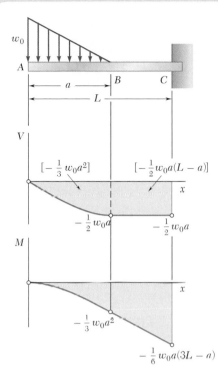

SAMPLE PROBLEM 7.6

Sketch the shear and bending-moment diagrams for the cantilever beam shown.

SOLUTION

Shear Diagram. At the free end of the beam, we find $V_A = 0$. Between A and B, the area under the load curve is $\frac{1}{2}w_0a$, we find V_B by writing

$$V_B - V_A = -\tfrac{1}{2}w_0a \qquad V_B = -\tfrac{1}{2}w_0a$$

Between B and C, the beam is not loaded; thus $V_C = V_B$. At A, we have $w = w_0$, and, according to Eq. (7.1), the slope of the shear curve is $dV/dx = -w_0$, while at B the slope is $dV/dx = 0$. Between A and B, the loading decreases linearly, and the shear diagram is parabolic. Between B and C, $w = 0$, and the shear diagram is a horizontal line.

Bending-Moment Diagram. We note that $M_A = 0$ at the free end of the beam. We compute the area under the shear curve and write

$$M_B - M_A = -\tfrac{1}{3}w_0a^2 \qquad M_B = -\tfrac{1}{3}w_0a^2$$
$$M_C - M_B = -\tfrac{1}{2}w_0a(L - a)$$
$$M_C = -\tfrac{1}{6}w_0a(3L - a)$$

The sketch of the bending-moment diagram is completed by recalling that $dM/dx = V$. We find that between A and B the diagram is represented by a cubic curve with zero slope at A, and between B and C the diagram is represented by a straight line.

SAMPLE PROBLEM 7.7

The simple beam AC is loaded by a couple of magnitude T applied at point B. Draw the shear and bending-moment diagrams for the beam.

SOLUTION

Free Body: Entire Beam. The entire beam is taken as a free body, and we obtain

$$\mathbf{R}_A = \frac{T}{L}\uparrow \qquad \mathbf{R}_C = \frac{T}{L}\downarrow$$

Shear and Bending-Moment Diagrams. The shear at any section is constant and equal to T/L. Since a couple is applied at B, the bending-moment diagram is discontinuous at B; the bending moment decreases suddenly by an amount equal to T.

In this lesson you learned how to use the relations existing among load, shear, and bending moment to simplify the drawing of the shear and bending-moment diagrams. These relations are

$$\frac{dV}{dx} = -w \qquad (7.1)$$

$$\frac{dM}{dx} = V \qquad (7.3)$$

$$V_D - V_C = -(\text{area under load curve between } C \text{ and } D) \qquad (7.2')$$

$$M_D - M_C = (\text{area under shear curve between } C \text{ and } D) \qquad (7.4')$$

Taking into account these relations, you can use the following procedure to draw the shear and bending-moment diagrams for a beam.

1. Draw a free-body diagram of the entire beam, and use it to determine the reactions at the beam supports.

2. Draw the shear diagram. This can be done as in the preceding lesson by cutting the beam at various points and considering the free-body diagram of one of the two portions of the beam that you have obtained [Sample Prob. 7.3]. You can, however, consider one of the following alternative procedures.

a. The shear V at any point of the beam is the sum of the reactions and loads to the left of that point; an upward force is counted as positive, and a downward force is counted as negative.

b. For a beam carrying a distributed load, you can start from a point where you know V and use Eq. (7.2′) repeatedly to find V at all the other points of interest.

3. Draw the bending-moment diagram, using the following procedure.

a. Compute the area under each portion of the shear curve, assigning a positive sign to areas located above the x axis and a negative sign to areas located below the x axis.

b. Apply Eq. (7.4′) repeatedly [Sample Probs. 7.4 and 7.5], starting from the left end of the beam, where $M = 0$ (except if a couple is applied at that end, or if the beam is a cantilever beam with a fixed left end).

c. Where a couple is applied to the beam, be careful to show a discontinuity in the bending-moment diagram by *increasing* the value of M at that point by an amount equal to the magnitude of the couple if the couple is *clockwise,* or *decreasing* the value of M by that amount if the couple is *counterclockwise* [Sample Prob. 7.7].

4. Determine the location and magnitude of $|M|_{max}$. The maximum absolute value of the bending moment occurs at one of the points where $dM/dx = 0$, that is, according to Eq. (7.3), at a point where V is equal to zero or changes sign. You should, therefore:

  a. Determine the values of $|M|$, using the area under the shear curve, at the points where V changes sign; these points will occur under concentrated loads [Sample Prob. 7.4].

  b. Determine the points where $V = 0$ and the corresponding values of $|M|$; these points will occur under a distributed load. To find the distance x between point C, where the distributed load starts, and point D, where the shear is zero, use Eq. (7.2′); for V_C use the known value of the shear at point C, for V_D use zero, and express the area under the load curve as a function of x [Sample Prob. 7.5].

5. You can improve the quality of your drawings by keeping in mind that at any given point, according to Eqs. (7.1) and (7.3), the slope of the V curve is equal to $-w$ and the slope of the M curve is equal to V. In particular, your drawings should clearly show where the slope of the curves is zero.

6. Finally, for beams supporting a distributed load expressed as a function $w(x)$, remember that the shear V can be obtained by integrating the function $-w(x)$, and the bending moment M can be obtained by integrating $V(x)$ [Eqs. (7.3) and (7.4)].

Problems

Fig. P7.64

Fig. P7.65

Fig. *P7.70*

Fig. P7.71

7.58 Using the method of Sec. 7.6, solve Prob. 7.29.

7.59 Using the method of Sec. 7.6, solve Prob. 7.30.

7.60 Using the method of Sec. 7.6, solve Prob. 7.31.

7.61 Using the method of Sec. 7.6, solve Prob. 7.32.

7.62 Using the method of Sec. 7.6, solve Prob. 7.33.

7.63 Using the method of Sec. 7.6, solve Prob. 7.36.

7.64 and 7.65 For the beam and loading shown, (*a*) draw the shear and bending-moment diagrams, (*b*) determine the maximum absolute values of the shear and bending moment.

7.66 Using the method of Sec. 7.6, solve Prob. 7.37.

7.67 Using the method of Sec. 7.6, solve Prob. 7.38.

7.68 Using the method of Sec. 7.6, solve Prob. 7.39.

7.69 Using the method of Sec. 7.6, solve Prob. 7.40.

7.70 and 7.71 For the beam and loading shown, (*a*) draw the shear and bending-moment diagrams, (*b*) determine the maximum absolute values of the shear and bending moment.

Fig. P7.72   **Fig. P7.73**

7.72 and 7.73 For the beam and loading shown, (*a*) draw the shear and bending-moment diagrams, (*b*) determine the magnitude and location of the maximum bending moment.

7.74 For the beam shown, draw the shear and bending-moment diagrams, and determine the maximum absolute value of the bending moment knowing that (*a*) $P = 14$ kN, (*b*) $P = 20$ kN.

Fig. P7.74

7.75 For the beam shown, draw the shear and bending-moment diagrams, and determine the magnitude and location of the maximum absolute value of the bending moment knowing that (*a*) $M = 0$, (*b*) $M = 12$ kN · m.

Fig. P7.75

7.76 For the beam and loading shown, (*a*) draw the shear and bending-moment diagrams, (*b*) determine the magnitude and location of the maximum absolute value of the bending moment.

Fig. P7.76

7.77 Solve Prob. 7.76 assuming that the 30 kN · m couple applied at *B* is counterclockwise.

7.78 For beam *AB*, (*a*) draw the shear and bending-moment diagrams, (*b*) determine the magnitude and location of the maximum absolute value of the bending moment.

7.79 Solve Prob. 7.78 assuming that the 4-kN force applied at *E* is directed upward.

Fig. P7.78

7.80 and 7.81 For the beam and loading shown, (*a*) derive the equations of the shear and bending-moment curves, (*b*) draw the shear and bending-moment diagrams, (*c*) determine the magnitude and location of the maximum bending moment.

Fig. P7.80

Fig. P7.81

Fig. P7.83

Fig. P7.82

7.82 For the beam shown, (*a*) draw the shear and bending-moment diagrams, (*b*) determine the magnitude and location of the maximum bending moment. (*Hint:* Derive the equations of the shear and bending-moment curves for portion *CD* of the beam.)

7.83 Beam *AB*, which lies on the ground, supports the parabolic load shown. Assuming the upward reaction of the ground to be uniformly distributed, (*a*) write the equations of the shear and bending-moment curves, (*b*) determine the maximum bending moment.

7.84 The beam *AB* is subjected to the uniformly distributed load shown and to two unknown forces **P** and **Q**. Knowing that it has been experimentally determined that the bending moment is $+325$ lb · ft at *D* and $+800$ lb · ft at *E*, (*a*) determine **P** and **Q**, (*b*) draw the shear and bending-moment diagrams for the beam.

7.85 Solve Prob. 7.84 assuming that the bending moment was found to be $+260$ lb · ft at *D* and $+860$ lb · ft at *E*.

Fig. P7.84

*7.86** The beam *AB* is subjected to the uniformly distributed load shown and to two unknown forces **P** and **Q**. Knowing that it has been experimentally determined that the bending moment is $+7$ kN · m at *D* and $+5$ kN · m at *E*, (*a*) determine **P** and **Q**, (*b*) draw the shear and bending-moment diagrams for the beam.

Fig. P7.86

*7.87** Solve Prob. 7.86 assuming that the bending moment was found to be $+3.6$ kN · m at *D* and $+4.14$ kN · m at *E*.

*7.7. CABLES WITH CONCENTRATED LOADS

Cables are used in many engineering applications, such as suspension bridges, transmission lines, aerial tramways, guy wires for high towers, etc. Cables may be divided into two categories according to their loading: (1) cables supporting concentrated loads, (2) cables supporting distributed loads. In this section, cables of the first category are examined.

Consider a cable attached to two fixed points A and B and supporting n vertical concentrated loads $\mathbf{P}_1, \mathbf{P}_2, \ldots, \mathbf{P}_n$ (Fig. 7.13a). We assume that the cable is *flexible*, that is, that its resistance to bending is small and can be neglected. We further assume that the *weight of the cable is negligible* compared with the loads supported by the cable. Any portion of cable between successive loads can therefore be considered as a two-force member, and the internal forces at any point in the cable reduce to a *force of tension directed along the cable*.

We assume that each of the loads lies in a given vertical line, that is, that the horizontal distance from support A to each of the loads is known; we also assume that the horizontal and vertical distances between the supports are known. We propose to determine the shape of the cable, that is, the vertical distance from support A to each of the points $C_1, C_2, \ldots, C_n$, and also the tension T in each portion of the cable.

Photo 7.3 Since the weight of the cable of the chairlift shown is negligible compared to the weights of the chairs and skiers, the methods of this section can be used to determine the force at any point in the cable.

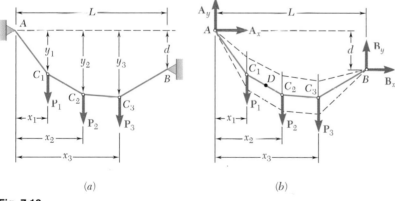

(a) $\qquad\qquad\qquad\qquad (b)$

Fig. 7.13

We first draw the free-body diagram of the entire cable (Fig. 7.13b). Since the slope of the portions of cable attached at A and B is not known, the reactions at A and B must be represented by two components each. Thus, four unknowns are involved, and the three equations of equilibrium are not sufficient to determine the reactions at A and B.† We must therefore obtain an additional equation by considering the equilibrium of a portion of the cable. This is possible if

---

†Clearly, the cable is not a rigid body; the equilibrium equations represent, therefore, *necessary but not sufficient conditions* (see Sec. 6.11).

(a)

Fig. 7.13 (*repeated*)

(a)

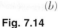

(b)

Fig. 7.14

we know the coordinates x and y of a point D of the cable. Drawing the free-body diagram of the portion of cable AD (Fig. 7.14a) and writing $\Sigma M_D = 0$, we obtain an additional relation between the scalar components A_x and A_y and can determine the reactions at A and B. The problem would remain indeterminate, however, if we did not know the coordinates of D, unless some other relation between A_x and A_y (or between B_x and B_y) were given. The cable might hang in any of various possible ways, as indicated by the dashed lines in Fig. 7.13b.

Once A_x and A_y have been determined, the vertical distance from A to any point of the cable can easily be found. Considering point C_2, for example, we draw the free-body diagram of the portion of cable AC_2 (Fig. 7.14b). Writing $\Sigma M_{C_2} = 0$, we obtain an equation which can be solved for y_2. Writing $\Sigma F_x = 0$ and $\Sigma F_y = 0$, we obtain the components of the force **T** representing the tension in the portion of cable to the right of C_2. We observe that $T \cos \theta = -A_x$; *the horizontal component of the tension force is the same at any point of the cable.* It follows that the tension T is maximum when $\cos \theta$ is minimum, that is, in the portion of cable which has the largest angle of inclination θ. Clearly, this portion of cable must be adjacent to one of the two supports of the cable.

*7.8. CABLES WITH DISTRIBUTED LOADS

Consider a cable attached to two fixed points A and B and carrying a *distributed load* (Fig. 7.15a). We saw in the preceding section that for a cable supporting concentrated loads, the internal force at any point is a force of tension directed along the cable. In the case of a cable carrying a distributed load, the cable hangs in the shape of a curve, and the internal force at a point D is a force of tension **T** *directed along the tangent to the curve.* In this section, you will learn to determine the tension at any point of a cable supporting a given distributed load. In the following sections, the shape of the cable will be determined for two particular types of distributed loads.

Considering the most general case of distributed load, we draw the free-body diagram of the portion of cable extending from the lowest point C to a given point D of the cable (Fig. 7.15b). The forces

Fig. 7.15

Fig. 7.16

acting on the free body are the tension force $\mathbf{T}_0$ at C, which is horizontal, the tension force $\mathbf{T}$ at D, directed along the tangent to the cable at D, and the resultant $\mathbf{W}$ of the distributed load supported by the portion of cable CD. Drawing the corresponding force triangle (Fig. 7.15c), we obtain the following relations:

$$T \cos \theta = T_0 \qquad T \sin \theta = W \qquad (7.5)$$

$$T = \sqrt{T_0^2 + W^2} \qquad \tan \theta = \frac{W}{T_0} \qquad (7.6)$$

From the relations (7.5), it appears that the horizontal component of the tension force $\mathbf{T}$ is the same at any point and that the vertical component of $\mathbf{T}$ is equal to the magnitude W of the load measured from the lowest point. Relations (7.6) show that the tension T is minimum at the lowest point and maximum at one of the two points of support.

*7.9. PARABOLIC CABLE

Let us assume, now, that the cable AB carries a load *uniformly distributed along the horizontal* (Fig. 7.16a). Cables of suspension bridges may be assumed loaded in this way, since the weight of the cables is small compared with the weight of the roadway. We denote by w the load per unit length (*measured horizontally*) and express it in N/m or in lb/ft. Choosing coordinate axes with origin at the lowest point C of the cable, we find that the magnitude W of the total load carried by the portion of cable extending from C to the point D of coordinates x and y is $W = wx$. The relations (7.6) defining the magnitude and direction of the tension force at D become

$$T = \sqrt{T_0^2 + w^2 x^2} \qquad \tan \theta = \frac{wx}{T_0} \qquad (7.7)$$

Moreover, the distance from D to the line of action of the resultant $\mathbf{W}$ is equal to half the horizontal distance from C to D (Fig. 7.16b). Summing moments about D, we write

$$+\gamma \Sigma M_D = 0: \qquad wx\frac{x}{2} - T_0 y = 0$$

and, solving for y,

$$y = \frac{wx^2}{2T_0} \tag{7.8}$$

This is the equation of a *parabola* with a vertical axis and its vertex at the origin of coordinates. The curve formed by cables loaded uniformly along the horizontal is thus a parabola.†

When the supports A and B of the cable have the same elevation, the distance L between the supports is called the *span* of the cable and the vertical distance h from the supports to the lowest point is called the *sag* of the cable (Fig. 7.17*a*). If the span and sag of a cable are known, and if the load w per unit horizontal length is given, the minimum tension T_0 may be found by substituting $x = L/2$ and $y = h$ in Eq. (7.8). Equations (7.7) will then yield the tension and the slope at any point of the cable, and Eq. (7.8) will define the shape of the cable.

When the supports have different elevations, the position of the lowest point of the cable is not known and the coordinates x_A, y_A and x_B, y_B of the supports must be determined. To this effect, we express that the coordinates of A and B satisfy Eq. (7.8) and that $x_B - x_A = L$ and $y_B - y_A = d$, where L and d denote, respectively, the horizontal and vertical distances between the two supports (Fig. 7.17*b* and *c*).

The length of the cable from its lowest point C to its support B can be obtained from the formula

$$s_B = \int_0^{x_B} \sqrt{1 + \left(\frac{dy}{dx}\right)^2}\, dx \tag{7.9}$$

Differentiating (7.8), we obtain the derivative $dy/dx = wx/T_0$; substituting into (7.9) and using the binomial theorem to expand the radical in an infinite series, we have

$$s_B = \int_0^{x_B} \sqrt{1 + \frac{w^2 x^2}{T_0^2}}\, dx = \int_0^{x_B} \left(1 + \frac{w^2 x^2}{2T_0^2} - \frac{w^4 x^4}{8T_0^4} + \cdots\right) dx$$

$$s_B = x_B \left(1 + \frac{w^2 x_B^2}{6T_0^2} - \frac{w^4 x_B^4}{40T_0^4} + \cdots\right)$$

and, since $wx_B^2/2T_0 = y_B$,

$$s_B = x_B \left[1 + \frac{2}{3}\left(\frac{y_B}{x_B}\right)^2 - \frac{2}{5}\left(\frac{y_B}{x_B}\right)^4 + \cdots\right] \tag{7.10}$$

The series converges for values of the ratio y_B/x_B less than 0.5; in most cases, this ratio is much smaller, and only the first two terms of the series need be computed.

(*a*)

(*b*)

(*c*)

Fig. 7.17

Photo 7.4 Suspension bridges, in which cables support the roadway, are used to span rivers and estuaries. The Verrazano-Narrows Bridge, which connects Staten Island and Brooklyn in New York City, has the longest span of all bridges in the United States.

†Cables hanging under their own weight are not loaded uniformly along the horizontal, and they do not form a parabola. The error introduced by assuming a parabolic shape for cables hanging under their weight, however, is small when the cable is sufficiently taut. A complete discussion of cables hanging under their own weight is given in the next section.

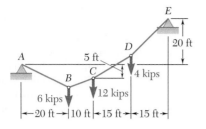

SAMPLE PROBLEM 7.8

The cable AE supports three vertical loads from the points indicated. If point C is 5 ft below the left support, determine (a) the elevation of points B and D, (b) the maximum slope and the maximum tension in the cable.

SOLUTION

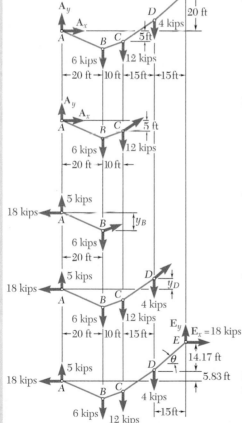

Reactions at Supports. The reaction components $\mathbf{A}_x$ and $\mathbf{A}_y$ are determined as follows:

Free Body: Entire Cable

$+\uparrow\Sigma M_E = 0$:

$A_x(20\text{ ft}) - A_y(60\text{ ft}) + (6\text{ kips})(40\text{ ft}) + (12\text{ kips})(30\text{ ft}) + (4\text{ kips})(15\text{ ft}) = 0$
$$20A_x - 60A_y + 660 = 0$$

Free Body: ABC

$+\uparrow\Sigma M_C = 0$: $\quad -A_x(5\text{ ft}) - A_y(30\text{ ft}) + (6\text{ kips})(10\text{ ft}) = 0$
$$-5A_x - 30A_y + 60 = 0$$

Solving the two equations simultaneously, we obtain

$$A_x = -18\text{ kips} \qquad \mathbf{A}_x = 18\text{ kips} \leftarrow$$
$$A_y = +5\text{ kips} \qquad \mathbf{A}_y = 5\text{ kips} \uparrow$$

a. Elevation of Points B and D.

Free Body: AB Considering the portion of cable AB as a free body, we write

$+\uparrow\Sigma M_B = 0$: $\quad (18\text{ kips})y_B - (5\text{ kips})(20\text{ ft}) = 0$
$$y_B = 5.56\text{ ft below } A \quad \blacktriangleleft$$

Free Body: ABCD. Using the portion of cable $ABCD$ as a free body, we write

$+\uparrow\Sigma M_D = 0$:

$-(18\text{ kips})y_D - (5\text{ kips})(45\text{ ft}) + (6\text{ kips})(25\text{ ft}) + (12\text{ kips})(15\text{ ft}) = 0$
$$y_D = 5.83\text{ ft above } A \quad \blacktriangleleft$$

b. Maximum Slope and Maximum Tension. We observe that the maximum slope occurs in portion DE. Since the horizontal component of the tension is constant and equal to 18 kips, we write

$$\tan\theta = \frac{14.17}{15\text{ ft}} \qquad\qquad \theta = 43.4° \quad \blacktriangleleft$$

$$T_{max} = \frac{18\text{ kips}}{\cos\theta} \qquad\qquad T_{max} = 24.8\text{ kips} \quad \blacktriangleleft$$

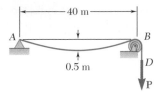

SAMPLE PROBLEM 7.9

A light cable is attached to a support at A, passes over a small pulley at B, and supports a load $\mathbf{P}$. Knowing that the sag of the cable is 0.5 m and that the mass per unit length of the cable is 0.75 kg/m, determine (a) the magnitude of the load $\mathbf{P}$, (b) the slope of the cable at B, (c) the total length of the cable from A to B. Since the ratio of the sag to the span is small, assume the cable to be parabolic. Also, neglect the weight of the portion of cable from B to D.

SOLUTION

a. Load P. We denote by C the lowest point of the cable and draw the free-body diagram of the portion CB of cable. Assuming the load to be uniformly distributed along the horizontal, we write

$$w = (0.75 \text{ kg/m})(9.81 \text{ m/s}^2) = 7.36 \text{ N/m}$$

The total load for the portion CB of cable is

$$W = wx_B = (7.36 \text{ N/m})(20 \text{ m}) = 147.2 \text{ N}$$

and is applied halfway between C and B. Summing moments about B, we write

$$+\curvearrowleft\Sigma M_B = 0: \qquad (147.2 \text{ N})(10 \text{ m}) - T_0(0.5 \text{ m}) = 0 \qquad\qquad T_0 = 2944 \text{ N}$$

From the force triangle we obtain

$$T_B = \sqrt{T_0^2 + W^2}$$
$$= \sqrt{(2944 \text{ N})^2 + (147.2 \text{ N})^2} = 2948 \text{ N}$$

Since the tension on each side of the pulley is the same, we find

$$P = T_B = 2948 \text{ N} \qquad \blacktriangleleft$$

b. Slope of Cable at B. We also obtain from the force triangle

$$\tan\theta = \frac{W}{T_0} = \frac{147.2 \text{ N}}{2944 \text{ N}} = 0.05$$

$$\theta = 2.9° \qquad \blacktriangleleft$$

c. Length of Cable. Applying Eq. (7.10) between C and B, we write

$$s_B = x_B\left[1 + \frac{2}{3}\left(\frac{y_B}{x_B}\right)^2 + \cdots\right]$$

$$= (20 \text{ m})\left[1 + \frac{2}{3}\left(\frac{0.5 \text{ m}}{20 \text{ m}}\right)^2 + \cdots\right] = 20.00833 \text{ m}$$

The total length of the cable between A and B is twice this value,

$$\text{Length} = 2s_B = 40.0167 \text{ m} \qquad \blacktriangleleft$$

In the problems of this section you will apply the equations of equilibrium to *cables that lie in a vertical plane*. We assume that a cable cannot resist bending, so that the force of tension in the cable is always directed along the cable.

A. *In the first part of this lesson we considered cables subjected to concentrated loads.* Since the weight of the cable is neglected, the cable is straight between loads.

Your solution will consist of the following steps:

1. *Draw a free-body diagram of the entire cable* showing the loads and the horizontal and vertical components of the reaction at each support. Use this free-body diagram to write the corresponding equilibrium equations.

2. *You will be confronted with four unknown components and only three equations of equilibrium* (see Fig. 7.13). You must therefore find an additional piece of information, such as the *position* of a point on the cable or the *slope* of the cable at a given point.

3. *After you have identified the point of the cable where the additional information exists,* cut the cable at that point, and draw a free-body diagram of one of the two portions of the cable you have obtained.

   a. If you know the position of the point where you have cut the cable, writing $\Sigma M = 0$ about that point for the new free body will yield the additional equation required to solve for the four unknown components of the reactions. [Sample Prob. 7.8].

   b. If you know the slope of the portion of the cable you have cut, writing $\Sigma F_x = 0$ and $\Sigma F_y = 0$ for the new free body will yield two equilibrium equations which, together with the original three, can be solved for the four reaction components and for the tension in the cable where it has been cut.

4. *To find the elevation of a given point of the cable and the slope and tension at that point* once the reactions at the supports have been found, you should cut the cable at that point and draw a free-body diagram of one of the two portions of the cable you have obtained. Writing $\Sigma M = 0$ about the given point yields its elevation. Writing $\Sigma F_x = 0$ and $\Sigma F_y = 0$ yields the components of the tension force, from which its magnitude and direction can easily be found.

(continued)

5. *For a cable supporting vertical loads only,* you will observe that *the horizontal component of the tension force is the same at any point.* It follows that, for such a cable, the *maximum tension occurs in the steepest portion of the cable.*

B. *In the second portion of this lesson we considered cables carrying a load uniformly distributed along the horizontal.* The shape of the cable is then parabolic.

Your solution will use one or more of the following concepts:

1. *Placing the origin of coordinates at the lowest point of the cable* and directing the x and y axes to the right and upward, respectively, we find that *the equation of the parabola* is

$$y = \frac{wx^2}{2T_0} \tag{7.8}$$

The minimum cable tension occurs at the origin, where the cable is horizontal, and the maximum tension is at the support where the slope is maximum.

2. *If the supports of the cable have the same elevation,* the sag h of the cable is the vertical distance from the lowest point of the cable to the horizontal line joining the supports. To solve a problem involving such a parabolic cable, you should write Eq. (7.8) for one of the supports; this equation can be solved for one unknown.

3. *If the supports of the cable have different elevations,* you will have to write Eq. (7.8) for each of the supports (see Fig. 7.17).

4. *To find the length of the cable* from the lowest point to one of the supports, you can use Eq. (7.10). In most cases, you will need to compute only the first two terms of the series.

In Sample Prob. 7.9 we considered a cable hanging under its own weight. We observe that for taut cables their weights per unit length in the horizontal direction are approximately constant. This was the external loading condition of the cables analyzed in Sec. 7.9. Thus, we are able to apply the results obtained for this type of external loading to the special case of *taut* cables hanging under their own weights if w is now taken as the weight per unit length of the cables.

Problems

7.88 Two loads are suspended as shown from cable $ABCD$. Knowing that $d_C = 1.5$ ft, determine (a) the distance d_B, (b) the components of the reaction at A, (c) the maximum tension in the cable.

7.89 Two loads are suspended as shown from cable $ABCD$. Knowing that the maximum tension in the cable is 720 lb, determine (a) the distance d_B, (b) the distance d_C.

7.90 Knowing that $d_C = 4$ m, determine (a) the reaction at A, (b) the reaction at E.

Fig. P7.88 and *P7.89*

Fig. P7.90 and P7.91

7.91 Knowing that $d_C = 2.25$ m, determine (a) the reaction at A, (b) the reaction at E.

7.92 Cable $ABCDE$ supports three loads as shown. Knowing that $d_C = 3.6$ ft, determine (a) the reaction at E, (b) the distances d_B and d_D.

7.93 Cable $ABCDE$ supports three loads as shown. Determine (a) the distance d_C for which portion CD of the cable is horizontal, (b) the corresponding reactions at the supports.

7.94 An oil pipeline is supported at 6-m intervals by vertical hangers attached to the cable shown. Due to the combined weight of the pipe and its contents, the tension in each hanger is 4 kN. Knowing that $d_C = 12$ m, determine (a) the maximum tension in the cable, (b) the distance d_D.

Fig. P7.92 and *P7.93*

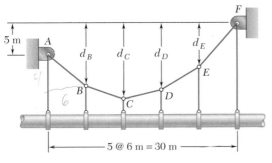

Fig. P7.94

7.95 Solve Prob. 7.94 assuming that $d_C = 9$ m.

391

Fig. P7.96 and *P7.97*

Fig. P7.100 and *P7.101*

Fig. *P7.103*

7.96 Cable *ABC* supports two boxes as shown. Knowing that *b* = 3.6 m, determine (*a*) the required magnitude of the horizontal force **P**, (*b*) the corresponding distance *a*.

7.97 Cable *ABC* supports two boxes as shown. Determine the distances *a* and *b* when a horizontal force **P** of magnitude 100 N is applied at *C*.

7.98 Knowing that W_B = 150 lb and W_C = 50 lb, determine the magnitude of the force **P** required to maintain equilibrium.

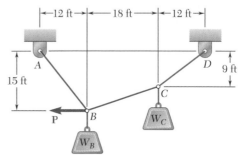

Fig. P7.98 and P7.99

7.99 Knowing that W_B = 40 lb and W_C = 22 lb, determine the magnitude of the force **P** required to maintain equilibrium.

7.100 If *a* = 4.5 m, determine the magnitudes of **P** and **Q** required to maintain the cable in the shape shown.

7.101 If *a* = 6 m, determine the magnitudes of **P** and **Q** required to maintain the cable in the shape shown.

7.102 A transmission cable having a mass per unit length of 1 kg/m is strung between two insulators at the same elevation that are 60 m apart. Knowing that the sag of the cable is 1.2 m, determine (*a*) the maximum tension in the cable, (*b*) the length of the cable.

7.103 Two cables of the same gauge are attached to a transmission tower at *B*. Since the tower is slender, the horizontal component of the resultant of the forces exerted by the cables at *B* is to be zero. Knowing that the mass per unit length of the cables is 0.4 kg/m, determine (*a*) the required sag *h*, (*b*) the maximum tension in each cable.

7.104 The center span of the George Washington Bridge, as originally constructed, consisted of a uniform roadway suspended from four cables. The uniform load supported by each cable was *w* = 9.75 kips/ft along the horizontal. Knowing that the span *L* is 3500 ft and that the sag *h* is 316 ft, determine for the original configuration (*a*) the maximum tension in each cable, (*b*) the length of each cable.

7.105 Each cable of the Golden Gate Bridge supports a load *w* = 11.1 kips/ft along the horizontal. Knowing that the span *L* is 4150 ft and that the sag *h* is 464 ft, determine (*a*) the maximum tension in each cable, (*b*) the length of each cable.

7.106 To mark the positions of the rails on the posts of a fence, a homeowner ties a cord to the post at A, passes the cord over a short piece of pipe attached to the post at B, and ties the free end of the cord to a bucket filled with bricks having a total mass of 20 kg. Knowing that the mass per unit length of the rope is 0.02 kg/m and assuming that A and B are at the same elevation, determine (a) the sag h, (b) the slope of the cable at B. Neglect the effect of friction.

Fig. P7.106

7.107 A small ship is tied to a pier with a 5-m length of rope as shown. Knowing that the current exerts on the hull of the ship a 300-N force directed from the bow to the stern and that the mass per unit length of the rope is 2.2 kg/m, determine (a) the maximum tension in the rope, (b) the sag h. [*Hint:* Use only the first two terms of Eq. (7.10).]

7.108 The center span of the Verrazano-Narrows Bridge consists of two uniform roadways suspended from four cables. The design of the bridge allowed for the effect of extreme temperature changes which cause the sag of the center span to vary from $h_w = 386$ ft in winter to $h_s = 394$ ft in summer. Knowing that the span is $L = 4260$ ft, determine the change in length of the cables due to extreme temperature changes.

Fig. P7.107

7.109 A cable of length $L + \Delta$ is suspended between two points which are at the same elevation and a distance L apart. (a) Assuming that Δ is small compared to L and that the cable is parabolic, determine the approximate sag in terms of L and Δ. (b) If $L = 30$ m and $\Delta = 1.2$ m, determine the approximate sag. [*Hint:* Use only the first two terms of Eq. (7.10).]

7.110 Each cable of the side spans of the Golden Gate Bridge supports a load $w = 10.2$ kips/ft along the horizontal. Knowing that for the side spans the maximum vertical distance h from each cable to the chord AB is 30 ft and occurs at midspan, determine (a) the maximum tension in each cable, (b) the slope at B.

Fig. P7.110

Fig. P7.111

7.111 A steam pipe weighing 50 lb/ft that passes between two buildings 60 ft apart is supported by a system of cables as shown. Assuming that the weight of the cable is equivalent to a uniformly distributed loading of 7.5 lb/ft, determine (a) the location of the lowest point C of the cable, (b) the maximum tension in the cable.

7.112 Chain AB supports a horizontal, uniform steel beam having a mass per unit length of 85 kg/m. If the maximum tension in the cable is not to exceed 8 kN, determine (a) the horizontal distance a from A to the lowest point C of the chain, (b) the approximate length of the chain.

7.113 Chain AB of length 6.4 m supports a horizontal, uniform steel beam having a mass per unit length of 85 kg/m. Determine (a) the horizontal distance a from A to the lowest point C of the chain, (b) the maximum tension in the chain.

Fig. P7.112 and P7.113

Fig. P7.114

Fig. P7.121

*7.114   A cable AB of span L and a simple beam $A'B'$ of the same span are subjected to identical vertical loadings as shown. Show that the magnitude of the bending moment at a point C' in the beam is equal to the product T_0h, where T_0 is the magnitude of the horizontal component of the tension force in the cable and h is the vertical distance between point C and the chord joining the points of support A and B.

7.115 through 7.118   Making use of the property established in Prob. 7.114, solve the problem indicated by first solving the corresponding beam problem.

> **7.115**   Prob. 7.89a.
> **7.116**   Prob. 7.92b.
> **7.117**   Prob. 7.94b.
> **7.118**   Prob. 7.95b.

*7.119   Show that the curve assumed by a cable that carries a distributed load $w(x)$ is defined by the differential equation $d^2y/dx^2 = w(x)/T_0$, where T_0 is the tension at the lowest point.

*7.120   Using the property indicated in Prob. 7.119, determine the curve assumed by a cable of span L and sag h carrying a distributed load $w = w_0 \cos(\pi x/L)$, where x is measured from mid-span. Also determine the maximum and minimum values of the tension in the cable.

*7.121   If the weight per unit length of the cable AB is $w_0/\cos^2 \theta$, prove that the curve formed by the cable is a circular arc. (*Hint:* Use the property indicated in Prob. 7.119.)

---

*7.10. CATENARY

Let us now consider a cable AB carrying a load *uniformly distributed along the cable itself* (Fig. 7.18a). Cables hanging under their own weight are loaded in this way. We denote by w the load per unit length (*measured along the cable*) and express it in N/m or in lb/ft. The magnitude W of the total load carried by a portion of cable of length s extending from the lowest point C to a point D is $W = ws$. Substituting this value for W in formula (7.6), we obtain the tension at D:

$$T = \sqrt{T_0^2 + w^2s^2}$$

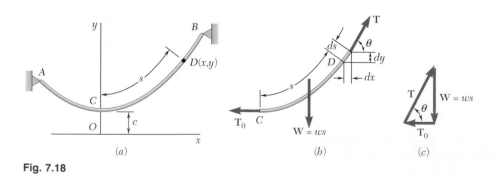

Fig. 7.18

In order to simplify the subsequent computations, we introduce the constant $c = T_0/w$. We thus write

$$T_0 = wc \qquad W = ws \qquad T = w\sqrt{c^2 + s^2} \qquad (7.11)$$

The free-body diagram of the portion of cable CD is shown in Fig. 7.18b. This diagram, however, cannot be used to obtain directly the equation of the curve assumed by the cable, since we do not know the horizontal distance from D to the line of action of the resultant **W** of the load. To obtain this equation, we first write that the horizontal projection of a small element of cable of length ds is $dx = ds \cos \theta$. Observing from Fig. 7.18c that $\cos \theta = T_0/T$ and using (7.11), we write

$$dx = ds \cos \theta = \frac{T_0}{T} ds = \frac{wc\,ds}{w\sqrt{c^2 + s^2}} = \frac{ds}{\sqrt{1 + s^2/c^2}}$$

Selecting the origin O of the coordinates at a distance c directly below C (Fig. 7.18a) and integrating from $C(0, c)$ to $D(x, y)$, we obtain†

$$x = \int_0^s \frac{ds}{\sqrt{1 + s^2/c^2}} = c\left[\sinh^{-1} \frac{s}{c} \right]_0^s = c \sinh^{-1} \frac{s}{c}$$

Photo 7.5 The forces on the supports and the internal forces in the cables of the power line shown are discussed in this section.

This equation, which relates the length s of the portion of cable CD and the horizontal distance x, can be written in the form

$$s = c \sinh \frac{x}{c} \qquad (7.15)$$

The relation between the coordinates x and y can now be obtained by writing $dy = dx \tan \theta$. Observing from Fig. 7.18c that $\tan \theta = W/T_0$ and using (7.11) and (7.15), we write

$$dy = dx \tan \theta = \frac{W}{T_0} dx = \frac{s}{c} dx = \sinh \frac{x}{c} dx$$

---

†This integral can be found in all standard integral tables. The function

$$z = \sinh^{-1} u$$

(read "arc hyperbolic sine u") is the *inverse* of the function $u = \sinh z$ (read "hyperbolic sine z"). This function and the function $v = \cosh z$ (read "hyperbolic cosine z") are defined as follows:

$$u = \sinh z = \tfrac{1}{2}(e^z - e^{-z}) \qquad v = \cosh z = \tfrac{1}{2}(e^z + e^{-z})$$

Numerical values of the functions $\sinh z$ and $\cosh z$ are found in *tables of hyperbolic functions*. They may also be computed on most calculators either directly or from the above definitions. The student is referred to any calculus text for a complete description of the properties of these functions. In this section, we use only the following properties, which are easily derived from the above definitions:

$$\frac{d \sinh z}{dz} = \cosh z \qquad \frac{d \cosh z}{dz} = \sinh z \qquad (7.12)$$

$$\sinh 0 = 0 \qquad \cosh 0 = 1 \qquad (7.13)$$

$$\cosh^2 z - \sinh^2 z = 1 \qquad (7.14)$$

Integrating from $C(0, c)$ to $D(x, y)$ and using (7.12) and (7.13), we obtain

$$y - c = \int_0^x \sinh \frac{x}{c} \, dx = c \left[\cosh \frac{x}{c} \right]_0^x = c \left(\cosh \frac{x}{c} - 1 \right)$$

$$y - c = c \cosh \frac{x}{c} - c$$

which reduces to

$$y = c \cosh \frac{x}{c} \qquad (7.16)$$

This is the equation of a *catenary* with a vertical axis. The ordinate c of the lowest point C is called the *parameter* of the catenary. Squaring both sides of Eqs. (7.15) and (7.16), subtracting, and taking (7.14) into account, we obtain the following relation between y and s:

$$y^2 - s^2 = c^2 \qquad (7.17)$$

Solving (7.17) for s^2 and carrying into the last of the relations (7.11), we write these relations as follows:

$$T_0 = wc \qquad W = ws \qquad T = wy \qquad (7.18)$$

The last relation indicates that the tension at any point D of the cable is proportional to the vertical distance from D to the horizontal line representing the x axis.

When the supports A and B of the cable have the same elevation, the distance L between the supports is called the *span* of the cable and the vertical distance h from the supports to the lowest point C is called the *sag* of the cable. These definitions are the same as those given in the case of parabolic cables, but it should be noted that because of our choice of coordinate axes, the sag h is now

$$h = y_A - c \qquad (7.19)$$

It should also be observed that certain catenary problems involve transcendental equations which must be solved by successive approximations (see Sample Prob. 7.10). When the cable is fairly taut, however, the load can be assumed uniformly distributed *along the horizontal* and the catenary can be replaced by a parabola. This greatly simplifies the solution of the problem, and the error introduced is small.

When the supports A and B have different elevations, the position of the lowest point of the cable is not known. The problem can then be solved in a manner similar to that indicated for parabolic cables, by expressing that the cable must pass through the supports and that $x_B - x_A = L$ and $y_B - y_A = d$, where L and d denote, respectively, the horizontal and vertical distances between the two supports.

SAMPLE PROBLEM 7.10

A uniform cable weighing 3 lb/ft is suspended between two points A and B as shown. Determine (a) the maximum and minimum values of the tension in the cable, (b) the length of the cable.

SOLUTION

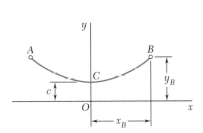

Equation of Cable. The origin of coordinates is placed at a distance c below the lowest point of the cable. The equation of the cable is given by Eq. (7.16),

$$y = c \cosh \frac{x}{c}$$

The coordinates of point B are

$$x_B = 250 \text{ ft} \qquad y_B = 100 + c$$

Substituting these coordinates into the equation of the cable, we obtain

$$100 + c = c \cosh \frac{250}{c}$$

$$\frac{100}{c} + 1 = \cosh \frac{250}{c}$$

The value of c is determined by assuming successive trial values, as shown in the following table:

c	$\dfrac{250}{c}$	$\dfrac{100}{c}$	$\dfrac{100}{c} + 1$	$\cosh \dfrac{250}{c}$
300	0.833	0.333	1.333	1.367
350	0.714	0.286	1.286	1.266
330	0.758	0.303	1.303	1.301
328	0.762	0.305	1.305	1.305

Taking $c = 328$, we have

$$y_B = 100 + c = 428 \text{ ft}$$

_a_. Maximum and Minimum Values of the Tension. Using Eqs. (7.18), we obtain

$$T_{\min} = T_0 = wc = (3 \text{ lb/ft})(328 \text{ ft}) \qquad T_{\min} = 984 \text{ lb} \quad \blacktriangleleft$$
$$T_{\max} = T_B = wy_B = (3 \text{ lb/ft})(428 \text{ ft}) \qquad T_{\max} = 1284 \text{ lb} \quad \blacktriangleleft$$

_b_. Length of Cable. One-half the length of the cable is found by solving Eq. (7.17):

$$y_B^2 - s_{CB}^2 = c^2 \qquad s_{CB}^2 = y_B^2 - c^2 = (428)^2 - (328)^2 \qquad s_{CB} = 275 \text{ ft}$$

The total length of the cable is therefore

$$s_{AB} = 2s_{CB} = 2(275 \text{ ft}) \qquad s_{AB} = 550 \text{ ft} \quad \blacktriangleleft$$

397

In the last section of this chapter you learned to solve problems involving a *cable carrying a load uniformly distributed along the cable*. The shape assumed by the cable is a catenary and is defined by the equation:

$$y = c \cosh \frac{x}{c} \qquad (7.16)$$

1. You should keep in mind that the origin of coordinates for a catenary is located at a distance c directly below the lowest point of the catenary. The length of the cable from the origin to any point is expressed as

$$s = c \sinh \frac{x}{c} \qquad (7.15)$$

2. You should first identify all of the known and unknown quantities. Then consider each of the equations listed in the text (Eqs. 7.15 through 7.19), and solve an equation that contains only one unknown. Substitute the value found into another equation, and solve that equation for another unknown.

3. If the sag h is given, use Eq. (7.19) to replace y by $h + c$ in Eq. (7.16) if x is known [Sample Prob. 7.10], or in Eq. (7.17) if s is known, and solve the equation obtained for the constant c.

4. Many of the problems that you will encounter will involve the solution by trial and error of an equation involving a hyperbolic sine or cosine. You can make your work easier by keeping track of your calculations in a table, as in Sample Prob. 7.10. Of course, you also can use a computer- or calculator-based numerical solution technique.

5. It is important to recognize that the defining equation may have more than one solution. To help you visualize this possibility, we encourage you first to graph the equation and then to solve for the root(s). If there is more than one root, you must determine which of the roots correspond to physically possible situations.

Problems

7.122 Two hikers are standing 30-ft apart and are holding the ends of a 35-ft length of rope as shown. Knowing that the weight per unit length of the rope is 0.05 lb/ft, determine (*a*) the sag *h*, (*b*) the magnitude of the force exerted on the hand of a hiker.

7.123 A 60-ft chain weighing 120 lb is suspended between two points at the same elevation. Knowing that the sag is 24 ft, determine (*a*) the distance between the supports, (*b*) the maximum tension in the chain.

7.124 A 200-ft steel surveying tape weighs 4 lb. If the tape is stretched between two points at the same elevation and pulled until the tension at each end is 16 lb, determine the horizontal distance between the ends of the tape. Neglect the elongation of the tape due to the tension.

7.125 An electric transmission cable of length 130 m and mass per unit length of 3.4 kg/m is suspended between two points at the same elevation. Knowing that the sag is 30 m, determine the horizontal distance between the supports and the maximum tension.

7.126 A 30 m length of wire having a mass per unit length of 0.3 kg/m is attached to a fixed support at *A* and to a collar at *B*. Neglecting the effect of friction, determine (*a*) the force **P** for which *h* = 12 m, (*b*) the corresponding span *L*.

7.127 A 30-m length of wire having a mass per unit length of 0.3 kg/m is attached to a fixed support at *A* and to a collar at *B*. Knowing that the magnitude of the horizontal force applied to the collar is *P* = 30 N, determine (*a*) the sag *h*, (*b*) the corresponding span *L*.

7.128 A 30-m length of wire having a mass per unit length of 0.3 kg/m is attached to a fixed support at *A* and to a collar at *B*. Neglecting the effect of friction, determine (*a*) the sag *h* for which *L* = 22.5 m, (*b*) the corresponding force **P**.

7.129 A 30-ft wire is suspended from two points at the same elevation that are 20 ft apart. Knowing that the maximum tension is 80 lb, determine (*a*) the sag of the wire, (*b*) the total weight of the wire.

7.130 Determine the sag of a 45-ft chain which is attached to two points at the same elevation that are 20 ft apart.

7.131 A 10-m rope is attached to two supports *A* and *B* as shown. Determine (*a*) the span of the rope for which the span is equal to the sag, (*b*) the corresponding angle θ_B.

7.132 A cable having a mass per unit length of 3 kg/m is suspended between two points at the same elevation that are 48 m apart. Determine the smallest allowable sag of the cable if the maximum tension is not to exceed 1800 N.

Fig. P7.122

Fig. P7.126, *P7.127,* and P7.128

Fig. P7.131

7.133 An 8-m length of chain having a mass per unit length of 3.72 kg/m is attached to a beam at *A* and passes over a small pulley at *B* as shown. Neglecting the effect of friction, determine the values of distance *a* for which the chain is in equilibrium.

Fig. P7.133

7.134 A motor *M* is used to slowly reel in the cable shown. Knowing that the weight per unit length of the cable is 0.5 lb/ft, determine the maximum tension in the cable when $h = 15$ ft.

Fig. P7.134 and *P7.135*

7.135 A motor *M* is used to slowly reel in the cable shown. Knowing that the weight per unit length of the cable is 0.5 lb/ft, determine the maximum tension in the cable when $h = 9$ ft.

7.136 To the left of point *B* the long cable *ABDE* rests on the rough horizontal surface shown. Knowing that the weight per unit length of the cable is 1.5 lb/ft, determine the force **F** when $a = 10.8$ ft.

Fig. P7.136 and P7.137

7.137 To the left of point *B* the long cable *ABDE* rests on the rough horizontal surface shown. Knowing that the weight per unit length of the cable is 1.5 lb/ft, determine the force **F** when $a = 18$ ft.

7.138 The uniform cable shown in Fig. P7.138 and P7.139 has a mass per unit length of 4 kg/m and is held in the position shown by a horizontal force **P** applied at *B*. Knowing that $P = 800$ N and $\theta_A = 60°$, determine (*a*) the location of point *B*, (*b*) the length of the cable.

Fig. P7.138 and P7.139

7.139 A uniform cable having a mass per unit length of 4 kg/m is held in the position shown by a horizontal force **P** applied at B. Knowing that $P = 600$ N and $\theta_A = 60°$, determine (a) the location of point B, (b) the length of the cable.

7.140 The cable ACB weighs 0.3 lb/ft. Knowing that the lowest point of the cable is located at a distance $a = 1.8$ ft below the support A, determine (a) the location of the lowest point C, (b) the maximum tension in the cable.

Fig. P7.140 and P7.141

7.141 The cable ACB weighs 0.3 lb/ft. Knowing that the lowest point of the cable is located at a distance $a = 6$ ft below the support A, determine (a) the location of the lowest point C, (b) the maximum tension in the cable.

7.142 Denoting by θ the angle formed by a uniform cable and the horizontal, show that at any point (a) $s = c \tan \theta$, (b) $y = c \sec \theta$.

7.143 (a) Determine the maximum allowable horizontal span for a uniform cable of mass per unit length m' if the tension in the cable is not to exceed a given value T_m. (b) Using the result of part a, determine the maximum span of a steel wire for which $m' = 0.34$ kg/m and $T_m = 32$ kN.

***7.144** A cable has a weight per unit length of 2 lb/ft and is supported as shown. Knowing that the span L is 18 ft, determine the two values of the sag h for which the maximum tension is 80 lb.

***7.145** Determine the sag-to-span ratio for which the maximum tension in the cable is equal to the total weight of the entire cable AB.

***7.146** A cable of weight w per unit length is suspended between two points at the same elevation that are a distance L apart. Determine (a) the sag-to-span ratio for which the maximum tension is as small as possible, (b) the corresponding values of θ_B and T_m.

Fig. P7.144, P7.145, and P7.146

In this chapter you learned to determine the internal forces which hold together the various parts of a given member in a structure.

Forces in straight two-force members

Considering first a *straight two-force member AB* [Sec. 7.2], we recall that such a member is subjected at A and B to equal and opposite forces $\mathbf{F}$ and $-\mathbf{F}$ directed along AB (Fig. 7.19a). Cutting member AB at C and drawing the free-body diagram of portion AC, we conclude that the internal forces which existed at C in member AB are equivalent to an *axial force* $-\mathbf{F}$ equal and opposite to $\mathbf{F}$ (Fig. 7.19b). We note that in the case of a two-force member which is not straight, the internal forces reduce to a force-couple system and not to a single force.

(a)   *(b)*

Fig. 7.19

(a)

Fig. 7.20

Forces in multiforce members

Considering next a *multiforce member AD* (Fig. 7.20a), cutting it at J, and drawing the free-body diagram of portion JD, we conclude that the internal forces at J are equivalent to a force-couple system consisting of the *axial force* $\mathbf{F}$, the *shearing force* $\mathbf{V}$, and a couple $\mathbf{M}$ (Fig. 7.20b). The magnitude of the shearing force measures the *shear* at point J, and the moment of the couple is referred to as the *bending moment* at J. Since an equal and opposite force-couple system would have been obtained by considering the free-body diagram of portion AJ, it is necessary to specify which portion of member AD was used when recording the answers [Sample Prob. 7.1].

Forces in beams

Most of the chapter was devoted to the analysis of the internal forces in two important types of engineering structures: *beams* and *cables*. *Beams* are usually long, straight prismatic members designed to support loads applied at various points along the member. In general the loads are perpendicular to the axis of the beam and produce only *shear and bending* in the beam. The loads may be either *concentrated* at specific points, or *distributed* along the entire length or a portion of the beam. The beam itself may be supported in various ways; since only statically determinate beams are considered in this text, we limited our analysis to that of *simply supported beams*, *overhanging beams*, and *cantilever beams* [Sec. 7.3].

To obtain the *shear V* and *bending moment M* at a given point C of a beam, we first determine the reactions at the supports by considering the entire beam as a free body. We then cut the beam at C and use the free-body diagram of one of the two portions obtained in this fashion to determine V and M. In order to avoid any confusion regarding the sense of the shearing force **V** and couple **M** (which act in opposite directions on the two portions of the beam), the sign convention illustrated in Fig. 7.21 was adopted [Sec. 7.4]. Once the values of the shear and bending moment have been determined at a few selected points of the beam, it is usually possible to draw a *shear diagram* and a *bending-moment diagram* representing, respectively, the shear and bending moment at any point of the beam [Sec. 7.5]. When a beam is subjected to concentrated loads only, the shear is of constant value between loads and the bending moment varies linearly between loads [Sample Prob. 7.2]. On the other hand, when a beam is subjected to distributed loads, the shear and bending moment vary quite differently [Sample Prob. 7.3].

The construction of the shear and bending-moment diagrams is facilitated if the following relations are taken into account. Denoting by w the distributed load per unit length (assumed positive if directed downward), we have [Sec. 7.5]:

$$\frac{dV}{dx} = -w \tag{7.1}$$

$$\frac{dM}{dx} = V \tag{7.3}$$

or, in integrated form,

$V_D - V_C = -$(area under load curve between C and D) (7.2′)
$M_D - M_C = $ area under shear curve between C and D   (7.4′)

Equation (7.2′) makes it possible to draw the shear diagram of a beam from the curve representing the distributed load on that beam and the value of V at one end of the beam. Similarly, Eq. (7.4′) makes it possible to draw the bending-moment diagram from the shear diagram and the value of M at one end of the beam. However, concentrated loads introduce discontinuities in the shear diagram and concentrated couples introduce discontinuities in the bending-moment diagram, none of which are accounted for in these equations [Sample Probs. 7.4 and 7.7]. Finally, we note from Eq. (7.3) that the points of the beam where the bending moment is maximum or minimum are also the points where the shear is zero [Sample Prob. 7.5].

Shear and bending moment in a beam

Internal forces at section
(positive shear and positive bending moment)
Fig. 7.21

Relations among load, shear, and bending moment

Cables with concentrated loads

Fig. 7.22

Cables with distributed loads

Fig. 7.23

Parabolic cable

Fig. 7.24

Catenary

Fig. 7.25

The second half of the chapter was devoted to the analysis of *flexible cables*. We first considered a cable of negligible weight supporting *concentrated loads* [Sec. 7.7]. Using the entire cable AB as a free body (Fig. 7.22), we noted that the three available equilibrium equations were not sufficient to determine the four unknowns representing the reactions at the supports A and B. However, if the coordinates of a point D of the cable are known, an additional equation can be obtained by considering the free-body diagram of the portion AD or DB of the cable. Once the reactions at the supports have been determined, the elevation of any point of the cable and the tension in any portion of the cable can be found from the appropriate free-body diagram [Sample Prob. 7.8]. It was noted that the horizontal component of the force **T** representing the tension is the same at any point of the cable.

We next considered cables carrying *distributed loads* [Sec. 7.8]. Using as a free body a portion of cable CD extending from the lowest point C to an arbitrary point D of the cable (Fig. 7.23), we observed that the horizontal component of the tension force **T** at D is constant and equal to the tension T_0 at C, while its vertical component is equal to the weight W of the portion of cable CD. The magnitude and direction of **T** were obtained from the force triangle:

$$T = \sqrt{T_0^2 + W^2} \qquad \tan \theta = \frac{W}{T_0} \qquad (7.6)$$

In the case of a load *uniformly distributed along the horizontal*—as in a suspension bridge (Fig. 7.24)—the load supported by portion CD is $W = wx$, where w is the constant load per unit horizontal length [Sec. 7.9]. We also found that the curve formed by the cable is a *parabola* of equation

$$y = \frac{wx^2}{2T_0} \qquad (7.8)$$

and that the length of the cable can be found by using the expansion in series given in Eq. (7.10) [Sample Prob. 7.9].

In the case of a load *uniformly distributed along the cable itself* [for example, a cable hanging under its own weight (Fig. 7.25)] the load supported by portion CD is $W = ws$, where s is the length measured along the cable and w is the constant load per unit length [Sec. 7.10]. Choosing the origin O of the coordinate axes at a distance $c = T_0/w$ below C, we derived the relations

$$s = c \sinh \frac{x}{c} \qquad (7.15)$$

$$y = c \cosh \frac{x}{c} \qquad (7.16)$$

$$y^2 - s^2 = c^2 \qquad (7.17)$$

$$T_0 = wc \qquad W = ws \qquad T = wy \qquad (7.18)$$

which can be used to solve problems involving cables hanging under their own weight [Sample Prob. 7.10]. Equation (7.16), which defines the shape of the cable, is the equation of a *catenary*.

7.147 For the beam and loading shown, (*a*) draw the shear and bending moment diagrams, (*b*) determine the maximum absolute values of the shear and bending moment.

Fig. P7.147

7.148 For the beam and loading shown, (*a*) draw the shear and bending moment diagrams, (*b*) determine the maximum absolute values of the shear and bending moment.

Fig. *P7.148*

7.149 Two loads are suspended as shown from the cable *ABCD*. Knowing that $h_B = 1.8$ m, determine (*a*) the distance h_C, (*b*) the components of the reaction at *D*, (*c*) the maximum tension in the cable.

Fig. P7.149 and *P7.150*

7.150 Knowing that the maximum tension in cable *ABCD* is 15 kN, determine (*a*) the distance h_B, (*b*) the distance h_C.

7.151 A semicircular rod of weight W and uniform cross section is supported as shown. Determine the bending moment at point J when $\theta = 60°$.

7.152 A semicircular rod of weight W and uniform cross section is supported as shown. Determine the bending moment at point J when $\theta = 150°$.

7.153 Determine the internal forces at point J of the structure shown.

Fig. P7.151 and P7.152

Fig. P7.153 and *P7.154*

7.154 Determine the internal forces at point K of the structure shown.

7.155 Two small channel sections DF and EH have been welded to the uniform beam AB of weight $W = 3$ kN to form the rigid structural member shown. This member is being lifted by two cables attached at D and E. Knowing that $\theta = 30°$ and neglecting the weight of the channel sections, (*a*) draw the shear and bending-moment diagrams for beam AB, (*b*) determine the maximum absolute values of the shear and bending moment in the beam.

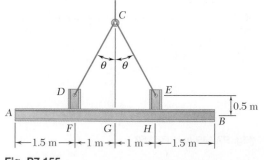

Fig. P7.155

7.156 (a) Draw the shear and bending moment diagrams for beam AB, (b) determine the magnitude and location of the maximum absolute value of the bending momkent.

Fig. P7.156

7.157 Cable ABC supports two loads as shown. Knowing that $b = 4$ ft, determine (a) the required magnitude of the horizontal force **P**, (b) the corresponding distance a.

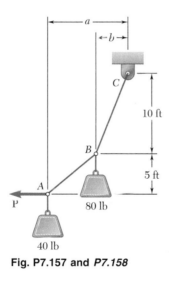

Fig. P7.157 and *P7.158*

7.158 Cable ABC supports two loads as shown. Determine the distances a and b when a horizontal force **P** of magnitude 60 lb is applied at A.

Computer Problems

Fig. P7.C1

Fig. P7.C2

7.C1 A semicircular member of radius 20 in. is loaded as shown. Plot the internal forces (axial force, shearing force, and bending moment) as functions of θ for $0 < \theta < 90°$.

7.C2 The axis of the curved member AB is a parabola for which the vertex is at A and $L = nh$. Knowing that a vertical load **P** of magnitude 1.2 kN is applied at A, determine the internal forces and bending moment at an arbitrary point J. Plot the internal forces and bending moment as functions of x for $0.25L \le x \le 0.75L$ when $L = 400$ mm and (*a*) $n = 2$, (*b*) $n = 3$, (*c*) $n = 4$.

7.C3 The floor of a bridge will consist of narrow planks resting on two simply supported beams, one of which is shown in the figure. As part of the design of the bridge, it is desired to simulate the effect that driving a 12-kN truck over the bridge will have on this beam. The distance between the truck's axles is 1.8 m, and it assumed that the weight of the truck is equally distributed over its four wheels. (*a*) Using computational software, calculate, using 0.15-m increments, and plot the magnitude and location of the maximum bending moment in the beam as a function of x for -0.9 m $\le x \le 3$ m. (*b*) Using smaller increments if necessary, determine the largest value of the bending moment that occurs in the beam as the truck is driven over the bridge, and determine the corresponding value of x.

Flg. P7.C3

7.C4 Determine the equations for the shear and bending moment curves for beams 1 and 2 shown assuming that $w_0 = 15$ kN/m and $L = 3$ m. Plot the shear and bending moment diagrams for each beam.

Beam 1

Beam 2

Fig. P7.C4

7.C5 A beam *AB* hinged at *B* and supported by a roller at *D* is to be designed to carry a load uniformly distributed from its end *A* to its midpoint *C* with maximum efficiency. As part of the design process, use computational software to determine the distance *a* from end *A* to the point *D* where the roller should be placed to minimize the absolute value of the bending moment *M* in the beam. (*Hint:* A short preliminary analysis will show that the roller should be placed under the load and that the largest negative value of *M* will occur at *D*, while its largest positive value will occur somewhere between *D* and *C*. Thus, you should draw the bending-moment diagram and then equate the absolute values of the largest positive and negative bending moments obtained.)

Fig. P7.C5

7.C6 Cable *AB* supports a load distributed uniformly along the horizontal as shown. The lowest portion of the cable is located at a distance *a* = 3.6 m below support *A*, and support *B* is located a distance *b* = *na* above *A*. (*a*) Determine the maximum tension in the cable as a function of *n*. (*b*) Plot the maximum tension in the cable for $2 \leq n \leq 6$.

7.C7 Using computational software, solve Prob. 7.127 for values of *P* from 0 to 75 N using 5-N increments.

Fig. P7.C6

7.C8 A typical transmission-line installation consists of a cable of length s_{AB} and mass per unit length *m′* suspended as shown between two points at the same elevation. Using computational software, construct a table that can be used in the design of future installations. The table should present the dimensionless quantities h/L, s_{AB}/L, $T_0/m'gL$ and $T_{max}/m'gL$ for values of c/L from 0.2 to 0.5 using 0.025 increments and from 1 to 4 using 0.5 increments.

7.C9 The electrician shown is installing an electric cable weighing 0.2 lb/ft. As the electrician slowly climbs the ladder, the cable unwinds from the spool so that the tension T_B in the cable at *B* is 1.2 lb. Knowing that *h* = 4 ft when the electrician is standing on the ground, plot the magnitude of the force exerted on the hand of the electrician by the cable as a function of *h* for 4 ft $\leq h \leq$ 16 ft.

Fig. P7.C8

Fig. P7.C9

Friction

The operation of the bicycle brake shown depends upon the normal and friction forces exerted by the brake pads on the rim of the wheel as the pads are pressed against the rim.

FRICTION

8.1   Introduction
8.2   The Laws of Dry Friction.
      Coefficients of Friction
8.3   Angles of Friction
8.4   Problems Involving Dry Friction
8.5   Wedges
8.6   Square-Threaded Screws
8.7   Journal Bearings. Axle Friction
8.8   Thrust Bearings. Disk Friction
8.9   Wheel Friction. Rolling Resistance
8.10 Belt Friction

8.1. INTRODUCTION

In the preceding chapters, it was assumed that surfaces in contact were either *frictionless* or *rough.* If they were frictionless, the force each surface exerted on the other was normal to the surfaces and the two surfaces could move freely with respect to each other. If they were rough, it was assumed that tangential forces could develop to prevent the motion of one surface with respect to the other.

This view was a simplified one. Actually, no perfectly frictionless surface exists. When two surfaces are in contact, tangential forces, called *friction forces,* will always develop if one attempts to move one surface with respect to the other. On the other hand, these friction forces are limited in magnitude and will not prevent motion if sufficiently large forces are applied. The distinction between frictionless and rough surfaces is thus a matter of degree. This will be seen more clearly in the present chapter, which is devoted to the study of friction and of its applications to common engineering situations.

There are two types of friction: *dry friction,* sometimes called *Coulomb friction,* and *fluid friction.* Fluid friction develops between layers of fluid moving at different velocities. Fluid friction is of great importance in problems involving the flow of fluids through pipes and orifices or dealing with bodies immersed in moving fluids. It is also basic in the analysis of the motion of *lubricated mechanisms.* Such problems are considered in texts on fluid mechanics. The present study is limited to dry friction, that is, to problems involving rigid bodies which are in contact along *nonlubricated* surfaces.

In the first part of this chapter, the equilibrium of various rigid bodies and structures, assuming dry friction at the surfaces of contact, is analyzed. Later a number of specific engineering applications where dry friction plays an important role are considered: wedges, square-threaded screws, journal bearings, thrust bearings, rolling resistance, and belt friction.

8.2. THE LAWS OF DRY FRICTION. COEFFICIENTS OF FRICTION

The laws of dry friction are exemplified by the following experiment. A block of weight $\mathbf{W}$ is placed on a horizontal plane surface (Fig. 8.1a). The forces acting on the block are its weight $\mathbf{W}$ and the reaction of the surface. Since the weight has no horizontal component, the reaction of the surface also has no horizontal component; the reaction is therefore *normal* to the surface and is represented by $\mathbf{N}$ in Fig. 8.1a. Suppose, now, that a horizontal force $\mathbf{P}$ is applied to the block (Fig. 8.1b). If $\mathbf{P}$ is small, the block will not move; some other horizontal force must therefore exist, which balances $\mathbf{P}$. This other force is the *static-friction force* $\mathbf{F}$, which is actually the resultant of a great number of forces acting over the entire surface of contact between the block and the plane. The nature of these forces is not known exactly, but it is generally assumed that these forces are due to the irregularities of the surfaces in contact and, to a certain extent, to molecular attraction.

If the force **P** is increased, the friction force **F** also increases, continuing to oppose **P**, until its magnitude reaches a certain *maximum value* F_m (Fig. 8.1c). If **P** is further increased, the friction

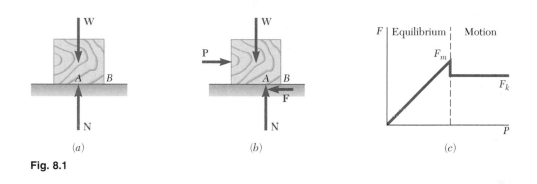

(a)   (b)   (c)

Fig. 8.1

force cannot balance it any more and the block starts sliding.† As soon as the block has been set in motion, the magnitude of **F** drops from F_m to a lower value F_k. This is because there is less interpenetration between the irregularities of the surfaces in contact when these surfaces move with respect to each other. From then on, the block keeps sliding with increasing velocity while the friction force, denoted by $\mathbf{F}_k$ and called the *kinetic-friction force*, remains approximately constant.

Experimental evidence shows that the maximum value F_m of the static-friction force is proportional to the normal component N of the reaction of the surface. We have

$$F_m = \mu_s N \qquad (8.1)$$

where μ_s is a constant called the *coefficient of static friction*. Similarly, the magnitude F_k of the kinetic-friction force may be put in the form

$$F_k = \mu_k N \qquad (8.2)$$

where μ_k is a constant called the *coefficient of kinetic friction*. The coefficients of friction μ_s and μ_k do not depend upon the area of the surfaces in contact. Both coefficients, however, depend strongly on the *nature* of the surfaces in contact. Since they also depend upon the exact condition of the surfaces, their value is seldom known with an accuracy greater than 5 percent. Approximate values of coefficients

---

†It should be noted that, as the magnitude F of the friction force increases from 0 to F_m, the point of application A of the resultant **N** of the normal forces of contact moves to the right, so that the couples formed, respectively, by **P** and **F** and by **W** and **N** remain balanced. If **N** reaches B before F reaches its maximum value F_m, the block will tip about B before it can start sliding (see Probs. 8.15 and 8.16).

of static friction for various dry surfaces are given in Table 8.1. The corresponding values of the coefficient of kinetic friction would be about 25 percent smaller. Since coefficients of friction are dimensionless quantities, the values given in Table 8.1 can be used with both SI units and U.S. customary units.

Table 8.1. Approximate Values of Coefficient of Static Friction for Dry Surfaces

Metal on metal	0.15–0.60
Metal on wood	0.20–0.60
Metal on stone	0.30–0.70
Metal on leather	0.30–0.60
Wood on wood	0.25–0.50
Wood on leather	0.25–0.50
Stone on stone	0.40–0.70
Earth on earth	0.20–1.00
Rubber on concrete	0.60–0.90

(a) No friction ($P_x = 0$)

$$F = 0$$
$$N = P + W$$

(b) No motion ($P_x < F_m$)

$$F = P_x$$
$$F < \mu_s N$$
$$N = P_y + W$$

(c) Motion impending $\longrightarrow$ ($P_x = F_m$)

$$F_m = P_x$$
$$F_m = \mu_s N$$
$$N = P_y + W$$

(d) Motion $\longrightarrow$ ($P_x > F_m$)

$$F_k < P_x$$
$$F_k = \mu_k N$$
$$N = P_y + W$$

Fig. 8.2

From the description given above, it appears that four different situations can occur when a rigid body is in contact with a horizontal surface:

1. The forces applied to the body do not tend to move it along the surface of contact; there is no friction force (Fig. 8.2a).

2. The applied forces tend to move the body along the surface of contact but are not large enough to set it in motion. The friction force **F** which has developed can be found by solving the equations of equilibrium for the body. Since there is no evidence that **F** has reached its maximum value, the equation $F_m = \mu_s N$ *cannot be used* to determine the friction force (Fig. 8.2b).

3. The applied forces are such that the body is just about to slide. We say that *motion is impending*. The friction force **F** has reached its maximum value F_m and, together with the normal force **N**, balances the applied forces. Both the equations of equilibrium and the equation $F_m = \mu_s N$ *can be used*. We also note that the friction force has a sense opposite to the sense of impending motion (Fig. 8.2c).

4. The body is sliding under the action of the applied forces, and the equations of equilibrium do not apply any more. However, **F** is now equal to F_k and the equation $F_k = \mu_k N$ may be used. The sense of F_k is opposite to the sense of motion (Fig. 8.2d).

It is sometimes convenient to replace the normal force **N** and the friction force **F** by their resultant **R**. Let us consider again a block of weight **W** resting on a horizontal plane surface. If no horizontal force is applied to the block, the resultant **R** reduces to the normal force **N** (Fig. 8.3*a*). However, if the applied force **P** has a horizontal component $\mathbf{P}_x$ which tends to move the block, the force **R** will have a horizontal component **F** and, thus, will form an angle ϕ with the normal to the surface (Fig. 8.3*b*). If $\mathbf{P}_x$ is increased until motion becomes impending, the angle between **R** and the vertical grows and reaches a maximum value (Fig. 8.3*c*). This value is called the *angle of static friction* and is denoted by ϕ_s. From the geometry of Fig. 8.3*c*, we note that

$$\tan \phi_s = \frac{F_m}{N} = \frac{\mu_s N}{N}$$

$$\tan \phi_s = \mu_s \tag{8.3}$$

If motion actually takes place, the magnitude of the friction force drops to F_k; similarly, the angle ϕ between **R** and **N** drops to a lower value ϕ_k, called the *angle of kinetic friction* (Fig. 8.3*d*). From the geometry of Fig. 8.3*d*, we write

$$\tan \phi_k - \frac{F_k}{N} = \frac{\mu_k N}{N}$$

$$\tan \phi_k = \mu_k \tag{8.4}$$

Another example will show how the angle of friction can be used to advantage in the analysis of certain types of problems. Consider a block resting on a board and subjected to no other force than its weight **W** and the reaction **R** of the board. The board can be given any desired inclination. If the board is horizontal, the force **R** exerted by the board on the block is perpendicular to the board and balances the weight **W** (Fig. 8.4*a*). If the board is given a small angle of inclina-

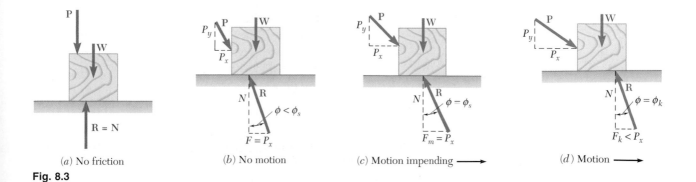

(*a*) No friction   (*b*) No motion   (*c*) Motion impending ⟶   (*d*) Motion ⟶

Fig. 8.3

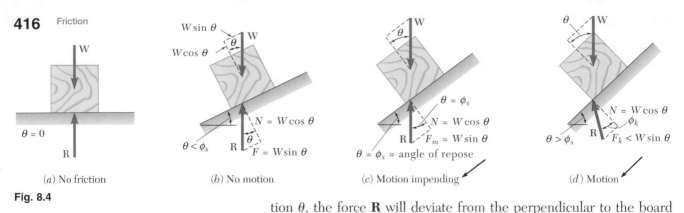

$\theta = 0$

R

(a) No friction

$W \sin \theta$ W

$W \cos \theta$

θ

$N = W \cos \theta$

θ

$\theta < \phi_s$ R $F = W \sin \theta$

(b) No motion

θ W

$\theta = \phi_s$

$N = W \cos \theta$

R $F_m = W \sin \theta$

$\theta = \phi_s = $ angle of repose

(c) Motion impending

θ W

$N = W \cos \theta$

ϕ_k

$\theta > \phi_s$ R $F_k < W \sin \theta$

(d) Motion

Fig. 8.4

Photo 8.1 The coefficient of static friction between a package and the inclined conveyer belt must be sufficiently large to enable the package to be transported without slipping.

tion θ, the force **R** will deviate from the perpendicular to the board by the angle θ and will keep balancing **W** (Fig. 8.4b); it will then have a normal component **N** of magnitude $N = W \cos \theta$ and a tangential component **F** of magnitude $F = W \sin \theta$.

If we keep increasing the angle of inclination, motion will soon become impending. At that time, the angle between **R** and the normal will have reached its maximum value ϕ_s (Fig. 8.4c). The value of the angle of inclination corresponding to impending motion is called the *angle of repose.* Clearly, the angle of repose is equal to the angle of static friction ϕ_s. If the angle of inclination θ is further increased, motion starts and the angle between **R** and the normal drops to the lower value ϕ_k (Fig. 8.4d). The reaction **R** is not vertical any more, and the forces acting on the block are unbalanced.

8.4. PROBLEMS INVOLVING DRY FRICTION

Problems involving dry friction are found in many engineering applications. Some deal with simple situations such as the block sliding on a plane described in the preceding sections. Others involve more complicated situations as in Sample Prob. 8.3; many deal with the stability of rigid bodies in accelerated motion and will be studied in dynamics. Also, a number of common machines and mechanisms can be analyzed by applying the laws of dry friction. These include wedges, screws, journal and thrust bearings, and belt transmissions. They will be studied in the following sections.

The *methods* which should be used to solve problems involving dry friction are the same that were used in the preceding chapters. If a problem involves only a motion of translation, with no possible rotation, the body under consideration can usually be treated as a particle, and the methods of Chap. 2 can be used. If the problem involves a possible rotation, the body must be considered as a rigid body, and the methods of Chap. 4 should be used. If the structure considered is made of several parts, the principle of action and reaction must be used as was done in Chap. 6.

If the body considered is acted upon by more than three forces (including the reactions at the surfaces of contact), the reaction at each surface will be represented by its components **N** and **F** and the problem will be solved from the equations of equilibrium. If only three forces act on the body under consideration, it may be more convenient to represent each reaction by a single force **R** and to solve the problem by drawing a force triangle.

Most problems involving friction fall into one of the following *three groups:* In the *first group* of problems, all applied forces are given and the coefficients of friction are known; we are to determine whether the body considered will remain at rest or slide. The friction force **F** *required to maintain equilibrium* is unknown (its magnitude is *not* equal to $\mu_s N$) and should be determined, together with the normal force **N**, by drawing a free-body diagram and *solving the equations of equilibrium* (Fig. 8.5a). The value found for the magnitude F of the friction force is then compared with the maximum value $F_m = \mu_s N$. If F is smaller than or equal to F_m, the body remains at rest. If the value found for F is larger than F_m, equilibrium cannot be maintained and motion takes place; the actual magnitude of the friction force is then $F_k = \mu_k N$.

In problems of the *second group*, all applied forces are given and the motion is known to be impending; we are to determine the value of the coefficient of static friction. Here again, we determine the friction force and the normal force by drawing a free-body diagram and solving the equations of equilibrium (Fig. 8.5b). Since we know that the value found for F is the maximum value F_m, the coefficient of friction may be found by writing and solving the equation $F_m = \mu_s N$.

In problems of the *third group*, the coefficient of static friction is given, and it is known that the motion is impending in a given direction; we are to determine the magnitude or the direction of one of the applied forces. The friction force should be shown in the free-body diagram with a *sense opposite to that of the impending motion* and with a magnitude $F_m = \mu_s N$ (Fig. 8.5c). The equations of equilibrium can then be written, and the desired force determined.

As noted above, when only three forces are involved it may be more convenient to represent the reaction of the surface by a single force **R** and to solve the problem by drawing a force triangle. Such a solution is used in Sample Prob. 8.2.

When two bodies A and B are in contact (Fig. 8.6a), the forces of friction exerted, respectively, by A on B and by B on A are equal and opposite (Newton's third law). In drawing the free-body diagram of one of the bodies, it is important to include the appropriate friction force with its correct sense. The following rule should then be observed: *The sense of the friction force acting on A is opposite to that of the motion (or impending motion) of A as observed from B* (Fig. 8.6b).† The sense of the friction force acting on B is determined in a similar way (Fig. 8.6c). Note that the motion of A as observed from B is a *relative motion*. For example, if body A is fixed and body B moves, body A will have a relative motion with respect to B. Also, if both B and A are moving down but B is moving faster than A, body A will be observed, from B, to be moving up.

†It is therefore *the same as that of the motion of B as observed from A.*

Fig. 8.5

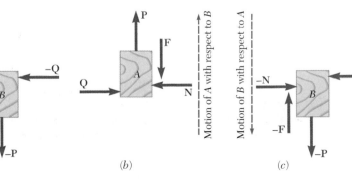

Fig. 8.6

SAMPLE PROBLEM 8.1

A 100-lb force acts as shown on a 300-lb block placed on an inclined plane. The coefficients of friction between the block and the plane are $\mu_s = 0.25$ and $\mu_k = 0.20$. Determine whether the block is in equilibrium, and find the value of the friction force.

SOLUTION

Force Required for Equilibrium. We first determine the value of the friction force *required to maintain equilibrium.* Assuming that **F** is directed down and to the left, we draw the free-body diagram of the block and write

$$+\nearrow \Sigma F_x = 0: \qquad 100 \text{ lb} - \tfrac{3}{5}(300 \text{ lb}) - F = 0$$
$$F = -80 \text{ lb} \qquad \mathbf{F} = 80 \text{ lb} \nearrow$$

$$+\nwarrow \Sigma F_y = 0: \qquad N - \tfrac{4}{5}(300 \text{ lb}) = 0$$
$$N = +240 \text{ lb} \qquad \mathbf{N} = 240 \text{ lb} \nwarrow$$

The force **F** required to maintain equilibrium is an 80-lb force directed up and to the right; the tendency of the block is thus to move down the plane.

Maximum Friction Force. The magnitude of the maximum friction force which can be developed is

$$F_m = \mu_s N \qquad F_m = 0.25(240 \text{ lb}) = 60 \text{ lb}$$

Since the value of the force required to maintain equilibrium (80 lb) is larger than the maximum value which can be obtained (60 lb), equilibrium will not be maintained and *the block will slide down the plane.*

Actual Value of Friction Force. The magnitude of the actual friction force is obtained as follows:

$$F_{\text{actual}} = F_k = \mu_k N$$
$$= 0.20(240 \text{ lb}) = 48 \text{ lb}$$

The sense of this force is opposite to the sense of motion; the force is thus directed up and to the right:

$$\mathbf{F}_{\text{actual}} = 48 \text{ lb} \nearrow \quad \blacktriangleleft$$

It should be noted that the forces acting on the block are not balanced; the resultant is

$$\tfrac{3}{5}(300 \text{ lb}) - 100 \text{ lb} - 48 \text{ lb} = 32 \text{ lb} \swarrow$$

SAMPLE PROBLEM 8.2

A support block is acted upon by two forces as shown. Knowing that the coefficients of friction between the block and the incline are $\mu_s = 0.35$ and $\mu_k = 0.25$, determine (*a*) the force **P** for which motion of the block up the incline is impending, (*b*) the friction force when the block is moving up, (*c*) the smallest force **P** required to prevent the block from sliding down.

800 N

SOLUTION

Free-Body Diagram. For each part of the problem we draw a free-body diagram of the block and a force triangle including the 800-N vertical force, the horizontal force **P**, and the force **R** exerted on the block by the incline. The direction of **R** must be determined in each separate case. We note that since **P** is perpendicular to the 800-N force, the force triangle is a right triangle, which can easily be solved for **P**. In most other problems, however, the force triangle will be an oblique triangle and should be solved by applying the law of sines.

*a. Force **P** for Impending Motion of the Block Up the Incline*

$\tan \phi_s = \mu_s$
$\quad = 0.35$
$\phi_s = 19.29°$
$25° + 19.29° = 44.29°$

$$P = (800\ N)\tan 44.29° \qquad \mathbf{P} = 780\ N \leftarrow \quad \blacktriangleleft$$

*b. Friction Force **F** when the Block Moves Up the Incline*

$\tan \phi_k = \mu_k$
$\quad = 0.25$
$\phi_k = 14.04°$
$25° + 14.04° = 39.04°$

$$R = (800\ N)/\cos 39.04° = 1029.9\ N$$
$$F = R \sin \phi_k = (1029.9\ N)\sin 14.04°$$
$$\mathbf{F} = 250\ N \searrow \quad \blacktriangleleft$$

*c. Force **P** for Impending Motion of the Block Down the Incline*

$\phi_s = 19.29°$
$25° - 19.29° = 5.71°$

$$P = (800\ N)\tan 5.71° \qquad \mathbf{P} = 80.0\ N \leftarrow \quad \blacktriangleleft$$

SAMPLE PROBLEM 8.3

The movable bracket shown may be placed at any height on the 3-in.-diameter pipe. If the coefficient of static friction between the pipe and bracket is 0.25, determine the minimum distance x at which the load **W** can be supported. Neglect the weight of the bracket.

SOLUTION

Free-Body Diagram. We draw the free-body diagram of the bracket. When **W** is placed at the minimum distance x from the axis of the pipe, the bracket is just about to slip, and the forces of friction at A and B have reached their maximum values:

$$F_A = \mu_s N_A = 0.25 N_A$$
$$F_B = \mu_s N_B = 0.25 N_B$$

Equilibrium Equations

$\xrightarrow{+} \Sigma F_x = 0:$ $N_B - N_A = 0$
$N_B = N_A$

$+\uparrow \Sigma F_y = 0:$ $F_A + F_B - W = 0$
$0.25 N_A + 0.25 N_B = W$

And, since N_B has been found equal to N_A,

$$0.50 N_A = W$$
$$N_A = 2W$$

$+\gamma \Sigma M_B = 0:$ $N_A(6 \text{ in.}) - F_A(3 \text{ in.}) - W(x - 1.5 \text{ in.}) = 0$
$6N_A - 3(0.25 N_A) - Wx + 1.5W = 0$
$6(2W) - 0.75(2W) - Wx + 1.5W = 0$

Dividing through by W and solving for x,

$x = 12 \text{ in.}$ ◀

In this lesson you studied and applied the *laws of dry friction.* Previously you had encountered only (*a*) frictionless surfaces that could move freely with respect to each other, (*b*) rough surfaces that allowed no motion relative to each other.

A. ***In solving problems involving dry friction,*** you should keep the following in mind.

1. ***The reaction* R *exerted by a surface on a free body*** can be resolved into a normal component **N** and a tangential component **F**. The tangential component is known as the *friction force.* When a body is in contact with a fixed surface the direction of the friction force **F** is opposite to that of the actual or impending motion of the body.

    a. No motion will occur as long as F does not exceed the maximum value $F_m = \mu_s N$, where μ_s is the *coefficient of static friction.*

    b. Motion will occur if a value of F larger than F_m is required to maintain equilibrium. As motion takes place, the actual value of F drops to $F_k = \mu_k N$, where μ_k is the *coefficient of kinetic friction* [Sample Prob. 8.1].

2. ***When only three forces are involved*** an alternative approach to the analysis of friction may be preferred [Sample Prob. 8.2]. The reaction **R** is defined by its magnitude R and the angle ϕ it forms with the normal to the surface. No motion will occur as long as ϕ does not exceed the maximum value ϕ_s, where $\tan \phi_s = \mu_s$. Motion will occur if a value of ϕ larger than ϕ_s is required to maintain equilibrium, and the actual value of ϕ will drop to ϕ_k, where $\tan \phi_k = \mu_k$.

3. ***When two bodies are in contact*** the sense of the actual or impending relative motion at the point of contact must be determined. On each of the two bodies a friction force **F** should be shown in a direction opposite to that of the actual or impending motion of the body as seen from the other body.

B. ***Methods of solution.*** The first step in your solution is to *draw a free-body diagram* of the body under consideration, resolving the force exerted on each surface where friction exists into a normal component **N** and a friction force **F**. If several bodies are involved, draw a free-body diagram of each of them, labeling and directing the forces at each surface of contact as you learned to do when analyzing frames in Chap. 6.

The problem you have to solve may fall into one of the following five categories:

1. ***All the applied forces and the coefficients of friction are known, and you must determine whether equilibrium is maintained.*** Note that in this situation the friction force is unknown and *cannot be assumed to be equal to* $\mu_s N$.

    a. Write the equations of equilibrium to determine N and F.

    b. Calculate the maximum allowable friction force, $F_m = \mu_s N$. If $F \leq F_m$, equilibrium is maintained. If $F > F_m$, motion occurs, and the magnitude of the friction force is $F_k = \mu_k N$ [Sample Prob. 8.1].

2. ***All the applied forces are known, and you must find the smallest allowable value of*** μ_s ***for which equilibrium is maintained.*** You will assume that motion is impending and determine the corresponding value of μ_s.

a. Write the equations of equilibrium to determine N and F.

b. Since motion is impending, $F = F_m$. Substitute the values found for N and F into the equation $F_m = \mu_s N$ and solve for μ_s.

3. *The motion of the body is impending and μ_s is known; you must find some unknown quantity,* such as a distance, an angle, the magnitude of a force, or the direction of a force.

a. Assume a possible impending motion of the body and, on the free-body diagram, draw the friction force in a direction opposite to that of the assumed motion.

b. Since motion is impending, $F = F_m = \mu_s N$. Substituting for μ_s its known value, you can express F in terms of N on the free-body diagram, thus eliminating one unknown.

c. Write and solve the equilibrium equations for the unknown you seek [Sample Prob. 8.3].

4. *The system consists of several bodies and more than one state of impending motion is possible; you must determine a force, a distance, or a coefficient of friction for which motion is impending.*

a. Count the number of unknowns (forces, angles, etc.) and the total number of independent equations of equilibrium; the difference is equal to the number of surfaces at which motion can be impending.

b. Assume where motion is impending, and on the free-body diagrams of the bodies, draw the friction forces in directions opposite to the assumed motion.

c. At those surfaces where motion is assumed to be impending, $F = F_m = \mu_s N$.

d. Write and solve the equations of equilibrium for all unknowns.

e. Check your assumption by computing F and F_m for the surfaces where motion was assumed not to be impending. If $F < F_m$, your assumption was correct; however, if $F > F_m$, your assumption was incorrect, and it is then necessary to make a new assumption as to where motion impends and to repeat the solution process. You should always include a statement of your assumption and the associated check.

5. *The impending motion of the body can be either sliding or tipping about a given point; you must find a force, a distance, or a coefficient of friction.*

a. Assume a possible impending sliding motion of the body and, on the free-body diagram, draw the friction force or forces in the direction opposite to that of the assumed motion.

b. Separately consider impending sliding and impending tipping.

> *i. For impending sliding, $F = F_m = \mu_s N$.* Write the equations of equilibrium, and solve for the required unknown.

> *ii. For impending tipping, assume that the external reactions go to zero at all supports except for the reactions at the point about which tipping is assumed to be impending.* Note for this case that $F \neq F_m$ and that the required unknown is determined from the equations of equilibrium.

The problems in this section require careful analysis, for they are some of the most difficult problems you will encounter in your statics course. You must always carefully draw free-body diagrams, paying particular attention to the directions of the friction forces. When it is necessary to assume the direction of the friction force, a positive answer implies a correct assumption, while a negative answer implies the friction force is directed opposite to the assumed direction. Also, the solutions of some problems are simplified using either the property of three-force bodies [Sec. 4.7] or the geometry of the system (for example, Prob. 8.23).

Problems

8.1 Determine whether the block shown is in equilibrium, and find the magnitude and direction of the friction force when $\theta = 30°$ and $P = 200$ N.

$\mu_s = 0.30$
$\mu_k = 0.20$

1 kN

P

θ

Fig. P8.1 and P8.2

8.2 Determine whether the block shown is in equilibrium, and find the magnitude and direction of the friction force when $\theta = 35°$ and $P = 400$ N.

8.3 Determine whether the 20-lb block shown is in equilibrium, and find the magnitude and direction of the friction force when $P = 8$ lb and $\theta = 20°$.

8.4 Determine whether the 20-lb block shown is in equilibrium, and find the magnitude and direction of the friction force when $P = 12.5$ lb and $\theta = 15°$.

8.5 Knowing that $\theta = 25°$, determine the range of values of P for which equilibrium is maintained.

$\mu_s = 0.30$
$\mu_k = 0.25$

θ

P

20°

Fig. P8.3, *P8.4,* and P8.5

8.6 Knowing that the coefficient of friction between the 60-lb block and the incline is $\mu_s = 0.25$, determine (*a*) the smallest value of P for which motion of the block up the incline is impending, (*b*) the corresponding value of β.

P

β

60 lb

30°

Fig. P8.6

8.7 Considering only values of θ less than 90°, determine the smallest value of θ for which motion of the block to the right is impending when (*a*) $m = 30$ kg, (*b*) $m = 40$ kg.

$\mu_s = 0.25$
$\mu_k = 0.20$

m

θ

120 N

Fig. P8.7

423

8.8 Knowing that the coefficient of friction between the 30-lb block and the incline is $\mu_s = 0.25$, determine (a) the smallest value of P required to maintain the block in equilibrium, (b) the corresponding value of β.

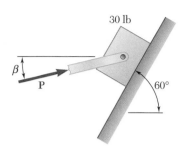

30 lb

β

P

60°

Fig. P8.8

40 N

θ

$\mu_s = 0.40$
$\mu_k = 0.35$

Fig. P8.9

$\mu_s = 0.45$
$\mu_k = 0.35$

18 lb

θ

P

Fig. P8.10

8.9 A 6-kg block is at rest as shown. Determine the positive range of values of θ for which the block is in equilibrium if (a) θ is less than 90°, (b) θ is between 90° and 180°.

8.10 Knowing that $P = 25$ lb, determine the range of values of θ for which equilibrium of the 18-lb block is maintained.

8.11 and 8.12 The coefficients of friction are $\mu_s = 0.40$ and $\mu_k = 0.30$ between all surfaces of contact. Determine the force **P** for which motion of the 60-lb block is impending if cable AB (a) is attached as shown, (b) is removed.

40 lb A

P

60 lb B

Fig. P8.11

40 lb A B

P

60 lb

Fig. P8.12

8.13 The 8-kg block A is attached to link AC and rests on the 12-kg block B. Knowing that the coefficient of static friction is 0.20 between all surfaces of contact and neglecting the mass of the link, determine the value of θ for which motion of block B is impending.

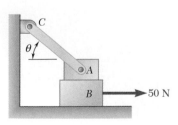

C

θ

A

B → 50 N

Fig. P8.13

A

B

θ

Fig. P8.14

8.14 The 8-kg block A and the 16-kg block B are at rest on an incline as shown. Knowing that the coefficient of static friction is 0.25 between all surfaces of contact, determine the value of θ for which motion is impending.

8.15 A 48-kg cabinet is mounted on casters which can be locked to prevent their rotation. The coefficient of static friction between the floor and each caster is 0.30. Knowing that $h = 640$ mm, determine the magnitude of the force **P** required for impending motion of the cabinet to the right (*a*) if all casters are locked, (*b*) if the casters at *B* are locked and the casters at *A* are free to rotate, (*c*) if the casters at *A* are locked and the casters at *B* are free to rotate.

8.16 A 48-kg cabinet is mounted on casters which can be locked to prevent their rotation. The coefficient of static friction between the floor and each caster is 0.30. Assuming that the casters at *A* and *B* are locked, determine (*a*) the force **P** required for impending motion of the cabinet to the right, (*b*) the largest allowable height h if the cabinet is not to tip over.

Fig. P8.15 and *P8.16*

8.17 The cylinder shown is of weight W and radius r, and the coefficient of static friction μ_s is the same at *A* and *B*. Determine the magnitude of the largest couple **M** which can be applied to the cylinder if it is not to rotate.

Fig. P8.17 and P8.18

8.18 The cylinder shown is of weight W and radius r. Express in terms of W and r the magnitude of the largest couple **M** which can be applied to the cylinder if it is not to rotate assuming that the coefficient of static friction is (*a*) zero at *A* and 0.36 at *B*, (*b*) 0.30 at *A* and 0.36 at *B*.

8.19 The hydraulic cylinder shown exerts a force of 680 lb directed to the right on point *B* and to the left on point *E*. Determine the magnitude of the couple **M** required to rotate the drum clockwise at a constant speed.

Fig. P8.19 and *P8.20*

8.20 A couple **M** of magnitude 70 lb · ft is applied to the drum as shown. Determine the smallest force which must be exerted by the hydraulic cylinder on joints *B* and *E* if the drum is not to rotate.

8.21 and 8.22 A 19.5-ft ladder *AB* leans against a wall as shown. Assuming that the coefficient of static friction μ_s is the same at *A* and *B*, determine the smallest value of μ_s for which equilibrium is maintained.

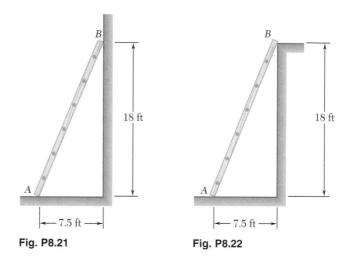

Fig. P8.21   **Fig. P8.22**

8.23 End *A* of a slender, uniform rod of weight *W* and length *L* bears on a horizontal surface as shown, while end *B* is supported by a cord *BC* of length *L*. Knowing that the coefficient of static friction is 0.40, determine (*a*) the value of θ for which motion is impending, (*b*) the corresponding value of the tension in the cord.

Fig. P8.23

8.24 A slender rod of length *L* is lodged between peg *C* and the vertical wall and supports a load **P** at end *A*. Knowing that the coefficient of static friction between the peg and the rod is 0.25 and neglecting friction at the roller, determine the range of values of the ratio *L/a* for which equilibrium is maintained.

8.25 The basic components of a clamping device are bar *AB*, locking plate *CD*, and lever *EFG*; the dimensions of the slot in *CD* are slightly larger than those of the cross section of *AB*. To engage the clamp, *AB* is pushed against the workpiece, and then force **P** is applied. Knowing that *P* = 160 N and neglecting the friction force between the lever and the plate, determine the smallest allowable value of the static coefficient of friction between the bar and the plate.

Fig. *P8.24*

Fig. P8.25

8.26   A window sash having a mass of 4 kg is normally supported by two 2-kg sash weights. Knowing that the window remains open after one sash cord has broken, determine the smallest possible value of the coefficient of static friction. (Assume that the sash is slightly smaller than the frame and will bind only at points A and D.)

Fig. P8.26

8.27   The steel-plate clamp shown is used to lift a steel plate H of mass 250 kg. Knowing that the normal force exerted on steel cam EG by pin D forms an angle of 40° with the horizontal and neglecting the friction force between the cam and the pin, determine the smallest allowable value of the coefficient of static friction.

8.28   The 5-in.-radius cam shown is used to control the motion of the plate CD. Knowing that the coefficient of static friction between the cam and the plate is 0.45 and neglecting friction at the roller supports, determine (*a*) the force **P** for which motion of the plate is impending knowing that the plate is 1 in. thick, (*b*) the largest thickness of the plate for which the mechanism is self-locking (that is, for which the plate cannot be moved however large the force **P** may be).

Fig. P8.27

Fig. P8.28

8.29   A child having a mass of 18 kg is seated halfway between the ends of a small, 16-kg table as shown. The coefficient of static friction is 0.20 between the ends of the table and the floor. If a second child pushes on edge B of the table top at a point directly opposite to the first child with a force **P** lying in a vertical plane parallel to the ends of the table and having a magnitude of 66 N, determine the range of values of θ for which the table will (*a*) tip, (*b*) slide.

Fig. P8.29

Fig. P8.30

8.30 A pipe of diameter 3 in. is gripped by the stillson wrench shown. Portions *AB* and *DE* of the wrench are rigidly attached to each other, and portion *CF* is connected by a pin at *D*. If the wrench is to grip the pipe and be self-locking, determine the required minimum coefficients of friction at *A* and *C*.

8.31 Solve Prob. 8.30 assuming that the diameter of the pipe is 1.5 in.

8.32 The 25-kg plate *ABCD* is attached at *A* and *D* to collars which can slide on the vertical rod. Knowing that the coefficient of static friction is 0.40 between both collars and the rod, determine whether the plate is in equilibrium in the position shown when the magnitude of the vertical force applied at *E* is (*a*) $P = 0$, (*b*) $P = 80$ N.

Fig. P8.32

8.33 In Prob. 8.32, determine the range of values of the magnitude *P* of the vertical force applied at *E* for which the plate will move downward.

8.34 and 8.35 A collar *B* of weight *W* is attached to the spring *AB* and can move along the rod shown. The constant of the spring is 1.5 kN/m and the spring is unstretched when $\theta = 0$. Knowing that the coefficient of static friction between the collar and the rod is 0.40, determine the range of values of *W* for which equilibrium is maintained when (*a*) $\theta = 20°$, (*b*) $\theta = 30°$.

Fig. P8.36

Fig. P8.34   **Fig. P8.35**

8.36 The slender rod *AB* of length $l = 30$ in. is attached to a collar at *B* and rests on a small wheel located at a horizontal distance $a = 4$ in. from the vertical rod on which the collar slides. Knowing that the coefficient of static friction between the collar and the vertical rod is 0.25 and neglecting the radius of the wheel, determine the range of values of *P* for which equilibrium is maintained when $Q = 25$ lb and $\theta = 30°$.

8.37   The 4.5-kg block A and the 3-kg block B are connected by a slender rod of negligible mass. The coefficient of static friction is 0.40 between all surfaces of contact. Knowing that for the position shown the rod is horizontal, determine the range of values of P for which equilibrium is maintained.

Fig. P8.37

8.38   Bar AB is attached to collars which can slide on the inclined rods shown. A force $\mathbf{P}$ is applied at point D located at a distance a from end A. Knowing that the coefficient of static friction μ_s between each collar and the rod upon which it slides is 0.30 and neglecting the weights of the bar and of the collars, determine the smallest value of the ratio a/L for which equilibrium is maintained.

Fig. P8.38

8.39   The 6-kg slender rod AB is pinned at A and rests on the 18-kg cylinder C. Knowing that the diameter of the cylinder is 250 mm and that the coefficient of static friction is 0.35 between all surfaces of contact, determine the largest magnitude of the force $\mathbf{P}$ for which equilibrium is maintained.

Fig. P8.39

8.40   Two rods are connected by a collar at B. A couple $\mathbf{M}_A$ of magnitude 12 lb · ft is applied to rod AB. Knowing that $\mu_s = 0.30$ between the collar and rod AB, determine the largest couple $\mathbf{M}_C$ for which equilibrium will be maintained.

8.41   In Prob. 8.40, determine the smallest couple $\mathbf{M}_C$ for which equilibrium will be maintained.

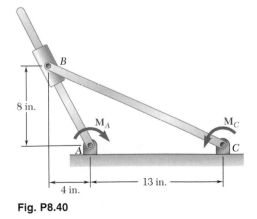

Fig. P8.40

8.42 Blocks *A*, *B*, and *C* having the masses shown are at rest on an incline. Denoting by μ_s the coefficient of static friction between all surfaces of contact, determine the smallest value of μ_s for which equilibrium is maintained.

2 kg

1.5 kg

4 kg

30°

Fig. P8.42

8.43 A slender steel rod of length 9 in. is placed inside a pipe as shown. Knowing that the coefficient of static friction between the rod and the pipe is 0.20, determine the largest value of θ for which the rod will not fall into the pipe.

B

θ

A

3 in.

Fig. P8.43

8.44 In Prob. 8.43, determine the smallest value of θ for which the rod will not fall out of the pipe.

8.45 Two slender rods of negligible weight are pin-connected at *C* and attached to blocks *A* and *B*, each of weight *W*. Knowing that $\theta = 70°$ and that the coefficient of static friction between the blocks and the horizontal surface is 0.30, determine the largest value of *P* for which equilibrium is maintained.

P

θ

C

A

30°

60°

B

W

W

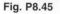

Fig. P8.45

8.5. WEDGES

Wedges are simple machines used to raise large stone blocks and other heavy loads. These loads can be raised by applying to the wedge a force usually considerably smaller than the weight of the load. In addition, because of the friction between the surfaces in contact, a properly shaped wedge will remain in place after being forced under the load. Wedges can thus be used advantageously to make small adjustments in the position of heavy pieces of machinery.

Consider the block A shown in Fig. 8.7a. This block rests against a vertical wall B and is to be raised slightly by forcing a wedge C between block A and a second wedge D. We want to find the minimum value of the force $\mathbf{P}$ which must be applied to the wedge C to move the block. It will be assumed that the weight $\mathbf{W}$ of the block is known, either given in pounds or determined in newtons from the mass of the block expressed in kilograms.

The free-body diagrams of block A and of wedge C have been drawn in Fig. 8.7b and c. The forces acting on the block include its weight and the normal and friction forces at the surfaces of contact with wall B and wedge C. The magnitudes of the friction forces $\mathbf{F}_1$ and $\mathbf{F}_2$ are equal, respectively, to $\mu_s N_1$ and $\mu_s N_2$ since the motion of the block is impending. It is important to show the friction forces with their correct sense. Since the block will move upward, the force $\mathbf{F}_1$ exerted by the wall on the block must be directed downward. On the other hand, since the wedge C will move to the right, the relative motion of A with respect to C is to the left and the force $\mathbf{F}_2$ exerted by C on A must be directed to the right.

Considering now the free body C in Fig. 8.7c, we note that the forces acting on C include the applied force $\mathbf{P}$ and the normal and friction forces at the surfaces of contact with A and D. The weight of the wedge is small compared with the other forces involved and can be neglected. The forces exerted by A on C are equal and opposite to the forces $\mathbf{N}_2$ and $\mathbf{F}_2$ exerted by C on A and are denoted, respectively, by $-\mathbf{N}_2$ and $-\mathbf{F}_2$; the friction force $-\mathbf{F}_2$ must therefore be directed to the left. We check that the force $\mathbf{F}_3$ exerted by D is also directed to the left.

The total number of unknowns involved in the two free-body diagrams can be reduced to four if the friction forces are expressed in terms of the normal forces. Expressing that block A and wedge C are in equilibrium will provide four equations which can be solved to obtain the magnitude of $\mathbf{P}$. It should be noted that in the example considered here, it will be more convenient to replace each pair of normal and friction forces by their resultant. Each free body is then subjected to only three forces, and the problem can be solved by drawing the corresponding force triangles (see Sample Prob. 8.4).

8.6. SQUARE-THREADED SCREWS

Square-threaded screws are frequently used in jacks, presses, and other mechanisms. Their analysis is similar to the analysis of a block sliding along an inclined plane.

(a)

(b)

(c)

Fig. 8.7

Photo 8.2 Wedges are used as shown to split tree trunks because the normal forces exerted by the wedges on the wood are much larger than the forces required to insert the wedges.

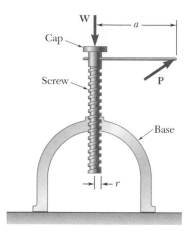

Fig. 8.8

Consider the jack shown in Fig. 8.8. The screw carries a load **W** and is supported by the base of the jack. Contact between the screw and the base takes place along a portion of their threads. By applying a force **P** on the handle, the screw can be made to turn and to raise the load **W**.

The thread of the base has been unwrapped and shown as a straight line in Fig. 8.9a. The correct slope was obtained by plotting horizontally the product $2\pi r$, where r is the mean radius of the thread, and vertically the *lead* L of the screw, that is, the distance through which the screw advances in one turn. The angle θ this line forms with the horizontal is the *lead angle*. Since the force of friction between two surfaces in contact does not depend upon the area of contact, a much smaller than actual area of contact between the two threads can be assumed, and the screw can be represented by the block shown in Fig. 8.9a. It should be noted, however, that in this analysis of the jack, the friction between the cap and the screw is neglected.

The free-body diagram of the block should include the load **W**, the reaction **R** of the base thread, and a horizontal force **Q** having the same effect as the force **P** exerted on the handle. The force **Q** should have the same moment as **P** about the axis of the screw and its magnitude should thus be $Q = Pa/r$. The force **Q**, and thus the force **P** required to raise the load **W**, can be obtained from the free-body diagram shown in Fig. 8.9a. The friction angle is taken equal to ϕ_s since the load will presumably be raised through a succession of short strokes. In mechanisms providing for the continuous rotation of a screw, it may be desirable to distinguish between the force required for impending motion (using ϕ_s) and that required to maintain motion (using ϕ_k).

(a) Impending motion upward

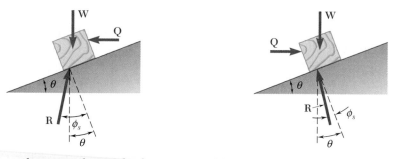

(b) Impending motion downward with $\phi_s > \theta$ (c) Impending motion downward with $\phi_s < \theta$

Fig. 8.9 Block-and-incline analysis of a screw.

If the friction angle ϕ_s is larger than the lead angle θ, the screw is said to be *self-locking;* it will remain in place under the load. To lower the load, we must then apply the force shown in Fig. 8.9b. If ϕ_s is smaller than θ, the screw will unwind under the load; it is then necessary to apply the force shown in Fig. 8.9c to maintain equilibrium.

The lead of a screw should not be confused with its *pitch.* The lead was defined as the distance through which the screw advances in one turn; the pitch is the distance measured between two consecutive threads. While lead and pitch are equal in the case of *single-threaded* screws, they are different in the case of *multiple-threaded* screws, that is, screws having several independent threads. It is easily verified that for double-threaded screws, the lead is twice as large as the pitch; for triple-threaded screws, it is three times as large as the pitch; etc.

SAMPLE PROBLEM 8.4

400 lb

B

8°   A   ← P

The position of the machine block B is adjusted by moving the wedge A. Knowing that the coefficient of static friction is 0.35 between all surfaces of contact, determine the force P for which motion of block B (a) is impending upward, (b) is impending downward.

[handwritten notes:]

$W = 210 kN$

69   $\theta = \tan^{-1} \dfrac{3mm}{(22.6)\pi} = 2.4195°$

$\phi_s = \tan^{-1} \mu_s = \tan^{-1} .4$

$\phi_s = 21.8014°$

$\pi d = \pi (22.6mm)$

$Q = (210 kN) \tan (21.8014 + 2.4195°)$

$Q = 94.47 kN$

$W = 210 kN$

$\theta + \phi_s =$

$Torque = \dfrac{d}{2} Q = \dfrac{22.6}{2} (94.47 kN)$

$= 1067.5 \ N \cdot m$

SOLUTION

For each part, the free-body diagrams of block B and wedge A are drawn, together with the corresponding force triangles, and the law of sines is used to find the desired forces. We note that since $\mu_s = 0.35$, the angle of friction is

$$\phi_s = \tan^{-1} 0.35 = 19.3°$$

a. Force P for Impending Motion of Block Upward

Free Body: Block B

$$\frac{R_1}{\sin 109.3°} = \frac{400 \text{ lb}}{\sin 43.4°}$$

$$R_1 = 549 \text{ lb}$$

Free Body: Wedge A

$$\frac{P}{\sin 46.6°} = \frac{549 \text{ lb}}{\sin 70.7°}$$

$$P = 423 \text{ lb} \qquad \mathbf{P = 423 \text{ lb}} \leftarrow \ \blacktriangleleft$$

b. Force P for Impending Motion of Block Downward

Free Body: Block B

$$\frac{R_1}{\sin 70.7°} = \frac{400 \text{ lb}}{\sin 98.0°}$$

$$R_1 = 381 \text{ lb}$$

Free Body: Wedge A

$$\frac{P}{\sin 30.6°} = \frac{381 \text{ lb}}{\sin 70.7°}$$

$$P = 206 \text{ lb} \qquad \mathbf{P = 206 \text{ lb}} \rightarrow \ \blacktriangleleft$$

[Figure labels, part a:]

400 lb

R_2 B $\phi_s = 19.3°$

$\phi_s = 19.3°$ 8° R_1

$8° + 19.3° = 27.3°$

400 lb   27.3° R_1

$180° - 27.3° - 109.3° = 43.4°$

R_2 $90° + 19.3° = 109.3°$

27.3°

$R_1 = 549 \text{ lb}$

A   P

R_3 19.3°

$90° - 19.3° = 70.7°$

P   19.3°

27.3° R_3

549 lb $27.3° + 19.3° = 46.6°$

[Figure labels, part b:]

400 lb

B $\phi_s = 19.3°$ R_2

$19.3° - 8° = 11.3°$ 8°

R_1 $\phi_s = 19.3°$

$90° - 19.3° = 70.7°$

R_2 $180° - 70.7° - 11.3° = 98.0°$

400 lb R_1 11.3°

11.3° $R_1 = 381 \text{ lb}$

A   P

19.3° R_3

$90° - 19.3° = 70.7°$

P   11.3°

19.3°   381 lb

R_3 $19.3° + 11.3° = 30.6°$

SAMPLE PROBLEM 8.5

A clamp is used to hold two pieces of wood together as shown. The clamp has a double square thread of mean diameter equal to 10 mm with a pitch of 2 mm. The coefficient of friction between threads is $\mu_s = 0.30$. If a maximum torque of 40 N · m is applied in tightening the clamp, determine (a) the force exerted on the pieces of wood, (b) the torque required to loosen the clamp.

SOLUTION

a. Force Exerted by Clamp. The mean radius of the screw is $r = 5$ mm. Since the screw is double-threaded, the lead L is equal to twice the pitch: $L = 2(2 \text{ mm}) = 4$ mm. The lead angle θ and the friction angle ϕ_s are obtained by writing

$$\tan \theta = \frac{L}{2\pi r} = \frac{4 \text{ mm}}{10\pi \text{ mm}} = 0.1273 \qquad \theta = 7.3°$$

$$\tan \phi_s = \mu_s = 0.30 \qquad\qquad\qquad \phi_s = 16.7°$$

The force **Q** which should be applied to the block representing the screw is obtained by expressing that its moment Qr about the axis of the screw is equal to the applied torque.

$$Q(5 \text{ mm}) = 40 \text{ N} \cdot \text{m}$$

$$Q = \frac{40 \text{ N} \cdot \text{m}}{5 \text{ mm}} = \frac{40 \text{ N} \cdot \text{m}}{5 \times 10^{-3} \text{ m}} = 8000 \text{ N} = 8 \text{ kN}$$

The free-body diagram and the corresponding force triangle can now be drawn for the block; the magnitude of the force **W** exerted on the pieces of wood is obtained by solving the triangle.

$$W = \frac{Q}{\tan (\theta + \phi_s)} = \frac{8 \text{ kN}}{\tan 24.0°}$$

$$W = 17.97 \text{ kN} \quad \blacktriangleleft$$

b. Torque Required to Loosen Clamp. The force **Q** required to loosen the clamp and the corresponding torque are obtained from the free-body diagram and force triangle shown.

$$Q = W \tan (\phi_s - \theta) = (17.97 \text{ kN}) \tan 9.4°$$
$$= 2.975 \text{ kN}$$

$$\text{Torque} = Qr = (2.975 \text{ kN})(5 \text{ mm})$$
$$= (2.975 \times 10^3 \text{ N})(5 \times 10^{-3} \text{ m}) = 14.87 \text{ N} \cdot \text{m}$$

$$\text{Torque} = 14.87 \text{ N} \cdot \text{m} \quad \blacktriangleleft$$

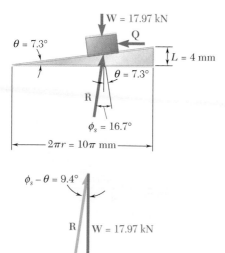

In this lesson you learned to apply the laws of friction to the solution of problems involving *wedges* and *square-threaded screws*.

1. Wedges. Keep the following in mind when solving a problem involving a wedge:

a. First draw a free-body diagram of the wedge and of all the other bodies involved. Carefully note the sense of the relative motion of all surfaces of contact and show each friction force acting in *a direction opposite* to the direction of that relative motion.

b. Show the maximum static friction force F_m *at each surface if the wedge is to be inserted or removed, since motion will be impending in each of these cases.*

c. The reaction R and the angle of friction, rather than the normal force and the friction force, can be used in many applications. You can then draw one or more force triangles and determine the unknown quantities either graphically or by trigonometry [Sample Prob. 8.4].

2. Square-Threaded Screws. The analysis of a square threaded screw is equivalent to the analysis of a block sliding on an incline. To draw the appropriate incline, you should unwrap the thread of the screw and represent it by a straight line [Sample Prob. 8.5]. When solving a problem involving a square-threaded screw, keep the following in mind:

a. Do not confuse the pitch of a screw with the lead of a screw. The *pitch* of a screw is the distance between two consecutive threads, while the *lead* of a screw is the distance the screw advances in one full turn. The lead and the pitch are equal only in single-threaded screws. In a double-threaded screw, the lead is twice the pitch.

b. The torque required to tighten a screw is different from the torque required to loosen it. Also, screws used in jacks and clamps are usually *self-locking;* that is, the screw will remain stationary as long as no torque is applied to it, and a torque must be applied to the screw to loosen it [Sample Prob. 8.5].

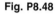

Fig. P8.46

8.46 A 40-lb weight is hung from a lever which rests against a 10° wedge at A and is supported by a frictionless hinge at C. Knowing that the coefficient of static friction is 0.25 at both surfaces of the wedge and that for the position shown the spring is stretched 4 in., determine (a) the magnitude of the force **P** for which motion of the wedge is impending, (b) the components of the corresponding reaction at C.

8.47 Solve Prob. 8.46 assuming that force **P** is directed to the left.

8.48 and 8.49 Two 8° wedges of negligible mass are used to move and position a 240-kg block. Knowing that the coefficient of static friction is 0.40 at all surfaces of contact, determine the magnitude of the force **P** for which motion of the block is impending.

Fig. P8.48

Fig. P8.49

8.50 and 8.51 The elevation of the end of the steel beam supported by a concrete floor is adjusted by means of the steel wedges E and F. The base plate CD has been welded to the lower flange of the beam, and the end reaction of the beam is known to be 150 kN. The coefficient of static friction is 0.30 between the two steel surfaces and 0.60 between the steel and the concrete. If the horizontal motion of the beam is prevented by the force **Q**, determine (a) the force **P** required to raise the beam, (b) the corresponding force **Q**.

Fig. P8.50

Fig. P8.51

8.52 Block *A* supports a pipe column and rests as shown on wedge *B*. Knowing that the coefficient of static friction at all surfaces of contact is 0.25 and that $\theta = 45°$, determine the smallest force **P** required to raise block *A*.

8.53 Block *A* supports a pipe column and rests as shown on wedge *B*. Knowing that the coefficient of static friction at all surfaces of contact is 0.25 and that $\theta = 45°$, determine the smallest force **P** for which equilibrium is maintained.

8.54 and 8.55 A 16° wedge *A* of negligible mass is placed between two 80-kg blocks *B* and *C* which are at rest on inclined surfaces as shown. The coefficient of static friction is 0.40 between both the wedge and the blocks and block *C* and the incline. Determine the magnitude of the force **P** for which motion of the wedge is impending when the coefficient of static friction between block *B* and the incline is (*a*) 0.40, (*b*) 0.60.

Fig. P8.52 and *P8.53*

Fig. P8.54

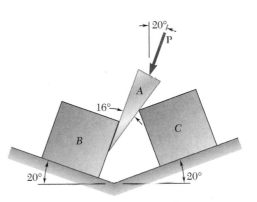

Fig. P8.55

8.56 A 10° wedge is to be forced under end *B* of the 12-lb rod *AB*. Knowing that the coefficient of static friction is 0.45 between the wedge and the rod and 0.25 between the wedge and the floor, determine the smallest force **P** required to raise end *B* of the rod.

Fig. *P8.56*

8.57 A small screwdriver is used to pry apart the two coils of a circular key ring. The wedge angle of the screwdriver blade is 16° and the coefficient of static friction is 0.12 between the coils and the blade. Knowing that a force **P** of magnitude 0.8 lb was required to insert the screwdriver to the equilibrium position shown, determine the magnitude of the forces exerted on the ring by the screwdriver immediately after force **P** is removed.

Fig. P8.57

Fig. P8.58

8.58 A conical wedge is placed between two horizontal plates that are then slowly moved toward each other. Indicate what will happen to the wedge (*a*) if $\mu_s = 0.20$, (*b*) if $\mu_s = 0.30$.

8.59 A 6° steel wedge is driven into the end of an ax handle to lock the handle to the ax head. The coefficient of static friction between the wedge and the handle is 0.35. Knowing that a force **P** of magnitude 250 N was required to insert the wedge to the equilibrium position shown, determine the magnitude of the forces exerted on the handle by the wedge after force **P** is removed.

Fig. P8.59

8.60 A 15° wedge is forced under a 100-lb pipe as shown. The coefficient of static friction at all surfaces is 0.20. Determine (*a*) at which surface slipping of the pipe will first occur, (*b*) the force **P** for which motion of the wedge is impending.

8.61 A 15° wedge is forced under a 100-lb pipe as shown. Knowing that the coefficient of static friction at both surfaces of the wedge is 0.20, determine the largest coefficient of static friction between the pipe and the vertical wall for which slipping is impending at *A*.

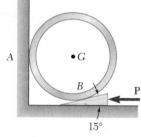

Fig. P8.60 and *P8.61*

8.62 Bags of grass seed are stored on a wooden plank as shown. To move the plank, a 9° wedge is driven under end *A*. Knowing that the weight of the grass seed can be represented by the distributed load shown and that the coefficient of static friction is 0.45 between all surfaces of contact, (*a*) determine the force **P** for which motion of the wedge is impending, (*b*) indicate whether the plank will slide on the floor.

Fig. P8.62

8.63 Solve Prob. 8.62 assuming that the wedge is driven under the plank at *B* instead of at *A*.

*8.64 The 20-lb block *A* is at rest against the 100-lb block *B* as shown. The coefficient of static friction μ_s is the same between blocks *A* and *B* and between block *B* and the floor, while friction between block *A* and the wall can be neglected. Knowing that $P = 30$ lb, determine the value of μ_s for which motion is impending.

Fig. P8.64

*8.65 Solve Prob. 8.64 assuming that μ_s is the coefficient of static friction between all surfaces of contact.

8.66 Derive the following formulas relating the load **W** and the force **P** exerted on the handle of the jack discussed in Sec. 8.6. (a) $P = (Wr/a) \tan (\theta + \phi_s)$, to raise the load; (b) $P = (Wr/a) \tan (\phi_s - \theta)$, to lower the load if the screw is self-locking; (c) $P = (Wr/a) \tan (\theta - \phi_s)$, to hold the load if the screw is not self-locking.

8.67 The square-threaded worm gear shown has a mean radius of 30 mm and a lead of 7.5 mm. The larger gear is subjected to a constant clockwise torque of 720 N · m. Knowing that the coefficient of static friction between the two gears is 0.12, determine the torque that must be applied to shaft AB in order to rotate the large gear counterclockwise. Neglect friction in the bearings at A, B, and C.

8.68 In Prob. 8.67, determine the torque that must be applied to shaft AB in order to rotate the gear clockwise.

8.69 High-strength bolts are used in the construction of many steel structures. For a 24-mm-nominal-diameter bolt the required minimum bolt tension is 210 kN. Assuming the coefficient of friction to be 0.40, determine the required torque that should be applied to the bolt and nut. The mean diameter of the thread is 22.6 mm, and the lead is 3 mm. Neglect friction between the nut and washer, and assume the bolt to be square-threaded.

Fig. P8.69

8.70 The ends of two fixed rods A and B are each made in the form of a single-threaded screw of mean radius 0.3 in. and pitch 0.1 in. Rod A has a right-handed thread and rod B a left-handed thread. The coefficient of static friction between the rods and the threaded sleeve is 0.12. Determine the magnitude of the couple that must be applied to the sleeve in order to draw the rods closer together.

Fig. P8.70

8.71 Assuming that in Prob. 8.70 a right-handed thread is used on *both* rods A and B, determine the magnitude of the couple that must be applied to the sleeve in order to rotate it.

8.72 The position of the automobile jack shown is controlled by a screw ABC that is single-threaded at each end (right-handed thread at A, left-handed thread at C). Each thread has a pitch of 2 mm and a mean diameter of 7.5 mm. If the coefficient of static friction is 0.15, determine the magnitude of the couple **M** that must be applied to raise the automobile.

8.73 For the jack of Prob. 8.72, determine the magnitude of the couple **M** that must be applied to lower the automobile.

Fig. P8.67

Fig. P8.72

8.74 In the gear-pulling assembly shown, the square-threaded screw *AB* has a mean radius of 22.5 mm and a lead of 6 mm. Knowing that the coefficient of static friction is 0.10, determine the torque which must be applied to the screw in order to produce a force of 4.5 kN on the gear. Neglect friction at end *A* of the screw.

Fig. P8.74

*8.7. JOURNAL BEARINGS. AXLE FRICTION

Journal bearings are used to provide lateral support to rotating shafts and axles. Thrust bearings, which will be studied in the next section, are used to provide axial support to shafts and axles. If the journal bearing is fully lubricated, the frictional resistance depends upon the speed of rotation, the clearance between the axle and the bearing, and the viscosity of the lubricant. As indicated in Sec. 8.1, such problems are studied in fluid mechanics. The methods of this chapter, however, can be applied to the study of axle friction when the bearing is not lubricated or is only partially lubricated. It can then be assumed that the axle and the bearing are in direct contact along a single straight line.

Consider two wheels, each of weight **W**, rigidly mounted on an axle supported symmetrically by two journal bearings (Fig. 8.10*a*). If the wheels rotate, we find that to keep them rotating at constant speed, it is necessary to apply to each of them a couple **M**. The free-body diagram in Fig. 8.10*c* represents one of the wheels and the corresponding half axle in projection on a plane perpendicular to the axle. The forces acting on the free body include the weight **W** of the wheel, the couple **M** required to maintain its motion, and a force **R** representing the reaction of the bearing. This force is vertical, equal, and opposite to **W** but does not pass through the center *O* of the axle; **R** is located to the right of *O* at a distance such that its moment about *O* balances the moment **M** of the couple. Therefore, contact between the axle and bearing does not take place at the lowest point *A* when the axle rotates. It takes place at point *B* (Fig. 8.10*b*) or, rather, along a straight line intersecting the plane of the figure at *B*. Physically, this is explained by the fact that when the wheels are set in motion, the axle "climbs" in the bearings until slippage occurs. After sliding back slightly, the axle settles more or less in the position shown. This position is such that the angle between the reaction **R** and the normal to the surface of the bearing is equal to the angle of kinetic friction ϕ_k.

(a)

(b)

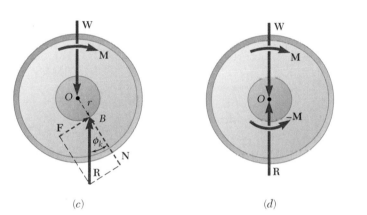

(c)   (d)   (e)

Fig. 8.10

The distance from O to the line of action of $\mathbf{R}$ is thus $r \sin \phi_k$, where r is the radius of the axle. Writing that $\Sigma M_O = 0$ for the forces acting on the free body considered, we obtain the magnitude of the couple $\mathbf{M}$ required to overcome the frictional resistance of one of the bearings:

$$M = Rr \sin \phi_k \qquad (8.5)$$

Observing that, for small values of the angle of friction, $\sin \phi_k$ can be replaced by $\tan \phi_k$, that is, by μ_k, we write the approximate formula

$$M \approx Rr\mu_k \qquad (8.6)$$

In the solution of certain problems, it may be more convenient to let the line of action of $\mathbf{R}$ pass through O, as it does when the axle does not rotate. A couple $-\mathbf{M}$ of the same magnitude as the couple $\mathbf{M}$ but of opposite sense must then be added to the reaction $\mathbf{R}$ (Fig. 8.10d). This couple represents the frictional resistance of the bearing.

In case a graphical solution is preferred, the line of action of $\mathbf{R}$ can be readily drawn (Fig. 8.10e) if we note that it must be tangent to a circle centered at O and of radius

$$r_f = r \sin \phi_k \approx r\mu_k \qquad (8.7)$$

This circle is called the *circle of friction* of the axle and bearing and is independent of the loading conditions of the axle.

*8.8. THRUST BEARINGS. DISK FRICTION

Two types of thrust bearings are used to provide axial support to rotating shafts and axles: (1) *end bearings* and (2) *collar bearings* (Fig. 8.11). In the case of collar bearings, friction forces develop between the two ring-shaped areas which are in contact. In the case of end bearings, friction takes place over full circular areas, or over ring-shaped areas when the end of the shaft is hollow. Friction between circular areas, called *disk friction,* also occurs in other mechanisms, such as *disk clutches.*

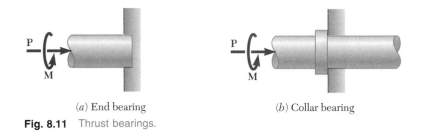

(*a*) End bearing   (*b*) Collar bearing

Fig. 8.11   Thrust bearings.

To obtain a formula which is valid in the most general case of disk friction, let us consider a rotating hollow shaft. A couple **M** keeps the shaft rotating at constant speed while a force **P** maintains it in contact with a fixed bearing (Fig. 8.12). Contact between the shaft and

Fig. 8.12

the bearing takes place over a ring-shaped area of inner radius R_1 and outer radius R_2. Assuming that the pressure between the two surfaces in contact is uniform, we find that the magnitude of the normal force $\Delta \mathbf{N}$ exerted on an element of area ΔA is $\Delta N = P \, \Delta A / A$, where $A = \pi(R_2^2 - R_1^2)$, and that the magnitude of the friction force $\Delta \mathbf{F}$ acting on ΔA is $\Delta F = \mu_k \, \Delta N$. Denoting by r the distance from the axis of the shaft to the element of area ΔA, we express the magnitude ΔM of the moment of $\Delta \mathbf{F}$ about the axis of the shaft as follows:

$$\Delta M = r \, \Delta F = \frac{r \mu_k P \, \Delta A}{\pi(R_2^2 - R_1^2)}$$

The equilibrium of the shaft requires that the moment **M** of the couple applied to the shaft be equal in magnitude to the sum of the moments of the friction forces $\Delta\mathbf{F}$. Replacing ΔA by the infinitesimal element $dA = r\,d\theta\,dr$ used with polar coordinates, and integrating over the area of contact, we thus obtain the following expression for the magnitude of the couple **M** required to overcome the frictional resistance of the bearing:

$$M = \frac{\mu_k P}{\pi(R_2^2 - R_1^2)} \int_0^{2\pi} \int_{R_1}^{R_2} r^2\,dr\,d\theta$$

$$= \frac{\mu_k P}{\pi(R_2^2 - R_1^2)} \int_0^{2\pi} \tfrac{1}{3}(R_2^3 - R_1^3)\,d\theta$$

$$M = \tfrac{2}{3}\mu_k P \frac{R_2^3 - R_1^3}{R_2^2 - R_1^2} \tag{8.8}$$

When contact takes place over a full circle of radius R, formula (8.8) reduces to

$$M = \tfrac{2}{3}\mu_k P R \tag{8.9}$$

The value of M is then the same as would be obtained if contact between the shaft and the bearing took place at a single point located at a distance $2R/3$ from the axis of the shaft.

The largest torque which can be transmitted by a disk clutch without causing slippage is given by a formula similar to (8.9), where μ_k is replaced by the coefficient of static friction μ_s.

*8.9. WHEEL FRICTION. ROLLING RESISTANCE

The wheel is one of the most important inventions of our civilization. Its use makes it possible to move heavy loads with relatively little effort. Because the point of the wheel in contact with the ground at any given instant has no relative motion with respect to the ground, the wheel eliminates the large friction forces which would arise if the load were in direct contact with the ground. However, some resistance to the wheel's motion exists. This resistance has two distinct causes. It is due (1) to a combined effect of axle friction and friction at the rim and (2) to the fact that the wheel and the ground deform, with the result that contact between the wheel and the ground takes place over a certain area, rather than at a single point.

To understand better the first cause of resistance to the motion of a wheel, let us consider a railroad car supported by eight wheels mounted on axles and bearings. The car is assumed to be moving to the right at constant speed along a straight horizontal track. The

(a) Effect of axle friction

(b) Free wheel

(c) Rolling resistance

Fig. 8.13

free-body diagram of one of the wheels is shown in Fig. 8.13a. The forces acting on the free body include the load **W** supported by the wheel and the normal reaction **N** of the track. Since **W** is drawn through the center O of the axle, the frictional resistance of the bearing should be represented by a counterclockwise couple **M** (see Sec. 8.7). To keep the free body in equilibrium, we must add two equal and opposite forces **P** and **F**, forming a clockwise couple of moment $-$**M**. The force **F** is the friction force exerted by the track on the wheel, and **P** represents the force which should be applied to the wheel to keep it rolling at constant speed. Note that the forces **P** and **F** would not exist if there were no friction between the wheel and the track. The couple **M** representing the axle friction would then be zero; thus, the wheel would slide on the track without turning in its bearing.

The couple **M** and the forces **P** and **F** also reduce to zero when there is no axle friction. For example, a wheel which is not held in bearings and rolls freely and at constant speed on horizontal ground (Fig. 8.13b) will be subjected to only two forces: its own weight **W** and the normal reaction **N** of the ground. Regardless of the value of the coefficient of friction between the wheel and the ground, no friction force will act on the wheel. A wheel rolling freely on horizontal ground should thus keep rolling indefinitely.

Experience, however, indicates that the wheel will slow down and eventually come to rest. This is due to the second type of resistance mentioned at the beginning of this section, known as the *rolling resistance*. Under the load **W**, both the wheel and the ground deform slightly, causing the contact between the wheel and the ground to take place over a certain area. Experimental evidence shows that the resultant of the forces exerted by the ground on the wheel over this area is a force **R** applied at a point B, which is not located directly under the center O of the wheel but slightly in front of it (Fig. 8.13c). To balance the moment of **W** about B and to keep the wheel rolling at constant speed, it is necessary to apply a horizontal force **P** at the center of the wheel. Writing $\Sigma M_B = 0$, we obtain

$$Pr = Wb \qquad (8.10)$$

where r = radius of wheel
b = horizontal distance between O and B

The distance b is commonly called the *coefficient of rolling resistance*. It should be noted that b is not a dimensionless coefficient since it represents a length; b is usually expressed in inches or in millimeters. The value of b depends upon several parameters in a manner which has not yet been clearly established. Values of the coefficient of rolling resistance vary from about 0.01 in. or 0.25 mm for a steel wheel on a steel rail to 5.0 in. or 125 mm for the same wheel on soft ground.

SAMPLE PROBLEM 8.6

A pulley of diameter 4 in. can rotate about a fixed shaft of diameter 2 in. The coefficient of static friction between the pulley and shaft is 0.20. Determine (a) the smallest vertical force **P** required to start raising a 500-lb load, (b) the smallest vertical force **P** required to hold the load, (c) the smallest horizontal force **P** required to start raising the same load.

SOLUTION

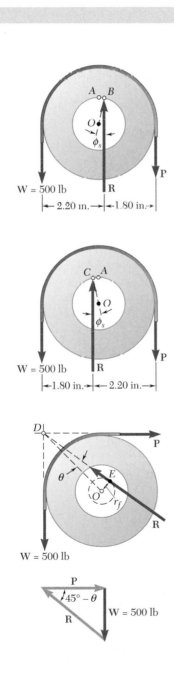

u. Vertical Force P Required to Start Raising the Load. When the forces in both parts of the rope are equal, contact between the pulley and shaft takes place at A. When **P** is increased, the pulley rolls around the shaft slightly and contact takes place at B. The free-body diagram of the pulley when motion is impending is drawn. The perpendicular distance from the center O of the pulley to the line of action of **R** is

$$r_f = r \sin \phi_s \approx r\mu_s \qquad r_f \approx (1 \text{ in.})0.20 = 0.20 \text{ in.}$$

Summing moments about B, we write

$$+\!\uparrow \Sigma M_B = 0: \qquad (2.20 \text{ in.})(500 \text{ lb}) - (1.80 \text{ in.})P = 0$$
$$P = 611 \text{ lb} \qquad\qquad\qquad \mathbf{P} = 611 \text{ lb} \downarrow \quad \blacktriangleleft$$

b. Vertical Force P to Hold the Load. As the force **P** is decreased, the pulley rolls around the shaft and contact takes place at C. Considering the pulley as a free body and summing moments about C, we write

$$+\!\uparrow \Sigma M_C = 0: \qquad (1.80 \text{ in.})(500 \text{ lb}) - (2.20 \text{ in.})P = 0$$
$$P = 409 \text{ lb} \qquad\qquad\qquad \mathbf{P} = 409 \text{ lb} \downarrow \quad \blacktriangleleft$$

c. Horizontal Force P to Start Raising the Load. Since the three forces **W**, **P**, and **R** are not parallel, they must be concurrent. The direction of **R** is thus determined from the fact that its line of action must pass through the point of intersection D of **W** and **P** and must be tangent to the circle of friction. Recalling that the radius of the circle of friction is $r_f = 0.20$ in., we write

$$\sin \theta = \frac{OE}{OD} = \frac{0.20 \text{ in.}}{(2 \text{ in.})\sqrt{2}} = 0.0707 \qquad \theta = 4.1°$$

From the force triangle, we obtain

$$P = W \cot (45° - \theta) = (500 \text{ lb}) \cot 40.9°$$
$$= 577 \text{ lb} \qquad\qquad\qquad \mathbf{P} = 577 \text{ lb} \rightarrow \quad \blacktriangleleft$$

In this lesson you learned about several additional engineering applications of the laws of friction.

1. Journal bearings and axle friction. In journal bearings, the *reaction does not pass through the center of the shaft or axle* which is being supported. The distance from the center of the shaft or axle to the line of action of the reaction (Fig. 8.10) is defined by the equation

$$r_f = r \sin \phi_k \approx r\mu_k$$

if motion is actually taking place, and by the equation

$$r_f = r \sin \phi_s \approx r\mu_s$$

if the motion is impending.

Once you have determined the line of action of the reaction, you can draw a *free-body diagram* and use the corresponding equations of equilibrium to complete your solution [Sample Prob. 8.6]. In some problems, it is useful to observe that the line of action of the reaction must be tangent to a circle of radius $r_f \approx r\mu_k$, or $r_f \approx r\mu_s$, known as the *circle of friction* [Sample Prob. 8.6, part c]. It is important to remember that the reaction is always directed to oppose the actual or impending rotation of the shaft or axle.

2. Thrust bearings and disk friction. In a *thrust bearing* the magnitude of the couple required to overcome frictional resistance is equal to the sum of the moments of the *kinetic friction forces* exerted on the elements of the end of the shaft [Eqs. (8.8) and (8.9)].

An example of disk friction is the *disk clutch*. It is analyzed in the same way as a thrust bearing, except that to determine the largest torque that can be transmitted, you must compute the sum of the moments of the *maximum static friction forces* exerted on the disk.

3. Wheel friction and rolling resistance. You saw that the rolling resistance of a wheel is caused by deformations of both the wheel and the ground. The line of action of the reaction **R** of the ground on the wheel intersects the ground at a horizontal distance b from the center of the wheel. The distance b is known as the *coefficient of rolling resistance* and is expressed in inches or millimeters.

4. In problems involving both rolling resistance and axle friction, your free-body diagram should show that the line of action of the reaction **R** of the ground on the wheel is tangent to the friction circle of the axle and intersects the ground at a horizontal distance from the center of the wheel equal to the coefficient of rolling resistance.

8.75 A 120-mm-radius pulley of mass 5 kg is attached to a 30-mm-radius shaft which fits loosely in a fixed bearing. It is observed that the pulley will just start rotating if a 0.5-kg mass is added to block A. Determine the coefficient of static friction between the shaft and the bearing.

Fig. P8.75

8.76 and 8.77 The double pulley shown is attached to a 0.5-in.-radius shaft which fits loosely in a fixed bearing. Knowing that the coefficient of static friction between the shaft and the poorly lubricated bearing is 0.40, determine the magnitude of the force **P** required to start raising the load.

Fig. P8.76 and *P8.78*

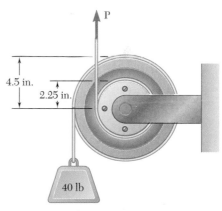

Fig. P8.77 and *P8.79*

8.78 and 8.79 The double pulley shown is attached to a 0.5-in.-radius shaft which fits loosely in a fixed bearing. Knowing that the coefficient of static friction between the shaft and the poorly lubricated bearing is 0.40, determine the magnitude of the force **P** required to maintain equilibrium.

8.80 Control lever ABC fits loosely on a 18-mm-diameter shaft at support B. Knowing that P = 130 N for impending clockwise rotation of the lever, determine (a) the coefficient of static friction between the pin and the lever, (b) the magnitude of the force **P** for which counterclockwise rotation of the lever is impending.

Fig. P8.80

Fig. P8.81 and P8.82

8.81 The block and tackle shown are used to raise a 600-N load. Each of the 60-mm-diameter pulleys rotates on a 10-mm-diameter axle. Knowing that the coefficient of kinetic friction is 0.20, determine the tension in each portion of the rope as the load is slowly raised.

8.82 The block and tackle shown are used to lower a 600-N load. Each of the 60-mm-diameter pulleys rotates on a 10-mm-diameter axle. Knowing that the coefficient of kinetic friction is 0.20, determine the tension in each portion of the rope as the load is slowly lowered.

8.83 The link arrangement shown is frequently used in highway bridge construction to allow for expansion due to changes in temperature. At each of the 3-in.-diameter pins A and B the coefficient of static friction is 0.20. Knowing that the vertical component of the force exerted by BC on the link is 50 kips, determine (a) the horizontal force which should be exerted on beam BC to just move the link, (b) the angle that the resulting force exerted by beam BC on the link will form with the vertical.

Fig. P8.83

8.84 and 8.85 A gate assembly consisting of a 24-kg gate ABC and a 66-kg counterweight D is attached to a 24-mm-diameter shaft B which fits loosely in a fixed bearing. Knowing that the coefficient of static friction is 0.20 between the shaft and the bearing, determine the magnitude of the force $\mathbf{P}$ for which counterclockwise rotation of the gate is impending.

Fig. P8.84 and *P8.86*

Fig. P8.85 and *P8.87*

8.86 and 8.87 A gate assembly consisting of a 24-kg gate ABC and a 66-kg counterweight D is attached to a 24-mm-diameter shaft B which fits loosely in a fixed bearing. Knowing that the coefficient of static friction is 0.20 between the shaft and the bearing, determine the magnitude of the force $\mathbf{P}$ for which clockwise rotation of the gate is impending.

8.88 A loaded railroad car has a weight of 35 tons and is supported by eight 32-in.-diameter wheels with 5-in.-diameter axles. Knowing that the coefficients of friction are $\mu_s = 0.020$ and $\mu_k = 0.015$, determine the horizontal force required (a) for impending motion of the car, (b) to keep the car moving at a constant speed. Neglect rolling resistance between the wheels and the track.

8.89 A scooter is designed to roll down a 2 percent slope at a constant speed. Assuming that the coefficient of kinetic friction between the 1-in.-diameter axles and the bearing is 0.10, determine the required diameter of the wheels. Neglect the rolling resistance between the wheels and the ground.

8.90 A 25-kg electric floor polisher is operated on a surface for which the coefficient of kinetic friction is 0.25. Assuming that the normal force per unit area between the disk and the floor is uniformly distributed, determine the magnitude Q of the horizontal forces required to prevent motion of the machine.

Fig. P8.90

8.91 The pivot for the seat of a desk chair consists of the steel plate A, which supports the seat, the solid steel shaft B which is welded to A and which turns freely in the tubular member C, and the nylon bearing D. If a person of weight $W = 180$ lb is seated directly above the pivot, determine the magnitude of the couple **M** for which rotation of the seat is impending knowing that the coefficient of static friction is 0.15 between the tubular member and the bearing.

*8.92 As the surfaces of a shaft and a bearing wear out, the frictional resistance of a thrust bearing decreases. It is generally assumed that the wear is directly proportional to the distance traveled by any given point of the shaft and thus to the distance r from the point to the axis of the shaft. Assuming, then, that the normal force per unit area is inversely proportional to r, show that the magnitude M of the couple required to overcome the frictional resistance of a worn-out end bearing (with contact over the full circular area) is equal to 75 percent of the value given by formula (8.9) for a new bearing.

*8.93 Assuming that bearings wear out as indicated in Prob. 8.92, show that the magnitude M of the couple required to overcome the frictional resistance of a worn-out collar bearing is

$$M = \frac{1}{2}\mu_k P(R_1 + R_2)$$

where P = magnitude of the total axial force
R_1, R_2 = inner and outer radii of collar

*8.94 Assuming that the pressure between the surfaces of contact is uniform, show that the magnitude M of the couple required to overcome frictional resistance for the conical bearing shown is

$$M = \frac{2}{3}\frac{\mu_k P}{\sin\theta}\frac{R_2^3 - R_1^3}{R_2^2 - R_1^2}$$

8.95 Solve Prob. 8.90 assuming that the normal force per unit area between the disk and the floor varies linearly from a maximum at the center to zero at the circumference of the disk.

Fig. P8.91

Fig. P8.94

Fig. P8.96

8.96 A 1-ton machine base is rolled along a concrete floor using a series of steel pipes with outside diameters of 5 in. Knowing that the coefficient of rolling resistance is 0.025 in. between the pipes and the base and 0.0625 in. between the pipes and the concrete floor, determine the magnitude of the force **P** required to slowly move the base along the floor.

8.97 Knowing that a 120-mm-diameter disk rolls at a constant velocity down a 2 percent incline, determine the coefficient of rolling resistance between the disk and the incline.

8.98 Determine the horizontal force required to move a 1-Mg automobile with 460-mm-diameter tires along a horizontal road at a constant speed. Neglect all forms of friction except rolling resistance, and assume the coefficient of rolling resistance to be 1 mm.

8.99 Solve Prob. 8.88 including the effect of a coefficient of rolling resistance of 0.02 in.

8.100 Solve Prob. 8.89 including the effect of a coefficient of rolling resistance of 0.07 in.

(a)

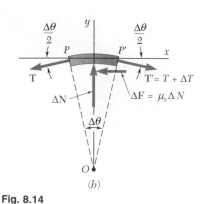

(b)

Fig. 8.14

8.10. BELT FRICTION

Consider a flat belt passing over a fixed cylindrical drum (Fig. 8.14a). We propose to determine the relation existing between the values T_1 and T_2 of the tension in the two parts of the belt when the belt is just about to slide toward the right.

Let us detach from the belt a small element PP' subtending an angle $\Delta\theta$. Denoting by T the tension at P and by $T + \Delta T$ the tension at P', we draw the free-body diagram of the element of the belt (Fig. 8.14b). Besides the two forces of tension, the forces acting on the free body are the normal component ΔN of the reaction of the drum and the friction force ΔF. Since motion is assumed to be impending, we have $\Delta F = \mu_s \Delta N$. It should be noted that if $\Delta\theta$ is made to approach zero, the magnitudes ΔN and ΔF, and the *difference* ΔT between the tension at P and the tension at P', will also approach zero; the value T of the tension at P, however, will remain unchanged. This observation helps in understanding our choice of notation.

Choosing the coordinate axes shown in Fig. 8.14b, we write the equations of equilibrium for the element PP':

$$\Sigma F_x = 0: \quad (T + \Delta T) \cos \frac{\Delta\theta}{2} - T \cos \frac{\Delta\theta}{2} - \mu_s \Delta N = 0 \quad (8.11)$$

$$\Sigma F_y = 0: \quad \Delta N - (T + \Delta T) \sin \frac{\Delta\theta}{2} - T \sin \frac{\Delta\theta}{2} = 0 \quad (8.12)$$

Solving Eq. (8.12) for ΔN and substituting into (8.11), we obtain after reductions

$$\Delta T \cos \frac{\Delta \theta}{2} - \mu_s(2T + \Delta T) \sin \frac{\Delta \theta}{2} = 0$$

Both terms are now divided by $\Delta \theta$. For the first term, this is done simply by dividing ΔT by $\Delta \theta$. The division of the second term is carried out by dividing the terms in the parentheses by 2 and the sine by $\Delta \theta/2$. We write

$$\frac{\Delta T}{\Delta \theta} \cos \frac{\Delta \theta}{2} - \mu_s \left(T + \frac{\Delta T}{2} \right) \frac{\sin(\Delta \theta/2)}{\Delta \theta/2} = 0$$

If we now let $\Delta \theta$ approach 0, the cosine approaches 1 and $\Delta T/2$ approaches zero, as noted above. The quotient of $\sin(\Delta \theta/2)$ over $\Delta \theta/2$ approaches 1, according to a lemma derived in all calculus textbooks. Since the limit of $\Delta T/\Delta \theta$ is by definition equal to the derivative $dT/d\theta$, we write

$$\frac{dT}{d\theta} - \mu_s T = 0 \qquad \frac{dT}{T} = \mu_s d\theta$$

Both members of the last equation (Fig. 8.14a) will now be integrated from P_1 to P_2. At P_1, we have $\theta = 0$ and $T = T_1$; at P_2, we have $\theta = \beta$ and $T = T_2$. Integrating between these limits, we write

$$\int_{T_1}^{T_2} \frac{dT}{T} = \int_0^\beta \mu_s \, d\theta$$

$$\ln T_2 - \ln T_1 = \mu_s \beta$$

or, noting that the left-hand member is equal to the natural logarithm of the quotient of T_2 and T_1,

$$\ln \frac{T_2}{T_1} = \mu_s \beta \tag{8.13}$$

This relation can also be written in the form

$$\frac{T_2}{T_1} = e^{\mu_s \beta} \tag{8.14}$$

The formulas we have derived apply equally well to problems involving flat belts passing over fixed cylindrical drums and to problems involving ropes wrapped around a post or capstan. They can also be used to solve problems involving band brakes. In such problems, it is the drum which is about to rotate, while the band remains fixed. The formulas can also be applied to problems involving belt drives. In

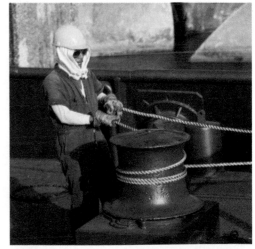

Photo 8.3 By wrapping the rope around the bollard, the force exerted by the worker to control the rope is much smaller than the tension in the taut portion of the rope.

these problems, both the pulley and the belt rotate; our concern is then to find whether the belt will slip, that is, whether it will move *with respect* to the pulley.

Formulas (8.13) and (8.14) should be used only if the belt, rope, or brake is *about to slip*. Formula (8.14) will be used if T_1 or T_2 is desired; formula (8.13) will be preferred if either μ_s or the angle of contact β is desired. We should note that T_2 is always larger than T_1; T_2 therefore represents the tension in that part of the belt or rope which *pulls*, while T_1 is the tension in the part which *resists*. We should also observe that the angle of contact β must be expressed in *radians*. The angle β may be larger than 2π; for example, if a rope is wrapped n times around a post, β is equal to $2\pi n$.

If the belt, rope, or brake is actually slipping, formulas similar to (8.13) and (8.14), but involving the coefficient of kinetic friction μ_k, should be used. If the belt, rope, or brake is not slipping and is not about to slip, none of these formulas can be used.

The belts used in belt drives are often V-shaped. In the V belt shown in Fig. 8.15a contact between the belt and the pulley takes

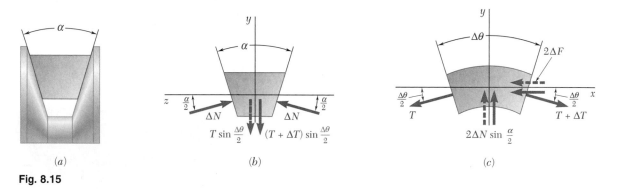

(a) (b) (c)

Fig. 8.15

Photo 8.4 Large normal and friction forces on the sloping sides of the V belt increase the power that can be transmitted before slipping occurs.

place along the sides of the groove. The relation existing between the values T_1 and T_2 of the tension in the two parts of the belt when the belt is just about to slip can again be obtained by drawing the free-body diagram of an element of belt (Fig. 8.15b and c). Equations similar to (8.11) and (8.12) are derived, but the magnitude of the total friction force acting on the element is now $2\,\Delta F$, and the sum of the y components of the normal forces is $2\,\Delta N\,\sin\,(\alpha/2)$. Proceeding as above, we obtain

or,

$$\ln \frac{T_2}{T_1} = \frac{\mu_s\beta}{\sin\,(\alpha/2)} \qquad (8.15)$$

$$\frac{T_2}{T_1} = e^{\mu_s\beta/\sin\,(\alpha/2)} \qquad (8.16)$$

150 N

7500 N

SAMPLE PROBLEM 8.7

A hawser thrown from a ship to a pier is wrapped two full turns around a bollard. The tension in the hawser is 7500 N; by exerting a force of 150 N on its free end, a dockworker can just keep the hawser from slipping. (*a*) Determine the coefficient of friction between the hawser and the bollard. (*b*) Determine the tension in the hawser that could be resisted by the 150-N force if the hawser were wrapped three full turns around the bollard.

SOLUTION

a. Coefficient of Friction. Since slipping of the hawser is impending, we use Eq. (8.13):

$$\ln \frac{T_2}{T_1} = \mu_s \beta$$

Since the hawser is wrapped two full turns around the bollard, we have

$$\beta = 2(2\pi \text{ rad}) = 12.57 \text{ rad}$$
$$T_1 = 150 \text{ N} \qquad T_2 = 7500 \text{ N}$$

Therefore,

$$\mu_s \beta = \ln \frac{T_2}{T_1}$$

$$\mu_s(12.57 \text{ rad}) = \ln \frac{7500 \text{ N}}{150 \text{ N}} = \ln 50 = 3.91$$

$$\mu_s = 0.311 \qquad\qquad \mu_s = 0.31 \quad \blacktriangleleft$$

b. Hawser Wrapped Three Turns around Bollard. Using the value of μ_s obtained in part *a*, we now have

$$\beta = 3(2\pi \text{ rad}) = 18.85 \text{ rad}$$
$$T_1 = 150 \text{ N} \qquad \mu_s = 0.311$$

Substituting these values into Eq. (8.14), we obtain

$$\frac{T_2}{T_1} = e^{\mu_s \beta}$$

$$\frac{T_2}{150 \text{ N}} = e^{(0.311)(18.85)} = e^{5.862} = 351.5$$

$$T_2 = 52\ 725 \text{ N}$$

$$T_2 = 52.7 \text{ kN} \quad \blacktriangleleft$$

$T_1 = 150$ N

T_2

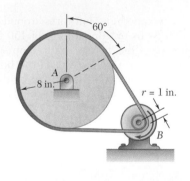

SAMPLE PROBLEM 8.8

A flat belt connects pulley A, which drives a machine tool, to pulley B, which is attached to the shaft of an electric motor. The coefficients of friction are $\mu_s = 0.25$ and $\mu_k = 0.20$ between both pulleys and the belt. Knowing that the maximum allowable tension in the belt is 600 lb, determine the largest torque which can be exerted by the belt on pulley A.

SOLUTION

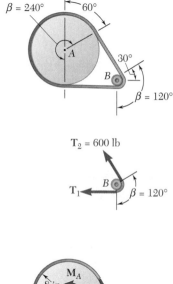

Since the resistance to slippage depends upon the angle of contact β between pulley and belt, as well as upon the coefficient of static friction μ_s, and since μ_s is the same for both pulleys, slippage will occur first on pulley B, for which β is smaller.

Pulley B.   Using Eq. (8.14) with $T_2 = 600$ lb, $\mu_s = 0.25$, and $\beta = 120° = 2\pi/3$ rad, we write

$$\frac{T_2}{T_1} = e^{\mu_s\beta} \qquad \frac{600\text{ lb}}{T_1} = e^{0.25(2\pi/3)} = 1.688$$

$$T_1 = \frac{600\text{ lb}}{1.688} = 355.4\text{ lb}$$

Pulley A.   We draw the free-body diagram of pulley A. The couple $\mathbf{M}_A$ is applied to the pulley by the machine tool to which it is attached and is equal and opposite to the torque exerted by the belt. We write

$$+\gamma\Sigma M_A = 0: \quad M_A - (600\text{ lb})(8\text{ in.}) + (355.4\text{ lb})(8\text{ in.}) = 0$$
$$M_A = 1957\text{ lb}\cdot\text{in.} \qquad\qquad M_A = 163.1\text{ lb}\cdot\text{ft} \quad\blacktriangleleft$$

Note.   We may check that the belt does not slip on pulley A by computing the value of μ_s required to prevent slipping at A and verifying that it is smaller than the actual value of μ_s. From Eq. (8.13) we have

$$\mu_s\beta = \ln\frac{T_2}{T_1} = \ln\frac{600\text{ lb}}{355.4\text{ lb}} = 0.524$$

and, since $\beta = 240° = 4\pi/3$ rad,

$$\frac{4\pi}{3}\mu_s = 0.524 \qquad \mu_s = 0.125 < 0.25$$

SOLVING PROBLEMS
ON YOUR OWN

In the preceding section you learned about *belt friction*. The problems you will have to solve include belts passing over fixed drums, band brakes in which the drum rotates while the band remains fixed, and belt drives.

1. ***Problems involving belt friction*** fall into one of the following two categories:

a. Problems in which slipping is impending. One of the following formulas, involving the *coefficient of static friction* μ_s, may then be used:

$$\ln \frac{T_2}{T_1} = \mu_s \beta \tag{8.13}$$

or

$$\frac{T_2}{T_1} = e^{\mu_s \beta} \tag{8.14}$$

b. Problems in which slipping is occurring. The formulas to be used can be obtained from Eqs. (8.13) and (8.14) by replacing μ_s with the *coefficient of kinetic friction* μ_k.

2. ***As you start solving a belt-friction problem,*** be sure to remember the following:

a. The angle β must be expressed in radians. In a belt-and-drum problem, this is the angle subtending the arc of the drum on which the belt is wrapped.

b. The larger tension is always denoted by T_2, and the smaller tension is denoted by T_1.

c. The larger tension occurs at the end of the belt which is in the direction of the motion, or impending motion, of the belt relative to the drum. You can also determine the direction of the larger tension by remembering that the friction force acting on the belt and the larger tension are always opposed.

3. ***In each of the problems you will be asked to solve,*** three of the four quantities T_1, T_2, β, and μ_s (or μ_k) will either be given or readily found, and you will then solve the appropriate equation for the fourth quantity. There are two kinds of problems that you will encounter:

a. Find μ_s between the belt and the drum knowing that slipping is impending. From the given data, determine T_1, T_2, and β; substitute these values into Eq. (8.13) and solve for μ_s [Sample Prob. 8.7, part *a*]. Follow the same procedure to find the *smallest value* of μ_s for which slipping will not occur.

b. Find the magnitude of a force or couple applied to the belt or drum knowing that slipping is impending. The given data should include μ_s and β. If it also includes T_1 or T_2, use Eq. (8.14) to find the other tension. If neither T_1 nor T_2 is known but some other data is given, use the free-body diagram of the belt-drum system to write an equilibrium equation that you will solve simultaneously with Eq. (8.14) for T_1 and T_2. You will then be able to find the magnitude of the specified force or couple from the free-body diagram of the system. Follow the same procedure to determine the *largest value* of a force or couple which can be applied to the belt or drum if no slipping is to occur [Sample Prob. 8.8].

8.101 A hawser is wrapped two full turns around a bollard. By exerting a 320-N force on the free end of the hawser, a dockworker can resist a force of 20 kN on the other end of the hawser. Determine (*a*) the coefficient of static friction between the hawser and the bollard, (*b*) the number of times the hawser should be wrapped around the bollard if a 80-kN force is to be resisted by the same 320-N force.

Fig. P8.102 and P8.103

8.102 Blocks *A* and *B* are connected by a cable that passes over support *C*. Friction between the blocks and the inclined surfaces can be neglected. Knowing that motion of block *B* up the incline is impending when $W_B = 16$ lb, determine (*a*) the coefficient of static friction between the rope and the support, (*b*) the largest value of W_B for which equilibrium is maintained. (*Hint:* See Prob. 8.128.)

8.103 Blocks *A* and *B* are connected by a cable that passes over support *C*. Friction between the blocks and the inclined surfaces can be neglected. Knowing that the coefficient of static friction between the rope and the support is 0.50, determine the range of values of W_B for which equilibrium is maintained. (*Hint:* See Prob. 8.128.)

8.104 A 120-kg block is supported by a rope which is wrapped $1\frac{1}{2}$ times around a horizontal rod. Knowing that the coefficient of static friction between the rope and the rod is 0.15, determine the range of values of *P* for which equilibrium is maintained.

Fig. P8.104

8.105 The coefficient of static friction between block *B* and the horizontal surface and between the rope and support *C* is 0.40. Knowing that $W_A = 30$ lb, determine the smallest weight of block *B* for which equilibrium is maintained.

8.106 The coefficient of static friction μ_s is the same between block *B* and the horizontal surface and between the rope and support *C*. Knowing that $W_A = W_B$, determine the smallest value of μ_s for which equilibrium is maintained.

Fig. P8.105 and P8.106

8.107 In the pivoted motor mount shown, the weight **W** of the 85-kg motor is used to maintain tension in the drive belt. Knowing that the coefficient of static friction between the flat belt and drums A and B is 0.40, and neglecting the weight of platform CD, determine the largest torque which can be transmitted to drum B when the drive drum A is rotating clockwise.

120 mm

200 mm

240 mm

Fig. P8.107

Handwritten notes:
109
flat
$T_2 = T_1 e^{\mu s \beta}$ $M_A = (6 in)(T_2 - T_1)$
$T_2 = T_1 e^{0.35\pi} = 3.0028 T_1$ ∴ max torque = $M_A = 338 lb\cdot in$
$\Sigma F_x = 0$: $T_1 + T_2 = 225 lb$ max tension: $T_2 = 168.8 lb$
$T_1 (1 + 3.0028) = 225 lb$ V-belt $T_2 = T_1 e^{\mu s / sin \frac{\alpha}{2}}$
$T_1 = 56.211 lb$ 131 $T_2 = T_1 e^{.35\pi/sin 18°} = 35.1015 T_1$
$T_2 = 3.0028 T_1$ $\rightarrow \Sigma F_x = 0$: $T_1 + T_2 - 225 lb = 0$
$T_2 = 168.79 lb$ $T_1 (1 + 35.1015) = 225 lb$
$T_1 = 6.2324 lb$
$T_2 = 218.768 lb = T_{max}$
$(\Sigma M_A = 0$: $M + (3 in)(T_1 - T_2) = 0$
$M = (3 in)(218.768 lb - 6.232 lb)$
$M = 638 lb\cdot in$
$T_{max} = 218.8 lb$

8.108 Solve Prob. 8.107 assuming that the drive drum A is rotating counterclockwise.

8.109 A flat belt is used to transmit a torque from pulley A to pulley B. The radius of each pulley is 3 in., and a force of magnitude $P = 225$ lb is applied as shown to the axle of pulley A. Knowing that the coefficient of static friction is 0.35, determine (*a*) the largest torque which can be transmitted, (*b*) the corresponding maximum value of the tension in the belt.

225 lb

P

12 in.

Fig. P8.109

8.110 Solve Prob. 8.109 assuming that the belt is looped around the pulleys in a figure eight.

8.111 A couple $\mathbf{M}_B$ of magnitude 2 lb · ft is applied to the drive drum B of a portable belt sander to maintain the sanding belt C at a constant speed. The total downward force exerted on the wooden workpiece E is 12 lb, and $\mu_k = 0.10$ between the belt and the sanding platen D. Knowing that $\mu_s = 0.35$ between the belt and the drive drum and that the radii of drums A and B are 1.00 in., determine (*a*) the minimum tension in the lower portion of the belt if no slipping is to occur between the belt and the drive drum, (*b*) the value of the coefficient of kinetic friction between the belt and the workpiece.

Fig. P8.111

Fig. P8.112

8.112 A band belt is used to control the speed of a flywheel as shown. Determine the magnitude of the couple being applied to the flywheel knowing that the coefficient of kinetic friction between the belt and the flywheel is 0.25 and that the flywheel is rotating clockwise at a constant speed. Show that the same result is obtained if the flywheel rotates counterclockwise.

$1/2 M_E = (7.5)(T_2 - T_1)$ $\Sigma M_c : (4\text{ in})(T_1 + T_2) = (16)(25)$ $T_1 + T_2 = 100\text{ lb}$ $T_2 = T_1 e^{\mu_s \beta}$ impending slip

8.113 The drum brake shown permits clockwise rotation of the drum but prevents rotation in the counterclockwise direction. Knowing that the maximum allowed tension in the belt is 7.2 kN, determine (a) the magnitude of the largest counterclockwise couple that can be applied to the drum, (b) the smallest value of the coefficient of static friction between the belt and the drum for which the drum will not rotate counterclockwise.

$T_2 = T_1 e^{.25(3\pi/2)} = 3.2482 T_1$

$So \ T_1(1 + 3.2482) = 100 \text{ lb}$

$T_1 = 23.539 \text{ lb}$

$M_E = (7.5)(3.2482 - 1)(25.539 \text{ lb})$

$= 396.9 \text{ lb} \cdot \text{in}$

Changing Direction of M_E will switch Mag of $T_1 + T_2$

Mag of Couple applied does change

Fig. P8.113

$(\Sigma M_B = 0: (25\text{mm}) T_C - (62.5\text{mm}) T_A$

$T_C = 2.5 T_A$

Impend CCW rotation

a.) $T_C = T_{max} = 7.2 \text{ kN}$

But $T_C = 2.5 T_A$

$T_A = \dfrac{7.2 \text{ kN}}{2.5} = 2.88 \text{ kN}$

$(\Sigma M_D = 0 = M_D + (100\text{mm})(T_A - T_C) = 0$

$M_D = (100\text{mm})(7.2 - 2.88) \text{kN}$

$M_D = 432 \text{ N} \cdot \text{m}$

Also impending slip: $\mu_s \beta = \ln \dfrac{T_C}{T_A}$

$\mu_{s,min} = .219$ $\mu_s = \dfrac{1}{\beta} \ln \dfrac{T_C}{T_A} = \dfrac{1}{\pi} \ln 2.5 = .2187$

8.114 A differential band brake is used to control the speed of a drum which rotates at a constant speed. Knowing that the coefficient of kinetic friction between the belt and the drum is 0.30 and that a couple of magnitude 150 N · m is applied to the drum, determine the corresponding magnitude of the force **P** that is exerted on end D of the lever when the drum is rotating (a) clockwise, (b) counterclockwise.

Fig. P8.114 and P8.115

8.115 A differential band brake is used to control the speed of a drum. Determine the minimum value of the coefficient of static friction for which the brake is self-locking when the drum rotates counterclockwise.

8.116 Bucket A and block C are connected by a cable that passes over drum B. Knowing that drum B rotates slowly counterclockwise and that the coefficients of friction at all surfaces are $\mu_s = 0.35$ and $\mu_k = 0.25$, determine the smallest combined weight W of the bucket and its contents for which block C will (a) remain at rest, (b) be about to move up the incline, (c) continue moving up the incline at a constant speed.

Fig. P8.116

8.117 Solve Prob. 8.116 assuming that drum B is frozen and cannot rotate.

8.118 and 8.120   A cable passes around three 30-mm-radius pulleys and supports two blocks as shown. Pulleys C and E are locked to prevent rotation, and the coefficients of friction between the cable and the pulleys are $\mu_s = 0.20$ and $\mu_k = 0.15$. Determine the range of values of the mass of block A for which equilibrium is maintained (*a*) if pulley D is locked, (*b*) if pulley D is free to rotate.

Fig. P8.118 and P8.119

Fig. P8.120 and P8.121

8.119 and 8.121   A cable passes around three 30-mm-radius pulleys and supports two blocks as shown. Two of the pulleys are locked to prevent rotation, while the third pulley is rotated slowly at a constant speed. Knowing that the coefficients of friction between the cable and the pulleys are $\mu_s = 0.20$ and $\mu_k = 0.15$, determine the largest mass m_A which can be raised (*a*) if pulley C is rotated, (*b*) if pulley E is rotated.

8.122   A recording tape passes over the 1-in.-radius drive drum B and under the idler drum C. Knowing that the coefficients of friction between the tape and the drums are $\mu_s = 0.40$ and $\mu_k = 0.30$ and that drum C is free to rotate, determine the smallest allowable value of P if slipping of the tape on drum B is not to occur.

Fig. P8.122

8.123   Solve Prob. 8.122 assuming that the idler drum C is frozen and cannot rotate.

$T_1 = T_2$ $\uparrow \Sigma F_y = 0:$

$B - 2\left(\dfrac{\sqrt{2}}{2} T_1\right) = 0$

1.125

$T_2 = T_4\left(e^{-\mu_s \beta_R}\right)$ or $B = \sqrt{2}\,T_1 = \sqrt{2}\,T_2$

$T_3 = T_1 e^{-\mu_s \beta_L}$ for impending Rotation $\circlearrowright$:

$T_3 > T_1 = T_2 > T_4$ so $T_3 = T_{max} = 5.6\,kN$

$T_1 = T_3 e^{-\mu_s \beta_L} = (5.6\,kN)e^{-.25\left(\frac{\pi}{4} + \frac{\pi}{2}\right)}$

$T_1 = 4.03706\,kN = T_2$

& $T_4 = T_2 e^{-\mu_s \beta_R} = (4.03706\,kN)e^{-.25\left(\frac{3\pi}{4}\right)}$

$T_4 = 2.23998\,kN$

$\uparrow \Sigma M_F = 0:$ $M_0 + r\left(T_4 - T_3 + T_2 - T_1\right) = 0$

$M_0 = (.16m)(5.6\,kN - 2.23998\,kN) = .5376\,kN \cdot m$

$M_0 = 538\,N \cdot m \circlearrowright$

Level D_x

$B = \sqrt{2}\,T_1 = \sqrt{2}(4.03706)\,kN$

$= 5.709\,kN$

$\curvearrowleft \Sigma M_D = 0: (.05)(5.709\,kN) -$

$(.25)P = 0$

$P = 1.142\,kN \downarrow$

8.124 For the band brake shown, the maximum allowed tension in either belt is 5.6 kN. Knowing that the coefficient of static friction between the belt and the 160-mm-radius drum is 0.25, determine (a) the largest clockwise moment $\mathbf{M}_0$ that can be applied to the drum if slipping is not to occur, (b) the corresponding force $\mathbf{P}$ exerted on end E of the lever.

Fig. P8.124

8.125 Solve Prob. 8.124 assuming that a counterclockwise moment is applied to the drum.

8.126 The strap wrench shown is used to grip the pipe firmly without marring the surface of the pipe. Knowing that the coefficient of static friction is the same for all surfaces of contact, determine the smallest value of μ_s for which the wrench will be self-locking when $a = 10$ in., $r = 1.5$ in., and $\theta = 65°$.

Fig. P8.126

8.127 Solve Prob. 8.126 assuming that $\theta = 75°$.

8.128 Prove that Eqs. (8.13) and (8.14) are valid for any shape of surface provided that the coefficient of friction is the same at all points of contact.

8.129 Complete the derivation of Eq. (8.15), which relates the tension in both parts of a V belt.

8.130 Solve Prob. 8.107 assuming that the flat belt and drums are replaced by a V belt and V pulleys with $\alpha = 36°$. (The angle α is as shown in Fig. 8.15a.)

8.131 Solve Prob. 8.109 assuming that the flat belt and drums are replaced by a V belt and V pulleys with $\alpha = 36°$. (The angle α is as shown in Fig. 8.15a.)

Fig. P8.128

This chapter was devoted to the study of *dry friction,* that is, to problems involving rigid bodies which are in contact along *nonlubricated surfaces*.

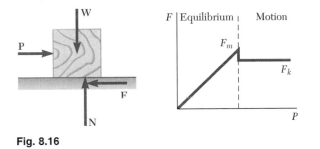

Fig. 8.16

Static and kinetic friction

Applying a horizontal force **P** to a block resting on a horizontal surface [Sec. 8.2], we note that the block at first does not move. This shows that a *friction force* **F** must have developed to balance **P** (Fig. 8.16). As the magnitude of **P** is increased, the magnitude of **F** also increases until it reaches a maximum value F_m. If **P** is further increased, the block starts sliding and the magnitude of **F** drops from F_m to a lower value F_k. Experimental evidence shows that F_m and F_k are proportional to the normal component N of the reaction of the surface. We have

$$F_m = \mu_s N \qquad F_k = \mu_k N \qquad (8.1, 8.2)$$

where μ_s and μ_k are called, respectively, the *coefficient of static friction* and the *coefficient of kinetic friction*. These coefficients depend on the nature and the condition of the surfaces in contact. Approximate values of the coefficients of static friction were given in Table 8.1.

Angles of friction

It is sometimes convenient to replace the normal force **N** and the friction force **F** by their resultant **R** (Fig. 8.17). As the friction force increases and reaches its maximum value $F_m = \mu_s N$, the angle ϕ that **R** forms with the normal to the surface increases and reaches a maximum value ϕ_s, called the *angle of static friction*. If motion actually takes place, the magnitude of **F** drops to F_k; similarly the angle ϕ drops to a lower value ϕ_k, called the *angle of kinetic friction*. As shown in Sec. 8.3, we have

$$\tan \phi_s = \mu_s \qquad \tan \phi_k = \mu_k \qquad (8.3, 8.4)$$

Fig. 8.17

When solving equilibrium problems involving friction, we should keep in mind that the magnitude F of the friction force is equal to $F_m = \mu_s N$ only if *the body is about to slide* [Sec. 8.4]. *If motion is not impending, F and N should be considered as independent unknowns* to be determined from the equilibrium equa-

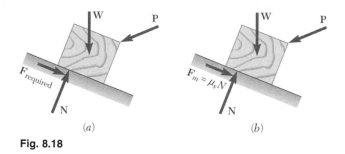

Fig. 8.18

tions (Fig. 8.18*a*). We should also check that the value of F required to maintain equilibrium is not larger than F_m; if it were, the body would move and the magnitude of the friction force would be $F_k = \mu_k N$ [Sample Prob. 8.1]. On the other hand, *if motion is known to be impending*, F has reached its maximum value $F_m = \mu_s N$ (Fig. 8.18*b*), and this expression may be substituted for F in the equilibrium equations [Sample Prob. 8.3]. When only three forces are involved in a free-body diagram, including the reaction **R** of the surface in contact with the body, it is usually more convenient to solve the problem by drawing a force triangle [Sample Prob. 8.2].

When a problem involves the analysis of the forces exerted on each other by *two bodies A and B*, it is important to show the friction forces with their correct sense. The correct sense for the friction force exerted by B on A, for instance, is opposite to that of the *relative motion* (or impending motion) of A with respect to B [Fig. 8.6].

In the second part of the chapter we considered a number of specific engineering applications where dry friction plays an important role. In the case of *wedges*, which are simple machines used to raise heavy loads [Sec. 8.5], two or more free-body diagrams were drawn and care was taken to show each friction force with its correct sense [Sample Prob. 8.4]. The analysis of *square-threaded screws*, which are frequently used in jacks, presses, and other mechanisms, was reduced to the analysis of a block sliding on an incline by unwrapping the thread of the screw and showing it as a straight line [Sec. 8.6]. This is done again in Fig. 8.19, where r denotes the *mean radius* of the thread, L is the *lead* of the screw, that is, the distance through which the screw advances in one turn, **W** is the load, and Qr is equal to the torque exerted on the screw. It was noted that in the case of multiple-threaded screws the lead L of the screw is *not* equal to its pitch, which is the distance measured between two consecutive threads.

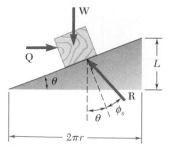

Fig. 8.19

Other engineering applications considered in this chapter were *journal bearings* and *axle friction* [Sec. 8.7], *thrust bearings* and *disk friction* [Sec. 8.8], *wheel friction* and *rolling resistance* [Sec. 8.9], and *belt friction* [Sec. 8.10].

In solving a problem involving a *flat belt* passing over a fixed cylinder, it is important to first determine the direction in which the belt slips or is about to slip. If the drum is rotating, the motion or impending motion of the belt should be determined *relative* to the rotating drum. For instance, if the belt shown in Fig. 8.20 is

Belt friction

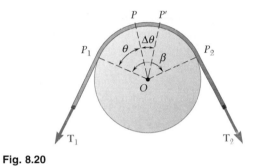

Fig. 8.20

about to slip to the right relative to the drum, the friction forces exerted by the drum on the belt will be directed to the left and the tension will be larger in the right-hand portion of the belt than in the left-hand portion. Denoting the larger tension by T_2, the smaller tension by T_1, the coefficient of static friction by μ_s, and the angle (in radians) subtended by the belt by β, we derived in Sec. 8.10 the formulas

$$\ln \frac{T_2}{T_1} = \mu_s \beta \qquad (8.13)$$

$$\frac{T_2}{T_1} = e^{\mu_s \beta} \qquad (8.14)$$

which were used in solving Sample Probs. 8.7 and 8.8. If the belt actually slips on the drum, the coefficient of static friction μ_s should be replaced by the coefficient of kinetic friction μ_k in both of these formulas.

Review Problems

$\mu_s = 0.25$
$\mu_k = 0.20$

W

θ 30 lb

Fig. *P8.132*

8.132 Considering only values of θ less than 90°, determine the smallest value of θ required to start the block moving to the right when (*a*) *W* = 75 lb, (*b*) *W* = 100 lb.

8.133 The machine base shown has a mass of 75 kg and is fitted with skids at *A* and *B*. The coefficient of static friction between the skids and the floor is 0.30. If a force **P** of magnitude 500 N is applied at corner *C*, determine the range of values of θ for which the base will not move.

Fig. P8.133

8.134 Knowing that a couple of magnitude 30 N · m is required to start the vertical shaft rotating, determine the coefficient of static friction between the annular surfaces of contact.

4 kN

M

50 mm

120 mm

Fig. *P8.134*

8.135 The 20-lb block *A* and the 30-lb block *B* are supported by an incline which is held in the position shown. Knowing that the coefficient of static friction is 0.15 between the two blocks and zero between block *B* and the incline, determine the value of θ for which motion is impending.

Fig. P8.135 and *P8.136*

8.136 The 20-lb block *A* and the 30-lb block *B* are supported by an incline which is held in the position shown. Knowing that the coefficient of static friction is 0.15 between all surfaces of contact, determine the value of θ for which motion is impending.

8.137 Two cylinders are connected by a rope that passes over two fixed rods as shown. Knowing that the coefficient of static friction between the rope and the rods is 0.40, determine the range of values of the mass m of cylinder D for which equilibrium is maintained.

8.138 Two cylinders are connected by a rope that passes over two fixed rods as shown. Knowing that for cylinder D upward motion impends when $m = 20$ kg, determine (a) the coefficient of static friction between the rope and the rods, (b) the corresponding tension in portion BC of the rope.

8.139 A 10° wedge is used to split a section of a log. The coefficient of static friction between the wedge and the log is 0.35. Knowing that a force **P** of magnitude 600 lb was required to insert the wedge, determine the magnitude of the forces exerted on the wood by the wedge after insertion.

Fig. P8.137 and *P8.138*

Fig. P8.139

8.140 A flat belt is used to transmit a torque from drum B to drum A. Knowing that the coefficient of static friction is 0.40 and that the allowable belt tension is 450 N, determine the largest torque that can be exerted on drum A.

Fig. P8.140

8.141   Two 10-lb blocks A and B are connected by a slender rod of negligible weight. The coefficient of static friction is 0.30 between all surfaces of contact, and the rod forms an angle $\theta = 30°$ with the vertical. (*a*) Show that the system is in equilibrium when $P = 0$. (*b*) Determine the largest value of P for which equilibrium is maintained.

Fig. P8.141

8.142   Determine the range of values of P for which equilibrium of the block shown is maintained.

Fig. P8.142

8.143   Two identical uniform boards, each of weight 40 lb, are temporarily leaned against each other as shown. Knowing that the coefficient of static friction between all surfaces is 0.40, determine (*a*) the largest magnitude of the force **P** for which equilibrium will be maintained, (*b*) the surface at which motion will impend.

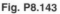

Fig. P8.143

Computer Problems

8.C1 Two blocks are connected by a cable as shown. Knowing that the coefficient of static friction between block A and the horizontal surface varies between 0 and 0.6, determine the values of θ for which motion is impending and plot these values as a function of the coefficient of static friction.

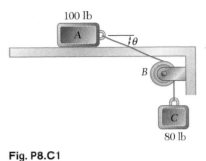

Fig. P8.C1

8.C2 The slender rod AB of length $l = 15$ in. is attached to a collar A and rests on a wheel located at a vertical distance $a = 2.5$ in. from the horizontal rod on which the collar slides. Neglecting friction at C, use computational software to determine the range of values of the magnitude of $\mathbf{Q}$ for which equilibrium is maintained. Plot the maximum and minimum values of $\mathbf{Q}$ as functions of the coefficient of static friction for $0.10 \leq \mu_s \leq 0.55$ when $P = 15$ lb and $\theta = 20°$.

Fig. P8.C2

8.C3 Blocks A and B are supported by an incline which is held in the position shown. Knowing that the mass of block A is 10 kg and that the coefficient of static friction between all surfaces of contact is 0.15, use computational software to determine the value of θ for which motion is impending for values of the mass of block B from 0 to 50 kg using 5-kg increments.

Fig. P8.C3

467

8.C4 The position of the 20-lb rod AB is controlled by the 4-lb block shown, which is slowly moved to the left by the force **P**. Knowing that the coefficient of kinetic friction between all surfaces of contact is 0.25, use computational software to plot the magnitude P of the force as a function of x for values of x from 45 in. to 5 in. Determine the maximum value of P and the corresponding value of x.

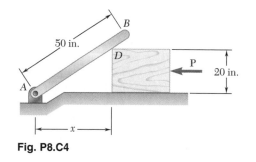

Fig. P8.C4

8.C5 A 0.6-lb cylinder C rests on cylinder D as shown. Knowing that the coefficient of static friction μ_s is the same at A and B, use computational software to determine, for values of μ_s from 0 to 0.40 and using 0.05 increments, the largest counterclockwise couple **M** which can be applied to cylinder D if it is not to rotate.

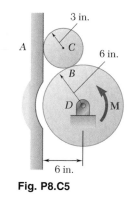

Fig. P8.C5

8.C6 Two identical wedges of negligible mass are used to move and position a 200-kg block. Knowing that the coefficient of static friction μ_s is the same between all surfaces of contact, plot the magnitude of the force **P** for which motion of the block is impending as a function of μ_s for $0.2 \leq \mu_s \leq 0.8$ when (a) $\theta = 8°$, (b) $\theta = 10°$, (c) $\theta = 12°$.

Fig. P8.C6

8.C7 The axle of the pulley shown is frozen and cannot rotate with respect to the block. Knowing that the coefficient of static friction between cable *ABCD* and the pulley varies between 0 and 0.55, determine (*a*) the corresponding values of θ for the system to remain in equilibrium, (*b*) the corresponding reactions at *A* and *D*, and (*c*) plot the values of θ as a function of the coefficient of friction.

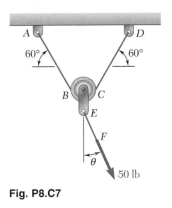

Fig. P8.C7

8.C8 A cable passes around two 50-mm-radius pulleys and supports two blocks as shown. Pulley *B* can only rotate clockwise and pulley *C* can only rotate counterclockwise. Knowing that the coefficient of static friction between the cable and the pulleys is 0.25, plot the minimum and maximum values of the mass of block *D* for which equilibrium is maintained as functions of θ for $15° \le \theta \le 40°$.

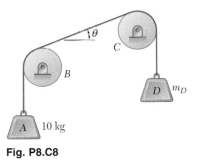

Fig. P8.C8

8.C9 A flat belt is used to transmit a torque from drum *A* to drum *B*. The radius of each drum is 4 in., and the system is fitted with an idler wheel *C* that is used to increase the contact between the belt and the drums. The allowable belt tension is 50 lb, and the coefficient of static friction between the belt and the drums is 0.30. Using computational software, calculate and plot the largest torque that can be transmitted as a function of θ for $0 \le \theta \le 30°$.

Fig. P8.C9

Distributed Forces: Moments of Inertia

The strength of structural members depends to a large extent on the properties of their cross sections, particularly on the second moments, or moments of inertia, of their areas.

DISTRIBUTED FORCES: MOMENTS OF INERTIA

9.1 Introduction
   Moments of Inertia of Areas
9.2 Second Moment, or Moment of Inertia, of an Area
9.3 Determination of the Moment of Inertia of an Area by Integration
9.4 Polar Moment of Inertia
9.5 Radius of Gyration of an Area
9.6 Parallel-Axis Theorem
9.7 Moments of Inertia of Composite Areas
9.8 Product of Inertia
9.9 Principal Axes and Principal Moments of Inertia
9.10 Mohr's Circle for Moments and Products of Inertia
   Moments of Inertia of Masses
9.11 Moment of Inertia of a Mass
9.12 Parallel-Axis Theorem
9.13 Moments of Inertia of Thin Plates
9.14 Determination of the Moment of Inertia of a Three-Dimensional Body by Integration
9.15 Moments of Inertia of Composite Bodies
9.16 Moment of Inertia of a Body with Respect to an Arbitrary Axis through O. Mass Products of Inertia
9.17 Ellipsoid of Inertia. Principal Axes of Inertia
9.18 Determination of the Principal Axes and Principal Moments of Inertia of a Body of Arbitrary Shape

9.1. INTRODUCTION

In Chap. 5, we analyzed various systems of forces distributed over an area or volume. The three main types of forces considered were (1) weights of homogeneous plates of uniform thickness (Secs. 5.3 through 5.6), (2) distributed loads on beams (Sec. 5.8) and hydrostatic forces (Sec. 5.9), and (3) weights of homogeneous three-dimensional bodies (Secs. 5.10 and 5.11). In the case of homogeneous plates, the magnitude ΔW of the weight of an element of a plate was proportional to the area ΔA of the element. For distributed loads on beams, the magnitude ΔW of each elemental weight was represented by an element of area $\Delta A = \Delta W$ under the load curve; in the case of hydrostatic forces on submerged rectangular surfaces, a similar procedure was followed. In the case of homogeneous three-dimensional bodies, the magnitude ΔW of the weight of an element of the body was proportional to the volume ΔV of the element. Thus, in all cases considered in Chap. 5, the distributed forces were proportional to the elemental areas or volumes associated with them. The resultant of these forces, therefore, could be obtained by summing the corresponding areas or volumes, and the moment of the resultant about any given axis could be determined by computing the first moments of the areas or volumes about that axis.

In the first part of this chapter, we consider distributed forces $\Delta \mathbf{F}$ whose magnitudes depend not only upon the elements of area ΔA on which these forces act but also upon the distance from ΔA to some given axis. More precisely, the magnitude of the force per unit area $\Delta F/\Delta A$ is assumed to vary linearly with the distance to the axis. As indicated in the next section, forces of this type are found in the study of the bending of beams and in problems involving submerged non-rectangular surfaces. Assuming that the elemental forces involved are distributed over an area A and vary linearly with the distance y to the x axis, it will be shown that while the magnitude of their resultant $\mathbf{R}$ depends upon the first moment $Q_x = \int y\,dA$ of the area A, the location of the point where $\mathbf{R}$ is applied depends upon the *second moment, or moment of inertia, $I_x = \int y^2\,dA$* of the same area with respect to the x axis. You will learn to compute the moments of inertia of various areas with respect to given x and y axes. Also introduced in the first part of this chapter is the *polar moment of inertia $J_O = \int r^2\,dA$* of an area, where r is the distance from the element of area dA to the point O. To facilitate your computations, a relation will be established between the moment of inertia I_x of an area A with respect to a given x axis and the moment of inertia $I_{x'}$ of the same area with respect to the parallel centroidal x' axis (parallel-axis theorem). You will also study the transformation of the moments of inertia of a given area when the coordinate axes are rotated (Secs. 9.9 and 9.10).

In the second part of the chapter, you will learn how to determine the moments of inertia of various *masses* with respect to a given axis. As you will see in Sec. 9.11, the moment of inertia of a given mass about an axis AA' is defined as $I = \int r^2\,dm$, where r is the distance from the axis AA' to the element of mass dm. Moments of inertia of masses are encountered in dynamics in problems involving the rotation of a rigid body about an axis. To facilitate the computation of mass moments of inertia, the parallel-axis theorem will be introduced (Sec. 9.12). Finally, you will learn to analyze the transformation of moments of inertia of masses when the coordinate axes are rotated (Secs. 9.16 through 9.18).

9.2. SECOND MOMENT, OR MOMENT OF INERTIA, OF AN AREA

In the first part of this chapter, we consider distributed forces $\Delta\mathbf{F}$ whose magnitudes ΔF are proportional to the elements of area ΔA on which the forces act and at the same time vary linearly with the distance from ΔA to a given axis.

Consider, for example, a beam of uniform cross section which is subjected to two equal and opposite couples applied at each end of the beam. Such a beam is said to be in *pure bending*, and it is shown in mechanics of materials that the internal forces in any section of the beam are distributed forces whose magnitudes $\Delta F = ky\,\Delta A$ vary linearly with the distance y between the element of area ΔA and an axis passing through the centroid of the section. This axis, represented by the x axis in Fig. 9.1, is known as the *neutral axis* of the section. The forces on one side of the neutral axis are forces of compression, while those on the other side are forces of tension; on the neutral axis itself the forces are zero.

The magnitude of the resultant $\mathbf{R}$ of the elemental forces $\Delta\mathbf{F}$ which act over the entire section is

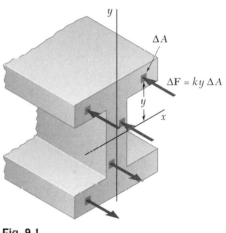

Fig. 9.1

$$R = \int ky\,dA = k\int y\,dA$$

The last integral obtained is recognized as the *first moment* Q_x of the section about the x axis; it is equal to $\bar{y}A$ and is thus equal to zero, since the centroid of the section is located on the x axis. The system of the forces $\Delta\mathbf{F}$ thus reduces to a couple. The magnitude M of this couple (bending moment) must be equal to the sum of the moments $\Delta M_x = y\,\Delta F = ky^2\,\Delta A$ of the elemental forces. Integrating over the entire section, we obtain

$$M = \int ky^2\,dA = k\int y^2\,dA$$

The last integral is known as the *second moment*, or *moment of iner-tia*,† of the beam section with respect to the x axis and is denoted by I_x. It is obtained by multiplying each element of area dA by the *square of its distance* from the x axis and integrating over the beam section. Since each product $y^2\,dA$ is positive, regardless of the sign of y, or zero (if y is zero), the integral I_x will always be positive.

Another example of a second moment, or moment of inertia, of an area is provided by the following problem from hydrostatics: A ver-tical circular gate used to close the outlet of a large reservoir is sub-merged under water as shown in Fig. 9.2. What is the resultant of the forces exerted by the water on the gate, and what is the moment of the resultant about the line of intersection of the plane of the gate and the water surface (x axis)?

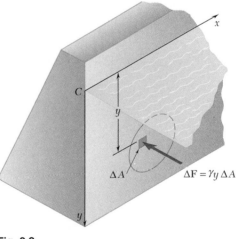

Fig. 9.2

---

†The term *second moment* is more proper than the term *moment of inertia*, since, log-ically, the latter should be used only to denote integrals of mass (see Sec. 9.11). In engi-neering practice, however, moment of inertia is used in connection with areas as well as masses.

If the gate were rectangular, the resultant of the forces of pressure could be determined from the pressure curve, as was done in Sec. 5.9. Since the gate is circular, however, a more general method must be used. Denoting by y the depth of an element of area ΔA and by γ the specific weight of water, the pressure at the element is $p = \gamma y$, and the magnitude of the elemental force exerted on ΔA is $\Delta F = p \, \Delta A = \gamma y \, \Delta A$. The magnitude of the resultant of the elemental forces is thus

$$R = \int \gamma y \, dA = \gamma \int y \, dA$$

and can be obtained by computing the first moment $Q_x = \int y \, dA$ of the area of the gate with respect to the x axis. The moment M_x of the resultant must be equal to the sum of the moments $\Delta M_x = y \, \Delta F = \gamma y^2 \, \Delta A$ of the elemental forces. Integrating over the area of the gate, we have

$$M_x = \int \gamma y^2 \, dA = \gamma \int y^2 \, dA$$

Here again, the integral obtained represents the second moment, or moment of inertia, I_x of the area with respect to the x axis.

9.3. DETERMINATION OF THE MOMENT OF INERTIA OF AN AREA BY INTEGRATION

We defined in the preceding section the second moment, or moment of inertia, of an area A with respect to the x axis. Defining in a similar way the moment of inertia I_y of the area A with respect to the y axis, we write (Fig. 9.3a)

$$I_x = \int y^2 \, dA \qquad I_y = \int x^2 \, dA \qquad (9.1)$$

These integrals, known as the *rectangular moments of inertia* of the area A, can be more easily evaluated if we choose dA to be a thin strip parallel to one of the coordinate axes. To compute I_x, the strip is chosen parallel to the x axis, so that all of the points of the strip are at the same distance y from the x axis (Fig. 9.3b); the moment of inertia dI_x of the strip is then obtained by multiplying the area dA of the strip by y^2. To compute I_y, the strip is chosen parallel to the y axis so that all of the points of the strip are at the same distance x from the y axis (Fig. 9.3c); the moment of inertia dI_y of the strip is $x^2 \, dA$.

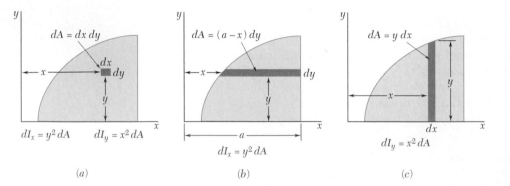

Fig. 9.3   (a)   (b)   (c)

Moment of Inertia of a Rectangular Area. As an example, let us determine the moment of inertia of a rectangle with respect to its base (Fig. 9.4). Dividing the rectangle into strips parallel to the x axis, we obtain

$$dA = b \, dy \qquad dI_x = y^2 b \, dy$$

$$I_x = \int_0^h by^2 \, dy = \tfrac{1}{3}bh^3 \qquad (9.2)$$

Computing I_x and I_y Using the Same Elemental Strip. The formula just derived can be used to determine the moment of inertia dI_x with respect to the x axis of a rectangular strip which is parallel to the y axis, such as the strip shown in Fig. 9.3c. Setting $b = dx$ and $h = y$ in formula (9.2), we write

$$dI_x = \tfrac{1}{3}y^3 \, dx$$

On the other hand, we have

$$dI_y = x^2 \, dA = x^2 y \, dx$$

The same element can thus be used to compute the moments of inertia I_x and I_y of a given area (Fig. 9.5a). The analogous results for the area of Fig. 9.3b are shown in Fig. 9.5b.

Fig. 9.4

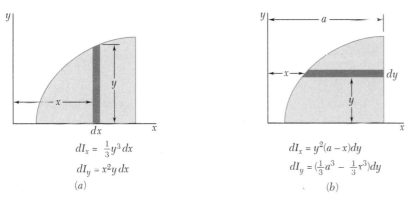

Fig. 9.5

9.4. POLAR MOMENT OF INERTIA

An integral of great importance in problems concerning the torsion of cylindrical shafts and in problems dealing with the rotation of slabs is

$$J_O = \int r^2 \, dA \qquad (9.3)$$

where r is the distance from O to the element of area dA (Fig. 9.6). This integral is the *polar moment of inertia* of the area A with respect to the "pole" O.

The polar moment of inertia of a given area can be computed from the rectangular moments of inertia I_x and I_y of the area if these quantities are already known. Indeed, noting that $r^2 = x^2 + y^2$, we write

$$J_O = \int r^2 \, dA = \int (x^2 + y^2) \, dA = \int y^2 \, dA + \int x^2 \, dA$$

Fig. 9.6

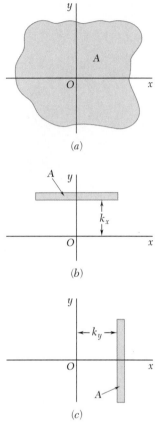

(a)

(b)

(c)

(d)

Fig. 9.7

that is,

$$J_O = I_x + I_y \qquad (9.4)$$

9.5. RADIUS OF GYRATION OF AN AREA

Consider an area A which has a moment of inertia I_x with respect to the x axis (Fig. 9.7a). Let us imagine that we concentrate this area into a thin strip parallel to the x axis (Fig. 9.7b). If the area A, thus concentrated, is to have the same moment of inertia with respect to the x axis, the strip should be placed at a distance k_x from the x axis, where k_x is defined by the relation

$$I_x = k_x^2 A$$

Solving for k_x, we write

$$k_x = \sqrt{\frac{I_x}{A}} \qquad (9.5)$$

The distance k_x is referred to as the *radius of gyration* of the area with respect to the x axis. In a similar way, we can define the radii of gyration k_y and k_O (Fig. 9.7c and d); we write

$$I_y = k_y^2 A \qquad k_y = \sqrt{\frac{I_y}{A}} \qquad (9.6)$$

$$J_O = k_O^2 A \qquad k_O = \sqrt{\frac{J_O}{A}} \qquad (9.7)$$

If we rewrite Eq. (9.4) in terms of the radii of gyration, we find that

$$k_O^2 = k_x^2 + k_y^2 \qquad (9.8)$$

Example.   For the rectangle shown in Fig. 9.8, let us compute the radius of gyration k_x with respect to its base. Using formulas (9.5) and (9.2), we write

$$k_x^2 = \frac{I_x}{A} = \frac{\frac{1}{3}bh^3}{bh} = \frac{h^2}{3} \qquad k_x = \frac{h}{\sqrt{3}}$$

The radius of gyration k_x of the rectangle is shown in Fig. 9.8. It should not be confused with the ordinate $\bar{y} = h/2$ of the centroid of the area. While k_x depends upon the *second moment*, or moment of inertia, of the area, the ordinate $\bar{y}$ is related to the *first moment* of the area.

Fig. 9.8

SAMPLE PROBLEM 9.1

Determine the moment of inertia of a triangle with respect to its base.

SOLUTION

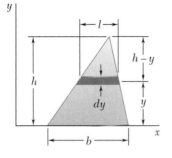

A triangle of base b and height h is drawn; the x axis is chosen to coincide with the base. A differential strip parallel to the x axis is chosen to be dA. Since all portions of the strip are at the same distance from the x axis, we write

$$dI_x = y^2\, dA \qquad dA = l\, dy$$

Using similar triangles, we have

$$\frac{l}{b} = \frac{h-y}{h} \qquad l = b\,\frac{h-y}{h} \qquad dA = b\,\frac{h-y}{h}\, dy$$

Integrating dI_x from $y = 0$ to $y = h$, we obtain

$$I_x = \int y^2\, dA = \int_0^h y^2 b\,\frac{h-y}{h}\, dy = \frac{b}{h}\int_0^h (hy^2 - y^3)\, dy$$

$$= \frac{b}{h}\left[h\frac{y^3}{3} - \frac{y^4}{4}\right]_0^h \qquad\qquad I_x = \frac{bh^3}{12} \quad \blacktriangleleft$$

SAMPLE PROBLEM 9.2

(a) Determine the centroidal polar moment of inertia of a circular area by direct integration. (b) Using the result of part a, determine the moment of inertia of a circular area with respect to a diameter.

SOLUTION

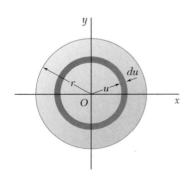

a. **Polar Moment of Inertia.** An annular differential element of area is chosen to be dA. Since all portions of the differential area are at the same distance from the origin, we write

$$dJ_O = u^2\, dA \qquad dA = 2\pi u\, du$$

$$J_O = \int dJ_O = \int_0^r u^2(2\pi u\, du) = 2\pi \int_0^r u^3\, du$$

$$J_O = \frac{\pi}{2}r^4 \quad \blacktriangleleft$$

b. **Moment of Inertia with Respect to a Diameter.** Because of the symmetry of the circular area, we have $I_x = I_y$. We then write

$$J_O = I_x + I_y = 2I_x \qquad \frac{\pi}{2}r^4 = 2I_x \qquad I_{\text{diameter}} = I_x = \frac{\pi}{4}r^4 \quad \blacktriangleleft$$

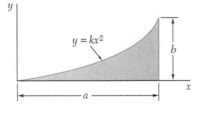

SAMPLE PROBLEM 9.3

(a) Determine the moment of inertia of the shaded area shown with respect to each of the coordinate axes. (Properties of this area were considered in Sample Prob. 5.4.) (b) Using the results of part a, determine the radius of gyration of the shaded area with respect to each of the coordinate axes.

SOLUTION

Referring to Sample Prob. 5.4, we obtain the following expressions for the equation of the curve and the total area:

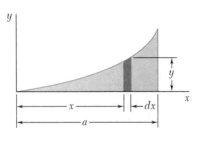

$$y = \frac{b}{a^2}x^2 \qquad A = \tfrac{1}{3}ab$$

Moment of Inertia I_x. A vertical differential element of area is chosen to be dA. Since all portions of this element are *not* at the same distance from the x axis, we must treat the element as a thin rectangle. The moment of inertia of the element with respect to the x axis is then

$$dI_x = \tfrac{1}{3}y^3\, dx = \frac{1}{3}\left(\frac{b}{a^2}x^2\right)^3 dx = \frac{1}{3}\frac{b^3}{a^6}x^6\, dx$$

$$I_x = \int dI_x = \int_0^a \frac{1}{3}\frac{b^3}{a^6}x^6\, dx = \left[\frac{1}{3}\frac{b^3}{a^6}\frac{x^7}{7}\right]_0^a$$

$$I_x = \frac{ab^3}{21} \qquad \blacktriangleleft$$

Moment of Inertia I_y. The same vertical differential element of area is used. Since all portions of the element are at the same distance from the y axis, we write

$$dI_y = x^2\, dA = x^2(y\, dx) = x^2\left(\frac{b}{a^2}x^2\right) dx = \frac{b}{a^2}x^4\, dx$$

$$I_y = \int dI_y = \int_0^a \frac{b}{a^2}x^4\, dx = \left[\frac{b}{a^2}\frac{x^5}{5}\right]_0^a$$

$$I_y = \frac{a^3b}{5} \qquad \blacktriangleleft$$

Radii of Gyration k_x and k_y. We have, by definition,

$$k_x^2 = \frac{I_x}{A} = \frac{ab^3/21}{ab/3} = \frac{b^2}{7} \qquad k_x = \sqrt{\tfrac{1}{7}}b \qquad \blacktriangleleft$$

and

$$k_y^2 = \frac{I_y}{A} = \frac{a^3b/5}{ab/3} = \tfrac{3}{5}a^2 \qquad k_y = \sqrt{\tfrac{3}{5}}a \qquad \blacktriangleleft$$

The purpose of this lesson was to introduce the *rectangular and polar moments of inertia of areas* and the corresponding *radii of gyration*. Although the problems you are about to solve may appear to be more appropriate for a calculus class than for one in mechanics, we hope that our introductory comments have convinced you of the relevance of the moments of inertia to your study of a variety of engineering topics.

1. **Calculating the rectangular moments of inertia I_x and I_y.** We defined these quantities as

$$I_x = \int y^2 \, dA \qquad I_y = \int x^2 \, dA \qquad (9.1)$$

where dA is a differential element of area $dx \, dy$. The moments of inertia are *the second moments of the area*; it is for that reason that I_x, for example, depends on the perpendicular distance y to the area dA. As you study Sec. 9.3, you should recognize the importance of carefully defining the shape and the orientation of dA. Further, you should note the following points.

   a. The moments of inertia of most areas can be obtained by means of a single integration. The expressions given in Figs. 9.3b and c and Fig. 9.5 can be used to calculate I_x and I_y. Regardless of whether you use a single or a double integration, be sure to show on your sketch the element dA that you have chosen.

   b. The moment of inertia of an area is always positive, regardless of the location of the area with respect to the coordinate axes. This is because it is obtained by integrating the product of dA and the *square* of a distance. (Note how this differs from the results for the first moment of the area.) Only when an area is *removed* (as in the case for a hole) will its moment of inertia be entered in your computations with a minus sign.

   c. As a partial check of your work, observe that the moments of inertia are equal to an area times the square of a length. Thus, every term in an expression for a moment of inertia must be a *length to the fourth power*.

2. **Computing the polar moment of inertia J_O.** We defined J_O as

$$J_O = \int r^2 \, dA \qquad (9.3)$$

where $r^2 = x^2 + y^2$. If the given area has circular symmetry (as in Sample Prob. 9.2), it is possible to express dA as a function of r and to compute J_O with a single integration. When the area lacks circular symmetry, it is usually easier first to calculate I_x and I_y and then to determine J_O from

$$J_O = I_x + I_y \qquad (9.4)$$

Lastly, if the equation of the curve that bounds the given area is expressed in polar coordinates, then $dA = r \, dr \, d\theta$ and a double integration is required to compute the integral for J_O [see Prob. 9.27].

3. **Determining the radii of gyration k_x and k_y and the polar radius of gyration k_O.** These quantities were defined in Sec. 9.5, and you should realize that they can be determined only after the area and the appropriate moments of inertia have been computed. It is important to remember that k_x is measured in the y direction, while k_y is measured in the x direction; you should carefully study Sec. 9.5 until you understand this point.

Problems

Fig. P9.1 and P9.5

9.1 through 9.4 Determine by direct integration the moment of inertia of the shaded area with respect to the y axis.

9.5 through 9.8 Determine by direct integration the moment of inertia of the shaded area with respect to the x axis.

Fig. P9.2 and *P9.6*

Fig. P9.3 and *P9.7*   **Fig. P9.4 and P9.8**

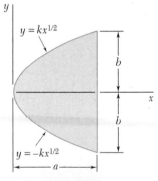

Fig. P9.9 and *P9.12*

9.9 through 9.11 Determine by direct integration the moment of inertia of the shaded area with respect to the x axis.

9.12 through 9.14 Determine by direct integration the moment of inertia of the shaded area with respect to the y axis.

Fig. P9.10 and P9.13

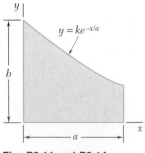

Fig. *P9.11* and P9.14

9.15 and *9.16* Determine the moment of inertia and the radius of gyration of the shaded area shown with respect to the x axis.

Fig. P9.15 and P9.17

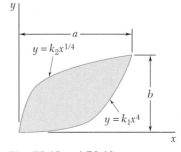

Fig. *P9.16* and P9.18

9.17 and 9.18 Determine the moment of inertia and the radius of gyration of the shaded area shown with respect to the y axis.

9.19 Determine the moment of inertia and the radius of gyration of the shaded area shown with respect to the x axis.

9.20 Determine the moment of inertia and the radius of gyration of the shaded area shown with respect to the y axis.

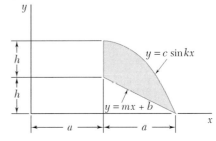

Fig. *P9.19* and P9.20

9.21 and 9.22 Determine the polar moment of inertia and the polar radius of gyration of the shaded area shown with respect to point P.

Fig. P9.21

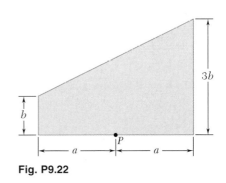

Fig. P9.22

9.23 and *9.24* Determine the polar moment of inertia and the polar radius of gyration of the shaded area shown with respect to point P.

Fig. P9.23

Fig. *P9.24*

9.25 (a) Determine by direct integration the polar moment of inertia of the area shown with respect to point O. (b) Using the result of part a, determine the moments of inertia of the given area with respect to the x and y axes.

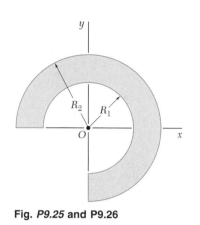

Fig. *P9.25* and P9.26

9.26 (a) Show that the polar radius of gyration k_O of the area shown is approximately equal to the mean radius $R_m = (R_1 + R_2)/2$ for small values of the thickness $t = R_2 - R_1$. (b) Determine the percentage error introduced by using R_m in place of k_O for the following values of t/R_m: $1, \frac{1}{4}, \frac{1}{16}$.

9.27 Determine the polar moment of inertia and the polar radius of gyration of the shaded area shown with respect to point O.

9.28 Determine the polar moment of inertia and the radius of gyration of the isosceles triangle shown with respect to point O.

Fig. P9.27

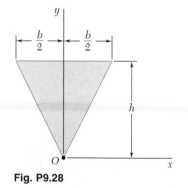

Fig. P9.28

*9.29 Using the polar moment of inertia of the isosceles triangle of Prob. 9.28, show that the centroidal polar moment of inertia of a circular area of radius r is $\pi r^4/2$. (*Hint:* As a circular area is divided into an increasing number of equal circular sectors, what is the approximate shape of each circular sector?)

*9.30 Prove that the centroidal polar moment of inertia of a given area A cannot be smaller than $A^2/2\pi$. (*Hint:* Compare the moment of inertia of the given area with the moment of inertia of a circle which has the same area and the same centroid.)

9.6. PARALLEL-AXIS THEOREM

Consider the moment of inertia I of an area A with respect to an axis AA' (Fig. 9.9). Denoting by y the distance from an element of area dA to AA', we write

$$I = \int y^2 \, dA$$

Let us now draw through the centroid C of the area an axis BB' parallel to AA'; this axis is called a *centroidal axis*. Denoting by y' the

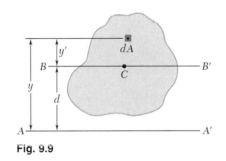

Fig. 9.9

distance from the element dA to BB', we write $y = y' + d$, where d is the distance between the axes AA' and BB'. Substituting for y in the above integral, we write

$$I = \int y^2 \, dA = \int (y' + d)^2 \, dA$$
$$= \int y'^2 \, dA + 2d \int y' \, dA + d^2 \int dA$$

The first integral represents the moment of inertia $\bar{I}$ of the area with respect to the centroidal axis BB'. The second integral represents the first moment of the area with respect to BB'; since the centroid C of the area is located on that axis, the second integral must be zero. Finally, we observe that the last integral is equal to the total area A. Therefore, we have

$$I = \bar{I} + Ad^2 \tag{9.9}$$

This formula expresses that the moment of inertia I of an area with respect to any given axis AA' is equal to the moment of inertia $\bar{I}$ of the area with respect to the centroidal axis BB' parallel to AA' *plus* the product of the area A and the square of the distance d between the two axes. This theorem is known as the *parallel-axis theorem*. Substituting k^2A for I and $\bar{k}^2A$ for $\bar{I}$, the theorem can also be expressed as

$$k^2 = \bar{k}^2 + d^2 \tag{9.10}$$

A similar theorem can be used to relate the polar moment of inertia J_O of an area about a point O to the polar moment of inertia $\bar{J}_C$ of the same area about its centroid C. Denoting by d the distance between O and C, we write

$$J_O = \bar{J}_C + Ad^2 \qquad \text{or} \qquad k_O^2 = \bar{k}_C^2 + d^2 \tag{9.11}$$

Fig. 9.10

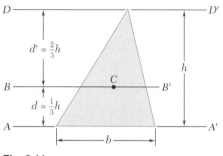

Fig. 9.11

Example 1.   As an application of the parallel-axis theorem, let us determine the moment of inertia I_T of a circular area with respect to a line tangent to the circle (Fig. 9.10). We found in Sample Prob. 9.2 that the moment of inertia of a circular area about a centroidal axis is $\bar{I} = \frac{1}{4}\pi r^4$. We can write, therefore,

$$I_T = \bar{I} + Ad^2 = \tfrac{1}{4}\pi r^4 + (\pi r^2)r^2 = \tfrac{5}{4}\pi r^4$$

Example 2.   The parallel-axis theorem can also be used to determine the centroidal moment of inertia of an area when the moment of inertia of the area with respect to a parallel axis is known. Consider, for instance, a triangular area (Fig. 9.11). We found in Sample Prob. 9.1 that the moment of inertia of a triangle with respect to its base AA' is equal to $\frac{1}{12}bh^3$. Using the parallel-axis theorem, we write

$$I_{AA'} = \bar{I}_{BB'} + Ad^2$$
$$I_{BB'} = I_{AA'} - Ad^2 = \tfrac{1}{12}bh^3 - \tfrac{1}{2}bh(\tfrac{1}{3}h)^2 = \tfrac{1}{36}bh^3$$

It should be observed that the product Ad^2 was *subtracted* from the given moment of inertia in order to obtain the centroidal moment of inertia of the triangle. Note that this product is *added* when transferring *from* a centroidal axis to a parallel axis, but it should be *subtracted* when transferring *to* a centroidal axis. In other words, the moment of inertia of an area is always smaller with respect to a centroidal axis than with respect to any parallel axis.

Returning to Fig. 9.11, we observe that the moment of inertia of the triangle with respect to the line DD' (which is drawn through a vertex) can be obtained by writing

$$I_{DD'} = \bar{I}_{BB'} + Ad'^2 = \tfrac{1}{36}bh^3 + \tfrac{1}{2}bh(\tfrac{2}{3}h)^2 = \tfrac{1}{4}bh^3$$

Note that $I_{DD'}$ could not have been obtained directly from $I_{AA'}$. The parallel-axis theorem can be applied only if one of the two parallel axes passes through the centroid of the area.

9.7. MOMENTS OF INERTIA OF COMPOSITE AREAS

Consider a composite area A made of several component areas A_1, A_2, A_3, . . . Since the integral representing the moment of inertia of A can be subdivided into integrals evaluated over $A_1, A_2, A_3, \ldots$, the moment of inertia of A with respect to a given axis is obtained by adding the moments of inertia of the areas $A_1, A_2, A_3, \ldots$ with respect to the same axis. The moment of inertia of an area consisting of several of the common shapes shown in Fig. 9.12 can thus be obtained by using the formulas given in that figure. Before adding the moments of inertia of the component areas, however, the parallel-axis theorem may have to be used to transfer each moment of inertia to the desired axis. This is shown in Sample Probs. 9.4 and 9.5.

The properties of the cross sections of various structural shapes are given in Fig. 9.13. As noted in Sec. 9.2, the moment of inertia of a beam section about its neutral axis is closely related to the computation of the bending moment in that section of the beam. The determination of moments of inertia is thus a prerequisite to the analysis and design of structural members.

Photo 9.1   Figure 9.13 tabulates data for a small sample of the rolled-steel shapes that are readily available. Shown above are two examples of wide-flange shapes that are commonly used in the construction of buildings.

Rectangle		$\overline{I}_{x'} = \frac{1}{12}bh^3$ $\overline{I}_{y'} = \frac{1}{12}b^3h$ $I_x = \frac{1}{3}bh^3$ $I_y = \frac{1}{3}b^3h$ $J_C = \frac{1}{12}bh(b^2 + h^2)$
Triangle		$\overline{I}_{x'} = \frac{1}{36}bh^3$ $I_x = \frac{1}{12}bh^3$
Circle		$\overline{I}_x = \overline{I}_y = \frac{1}{4}\pi r^4$ $J_O = \frac{1}{2}\pi r^4$
Semicircle		$I_x = I_y = \frac{1}{8}\pi r^4$ $J_O = \frac{1}{4}\pi r^4$
Quarter circle		$I_x = I_y = \frac{1}{16}\pi r^4$ $J_O = \frac{1}{8}\pi r^4$
Ellipse		$\overline{I}_x = \frac{1}{4}\pi ab^3$ $\overline{I}_y = \frac{1}{4}\pi a^3b$ $J_O = \frac{1}{4}\pi ab(a^2 + b^2)$

Fig. 9.12 Moments of inertia of common geometric shapes.

It should be noted that the radius of gyration of a composite area is *not* equal to the sum of the radii of gyration of the component areas. In order to determine the radius of gyration of a composite area, it is first necessary to compute the moment of inertia of the area.

		Area in²	Depth in.	Width in.	Axis X-X			Axis Y-Y		
	Designation				$\bar{I}_x$, in⁴	$\bar{k}_x$, in.	$\bar{y}$, in.	$\bar{I}_y$, in⁴	$\bar{k}_y$, in.	$\bar{x}$, in.
W Shapes (Wide-Flange Shapes)	W18 × 50†	14.7	17.99	7.495	800	7.38		40.1	1.65	
	W16 × 40	11.8	16.01	6.995	518	6.63		28.9	1.57	
	W14 × 30	8.85	13.84	6.730	291	5.73		19.6	1.49	
	W8 × 24	7.08	7.93	6.495	82.8	3.42		18.3	1.61	
S Shapes (American Standard Shapes)	S18 × 70†	20.6	18.00	6.251	926	6.71		24.1	1.08	
	S12 × 50	14.7	12.00	5.477	305	4.55		15.7	1.03	
	S10 × 35	10.3	10.00	4.944	147	3.78		8.36	0.901	
	S6 × 17.25	5.07	6.00	3.565	26.3	2.28		2.31	0.675	
C Shapes (American Standard Channels)	C12 × 25†	7.35	12.00	3.047	144	4.43		4.47	0.780	0.674
	C10 × 20	5.88	10.00	2.739	78.9	3.66		2.81	0.692	0.606
	C8 × 13.75	4.04	8.00	2.343	36.1	2.99		1.53	0.615	0.553
	C6 × 10.5	3.09	6.00	2.034	15.2	2.22		0.866	0.529	0.499
Angles	L6 × 6 × $\frac{3}{4}$‡	8.44			28.2	1.83	1.78	28.2	1.83	1.78
	L4 × 4 × $\frac{1}{2}$	3.75			5.56	1.22	1.18	5.56	1.22	1.18
	L3 × 3 × $\frac{1}{4}$	1.44			1.24	0.930	0.842	1.24	0.930	0.842
	L6 × 4 × $\frac{1}{2}$	4.75			17.4	1.91	1.99	6.27	1.15	0.987
	L5 × 3 × $\frac{1}{2}$	3.75			9.45	1.59	1.75	2.58	0.829	0.750
	L3 × 2 × $\frac{1}{4}$	1.19			1.09	0.957	0.993	0.392	0.574	0.493

Fig. 9.13A Properties of Rolled-Steel Shapes (U.S. Customary Units).*

*Courtesy of the American Institute of Steel Construction, Chicago, Illinois

†Nominal depth in inches and weight in pounds per foot

‡Depth, width, and thickness in inches

$$.50 \quad 2\,I_{xw} = 2\left[18.3 + (7.08)\left(\frac{6.495}{2}\right)^2\right] + I_{xplate}\left\{\frac{(7.93)(.3)^3}{12} + (7.93)(.3)\left[6.495 + .15\right]^2\right\}$$

$$I_{y\,plate}\left[\frac{(.3)(7.93)^3}{12}\right]$$

486

	Designation	Area mm²	Depth mm	Width mm	Axis X-X			Axis Y-Y		
					$\overline{I}_x$ 10⁶ mm⁴	$\overline{k}_x$ mm	$\overline{y}$ mm	$\overline{I}_y$ 10⁶ mm⁴	$\overline{k}_y$ mm	$\overline{x}$ mm
W Shapes (Wide-Flange Shapes)	W460 × 74†	9450	457	190	333	188		16.6	41.9	
	W410 × 60	7580	407	178	216	169		12.1	40.0	
	W360 × 44	5730	352	171	122	146		8.18	37.8	
	W200 × 35.9	4580	201	165	34.4	86.7		7.64	40.8	
S Shapes (American Standard Shapes)	S460 × 104†	13300	457	159	385	170		10.4	27.5	
	S310 × 74	9480	305	139	126	115		6.69	26.1	
	S250 × 52	6670	254	126	61.2	95.8		3.59	22.9	
	S150 × 25.7	3270	152	91	10.8	57.5		1.00	17.2	
C Shapes (American Standard Channels)	C310 × 37†	5690	305	77	59.7	112		1.83	19.7	17.0
	C250 × 30	3780	254	69	32.6	92.9		1.14	17.4	15.3
	C200 × 27.9	3560	203	64	18.2	71.5		0.817	15.1	14.3
	C150 × 15.6	1980	152	51	6.21	56.0		0.347	13.2	12.5
Angles	L152 × 152 × 19.0‡	5420			11.6	46.3	44.9	11.6	46.3	44.9
	L102 × 102 × 12.7	2430			2.34	31.0	30.2	2.34	31.0	30.2
	L76 × 76 × 6.4	932			0.517	23.6	21.4	0.517	23.6	21.4
	L152 × 102 × 12.7	3060			7.20	48.5	50.3	2.64	29.4	25.3
	L127 × 76 × 12.7	2420			3.93	40.3	44.4	1.06	20.9	19.0
	L76 × 51 × 6.4	772			0.453	24.2	25.1	0.166	14.7	12.6

Fig. 9.13B Properties of Rolled-Steel Shapes (SI Units).
†Nominal depth in millimeters and mass in kilograms per meter
‡Depth, width, and thickness in millimeters

32

$I_{x_{1-2-3}}$

$I_{x_1} = \frac{1}{12}(100)(120)^3$ $I_{x_2} = \frac{1}{12}(80)(40)^3 + (3200)(40)^2$

$I_{x_3} = \frac{1}{12}(80)(20)^3 + (1600)(30)^2$

31 $I_{x_1} = \frac{1}{12}(1.2)(.3)^3 + (.36)(1.36)^2 = .6588$

$I_{x_2} = \frac{1}{12}(.4)(2.4)^3$ $I_{x_3} = \frac{1}{12}(2.4)(.3)^3 + (.72)(1.35)^2$

$\frac{3}{4}$ in.

9 in.

C

13.84 in.

6.73 in.

SAMPLE PROBLEM 9.4

The strength of a W14 × 30 rolled-steel beam is increased by attaching a $9 \times \frac{3}{4}$-in. plate to its upper flange as shown. Determine the moment of inertia and the radius of gyration of the composite section with respect to an axis which is parallel to the plate and passes through the centroid C of the section.

$36. \; y_1 - y_2 - y_3$

$I_{y_1} \left[\frac{1}{12}(4a)(3a)^3 + (3a)(2a)\left(\frac{a}{2}\right)^2 \right]$

$I_{y_2} = -\left\{ \left[\frac{\pi}{8}a^4 - \frac{\pi}{2}a^2\left(\frac{4a}{3\pi}\right)^2 \right] + \frac{\pi}{2}a^2\left[2a - \frac{4a}{3\pi}\right]^2 \right\}$

SOLUTION

$I_{y_3} = -\left\{ \left[\frac{\pi}{8}a^4 - \frac{\pi}{2}a^2\left(\frac{4a}{3\pi}\right)^2 \right] + \frac{\pi}{2}a^2\left(a - \frac{4a}{3\pi}\right)^2 \right\}$

The origin O of the coordinates is placed at the centroid of the wide-flange shape, and the distance $\overline{Y}$ to the centroid of the composite section is computed using the methods of Chap. 5. The area of the wide-flange shape is found by referring to Fig. 9.13A. The area and the y coordinate of the centroid of the plate are

7.295 in.

C

O

d

x'

x

$\overline{Y}$

.49

$$A = (9 \text{ in.})(0.75 \text{ in.}) = 6.75 \text{ in}^2$$
$$\overline{y} = \tfrac{1}{2}(13.84 \text{ in.}) + \tfrac{1}{2}(0.75 \text{ in.}) = 7.295 \text{ in.}$$

$I_{x_s} + 2(I_x)_{plate}$

$I_{x \, plate} = \left[\frac{[8(1)^3}{12} + (8)(1)(6.5)^3 \right]$

$I_{y \, plate} = \left(\frac{1(8)^3}{12} \right)$

Section	Area, in²	$\overline{y}$, in.	$\overline{y}A$, in³
Plate	6.75	7.295	49.24
Wide-flange shape	8.85	0	0
	$\Sigma A = 15.60$		$\Sigma \overline{y}A = 49.24$

$$\overline{Y}\Sigma A = \Sigma \overline{y}A \qquad \overline{Y}(15.60) = 49.24 \qquad \overline{Y} = 3.156 \text{ in.}$$

Moment of Inertia. The parallel-axis theorem is used to determine the moments of inertia of the wide-flange shape and the plate with respect to the x' axis. This axis is a centroidal axis for the composite section but *not* for either of the elements considered separately. The value of $\overline{I}_x$ for the wide-flange shape is obtained from Fig. 9.13A.

For the wide-flange shape,

$$I_{x'} = \overline{I}_x + A\overline{Y}^2 = 291 + (8.85)(3.156)^2 = 379.1 \text{ in}^4$$

For the plate,

$$I_{x'} = \overline{I}_x + Ad^2 = (\tfrac{1}{12})(9)(\tfrac{3}{4})^3 + (6.75)(7.295 - 3.156)^2 = 116.0 \text{ in}^4$$

For the composite area,

$$I_{x'} = 379.1 + 116.0 = 495.1 \text{ in}^4 \qquad I_{x'} = 495 \text{ in}^4 \; \blacktriangleleft$$

Radius of Gyration. We have

$$k_{x'}^2 = \frac{I_{x'}}{A} = \frac{495.1 \text{ in}^4}{15.60 \text{ in}^2} \qquad k_{x'} = 5.63 \text{ in.} \; \blacktriangleleft$$

SAMPLE PROBLEM 9.5

Determine the moment of inertia of the shaded area with respect to the x axis.

SOLUTION

The given area can be obtained by subtracting a half circle from a rectangle. The moments of inertia of the rectangle and the half circle will be computed separately. $35 \quad I_{x_1} = \frac{1}{3}(2a)(4b)^3 \qquad I_{x_2} = \frac{1}{3}(a)(8a)^3$

Moment of Inertia of Rectangle. Referring to Fig. 9.12, we obtain

$$I_x = \tfrac{1}{3}bh^3 = \tfrac{1}{3}(240 \text{ mm})(120 \text{ mm})^3 = 138.2 \times 10^6 \text{ mm}^4$$

Moment of Inertia of Half Circle. Referring to Fig. 5.8, we determine the location of the centroid C of the half circle with respect to diameter AA'.

$$a = \frac{4r}{3\pi} = \frac{(4)(90 \text{ mm})}{3\pi} = 38.2 \text{ mm}$$

The distance b from the centroid C to the x axis is

$$b = 120 \text{ mm} - a = 120 \text{ mm} - 38.2 \text{ mm} = 81.8 \text{ mm}$$

Referring now to Fig. 9.12, we compute the moment of inertia of the half circle with respect to diameter AA'; we also compute the area of the half circle.

$$I_{AA'} = \tfrac{1}{8}\pi r^4 = \tfrac{1}{8}\pi(90 \text{ mm})^4 = 25.76 \times 10^6 \text{ mm}^4$$
$$A = \tfrac{1}{2}\pi r^2 = \tfrac{1}{2}\pi(90 \text{ mm})^2 = 12.72 \times 10^3 \text{ mm}^2$$

Using the parallel-axis theorem, we obtain the value of $\bar{I}_{x'}$:

$$I_{AA'} = \bar{I}_{x'} + Aa^2$$
$$25.76 \times 10^6 \text{ mm}^4 = \bar{I}_{x'} + (12.72 \times 10^3 \text{ mm}^2)(38.2 \text{ mm})^2$$
$$\bar{I}_{x'} = 7.20 \times 10^6 \text{ mm}^4$$

Again using the parallel-axis theorem, we obtain the value of I_x:

$$I_x = \bar{I}_{x'} + Ab^2 = 7.20 \times 10^6 \text{ mm}^4 + (12.72 \times 10^3 \text{ mm}^2)(81.8 \text{ mm})^2$$
$$= 92.3 \times 10^6 \text{ mm}^4$$

Moment of Inertia of Given Area. Subtracting the moment of inertia of the half circle from that of the rectangle, we obtain

$$I_x = 138.2 \times 10^6 \text{ mm}^4 - 92.3 \times 10^6 \text{ mm}^4$$
$$I_x = 45.9 \times 10^6 \text{ mm}^4 \quad \blacktriangleleft$$

In this lesson we introduced the *parallel-axis theorem* and illustrated how it can be used to simplify the computation of moments and polar moments of inertia of composite areas. The areas that you will consider in the following problems will consist of common shapes and rolled-steel shapes. You will also use the parallel-axis theorem to locate the point of application (the center of pressure) of the resultant of the hydrostatic forces acting on a submerged plane area.

1. Applying the parallel-axis theorem. In Sec. 9.6 we derived the parallel-axis theorem

$$I = \bar{I} + Ad^2 \tag{9.9}$$

which states that the moment of inertia I of an area A with respect to a given axis is equal to the sum of the moment of inertia $\bar{I}$ of that area with respect to the *parallel centroidal axis* and the product Ad^2, where d is the distance between the two axes. It is important that you remember the following points as you use the parallel-axis theorem.

   a. The centroidal moment of inertia $\bar{I}$ of an area A can be obtained by subtracting the product Ad^2 from the moment of inertia I of the area with respect to a parallel axis. Therefore, as we noted in Example 2 and illustrated in Sample Prob. 9.5, it follows that the moment of inertia $\bar{I}$ is *smaller* than the moment of inertia I of the same area with respect to any parallel axis.

   b. The parallel-axis theorem can be applied only if one of the two axes involved is a centroidal axis. Therefore, as we noted in Example 2, to compute the moment of inertia of an area with respect to a *noncentroidal axis* when the moment of inertia of the area is known with respect to *another noncentroidal axis*, it is necessary to *first compute* the moment of inertia of the area with respect to the *centroidal axis parallel to the two given axes.*

2. Computing the moments and polar moments of inertia of composite areas. Sample Probs. 9.4 and 9.5 illustrate the steps you should follow to solve problems of this type. As with all composite-area problems, you should show on your sketch the common shapes or rolled-steel shapes that constitute the various elements of the given area, as well as the distances between the centroidal axes of the elements and the axes about which the moments of inertia are to be computed. In addition, it is important that the following points be noted.

a. The moment of inertia of an area is always positive, regardless of the location of the axis with respect to which it is computed. As pointed out in the comments for the preceding lesson, it is only when an area is *removed* (as in the case of a hole) that its moment of inertia should be entered in your computations with a minus sign.

b. The moments of inertia of a semiellipse and a quarter ellipse can be determined by dividing the moment of inertia of an ellipse by 2 and 4, respectively. It should be noted, however, that the moments of inertia obtained in this manner are *with respect to the axes of symmetry of the ellipse.* To obtain the *centroidal* moments of inertia of these shapes, the parallel-axis theorem should be used. Note that this remark also applies to a semicircle and to a quarter circle and that the expressions given for these shapes in Fig. 9.12 are *not* centroidal moments of inertia.

c. To calculate the polar moment of inertia of a composite area, you can use either the expressions given in Fig. 9.12 for J_O or the relationship

$$J_O = I_x + I_y \tag{9.4}$$

depending on the shape of the given area.

d. Before computing the centroidal moments of inertia of a given area, you may find it necessary to first locate the centroid of the area using the methods of Chap. 5.

3. *Locating the point of application of the resultant of a system of hydrostatic forces.* In Sec. 9.2 we found that

$$R = \gamma \int y \, dA = \gamma \bar{y} A$$
$$M_x = \gamma \int y^2 \, dA = \gamma I_x$$

where $\bar{y}$ is the distance from the x axis to the centroid of the submerged plane area. Since $\mathbf{R}$ is equivalent to the system of elemental hydrostatic forces, it follows that

$$\Sigma M_x: \qquad y_P R = M_x$$

where y_P is the depth of the point of application of $\mathbf{R}$. Then

$$y_P(\gamma \bar{y} A) = \gamma I_x \qquad \text{or} \qquad y_P = \frac{I_x}{\bar{y} A}$$

In closing, we encourage you to carefully study the notation used in Fig. 9.13 for the rolled-steel shapes, as you will likely encounter it again in subsequent engineering courses.

Problems

9.31 and 9.32 Determine the moment of inertia and the radius of gyration of the shaded area with respect to the x axis.

Fig. P9.31 and P9.33

Fig. P9.32 and *P9.34*

9.33 and *9.34* Determine the moment of inertia and the radius of gyration of the shaded area with respect to the y axis.

9.35 and *9.36* Determine the moments of inertia of the shaded area shown with respect to the x and y axes.

Fig. P9.35

Fig. *P9.36*

9.37 For the 6-in^2 shaded area shown, determine the distance d_2 and the moment of inertia with respect to the centroidal axis parallel to AA' knowing that the moments of inertia with respect to AA' and BB' are 30 in^4 and 58 in^4, respectively, and that $d_1 = 1.25$ in.

9.38 Determine for the shaded region the area and the moment of inertia with respect to the centroidal axis parallel to BB' knowing that $d_1 = 1.25$ in. and $d_2 = 0.75$ in. and that the moments of inertia with respect to AA' and BB' are 20 in^4 and 15 in^4, respectively.

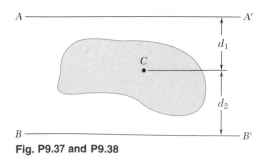

Fig. P9.37 and P9.38

9.39 The centroidal polar moment of inertia $\bar{J}_C$ of the 15.5×10^3 mm^2 shaded region is 250×10^6 mm^4. Determine the polar moments of inertia J_B and J_D of the shaded region knowing that $J_D = 2J_B$ and $d = 100$ mm.

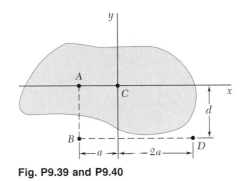

Fig. P9.39 and P9.40

9.40 Determine the centroidal polar moment of inertia $\bar{J}_C$ of the 10×10^3 mm^2 shaded area knowing that the polar moments of inertia of the area with respect to points A, B, and D are $J_A = 45 \times 10^6$ mm^4, $J_B = 130 \times 10^6$ mm^4, and $J_D = 252 \times 10^6$ mm^4, respectively.

9.41 through 9.44 Determine the moments of inertia $\bar{I}_x$ and $\bar{I}_y$ of the area shown with respect to centroidal axes respectively parallel and perpendicular to side AB.

Fig. P9.41

Fig. P9.42   **Fig. P9.43**

Fig. P9.44

9.45 and 9.46 Determine the polar moment of inertia of the area shown with respect to (a) point O, (b) the centroid of the area.

Fig. P9.45   **Fig. P9.46**

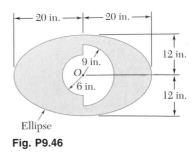

$J_0 = (J_0)_1 - (J_0)_2 = \frac{\pi}{8}(2.7)(1.8)[2.7^2 + (1.8)^2] - \frac{\pi}{4}(.9)^4 = 19.58$

$\bar{J}_C = ?$ $J_0 = \bar{J}_C + A(\bar{y})^2$; $19.5814 - (6.3618)(.84033) = 15.089$

9.47 and 9.48 Determine the polar moment of inertia of the area shown with respect to (*a*) point *O*, (*b*) the centroid of the area.

Semicircle

90 mm

O

←— 120 mm —→←— 120 mm —→

Fig. P9.47

120 mm

120 mm

57.5 mm

O

100 mm   150 mm   150 mm

Fig. P9.48

9.49 Two 1-in. steel plates are welded to a rolled S section as shown. Determine the moments of inertia and the radii of gyration of the section with respect to the centroidal *x* and *y* axes.

y

4 in.   4 in.

1 in.

C   *x*

S12 × 50

Fig. P9.49

y

W8 × 24

C   *x*

0.3 in.

←— 7.93 in. —→

Fig. P9.50

y

S250 × 52

C   *x*

C250 × 30

Fig. P9.51

9.50 To form a reinforced box section, two rolled W sections and two plates are welded together. Determine the moments of inertia and the radii of gyration of the combined section with respect to the centroidal axes shown.

9.51 Two C250 × 30 channels are welded to a 250 × 52 rolled S section as shown. Determine the moments of inertia and the radii of gyration of the combined section with respect to its centroidal *x* and *y* axes.

9.52 Two channels are welded to a *d* × 300-mm steel plate as shown. Determine the width *d* for which the ratio $\bar{I}_x/\bar{I}_y$ of the centroidal moments of inertia of the section is 16.

y

C250×30

150 mm

C   *x*

150 mm

←*d*→

Fig. P9.52

9.53 Two L3 × 3 × $\frac{1}{4}$-in. angles are welded to a C10 × 20 channel. Determine the moments of inertia of the combined section with respect to centroidal axes respectively parallel and perpendicular to the web of the channel.

L3 × 3 × $\frac{1}{4}$

C10 × 20

Fig. P9.53

9.54 To form an unsymmetrical girder, two L3 × 3 × $\frac{1}{4}$-in. angles and two L6 × 4 × $\frac{1}{2}$-in. angles are welded to a 0.8-in. steel plate as shown. Determine the moments of inertia of the combined section with respect to its centroidal x and y axes.

9.55 Two L127 × 76 × 12.7-mm angles are welded to a 10-mm steel plate. Determine the distance b and the centroidal moments of inertia $\bar{I}_x$ and $\bar{I}_y$ of the combined section knowing that $\bar{I}_y = 3\bar{I}_x$.

L3 × 3 × $\frac{1}{4}$

27 in.

C

0.8 in. L6 × 4 × $\frac{1}{2}$

Fig. P9.54

b b

L127 × 76 × 12.7

10 mm

C

100 mm 100 mm

Fig. P9.55

C150 × 15.6 L152 × 152 × 19.0

20 mm

C

a

Fig. P9.56

9.56 A channel and an angle are welded to an a × 20-mm steel plate. Knowing that the centroidal y axis is located as shown, determine (a) the width a, (b) the moments of inertia with respect to the centroidal x and y axes.

9.57 and 9.58 The panel shown forms the end of a trough which is filled with water to the line AA'. Referring to Sec. 9.2, determine the depth of the point of application of the resultant of the hydrostatic forces acting on the panel (the center of pressure).

A A'

r

Fig. P9.57

A a a A'

b

Semiellipse

Fig. P9.58

9.59 and 9.60   The panel shown forms the end of a trough which is filled with water to the line AA'. Referring to Sec. 9.2, determine the depth of the point of application of the resultant of the hydrostatic forces acting on the panel (the center of pressure).

Fig. P9.59

Fig. P9.60

9.61   The cover for a 10×22-in. access hole in an oil storage tank is attached to the outside of the tank with four bolts as shown. Knowing that the specific weight of the oil is 57.4 lb/ft³ and that the center of the cover is located 10 ft below the surface of the oil, determine the additional force on each bolt because of the pressure of the oil.

Fig. P9.61

9.62   A vertical trapezoidal gate that is used as an automatic valve is held shut by two springs attached to hinges located along edge AB. Knowing that each spring exerts a couple of magnitude 8.50 kN · m, determine the depth d of water for which the gate will open.

Fig. P9.62

***9.63**   Determine the x coordinate of the centroid of the volume shown. (*Hint:* The height y of the volume is proportional to the x coordinate; consider an analogy between this height and the water pressure on a submerged surface.)

Fig. P9.63

*9.64 Determine the x coordinate of the centroid of the volume shown; this volume was obtained by intersecting an elliptic cylinder with an oblique plane. (*Hint:* The height y of the volume is proportional to the x coordinate; consider an analogy between this height and the water pressure on a submerged surface.)

Fig. P9.64

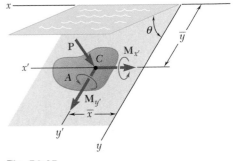

Fig. P9.65

*9.65 Show that the system of hydrostatic forces acting on a submerged plane area A can be reduced to a force $\mathbf{P}$ at the centroid C of the area and two couples. The force $\mathbf{P}$ is perpendicular to the area and is of magnitude $P = \gamma A\bar{y} \sin\theta$, where γ is the specific weight of the liquid, and the couples are $\mathbf{M}_{x'} = (\gamma \bar{I}_{x'} \sin\theta)\mathbf{i}$ and $\mathbf{M}_{y'} = (\gamma \bar{I}_{x'y'} \sin\theta)\mathbf{j}$, where $\bar{I}_{x'y'} = \int x'y'dA$ (see Sec. 9.8). Note that the couples are independent of the depth at which the area is submerged.

*9.66 Show that the resultant of the hydrostatic forces acting on a submerged plane area A is a force $\mathbf{P}$ perpendicular to the area and of magnitude $P = \gamma A\bar{y} \sin\theta = \bar{p}A$, where γ is the specific weight of the liquid and $\bar{p}$ is the pressure at the centroid C of the area. Show that $\mathbf{P}$ is applied at a point C_p, called the center of pressure, whose coordinates are $x_p = I_{xy}/A\bar{y}$ and $y_p = I_x/A\bar{y}$, where $I_{xy} = \int xy\, dA$ (see Sec. 9.8). Show also that the difference of ordinates $y_p - \bar{y}$ is equal to $\bar{k}_{x'}^2/\bar{y}$ and thus depends upon the depth at which the area is submerged.

Fig. P9.66

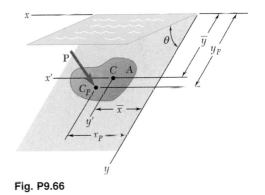

Fig. 9.14

*9.8. PRODUCT OF INERTIA

The integral

$$I_{xy} = \int xy\, dA \qquad (9.12)$$

which is obtained by multiplying each element dA of an area A by its coordinates x and y and integrating over the area (Fig. 9.14), is known as the *product of inertia* of the area A with respect to the x and y axes. Unlike the moments of inertia I_x and I_y, the product of inertia I_{xy} can be positive, negative, or zero.

When one or both of the x and y axes are axes of symmetry for the area A, the product of inertia I_{xy} is zero. Consider, for example, the channel section shown in Fig. 9.15. Since this section is symmetrical with respect to the x axis, we can associate with each element dA of coordinates x and y an element dA' of coordinates x and $-y$. Clearly, the contributions to I_{xy} of any pair of elements chosen in this way cancel out, and the integral (9.12) reduces to zero.

A parallel-axis theorem similar to the one established in Sec. 9.6 for moments of inertia can be derived for products of inertia.

Fig. 9.15

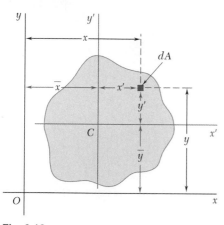

Fig. 9.16

Consider an area A and a system of rectangular coordinates x and y (Fig. 9.16). Through the centroid C of the area, of coordinates $\bar{x}$ and $\bar{y}$, we draw two *centroidal axes* x' and y' which are parallel, respectively, to the x and y axes. Denoting by x and y the coordinates of an element of area dA with respect to the original axes, and by x' and y' the coordinates of the same element with respect to the centroidal axes, we write $x = x' + \bar{x}$ and $y = y' + \bar{y}$. Substituting into (9.12), we obtain the following expression for the product of inertia I_{xy}:

$$I_{xy} = \int xy\, dA = \int (x' + \bar{x})(y' + \bar{y})\, dA$$
$$= \int x'y'\, dA + \bar{y}\int x'\, dA + \bar{x}\int y'\, dA + \bar{x}\bar{y}\int dA$$

The first integral represents the product of inertia $\bar{I}_{x'y'}$ of the area A with respect to the centroidal axes x' and y'. The next two integrals represent first moments of the area with respect to the centroidal axes; they reduce to zero, since the centroid C is located on these axes. Finally, we observe that the last integral is equal to the total area A. Therefore, we have

$$I_{xy} = \bar{I}_{x'y'} + \bar{x}\bar{y}\, A \tag{9.13}$$

*9.9. PRINCIPAL AXES AND PRINCIPAL MOMENTS OF INERTIA

Consider the area A and the coordinate axes x and y (Fig. 9.17). Assuming that the moments and product of inertia

$$I_x = \int y^2\, dA \qquad I_y = \int x^2\, dA \qquad I_{xy} = \int xy\, dA \tag{9.14}$$

of the area A are known, we propose to determine the moments and product of inertia $I_{x'}$, $I_{y'}$, and $I_{x'y'}$ of A with respect to new axes x' and y' which are obtained by rotating the original axes about the origin through an angle θ.

We first note the following relations between the coordinates x', y' and x, y of an element of area dA:

$$x' = x\cos\theta + y\sin\theta \qquad y' = y\cos\theta - x\sin\theta$$

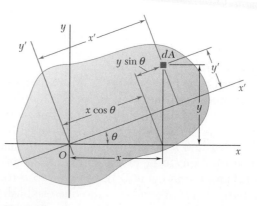

Fig. 9.17

Substituting for y' in the expression for $I_{x'}$, we write

$$I_{x'} = \int (y')^2 \, dA = \int (y \cos \theta - x \sin \theta)^2 \, dA$$

$$= \cos^2 \theta \int y^2 \, dA - 2 \sin \theta \cos \theta \int xy \, dA + \sin^2 \theta \int x^2 \, dA$$

Using the relations (9.14), we write

$$I_{x'} = I_x \cos^2 \theta - 2I_{xy} \sin \theta \cos \theta + I_y \sin^2 \theta \qquad (9.15)$$

Similarly, we obtain for $I_{y'}$ and $I_{x'y'}$ the expressions

$$I_{y'} = I_x \sin^2 \theta + 2I_{xy} \sin \theta \cos \theta + I_y \cos^2 \theta \qquad (9.16)$$

$$I_{x'y'} = (I_x - I_y) \sin \theta \cos \theta + I_{xy}(\cos^2 \theta - \sin^2 \theta) \qquad (9.17)$$

Recalling the trigonometric relations

$$\sin 2\theta = 2 \sin \theta \cos \theta \qquad \cos 2\theta = \cos^2 \theta - \sin^2 \theta$$

and

$$\cos^2 \theta = \frac{1 + \cos 2\theta}{2} \qquad \sin^2 \theta = \frac{1 - \cos 2\theta}{2}$$

we can write (9.15), (9.16), and (9.17) as follows:

$$I_{x'} = \frac{I_x + I_y}{2} + \frac{I_x - I_y}{2} \cos 2\theta - I_{xy} \sin 2\theta \qquad (9.18)$$

$$I_{y'} = \frac{I_x + I_y}{2} - \frac{I_x - I_y}{2} \cos 2\theta + I_{xy} \sin 2\theta \qquad (9.19)$$

$$I_{x'y'} = \frac{I_x - I_y}{2} \sin 2\theta + I_{xy} \cos 2\theta \qquad (9.20)$$

Adding (9.18) and (9.19) we observe that

$$I_{x'} + I_{y'} = I_x + I_y \qquad (9.21)$$

This result could have been anticipated, since both members of (9.21) are equal to the polar moment of inertia J_O.

Equations (9.18) and (9.20) are the parametric equations of a circle. This means that if we choose a set of rectangular axes and plot a point M of abscissa $I_{x'}$ and ordinate $I_{x'y'}$ for any given value of the parameter θ, all of the points thus obtained will lie on a circle. To establish this property, we eliminate θ from Eqs. (9.18) and (9.20); this is done by transposing $(I_x + I_y)/2$ in Eq. (9.18), squaring both members of Eqs. (9.18) and (9.20), and adding. We write

$$\left(I_{x'} - \frac{I_x + I_y}{2}\right)^2 + I_{x'y'}^2 = \left(\frac{I_x - I_y}{2}\right)^2 + I_{xy}^2 \qquad (9.22)$$

Setting

$$I_{\text{ave}} = \frac{I_x + I_y}{2} \qquad \text{and} \qquad R = \sqrt{\left(\frac{I_x - I_y}{2}\right)^2 + I_{xy}^2} \qquad (9.23)$$

we write the identity (9.22) in the form

$$(I_{x'} - I_{\text{ave}})^2 + I_{x'y'}^2 = R^2 \qquad (9.24)$$

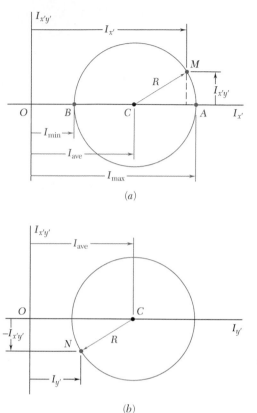

Fig. 9.18

which is the equation of a circle of radius R centered at the point C whose x and y coordinates are I_{ave} and 0, respectively (Fig. 9.18a). We observe that Eqs. (9.19) and (9.20) are the parametric equations of the same circle. Furthermore, because of the symmetry of the circle about the horizontal axis, the same result would have been obtained if instead of plotting M, we had plotted a point N of coordinates $I_{y'}$ and $-I_{x'y'}$ (Fig. 9.18b). This property will be used in Sec. 9.10.

The two points A and B where the above circle intersects the horizontal axis (Fig. 9.18a) are of special interest: Point A corresponds to the maximum value of the moment of inertia $I_{x'}$, while point B corresponds to its minimum value. In addition, both points correspond to a zero value of the product of inertia $I_{x'y'}$. Thus, the values θ_m of the parameter θ which correspond to the points A and B can be obtained by setting $I_{x'y'} = 0$ in Eq. (9.20). We obtain†

$$\tan 2\theta_m = -\frac{2I_{xy}}{I_x - I_y} \qquad (9.25)$$

This equation defines two values $2\theta_m$ which are 180° apart and thus two values θ_m which are 90° apart. One of these values corresponds to point A in Fig. 9.18a and to an axis through O in Fig. 9.17 with respect to which the moment of inertia of the given area is maximum; the other value corresponds to point B and to an axis through O with respect to which the moment of inertia of the area is minimum. The two axes thus defined, which are perpendicular to each other, are called the *principal axes of the area about O*, and the corresponding values I_{max} and I_{min} of the moment of inertia are called the *principal moments of inertia of the area about O*. Since the two values θ_m defined by Eq. (9.25) were obtained by setting $I_{x'y'} = 0$ in Eq. (9.20), it is clear that the product of inertia of the given area with respect to its principal axes is zero.

We observe from Fig. 9.18a that

$$I_{\text{max}} = I_{\text{ave}} + R \qquad I_{\text{min}} = I_{\text{ave}} - R \qquad (9.26)$$

Using the values for I_{ave} and R from formulas (9.23), we write

$$I_{\text{max,min}} = \frac{I_x + I_y}{2} \pm \sqrt{\left(\frac{I_x - I_y}{2}\right)^2 + I_{xy}^2} \qquad (9.27)$$

Unless it is possible to tell by inspection which of the two principal axes corresponds to I_{max} and which corresponds to I_{min}, it is necessary to substitute one of the values of θ_m into Eq. (9.18) in order to determine which of the two corresponds to the maximum value of the moment of inertia of the area about O.

Referring to Sec. 9.8, we note that if an area possesses an axis of symmetry through a point O, this axis must be a principal axis of the area about O. On the other hand, a principal axis does not need to be an axis of symmetry; whether or not an area possesses any axes of symmetry, it will have two principal axes of inertia about any point O.

The properties we have established hold for any point O located inside or outside the given area. If the point O is chosen to coincide with the centroid of the area, any axis through O is a centroidal axis; the two principal axes of the area about its centroid are referred to as the *principal centroidal axes of the area*.

---

†This relation can also be obtained by differentiating $I_{x'}$ in Eq. (9.18) and setting $dI_{x'}/d\theta = 0$.

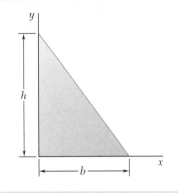

SAMPLE PROBLEM 9.6

Determine the product of inertia of the right triangle shown (*a*) with respect to the *x* and *y* axes and (*b*) with respect to centroidal axes parallel to the *x* and *y* axes.

SOLUTION

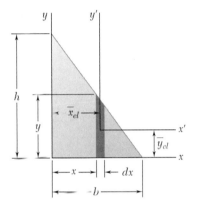

***a*. Product of Inertia** I_{xy}. A vertical rectangular strip is chosen as the differential element of area. Using the parallel-axis theorem, we write

$$dI_{xy} = dI_{x'y'} + \bar{x}_{el}\bar{y}_{el}\, dA$$

Since the element is symmetrical with respect to the x' and y' axes, we note that $dI_{x'y'} = 0$. From the geometry of the triangle, we obtain

$$y = h\left(1 - \frac{x}{b}\right) \qquad dA = y\, dx = h\left(1 - \frac{x}{b}\right) dx$$

$$\bar{x}_{el} = x \qquad \bar{y}_{el} = \tfrac{1}{2}y = \tfrac{1}{2}h\left(1 - \frac{x}{b}\right)$$

Integrating dI_{xy} from $x = 0$ to $x = b$, we obtain

$$I_{xy} = \int dI_{xy} = \int \bar{x}_{el}\bar{y}_{el}\, dA = \int_0^b x(\tfrac{1}{2})h^2\left(1 - \frac{x}{b}\right)^2 dx$$

$$= h^2 \int_0^b \left(\frac{x}{2} - \frac{x^2}{b} + \frac{x^3}{2b^2}\right) dx = h^2 \left[\frac{x^2}{4} - \frac{x^3}{3b} + \frac{x^4}{8b^2}\right]_0^b$$

$$I_{xy} = \tfrac{1}{24}b^2h^2 \quad \blacktriangleleft$$

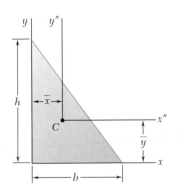

***b*. Product of Inertia** $\bar{I}_{x''y''}$. The coordinates of the centroid of the triangle relative to the *x* and *y* axes are

$$\bar{x} = \tfrac{1}{3}b \qquad \bar{y} = \tfrac{1}{3}h$$

Using the expression for I_{xy} obtained in part *a*, we apply the parallel-axis theorem and write

$$I_{xy} = \bar{I}_{x''y''} + \bar{x}\bar{y}A$$
$$\tfrac{1}{24}b^2h^2 = \bar{I}_{x''y''} + (\tfrac{1}{3}b)(\tfrac{1}{3}h)(\tfrac{1}{2}bh)$$
$$\bar{I}_{x''y''} = \tfrac{1}{24}b^2h^2 - \tfrac{1}{18}b^2h^2$$

$$\bar{I}_{x''y''} = -\tfrac{1}{72}b^2h^2 \quad \blacktriangleleft$$

SAMPLE PROBLEM 9.7

For the section shown, the moments of inertia with respect to the x and y axes have been computed and are known to be

$$I_x = 10.38 \text{ in}^4 \qquad I_y = 6.97 \text{ in}^4$$

Determine (a) the orientation of the principal axes of the section about O, (b) the values of the principal moments of inertia of the section about O.

SOLUTION

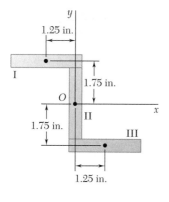

We first compute the product of inertia with respect to the x and y axes. The area is divided into three rectangles as shown. We note that the product of inertia $\bar{I}_{x'y'}$ with respect to centroidal axes parallel to the x and y axes is zero for each rectangle. Using the parallel-axis theorem $I_{xy} = \bar{I}_{x'y'} + \bar{x}\bar{y}A$, we find that I_{xy} reduces to $\bar{x}\bar{y}A$ for each rectangle.

Rectangle	Area, in²	$\bar{x}$, in.	$\bar{y}$, in.	$\bar{x}\bar{y}A$, in⁴
I	1.5	−1.25	+1.75	−3.28
II	1.5	0	0	0
III	1.5	+1.25	−1.75	−3.28
				$\Sigma\bar{x}\bar{y}A = -6.56$

$$I_{xy} = \Sigma\bar{x}\bar{y}A = -6.56 \text{ in}^4$$

***a*. Principal Axes.** Since the magnitudes of I_x, I_y, and I_{xy} are known, Eq. (9.25) is used to determine the values of θ_m:

$$\tan 2\theta_m = -\frac{2I_{xy}}{I_x - I_y} = -\frac{2(-6.56)}{10.38 - 6.97} = +3.85$$

$$2\theta_m = 75.4° \text{ and } 255.4°$$

$$\theta_m = 37.7° \qquad \text{and} \qquad \theta_m = 127.7° \quad \blacktriangleleft$$

***b*. Principal Moments of Inertia.** Using Eq. (9.27), we write

$$I_{max,min} = \frac{I_x + I_y}{2} \pm \sqrt{\left(\frac{I_x - I_y}{2}\right)^2 + I_{xy}^2}$$

$$= \frac{10.38 + 6.97}{2} \pm \sqrt{\left(\frac{10.38 - 6.97}{2}\right)^2 + (-6.56)^2}$$

$$I_{max} = 15.45 \text{ in}^4 \qquad I_{min} = 1.897 \text{ in}^4 \quad \blacktriangleleft$$

Noting that the elements of the area of the section are more closely distributed about the b axis than about the a axis, we conclude that $I_a = I_{max} = 15.45 \text{ in}^4$ and $I_b = I_{min} = 1.897 \text{ in}^4$. This conclusion can be verified by substituting $\theta = 37.7°$ into Eqs. (9.18) and (9.19).

In the problems for this lesson, you will continue your work with *moments of inertia* and will utilize various techniques for computing *products of inertia*. Although the problems are generally straightforward, several items are worth noting.

1. **Calculating the product of inertia I_{xy} by integration.** We defined this quantity as

$$I_{xy} = \int xy \, dA \tag{9.12}$$

and stated that its value can be positive, negative, or zero. The product of inertia can be computed directly from the above equation using double integration, or it can be determined using single integration as shown in Sample Prob. 9.6. When applying the latter technique and using the parallel-axis theorem, it is important to remember that $\bar{x}_{el}$ and $\bar{y}_{el}$ in the equation

$$dI_{xy} = dI_{x'y'} + x_{el}\bar{y}_{el} \, dA$$

are the coordinates of the centroid of the element of area dA. Thus, if dA is not in the first quadrant, one or both of these coordinates will be negative.

2. **Calculating the products of inertia of composite areas.** They can easily be computed from the products of inertia of their component parts by using the parallel-axis theorem

$$I_{xy} = \bar{I}_{x'y'} + \bar{x}\bar{y}A \tag{9.13}$$

The proper technique to use for problems of this type is illustrated in Sample Probs. 9.6 and 9.7. In addition to the usual rules for composite-area problems, it is essential that you remember the following points.

   a. If either of the centroidal axes of a component area is an axis of symmetry for that area, the product of inertia $\bar{I}_{x'y'}$ for that area is zero. Thus, $\bar{I}_{x'y'}$ is zero for component areas such as circles, semicircles, rectangles, and isosceles triangles which possess an axis of symmetry parallel to one of the coordinate axes.

   *b. Pay careful attention to the signs of the coordinates $\bar{x}$ and $\bar{y}$ of each component area when you use the parallel-axis theorem [Sample Prob. 9.7].

3. **Determining the moments of inertia and the product of inertia for rotated coordinate axes.** In Sec. 9.9 we derived Eqs. (9.18), (9.19), and (9.20), from which the moments of inertia and the product of inertia can be computed for coordinate axes which have been rotated about the origin O. To apply these equations, you must know a set of values I_x, I_y, and I_{xy} for a given orientation of the axes, and you must remember that θ is positive for counterclockwise rotations of the axes and negative for clockwise rotations of the axes.

4. **Computing the principal moments of inertia.** We showed in Sec. 9.9 that there is a particular orientation of the coordinate axes for which the moments of inertia attain their maximum and minimum values, I_{max} and I_{min}, and for which the product of inertia is zero. Equation (9.27) can be used to compute these values, known as the *principal moments of inertia* of the area about O. The corresponding axes are referred to as the *principal axes* of the area about O, and their orientation is defined by Eq. (9.25). *To determine which of the principal axes corresponds to I_{max} and which corresponds to I_{min},* you can either follow the procedure outlined in the text after Eq. (9.27) or observe about which of the two principal axes the area is more closely distributed; that axis corresponds to I_{min} [Sample Prob. 9.7].

Problems

Fig. P9.67

9.67 through 9.70 Determine by direct integration the product of inertia of the given area with respect to the x and y axes.

Fig. P9.68

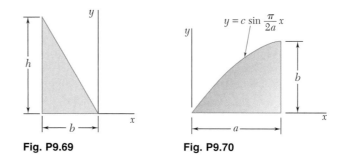

Fig. P9.69

Fig. P9.70

9.71 through 9.74 Using the parallel-axis theorem, determine the product of inertia of the area shown with respect to the centroidal x and y axes.

Fig. *P9.71*

Fig. P9.72

Fig. P9.73

Fig. P9.74

9.75 through 9.78 Using the parallel-axis theorem, determine the product of inertia of the area shown with respect to the centroidal x and y axes.

Fig. P9.75

$$I_{xy} = (I_{xy})_1 + \, \cdot \, + \,$$
$$= 0$$

Fig. P9.76

$$I_{xy} = \bar{I}_{x'y'} + \bar{x}\,\bar{y}\,A$$
$$W\!be \quad I_{x'y'} = 0$$
$$I_{xy} = (\bar{x}\bar{y}\,A)_2 + (\bar{x}\bar{y})A_3$$
$$(-69)(-25)(12)(38) + (69)(25)(12)(38)$$
$$I_{xy} = 1.573 \times 10^6 \,^4$$

Fig. P9.77

Fig. P9.78

9.79 Determine for the quarter ellipse of Prob. 9.67 the moments of inertia and the product of inertia with respect to new axes obtained by rotating the x and y axes about O (a) through 45° counterclockwise, (b) through 30° clockwise.

9.80 Determine the moments of inertia and the product of inertia of the area of Prob. 9.72 with respect to new centroidal axes obtained by rotating the x and y axes 45° clockwise.

9.81 Determine the moments of inertia and the product of inertia of the area of Prob. 9.73 with respect to new centroidal axes obtained by rotating the x and y axes through 30° clockwise.

9.82 Determine the moments of inertia and the product of inertia of the area of Prob. 9.75 with respect to new centroidal axes obtained by rotating the x and y axes through 60° counterclockwise.

9.83 Determine the moments of inertia and the product of inertia of the L76 × 51 × 6.4-mm angle cross section of Prob. 9.74 with respect to new centroidal axes obtained by rotating the x and y axes through 45° clockwise.

9.84 Determine the moments of inertia and the product of inertia of the L5 × 3 × $\frac{1}{2}$-in. angle cross section of Prob. 9.78 with respect to new centroidal axes obtained by rotating the x and y axes through 30° counterclockwise.

9.85 For the quarter ellipse of Prob. 9.67, determine the orientation of the principal axes at the origin and the corresponding values of the moments of inertia.

9.86 through 9.88 For the area indicated, determine the orientation of the principal axes at the origin and the corresponding values of the moments of inertia.

    9.86   Area of Prob. 9.72
    9.87   Area of Prob. 9.73
    9.88   Area of Prob. 9.75

9.89 and 9.90 For the angle cross section indicated, determine the orientation of the principal axes at the origin and the corresponding values of the moments of inertia.

    9.89   The L76 × 51 × 6.4-mm angle cross section of Prob. 9.74
    9.90   The L5 × 3 × $\frac{1}{2}$-in. angle cross section of Prob. 9.78

---

*9.10. MOHR'S CIRCLE FOR MOMENTS AND PRODUCTS OF INERTIA

The circle used in the preceding section to illustrate the relations existing between the moments and products of inertia of a given area with respect to axes passing through a fixed point O was first introduced by the German engineer Otto Mohr (1835–1918) and is known as *Mohr's circle*. It will be shown that if the moments and product of inertia of an area A are known with respect to two rectangular x and y axes which pass through a point O, Mohr's circle can be used to graphically determine (*a*) the principal axes and principal moments of inertia of the area about O and (*b*) the moments and product of inertia of the area with respect to any other pair of rectangular axes x' and y' through O.

Consider a given area A and two rectangular coordinate axes x and y (Fig. 9.19*a*). Assuming that the moments of inertia I_x and I_y and the product of inertia I_{xy} are known, we will represent them on a diagram by plotting a point X of coordinates I_x and I_{xy} and a point Y of coordinates I_y and $-I_{xy}$ (Fig. 9.19*b*). If I_{xy} is positive, as assumed in Fig. 9.19*a*, point X is located above the horizontal axis and point Y is located below, as shown in Fig. 9.19*b*. If I_{xy} is negative, X is located below the horizontal axis and Y is located above. Joining X and Y with a straight line, we denote by C the point of intersection of line XY with the horizontal axis and draw the circle of center C and diameter XY. Noting that the abscissa of C and the radius of the circle are respectively equal to the quantities I_{ave} and R defined by the formula (9.23), we conclude that the circle obtained is Mohr's circle for the given area about point O. Thus, the abscissas of the points A and B where the circle intersects the horizontal axis represent respectively the principal moments of inertia $I_{\max}$ and $I_{\min}$ of the area.

We also note that, since $\tan{(XCA)} = 2I_{xy}/(I_x - I_y)$, the angle XCA is equal in magnitude to one of the angles $2\theta_m$ which satisfy Eq. (9.25);

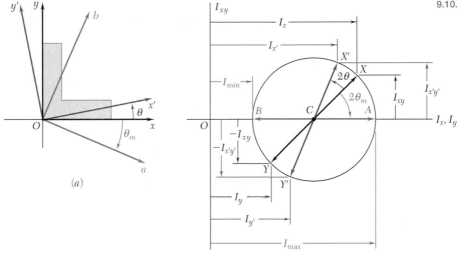

Fig. 9.19

(a)

(b)

thus, the angle θ_m, which defines in Fig. 9.19a the principal axis Oa corresponding to point A in Fig. 9.19b, is equal to half of the angle XCA of Mohr's circle. We further observe that if $I_x > I_y$ and $I_{xy} > 0$, as in the case considered here, the rotation which brings CX into CA is clockwise. Also, under these conditions, the angle θ_m obtained from Eq. (9.25), which defines the principal axis Oa in Fig. 9.19a, is negative; thus, the rotation which brings Ox into Oa is also clockwise. We conclude that the senses of rotation in both parts of Fig. 9.19 are the same. If a clockwise rotation through $2\theta_m$ is required to bring CX into CA on Mohr's circle, a clockwise rotation through θ_m will bring Ox into the corresponding principal axis Oa in Fig. 9.19a.

Since Mohr's circle is uniquely defined, the same circle can be obtained by considering the moments and product of inertia of the area A with respect to the rectangular axes x' and y' (Fig. 9.19a). The point X' of coordinates $I_{x'}$ and $I_{x'y'}$ and the point Y' of coordinates $I_{y'}$ and $-I_{x'y'}$ are thus located on Mohr's circle, and the angle $X'CA$ in Fig. 9.19b must be equal to twice the angle $x'Oa$ in Fig. 9.19a. Since, as noted above, the angle XCA is twice the angle xOa, it follows that the angle XCX' in Fig. 9.19b is twice the angle xOx' in Fig. 9.19a. The diameter $X'Y'$, which defines the moments and product of inertia $I_{x'}$, $I_{y'}$, and $I_{x'y'}$ of the given area with respect to rectangular axes x' and y' forming an angle θ with the x and y axes, can be obtained by rotating through an angle 2θ the diameter XY, which corresponds to the moments and product of inertia I_x, I_y, and I_{xy}. We note that the rotation which brings the diameter XY into the diameter $X'Y'$ in Fig. 9.19b has the same sense as the rotation which brings the x and y axes into the x' and y' axes in Fig. 9.19a.

It should be noted that the use of Mohr's circle is not limited to graphical solutions, that is, to solutions based on the careful drawing and measuring of the various parameters involved. By merely sketching Mohr's circle and using trigonometry, one can easily derive the various relations required for a numerical solution of a given problem (see Sample Prob. 9.8).

L152 × 102 × 12.7

SAMPLE PROBLEM 9.8

For the section shown, the moments and product of inertia with respect to the x and y axes are known to be

$$I_x = 7.24 \times 10^6 \text{ mm}^4 \qquad I_y = 2.61 \times 10^6 \text{ mm}^4 \qquad I_{xy} = -2.54 \times 10^6 \text{ mm}^4$$

Using Mohr's circle, determine (a) the principal axes of the section about O, (b) the values of the principal moments of inertia of the section about O, (c) the moments and product of inertia of the section with respect to the x' and y' axes which form an angle of $60°$ with the x and y axes.

SOLUTION

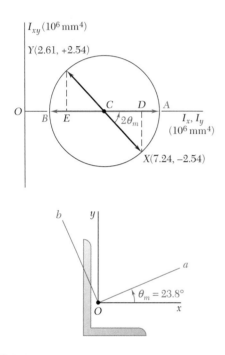

Drawing Mohr's Circle. We first plot point X of coordinates $I_x = 7.24$, $I_{xy} = -2.54$, and point Y of coordinates $I_y = 2.61$, $-I_{xy} = +2.54$. Joining X and Y with a straight line, we define the center C of Mohr's circle. The abscissa of C, which represents I_{ave}, and the radius R of the circle can be measured directly or calculated as follows:

$$I_{ave} = OC = \tfrac{1}{2}(I_x + I_y) = \tfrac{1}{2}(7.24 \times 10^6 + 2.61 \times 10^6) = 4.925 \times 10^6 \text{ mm}^4$$
$$CD = \tfrac{1}{2}(I_x - I_y) = \tfrac{1}{2}(7.24 \times 10^6 - 2.61 \times 10^6) = 2.315 \times 10^6 \text{ mm}^4$$
$$R = \sqrt{(CD)^2 + (DX)^2} = \sqrt{(2.315 \times 10^6)^2 + (2.54 \times 10^6)^2}$$
$$= 3.437 \times 10^6 \text{ mm}^4$$

a. Principal Axes. The principal axes of the section correspond to points A and B on Mohr's circle, and the angle through which we should rotate CX to bring it into CA defines $2\theta_m$. We have

$$\tan 2\theta_m = \frac{DX}{CD} = \frac{2.54}{2.315} = 1.097 \qquad 2\theta_m = 47.6° \,\text{↰} \qquad \theta_m = 23.8° \,\text{↰} \quad \blacktriangleleft$$

Thus, the principal axis Oa corresponding to the maximum value of the moment of inertia is obtained by rotating the x axis through $23.8°$ counterclockwise; the principal axis Ob corresponding to the minimum value of the moment of inertia is obtained by rotating the y axis through the same angle.

b. Principal Moments of Inertia. The principal moments of inertia are represented by the abscissas of A and B. We have

$$I_{max} = OA = OC + CA = I_{ave} + R = (4.925 + 3.437)10^6 \text{ mm}^4$$
$$I_{max} = 8.36 \times 10^6 \text{ mm}^4 \quad \blacktriangleleft$$
$$I_{min} = OB = OC - BC = I_{ave} - R = (4.925 - 3.437)10^6 \text{ mm}^4$$
$$I_{min} = 1.49 \times 10^6 \text{ mm}^4 \quad \blacktriangleleft$$

c. Moments and Product of Inertia with Respect to the x' and y' Axes. On Mohr's circle, the points X' and Y', which correspond to the x' and y' axes, are obtained by rotating CX and CY through an angle $2\theta = 2(60°) = 120°$ counterclockwise. The coordinates of X' and Y' yield the desired moments and product of inertia. Noting that the angle that CX' forms with the horizontal axis is $\phi = 120° - 47.6° = 72.4°$, we write

$$I_{x'} = OF = OC + CF = 4.925 \times 10^6 \text{ mm}^4 + (3.437 \times 10^6 \text{ mm}^4) \cos 72.4°$$
$$I_{x'} = 5.96 \times 10^6 \text{ mm}^4 \quad \blacktriangleleft$$
$$I_{y'} = OG = OC - GC = 4.925 \times 10^6 \text{ mm}^4 - (3.437 \times 10^6 \text{ mm}^4) \cos 72.4°$$
$$I_{y'} = 3.89 \times 10^6 \text{ mm}^4 \quad \blacktriangleleft$$
$$I_{x'y'} = FX' = (3.437 \times 10^6 \text{ mm}^4) \sin 72.4°$$
$$I_{x'y'} = 3.28 \times 10^6 \text{ mm}^4 \quad \blacktriangleleft$$

In the problems for this lesson, you will use *Mohr's circle* to determine the moments and products of inertia of a given area for different orientations of the coordinate axes. Although in some cases using Mohr's circle may not be as direct as substituting into the appropriate equations [Eqs. (9.18) through (9.20)], this method of solution has the advantage of providing a visual representation of the relationships among the various variables. Further, Mohr's circle shows all of the values of the moments and products of inertia which are possible for a given problem.

Using Mohr's circle. The underlying theory was presented in Sec. 9.9, and we discussed the application of this method in Sec. 9.10 and in Sample Prob. 9.8. In the same problem, we presented the steps you should follow to determine the *principal axes, the principal moments of inertia,* and the *moments and product of inertia with respect to a specified orientation of the coordinate axes.* When you use Mohr's circle to solve problems, it is important that you remember the following points.

a. Mohr's circle is completely defined by the quantities R and I_{ave}, which represent, respectively, the radius of the circle and the distance from the origin O to the center C of the circle. These quantities can be obtained from Eqs. (9.23) if the moments and product of inertia are known for a given orientation of the axes. However, Mohr's circle can be defined by other combinations of known values [Probs. 9.105, 9.108, and 9.109]. For these cases, it may be necessary to first make one or more assumptions, such as choosing an arbitrary location for the center when I_{ave} is unknown, assigning relative magnitudes to the moments of inertia (for example, $I_x > I_y$), or selecting the sign of the product of inertia.

b. Point X of coordinates (I_x, I_{xy}) and point Y of coordinates $(I_y, -I_{xy})$ are both located on Mohr's circle and are diametrically opposite.

c. Since moments of inertia must be positive, the entire Mohr's circle must lie to the right of the I_{xy} axis; it follows that $I_{ave} > R$ for all cases.

d. As the coordinate axes are rotated through an angle θ, the associated rotation of the diameter of Mohr's circle is equal to 2θ and is in the same sense (clockwise or counterclockwise). We strongly suggest that the known points on the circumference of the circle be labeled with the appropriate capital letter, as was done in Fig. 9.19b and for the Mohr's circles of Sample Prob. 9.8. This will enable you to determine, for each value of θ, the sign of the corresponding product of inertia and to determine which moment of inertia is associated with each of the coordinate axes [Sample Prob. 9.8, parts a and c].

Although we have introduced Mohr's circle within the specific context of the study of moments and products of inertia, the Mohr's circle technique is also applicable to the solution of analogous but physically different problems in mechanics of materials. This multiple use of a specific technique is not unique, and as you pursue your engineering studies, you will encounter several methods of solution which can be applied to a variety of problems.

Problems

9.91 Using Mohr's circle, determine for the quarter ellipse of Prob. 9.67 the moments of inertia and the product of inertia with respect to new axes obtained by rotating the x and y axes about O (*a*) through 45° counterclockwise, (*b*) through 30° clockwise.

9.92 Using Mohr's circle, determine the moments of inertia and the product of inertia of the area of Prob. 9.72 with respect to new centroidal axes obtained by rotating the x and y axes 45° clockwise.

9.93 Using Mohr's circle, determine the moments of inertia and the product of inertia of the area of Prob. 9.73 with respect to new centroidal axes obtained by rotating the x and y axes through 30° clockwise.

9.94 Using Mohr's circle, determine the moments of inertia and the product of inertia of the area of Prob. 9.75 with respect to new centroidal axes obtained by rotating the x and y axes through 60° counterclockwise.

9.95 Using Mohr's circle, determine the moments of inertia and the product of inertia of the L76 × 51 × 6.4-mm angle cross section of Prob. 9.74 with respect to new centroidal axes obtained by rotating the x and y axes through 45° clockwise.

9.96 Using Mohr's circle, determine the moments of inertia and the product of inertia of the L5 × 3 × $\frac{1}{2}$-in. angle cross section of Prob. 9.78 with respect to new centroidal axes obtained by rotating the x and y axes through 30° counterclockwise.

9.97 For the quarter ellipse of Prob. 9.67, use Mohr's circle to determine the orientation of the principal axes at the origin and the corresponding values of the moments of inertia.

9.98 through 9.104 Using Mohr's circle, determine for the area indicated the orientation of the principal centroidal axes and the corresponding values of the moments of inertia.

9.98 Area of Prob. 9.72
9.99 Area of Prob. 9.76
9.100 Area of Prob. 9.73
9.101 Area of Prob. 9.74
9.102 Area of Prob. 9.75
9.103 Area of Prob. 9.71
9.104 Area of Prob. 9.77

(The moments of inertia $\bar{I}_x$ and $\bar{I}_y$ of the area of Prob. 9.104 were determined in Prob. 9.44.)

9.105 The moments and product of inertia for an L102 × 76 × 6.4-mm angle cross section with respect to two rectangular axes x and y through C are, respectively, $\bar{I}_x = 0.166 \times 10^6$ mm^4, $\bar{I}_y = 0.453 \times 10^6$ mm^4, and $\bar{I}_{xy} < 0$, with the minimum value of the moment of inertia of the area with respect to any axis through C being $\bar{I}_{min} = 0.051 \times 10^6$ mm^4. Using Mohr's circle, determine (*a*) the product of inertia $\bar{I}_{xy}$ of the area, (*b*) the orientation of the principal axes, (*c*) the value of $\bar{I}_{max}$.

9.106 and 9.107   Using Mohr's circle, determine for the cross section of the rolled-steel angle shown the orientation of the principal centroidal axes and the corresponding values of the moments of inertia. (Properties of the cross sections are given in Fig. 9.13.)

Fig. P9.106

Fig. P9.107

*__9.108__   For a given area the moments of inertia with respect to two rectangular centroidal x and y axes are $\bar{I}_x = 640$ in^4 and $\bar{I}_y = 280$ in^4, respectively. Knowing that after rotating the x and y axes about the centroid 60° clockwise the product of inertia relative to the rotated axes is -180 in^4, use Mohr's circle to determine (*a*) the orientation of the principal axes, (*b*) the centroidal principal moments of inertia.

__9.109__   It is known that for a given area $\bar{I}_y = 300$ in^4 and $\bar{I}_{xy} = -125$ in^4, where the x and y axes are rectangular centroidal axes. If the axis corresponding to the maximum product of inertia is obtained by rotating the x axis 67.5° counterclockwise about C, use Mohr's circle to determine (*a*) the moment of inertia $\bar{I}_x$ of the area, (*b*) the principal centroidal moments of inertia.

__9.110__   Using Mohr's circle, show that for any regular polygon (such as a pentagon) (*a*) the moment of inertia with respect to every axis through the centroid is the same, (*b*) the product of inertia with respect to every pair of rectangular axes through the centroid is zero.

__9.111__   Using Mohr's circle, prove that the expression $I_{x'}I_{y'} - I_{x'y'}^2$ is independent of the orientation of the x' and y' axes, where $I_{x'}$, $I_{y'}$, and $I_{x'y'}$ represent the moments and product of inertia, respectively, of a given area with respect to a pair of rectangular axes x' and y' through a given point O. Also show that the given expression is equal to the square of the length of the tangent drawn from the origin of the coordinate system to Mohr's circle.

__9.112__   Using the invariance property established in the preceding problem, express the product of inertia I_{xy} of an area A with respect to a pair of rectangular axes through O in terms of the moments of inertia I_x and I_y of A and the principal moments of inertia I_{min} and I_{max} of A about O. Use the formula obtained to calculate the product of inertia $\bar{I}_{xy}$ of the $76 \times 51 \times$ 6.4-mm angle cross section shown in Fig. 9.13B knowing that its maximum moment of inertia is 524×10^3 mm^4.

MOMENTS OF INERTIA OF MASSES

9.11. MOMENT OF INERTIA OF A MASS

Consider a small mass Δm mounted on a rod of negligible mass which can rotate freely about an axis AA' (Fig. 9.20a). If a couple is applied to the system, the rod and mass, assumed to be initially at rest, will start rotating about AA'. The details of this motion will be studied later in dynamics. At present, we wish only to indicate that the time required for the system to reach a given speed of rotation is proportional to the mass Δm and to the square of the distance r. The product $r^2 \Delta m$ provides, therefore, a measure of the *inertia* of the system, that is, a measure of the resistance the system offers when we try to set it in motion. For this reason, the product $r^2 \Delta m$ is called the *moment of inertia* of the mass Δm with respect to the axis AA'.

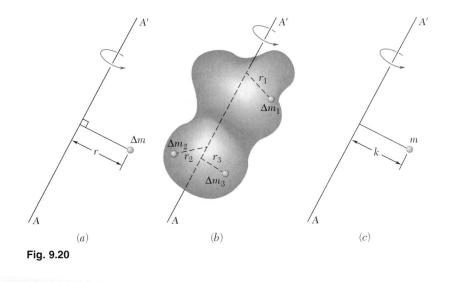

(a)   (b)   (c)

Fig. 9.20

Consider now a body of mass m which is to be rotated about an axis AA' (Fig. 9.20b). Dividing the body into elements of mass Δm_1, Δm_2, etc., we find that the body's resistance to being rotated is measured by the sum $r_1^2 \Delta m_1 + r_2^2 \Delta m_2 + \cdots$. This sum defines, therefore, the moment of inertia of the body with respect to the axis AA'. Increasing the number of elements, we find that the moment of inertia is equal, in the limit, to the integral

$$I = \int r^2 \, dm \qquad (9.28)$$

The *radius of gyration* k of the body with respect to the axis AA' is defined by the relation

$$I = k^2m \qquad \text{or} \qquad k = \sqrt{\frac{I}{m}} \qquad (9.29)$$

The radius of gyration k represents, therefore, the distance at which the entire mass of the body should be concentrated if its moment of inertia with respect to AA' is to remain unchanged (Fig. 9.20c). Whether it is kept in its original shape (Fig. 9.20b) or whether it is concentrated as shown in Fig. 9.20c, the mass m will react in the same way to a rotation, or *gyration,* about AA'.

If SI units are used, the radius of gyration k is expressed in meters and the mass m in kilograms, and thus the unit used for the moment of inertia of a mass is $kg \cdot m^2$. If U.S. customary units are used, the radius of gyration is expressed in feet and the mass in slugs (that is, in $lb \cdot s^2/ft$), and thus the derived unit used for the moment of inertia of a mass is $lb \cdot ft \cdot s^2$.†

The moment of inertia of a body with respect to a coordinate axis can easily be expressed in terms of the coordinates x, y, z of the element of mass dm (Fig. 9.21). Noting, for example, that the square of the distance r from the element dm to the y axis is $z^2 + x^2$, we express the moment of inertia of the body with respect to the y axis as

$$I_y = \int r^2 \, dm = \int (z^2 + x^2) \, dm$$

Similar expressions can be obtained for the moments of inertia with respect to the x and z axes. We write

$$I_x = \int (y^2 + z^2) \, dm$$

$$I_y = \int (z^2 + x^2) \, dm \qquad (9.30)$$

$$I_z = \int (x^2 + y^2) \, dm$$

Fig. 9.21

Photo 9.2 As you will discuss in your dynamics course, the rotational behavior of the camshaft shown is dependent upon the mass moment of inertia of the camshaft with respect to its axis of rotation.

†It should be kept in mind when converting the moment of inertia of a mass from U.S. customary units to SI units that the base unit *pound* used in the derived unit $lb \cdot ft \cdot s^2$ is a unit of force (*not* of mass) and should therefore be converted into newtons. We have

$$1 \, lb \cdot ft \cdot s^2 = (4.45 \, N)(0.3048 \, m)(1 \, s)^2 = 1.356 \, N \cdot m \cdot s^2$$

or, since $1 \, N = 1 \, kg \cdot m/s^2$,

$$1 \, lb \cdot ft \cdot s^2 = 1.356 \, kg \cdot m^2$$

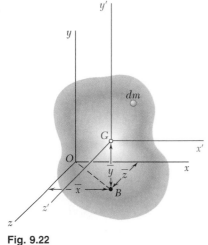

Fig. 9.22

9.12. PARALLEL-AXIS THEOREM

Consider a body of mass m. Let $Oxyz$ be a system of rectangular coordinates whose origin is at the arbitrary point O, and $Gx'y'z'$ a system of parallel *centroidal axes*, that is, a system whose origin is at the center of gravity G of the body† and whose axes x', y', z' are parallel to the x, y, and z axes, respectively (Fig. 9.22). Denoting by $\bar{x}, \bar{y}, \bar{z}$ the coordinates of G with respect to $Oxyz$, we write the following relations between the coordinates x, y, z of the element dm with respect to $Oxyz$ and its coordinates x', y', z' with respect to the centroidal axes $Gx'y'z'$:

$$x = x' + \bar{x} \qquad y = y' + \bar{y} \qquad z = z' + \bar{z} \qquad (9.31)$$

Referring to Eqs. (9.30), we can express the moment of inertia of the body with respect to the x axis as follows:

$$I_x = \int (y^2 + z^2)\, dm = \int [(y' + \bar{y})^2 + (z' + \bar{z})^2]\, dm$$
$$= \int (y'^2 + z'^2)\, dm + 2\bar{y} \int y'\, dm + 2\bar{z} \int z'\, dm + (\bar{y}^2 + \bar{z}^2) \int dm$$

The first integral in this expression represents the moment of inertia $I_{x'}$ of the body with respect to the centroidal axis x'; the second and third integrals represent the first moment of the body with respect to the $z'x'$ and $x'y'$ planes, respectively, and, since both planes contain G, the two integrals are *zero;* the last integral is equal to the total mass m of the body. We write, therefore,

$$I_x = \bar{I}_{x'} + m(\bar{y}^2 + \bar{z}^2) \qquad (9.32)$$

and, similarly,

$$I_y = \bar{I}_{y'} + m(\bar{z}^2 + \bar{x}^2) \qquad I_z = \bar{I}_{z'} + m(\bar{x}^2 + \bar{y}^2) \qquad (9.32')$$

We easily verify from Fig. 9.22 that the sum $\bar{z}^2 + \bar{x}^2$ represents the square of the distance OB, between the y and y' axes. Similarly, $\bar{y}^2 + \bar{z}^2$ and $\bar{x}^2 + \bar{y}^2$ represent the squares of the distance between the x and x' axes and the z and z' axes, respectively. Denoting by d the distance between an arbitrary axis AA' and the parallel centroidal axis BB' (Fig. 9.23), we can, therefore, write the following general relation between the moment of inertia I of the body with respect to AA' and its moment of inertia $\bar{I}$ with respect to BB':

$$I = \bar{I} + md^2 \qquad (9.33)$$

Expressing the moments of inertia in terms of the corresponding radii of gyration, we can also write

$$k^2 = \bar{k}^2 + d^2 \qquad (9.34)$$

where k and $\bar{k}$ represent the radii of gyration of the body about AA' and BB', respectively.

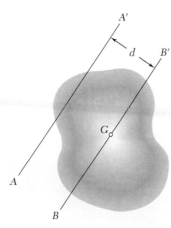

Fig. 9.23

†Note that the term *centroidal* is used here to define an axis passing through the center of gravity G of the body, whether or not G coincides with the centroid of the volume of the body.

9.13. MOMENTS OF INERTIA OF THIN PLATES

Consider a thin plate of uniform thickness t, which is made of a homogeneous material of density ρ (density = mass per unit volume). The mass moment of inertia of the plate with respect to an axis AA' *contained in the plane* of the plate (Fig. 9.24a) is

$$I_{AA', \text{ mass}} = \int r^2 \, dm$$

Since $dm = \rho t \, dA$, we write

$$I_{AA', \text{ mass}} = \rho t \int r^2 \, dA$$

But r represents the distance of the element of area dA to the axis

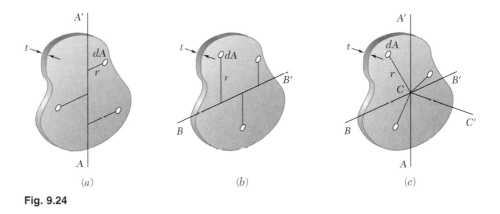

Fig. 9.24

AA'; the integral is therefore equal to the moment of inertia of the area of the plate with respect to AA'. We have

$$I_{AA', \text{ mass}} = \rho t I_{AA', \text{ area}} \qquad (9.35)$$

Similarly, for an axis BB' which is contained in the plane of the plate and is perpendicular to AA' (Fig. 9.24b), we have

$$I_{BB', \text{ mass}} = \rho t I_{BB', \text{ area}} \qquad (9.36)$$

Considering now the axis CC' which is *perpendicular* to the plate and passes through the point of intersection C of AA' and BB' (Fig. 9.24c), we write

$$I_{CC', \text{ mass}} = \rho t J_{C, \text{ area}} \qquad (9.37)$$

where J_C is the *polar* moment of inertia of the area of the plate with respect to point C.

Recalling the relation $J_C = I_{AA'} + I_{BB'}$ which exists between polar and rectangular moments of inertia of an area, we write the following relation between the mass moments of inertia of a thin plate:

$$I_{CC'} = I_{AA'} + I_{BB'} \qquad (9.38)$$

Fig. 9.25

Fig. 9.26

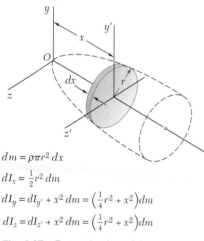

$dm = \rho\pi r^2\, dx$

$dI_x = \frac{1}{2}r^2\, dm$

$dI_y = dI_{y'} + x^2\, dm = \left(\frac{1}{4}r^2 + x^2\right)dm$

$dI_z = dI_{z'} + x^2\, dm = \left(\frac{1}{4}r^2 + x^2\right)dm$

Fig. 9.27 Determination of the moment of inertia of a body of revolution.

Rectangular Plate. In the case of a rectangular plate of sides a and b (Fig. 9.25), we obtain the following mass moments of inertia with respect to axes through the center of gravity of the plate:

$$I_{AA', \text{ mass}} = \rho t I_{AA', \text{ area}} = \rho t (\tfrac{1}{12}a^3 b)$$
$$I_{BB', \text{ mass}} = \rho t I_{BB', \text{ area}} = \rho t (\tfrac{1}{12}ab^3)$$

Observing that the product ρabt is equal to the mass m of the plate, we write the mass moments of inertia of a thin rectangular plate as follows:

$$I_{AA'} = \tfrac{1}{12}ma^2 \qquad I_{BB'} = \tfrac{1}{12}mb^2 \tag{9.39}$$
$$I_{CC'} = I_{AA'} + I_{BB'} = \tfrac{1}{12}m(a^2 + b^2) \tag{9.40}$$

Circular Plate. In the case of a circular plate, or disk, of radius r (Fig. 9.26), we write

$$I_{AA', \text{ mass}} = \rho t I_{AA', \text{ area}} = \rho t (\tfrac{1}{4}\pi r^4)$$

Observing that the product $\rho\pi r^2 t$ is equal to the mass m of the plate and that $I_{AA'} = I_{BB'}$, we write the mass moments of inertia of a circular plate as follows:

$$I_{AA'} = I_{BB'} = \tfrac{1}{4}mr^2 \tag{9.41}$$
$$I_{CC'} = I_{AA'} + I_{BB'} = \tfrac{1}{2}mr^2 \tag{9.42}$$

9.14. DETERMINATION OF THE MOMENT OF INERTIA OF A THREE-DIMENSIONAL BODY BY INTEGRATION

The moment of inertia of a three-dimensional body is obtained by evaluating the integral $I = \int r^2\, dm$. If the body is made of a homogeneous material of density ρ, the element of mass dm is equal to ρdV and we can write $I = \rho \int r^2\, dV$. This integral depends only upon the shape of the body. Thus, in order to compute the moment of inertia of a three-dimensional body, it will generally be necessary to perform a triple, or at least a double, integration.

However, if the body possesses two planes of symmetry, it is usually possible to determine the body's moment of inertia with a single integration by choosing as the element of mass dm a thin slab which is perpendicular to the planes of symmetry. In the case of bodies of revolution, for example, the element of mass would be a thin disk (Fig. 9.27). Using formula (9.42), the moment of inertia of the disk with respect to the axis of revolution can be expressed as indicated in Fig. 9.27. Its moment of inertia with respect to each of the other two coordinate axes is obtained by using formula (9.41) and the parallel-axis theorem. Integration of the expression obtained yields the desired moment of inertia of the body.

9.15. MOMENTS OF INERTIA OF COMPOSITE BODIES

The moments of inertia of a few common shapes are shown in Fig. 9.28. For a body consisting of several of these simple shapes, the moment of inertia of the body with respect to a given axis can be obtained by first computing the moments of inertia of its component parts about the desired axis and then adding them together. As was the case for areas, the radius of gyration of a composite body *cannot* be obtained by adding the radii of gyration of its component parts.

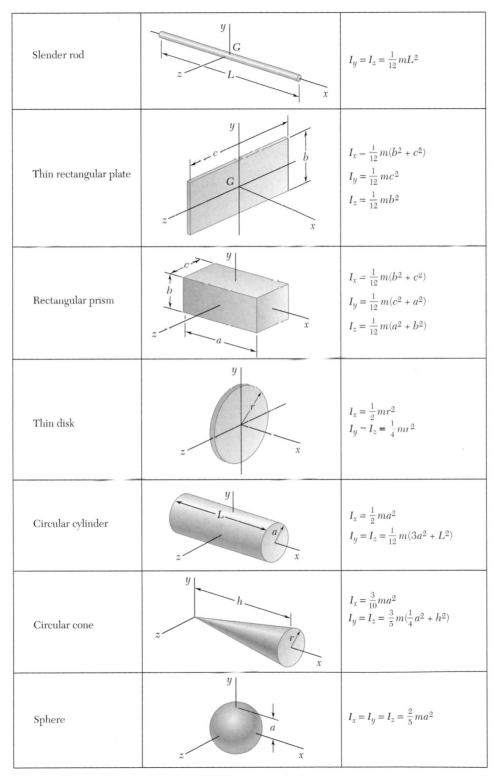

Slender rod		$I_y = I_z = \frac{1}{12} mL^2$
Thin rectangular plate		$I_x = \frac{1}{12} m(b^2 + c^2)$ $I_y = \frac{1}{12} mc^2$ $I_z = \frac{1}{12} mb^2$
Rectangular prism		$I_x = \frac{1}{12} m(b^2 + c^2)$ $I_y = \frac{1}{12} m(c^2 + a^2)$ $I_z = \frac{1}{12} m(a^2 + b^2)$
Thin disk		$I_x = \frac{1}{2} mr^2$ $I_y = I_z = \frac{1}{4} mr^2$
Circular cylinder		$I_x = \frac{1}{2} ma^2$ $I_y = I_z = \frac{1}{12} m(3a^2 + L^2)$
Circular cone		$I_x = \frac{3}{10} ma^2$ $I_y = I_z = \frac{3}{5} m(\frac{1}{4}a^2 + h^2)$
Sphere		$I_x = I_y = I_z = \frac{2}{5} ma^2$

Fig. 9.28 Mass moments of inertia of common geometric shapes.

SAMPLE PROBLEM 9.9

Determine the moment of inertia of a slender rod of length L and mass m with respect to an axis which is perpendicular to the rod and passes through one end of the rod.

SOLUTION

Choosing the differential element of mass shown, we write

$$dm = \frac{m}{L} dx$$

$$I_y = \int x^2 dm = \int_0^L x^2 \frac{m}{L} dx = \left[\frac{m}{L} \frac{x^3}{3} \right]_0^L \qquad I_y = \tfrac{1}{3}mL^2 \blacktriangleleft$$

SAMPLE PROBLEM 9.10

For the homogeneous rectangular prism shown, determine the moment of inertia with respect to the z axis.

SOLUTION

We choose as the differential element of mass the thin slab shown; thus

$$dm = \rho bc \, dx$$

Referring to Sec. 9.13, we find that the moment of inertia of the element with respect to the z' axis is

$$dI_{z'} = \tfrac{1}{12}b^2 \, dm$$

Applying the parallel-axis theorem, we obtain the mass moment of inertia of the slab with respect to the z axis.

$$dI_z = dI_{z'} + x^2 \, dm = \tfrac{1}{12}b^2 \, dm + x^2 \, dm = (\tfrac{1}{12}b^2 + x^2) \, \rho bc \, dx$$

Integrating from $x = 0$ to $x = a$, we obtain

$$I_z = \int dI_z = \int_0^a (\tfrac{1}{12}b^2 + x^2) \, \rho bc \, dx = \rho abc(\tfrac{1}{12}b^2 + \tfrac{1}{3}a^2)$$

Since the total mass of the prism is $m = \rho abc$, we can write

$$I_z = m(\tfrac{1}{12}b^2 + \tfrac{1}{3}a^2) \qquad I_z = \tfrac{1}{12}m(4a^2 + b^2) \blacktriangleleft$$

We note that if the prism is thin, b is small compared to a, and the expression for I_z reduces to $\tfrac{1}{3}ma^2$, which is the result obtained in Sample Prob. 9.9 when $L = a$.

SAMPLE PROBLEM 9.11

Determine the moment of inertia of a right circular cone with respect to (a) its longitudinal axis, (b) an axis through the apex of the cone and perpendicular to its longitudinal axis, (c) an axis through the centroid of the cone and perpendicular to its longitudinal axis.

SOLUTION

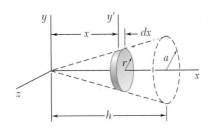

We choose the differential element of mass shown.

$$r = a\frac{x}{h} \qquad dm = \rho\pi r^2\, dx = \rho\pi\frac{a^2}{h^2}x^2\, dx$$

a. Moment of Inertia I_x. Using the expression derived in Sec. 9.13 for a thin disk, we compute the mass moment of inertia of the differential element with respect to the x axis.

$$dI_x = \tfrac{1}{2}r^2\, dm = \tfrac{1}{2}\Big(a\frac{x}{h}\Big)^2\Big(\rho\pi\frac{a^2}{h^2}x^2\, dx\Big) = \tfrac{1}{2}\rho\pi\frac{a^4}{h^4}x^4\, dx$$

Integrating from $x = 0$ to $x = h$, we obtain

$$I_x = \int dI_x = \int_0^h \tfrac{1}{2}\rho\pi\frac{a^4}{h^4}x^4\, dx = \tfrac{1}{2}\rho\pi\frac{a^4}{h^4}\frac{h^5}{5} = \tfrac{1}{10}\rho\pi a^4 h$$

Since the total mass of the cone is $m = \tfrac{1}{3}\rho\pi a^2 h$, we can write

$$I_x = \tfrac{1}{10}\rho\pi a^4 h = \tfrac{3}{10}a^2(\tfrac{1}{3}\rho\pi a^2 h) = \tfrac{3}{10}ma^2 \qquad I_x = \tfrac{3}{10}ma^2 \quad \blacktriangleleft$$

b. Moment of Inertia I_y. The same differential element is used. Applying the parallel-axis theorem and using the expression derived in Sec. 9.13 for a thin disk, we write

$$dI_y = dI_{y'} + x^2\, dm = \tfrac{1}{4}r^2\, dm + x^2\, dm = (\tfrac{1}{4}r^2 + x^2)\, dm$$

Substituting the expressions for r and dm into the equation, we obtain

$$dI_y = \Big(\frac{1}{4}\frac{a^2}{h^2}x^2 + x^2\Big)\Big(\rho\pi\frac{a^2}{h^2}x^2\, dx\Big) = \rho\pi\frac{a^2}{h^2}\Big(\frac{a^2}{4h^2} + 1\Big)x^4\, dx$$

$$I_y = \int dI_y = \int_0^h \rho\pi\frac{a^2}{h^2}\Big(\frac{a^2}{4h^2} + 1\Big)x^4\, dx = \rho\pi\frac{a^2}{h^2}\Big(\frac{a^2}{4h^2} + 1\Big)\frac{h^5}{5}$$

Introducing the total mass of the cone m, we rewrite I_y as follows:

$$I_y = \tfrac{3}{5}(\tfrac{1}{4}a^2 + h^2)\tfrac{1}{3}\rho\pi a^2 h \qquad I_y = \tfrac{3}{5}m(\tfrac{1}{4}a^2 + h^2) \quad \blacktriangleleft$$

c. Moment of Inertia $\overline{I}_{y''}$. We apply the parallel-axis theorem and write

$$I_y = \overline{I}_{y''} + m\overline{x}^2$$

Solving for $\overline{I}_{y''}$ and recalling from Fig. 5.21 that $\overline{x} = \tfrac{3}{4}h$, we have

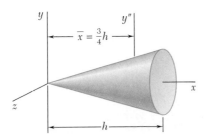

$$\overline{I}_{y''} = I_y - m\overline{x}^2 = \tfrac{3}{5}m(\tfrac{1}{4}a^2 + h^2) - m(\tfrac{3}{4}h)^2$$

$$\overline{I}_{y''} = \tfrac{3}{20}m(a^2 + \tfrac{1}{4}h^2) \quad \blacktriangleleft$$

SAMPLE PROBLEM 9.12

A steel forging consists of a 6 × 2 × 2-in. rectangular prism and two cylinders of diameter 2 in. and length 3 in. as shown. Determine the moments of inertia of the forging with respect to the coordinate axes knowing that the specific weight of steel is 490 lb/ft³.

SOLUTION

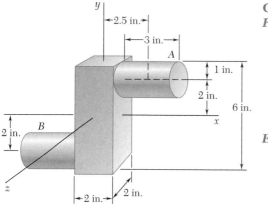

Computation of Masses
Prism

$$V = (2 \text{ in.})(2 \text{ in.})(6 \text{ in.}) = 24 \text{ in}^3$$

$$W = (24 \text{ in}^3)(490 \text{ lb/ft}^3)\left(\frac{1 \text{ ft}}{12 \text{ in.}}\right)^3 = 6.81 \text{ lb}$$

$$m = \frac{6.81 \text{ lb}}{32.2 \text{ ft/s}^2} = 0.211 \text{ lb} \cdot \text{s}^2/\text{ft}$$

Each Cylinder

$$V = \pi(1 \text{ in.})^2(3 \text{ in.}) = 9.42 \text{ in}^3$$

$$W = (9.42 \text{ in}^3)(490 \text{ lb/ft}^3)\left(\frac{1 \text{ ft}}{12 \text{ in.}}\right)^3 = 2.67 \text{ lb}$$

$$m = \frac{2.67 \text{ lb}}{32.2 \text{ ft/s}^2} = 0.0829 \text{ lb} \cdot \text{s}^2/\text{ft}$$

Moments of Inertia. The moments of inertia of each component are computed from Fig. 9.28, using the parallel-axis theorem when necessary. Note that all lengths are expressed in feet.

Prism

$$I_x - I_z = \tfrac{1}{12}(0.211 \text{ lb} \cdot \text{s}^2/\text{ft})[(\tfrac{6}{12} \text{ ft})^2 + (\tfrac{2}{12} \text{ ft})^2] = 4.88 \times 10^{-3} \text{ lb} \cdot \text{ft} \cdot \text{s}^2$$

$$I_y = \tfrac{1}{12}(0.211 \text{ lb} \cdot \text{s}^2/\text{ft})[(\tfrac{2}{12} \text{ ft})^2 + (\tfrac{2}{12} \text{ ft})^2] = 0.977 \times 10^{-3} \text{ lb} \cdot \text{ft} \cdot \text{s}^2$$

Each Cylinder

$$I_x = \tfrac{1}{2}ma^2 + m\overline{y}^2 = \tfrac{1}{2}(0.0829 \text{ lb} \cdot \text{s}^2/\text{ft})(\tfrac{1}{12} \text{ ft})^2 + (0.0829 \text{ lb} \cdot \text{s}^2/\text{ft})(\tfrac{2}{12} \text{ ft})^2$$
$$= 2.59 \times 10^{-3} \text{ lb} \cdot \text{ft} \cdot \text{s}^2$$

$$I_y = \tfrac{1}{12}m(3a^2 + L^2) + m\overline{x}^2 = \tfrac{1}{12}(0.0829 \text{ lb} \cdot \text{s}^2/\text{ft})[3(\tfrac{1}{12} \text{ ft})^2 + (\tfrac{3}{12} \text{ ft})^2]$$
$$+ (0.0829 \text{ lb} \cdot \text{s}^2/\text{ft})(\tfrac{2.5}{12} \text{ ft})^2 = 4.17 \times 10^{-3} \text{ lb} \cdot \text{ft} \cdot \text{s}^2$$

$$I_z = \tfrac{1}{12}m(3a^2 + L^2) + m(\overline{x}^2 + \overline{y}^2) = \tfrac{1}{12}(0.0829 \text{ lb} \cdot \text{s}^2/\text{ft})[3(\tfrac{1}{12} \text{ ft})^2 + (\tfrac{3}{12} \text{ ft})^2]$$
$$+ (0.0829 \text{ lb} \cdot \text{s}^2/\text{ft})[(\tfrac{2.5}{12} \text{ ft})^2 + (\tfrac{2}{12} \text{ ft})^2] = 6.48 \times 10^{-3} \text{ lb} \cdot \text{ft} \cdot \text{s}^2$$

Entire Body. Adding the values obtained,

$$I_x = 4.88 \times 10^{-3} + 2(2.59 \times 10^{-3}) \qquad I_x = 10.06 \times 10^{-3} \text{ lb} \cdot \text{ft} \cdot \text{s}^2 \blacktriangleleft$$
$$I_y = 0.977 \times 10^{-3} + 2(4.17 \times 10^{-3}) \qquad I_y = 9.32 \times 10^{-3} \text{ lb} \cdot \text{ft} \cdot \text{s}^2 \blacktriangleleft$$
$$I_z = 4.88 \times 10^{-3} + 2(6.48 \times 10^{-3}) \qquad I_z = 17.84 \times 10^{-3} \text{ lb} \cdot \text{ft} \cdot \text{s}^2 \blacktriangleleft$$

Dimensions in mm

SAMPLE PROBLEM 9.13

A thin steel plate which is 4 mm thick is cut and bent to form the machine part shown. Knowing that the density of steel is 7850 kg/m^3, determine the moments of inertia of the machine part with respect to the coordinate axes.

SOLUTION

We observe that the machine part consists of a semicircular plate and a rectangular plate from which a circular plate has been removed.

Computation of Masses. *Semicircular Plate*

$$V_1 = \tfrac{1}{2}\pi r^2 t = \tfrac{1}{2}\pi(0.08 \text{ m})^2(0.004 \text{ m}) = 40.21 \times 10^{-6} \text{ m}^3$$
$$m_1 = \rho V_1 = (7.85 \times 10^3 \text{ kg/m}^3)(40.21 \times 10^{-6} \text{ m}^3) = 0.3156 \text{ kg}$$

Rectangular Plate

$$V_2 = (0.200 \text{ m})(0.160 \text{ m})(0.004 \text{ m}) = 128 \times 10^{-6} \text{ m}^3$$
$$m_2 = \rho V_2 = (7.85 \times 10^3 \text{ kg/m}^3)(128 \times 10^{-6} \text{ m}^3) = 1.005 \text{ kg}$$

Circular Plate

$$V_3 = \pi a^2 t = \pi(0.050 \text{ m})^2(0.004 \text{ m}) = 31.42 \times 10^{-6} \text{ m}^3$$
$$m_3 = \rho V_3 = (7.85 \times 10^3 \text{ kg/m}^3)(31.42 \times 10^{-6} \text{ m}^3) = 0.2466 \text{ kg}$$

Moments of Inertia. Using the method presented in Sec. 9.13, we compute the moments of inertia of each component.

Semicircular Plate. From Fig. 9.28, we observe that for a circular plate of mass m and radius r

$$I_x = \tfrac{1}{2}mr^2 \qquad I_y = I_z = \tfrac{1}{4}mr^2$$

Because of symmetry, we note that for a semicircular plate

$$I_x = \tfrac{1}{2}(\tfrac{1}{2}mr^2) \qquad I_y = I_z = \tfrac{1}{2}(\tfrac{1}{4}mr^2)$$

Since the mass of the semicircular plate is $m_1 = \tfrac{1}{2}m$, we have

$$I_x = \tfrac{1}{2}m_1 r^2 = \tfrac{1}{2}(0.3156 \text{ kg})(0.08 \text{ m})^2 = 1.010 \times 10^{-3} \text{ kg} \cdot \text{m}^2$$
$$I_y = I_z = \tfrac{1}{4}(\tfrac{1}{2}mr^2) = \tfrac{1}{4}m_1 r^2 = \tfrac{1}{4}(0.3156 \text{ kg})(0.08 \text{ m})^2 = 0.505 \times 10^{-3} \text{ kg} \cdot \text{m}^2$$

Rectangular Plate

$$I_x = \tfrac{1}{12}m_2 c^2 = \tfrac{1}{12}(1.005 \text{ kg})(0.16 \text{ m})^2 = 2.144 \times 10^{-3} \text{ kg} \cdot \text{m}^2$$
$$I_z = \tfrac{1}{3}m_2 b^2 = \tfrac{1}{3}(1.005 \text{ kg})(0.2 \text{ m})^2 = 13.400 \times 10^{-3} \text{ kg} \cdot \text{m}^2$$
$$I_y = I_x + I_z = (2.144 + 13.400)(10^{-3}) = 15.544 \times 10^{-3} \text{ kg} \cdot \text{m}^2$$

Circular Plate

$$I_x = \tfrac{1}{4}m_3 a^2 = \tfrac{1}{4}(0.2466 \text{ kg})(0.05 \text{ m})^2 = 0.154 \times 10^{-3} \text{ kg} \cdot \text{m}^2$$
$$I_y = \tfrac{1}{2}m_3 a^2 + m_3 d^2$$
$$= \tfrac{1}{2}(0.2466 \text{ kg})(0.05 \text{ m})^2 + (0.2466 \text{ kg})(0.1 \text{ m})^2 = 2.774 \times 10^{-3} \text{ kg} \cdot \text{m}^2$$
$$I_z = \tfrac{1}{4}m_3 a^2 + m_3 d^2 = \tfrac{1}{4}(0.2466 \text{ kg})(0.05 \text{ m})^2 + (0.2466 \text{ kg})(0.1 \text{ m})^2$$
$$= 2.620 \times 10^{-3} \text{ kg} \cdot \text{m}^2$$

Entire Machine Part

$$I_x = (1.010 + 2.144 - 0.154)(10^{-3}) \text{ kg} \cdot \text{m}^2 \qquad I_x = 3.00 \times 10^{-3} \text{ kg} \cdot \text{m}^2 \blacktriangleleft$$
$$I_y = (0.505 + 15.544 - 2.774)(10^{-3}) \text{ kg} \cdot \text{m}^2 \qquad I_y = 13.28 \times 10^{-3} \text{ kg} \cdot \text{m}^2 \blacktriangleleft$$
$$I_z = (0.505 + 13.400 - 2.620)(10^{-3}) \text{ kg} \cdot \text{m}^2 \qquad I_z = 11.29 \times 10^{-3} \text{ kg} \cdot \text{m}^2 \blacktriangleleft$$

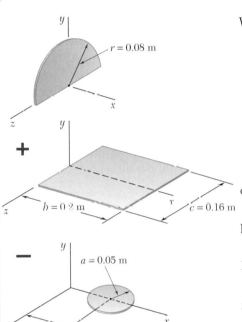

SOLVING PROBLEMS
ON YOUR OWN

In this lesson we introduced the *mass moment of inertia* and the *radius of gyration* of a three-dimensional body with respect to a given axis [Eqs. (9.28) and (9.29)]. We also derived a *parallel-axis theorem* for use with mass moments of inertia and discussed the computation of the mass moments of inertia of thin plates and three-dimensional bodies.

1. Computing mass moments of inertia. The mass moment of inertia I of a body with respect to a given axis can be calculated directly from the definition given in Eq. (9.28) for simple shapes [Sample Prob. 9.9]. In most cases, however, it is necessary to divide the body into thin slabs, compute the moment of inertia of a typical slab with respect to the given axis—using the parallel-axis theorem if necessary—and integrate the expression obtained.

2. Applying the parallel-axis theorem. In Sec. 9.12 we derived the parallel-axis theorem for mass moments of inertia

$$I = \bar{I} + md^2 \tag{9.33}$$

which states that the moment of inertia I of a body of mass m with respect to a given axis is equal to the sum of the moment of inertia $\bar{I}$ of that body with respect to the *parallel centroidal axis* and the product md^2, where d is the distance between the two axes. When the moment of inertia of a three-dimensional body is calculated with respect to one of the coordinate axes, d^2 can be replaced by the sum of the squares of distances measured along the other two coordinate axes [Eqs. (9.32) and (9.32′)].

3. Avoiding unit-related errors. To avoid errors, it is essential that you be consistent in your use of units. Thus, all lengths should be expressed in meters or feet, as appropriate, and for problems using U.S. customary units, masses should be given in $lb \cdot s^2/ft$. In addition, we strongly recommend that you include units as you perform your calculations [Sample Probs. 9.12 and 9.13].

4. Calculating the mass moment of inertia of thin plates. We showed in Sec. 9.13 that the mass moment of inertia of a thin plate with respect to a given axis can be obtained by multiplying the corresponding moment of inertia of the area of the plate by the density ρ and the thickness t of the plate [Eqs. (9.35) through (9.37)]. Note that since the axis CC' in Fig. 9.24c is *perpendicular to the plate*, $I_{CC', \text{mass}}$ is associated with the *polar* moment of inertia $J_{C, \text{area}}$.

Instead of calculating directly the moment of inertia of a thin plate with respect to a specified axis, you may sometimes find it convenient to first compute its moment of inertia with respect to an axis parallel to the specified axis and then apply the parallel-axis theorem. Further, to determine the moment of inertia of a thin plate with respect to an axis perpendicular to the plate, you may wish to first determine its moments of inertia with respect to two perpendicular in-plane axes and then use Eq. (9.38). Finally, remember that the mass of a plate of area A, thickness t, and density ρ is $m = \rho t A$.

5. Determining the moment of inertia of a body by direct single integration. We discussed in Sec. 9.14 and illustrated in Sample Probs. 9.10 and 9.11 how single integration can be used to compute the moment of inertia of a body that can be divided into a series of thin, parallel slabs. For such cases, you will often need to express the mass of the body in terms of the body's density and dimensions. Assuming that the body has been divided, as in the sample problems, into thin slabs perpendicular to the x axis, you will need to express the dimensions of each slab as functions of the variable x.

  a. In the special case of a body of revolution, the elemental slab is a thin disk, and the equations given in Fig. 9.27 should be used to determine the moments of inertia of the body [Sample Prob. 9.11].

  b. In the general case, when the body is not of revolution, the differential element is not a disk, but a thin slab of a different shape, and the equations of Fig. 9.27 cannot be used. See, for example, Sample Prob. 9.10, where the element was a thin, rectangular slab. For more complex configurations, you may want to use one or more of the following equations, which are based on Eqs. (9.32) and (9.32′) of Sec. 9.12.

$$dI_x = dI_{x'} + (\overline{y}_{el}^2 + \overline{z}_{el}^2)\,dm$$
$$dI_y = dI_{y'} + (\overline{z}_{el}^2 + \overline{x}_{el}^2)\,dm$$
$$dI_z = dI_{z'} + (\overline{x}_{el}^2 + \overline{y}_{el}^2)\,dm$$

where the primes denote the centroidal axes of each elemental slab, and where $\overline{x}_{el}$, $\overline{y}_{el}$, and $\overline{z}_{el}$ represent the coordinates of its centroid. The centroidal moments of inertia of the slab are determined in the manner described earlier for a thin plate: Referring to Fig. 9.12 on page 485, calculate the corresponding moments of inertia of the area of the slab and multiply the result by the density ρ and the thickness t of the slab. Also, assuming that the body has been divided into thin slabs perpendicular to the x axis, remember that you can obtain $dI_{x'}$ by adding $dI_{y'}$ and $dI_{z'}$ instead of computing it directly. Finally, using the geometry of the body, express the result obtained in terms of the single variable x and integrate in x.

6. Computing the moment of inertia of a composite body. As stated in Sec. 9.15, the moment of inertia of a composite body with respect to a specified axis is equal to the sum of the moments of its components with respect to that axis. Sample Probs. 9.12 and 9.13 illustrate the appropriate method of solution. You must also remember that the moment of inertia of a component will be negative only if the component is *removed* (as in the case of a hole).

Although the composite-body problems in this lesson are relatively straightforward, you will have to work carefully to avoid computational errors. In addition, if some of the moments of inertia that you need are not given in Fig. 9.28, you will have to derive your own formulas using the techniques of this lesson.

9.113 The quarter ring shown has a mass m and was cut from a thin, uniform plate. Knowing that $r_1 = \frac{1}{2}r_2$, determine the mass moment of inertia of the quarter ring with respect to (a) axis AA', (b) the centroidal axis CC' that is perpendicular to the plane of the quarter ring.

Fig. P9.113

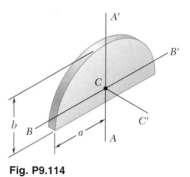

Fig. P9.114

9.114 A thin, semielliptical plate has a mass m. Determine the mass moment of inertia of the plate with respect to (a) the centroidal axis BB', (b) the centroidal axis CC' that is perpendicular to the plate.

9.115 The elliptical ring shown was cut from a thin, uniform plate. Denoting the mass of the ring by m, determine its moment of inertia with respect to (a) the centroidal axis BB', (b) the centroidal axis CC' that is perpendicular to the plane of the ring.

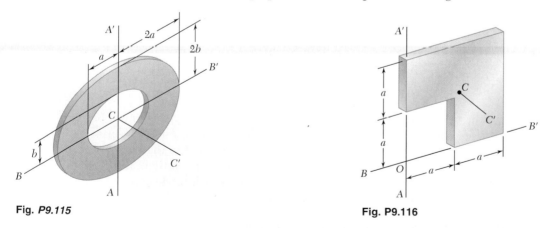

Fig. *P9.115*

Fig. P9.116

9.116 The machine component shown was cut from a thin, uniform plate. Denoting the mass of the component by m, determine its mass moment of inertia with respect to (a) the axis BB', (b) the centroidal axis CC' that is perpendicular to the plane of the component.

9.117 The rhombus shown has a mass m and was cut from a thin, uniform plate. Determine the mass moment of inertia of the rhombus with respect to (a) the x axis, (b) the y axis.

9.118 The rhombus shown has a mass m and was cut from a thin, uniform plate. Knowing that the AA' and BB' axes are parallel to the z axis and lie in a plane parallel to and at a distance a above the zx plane, determine the mass moment of inertia of the rhombus with respect to (a) the axis AA', (b) the axis BB'.

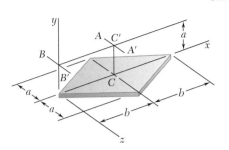

Fig. P9.117 and *P9.118*

9.119 A thin plate of mass m has the trapezoidal shape shown. Determine the mass moment of inertia of the plate with respect to (a) the x axis, (b) the y axis.

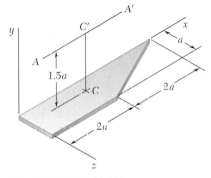

Fig. P9.119 and P9.120

9.120 A thin plate of mass m has the trapezoidal shape shown. Determine the mass moment of inertia of the plate with respect to (a) the centroidal axis CC' that is perpendicular to the plate, (b) the axis AA' which is parallel to the x axis and is located at a distance $1.5a$ from the plate.

9.121 The parabolic spandrel shown is revolved about the x axis to form a homogeneous solid of revolution of mass m. Using direct integration, express the moment of inertia of the solid with respect to the x axis in terms of m and b.

Fig. P9.121

9.122 Determine by direct integration the moment of inertia with respect to the z axis of the right circular cylinder shown assuming that it has a uniform density and a mass m.

Fig. *P9.122*

Fig. P9.123

9.123 The area shown is revolved about the x axis to form a homogeneous solid of revolution of mass m. Determine by direct integration the moment of inertia of the solid with respect to (a) the x axis, (b) the y axis. Express your answers in terms of m and a.

9.124 Determine by direct integration the moment of inertia with respect to the x axis of the tetrahedron shown assuming that it has a uniform density and a mass m.

Fig. P9.124 and *P9.125*

9.125 Determine by direct integration the moment of inertia with respect to the y axis of the tetrahedron shown assuming that it has a uniform density and a mass m.

***9.126** Determine by direct integration the moment of inertia with respect to the z axis of the semiellipsoid shown assuming that it has a uniform density and a mass m.

***9.127** A thin steel wire is bent into the shape shown. Denoting the mass per unit length of the wire by m', determine by direct integration the moment of inertia of the wire with respect to each of the coordinate axes.

9.128 A thin triangular plate of mass m is welded along its base AB to a block as shown. Knowing that the plate forms an angle θ with the y axis, determine by direct integration the mass moment of inertia of the plate with respect to (a) the x axis, (b) the y axis, (c) the z axis.

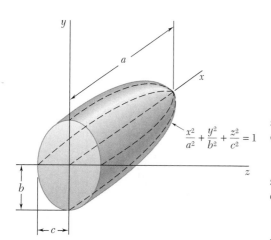

$$\frac{x^2}{a^2} + \frac{y^2}{b^2} + \frac{z^2}{c^2} = 1$$

Fig. P9.126

Fig. P9.127

Fig. P9.128

9.129 Shown is the cross section of a molded flat-belt pulley. Determine its mass moment of inertia and its radius of gyration with respect to the axis AA'. (The specific weight of brass is 0.306 lb/in³ and the specific weight of the fiber-reinforced polycarbonate used is 0.0433 lb/in³.)

Fig. P9.129

Fig. P9.130

9.130 Shown is the cross section of an idler roller. Determine its moment of inertia and its radius of gyration with respect to the axis AA'. (The density of bronze is 8580 kg/m³; of aluminum, 2770 kg/m³; and of neoprene, 1250 kg/m³.)

9.131 Given the dimensions and the mass m of the thin conical shell shown, determine the moment of inertia and the radius of gyration of the shell with respect to the x axis. (*Hint:* Assume that the shell was formed by removing a cone with a circular base of radius a from a cone with a circular base of radius a + t. In the resulting expressions, neglect terms containing t², t³, etc. Do not forget to account for the difference in the heights of the two cones.)

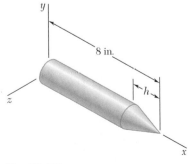

Fig. P9.131

9.132 A portion of an 8-in.-long steel rod of diameter 1.50 in. is turned to form the conical section shown. Knowing that the turning process reduces the moment of inertia of the rod with respect to the x axis by 20 percent, determine the height h of the cone.

Fig. P9.132

90 mm

120 mm

240 mm

80 mm

Fig. P9.133

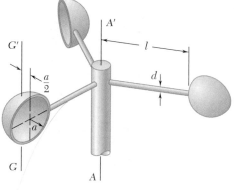

A'

l

d

G'

$\frac{a}{2}$

a

G

A

Fig. P9.135

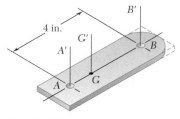

9.133   The steel machine component shown is formed by machining a hemisphere into the base of a truncated cone. Knowing that the density of steel is 7850 kg/m³, determine the mass moment of inertia of the component with respect to the y axis.

9.134   After a period of use, one of the blades of a shredder has been worn to the shape shown and is of weight 0.4 lb. Knowing that the moments of inertia of the blade with respect to the AA' and BB' axes are 0.6×10^{-3} lb · ft · s² and 1.26×10^{-3} lb · ft · s², respectively, determine (a) the location of the centroidal axis GG', (b) the radius of gyration with respect to axis GG'.

4 in.

B'

G'

A'

B

A

G

Fig. P9.134

9.135   The cups and the arms of an anemometer are fabricated from a material of density ρ. Knowing that the moment of inertia of a thin, hemispherical shell of mass m and thickness t with respect to its centroidal axis GG' is $5ma^2/12$, determine (a) the moment of inertia of the anemometer with respect to the axis AA', (b) the ratio of a to l for which the centroidal moment of inertia of the cups is equal to 1 percent of the moment of inertia of the cups with respect to the axis AA'.

9.136   A square hole is centered in and extends through the aluminum machine component shown. Determine (a) the value of a for which the mass moment of inertia of the component with respect to the axis AA', which bisects the top surface of the hole, is maximum, (b) the corresponding values of the mass moment of inertia and the radius of gyration with respect to the axis AA'. (The density of aluminum is 2800 kg/m³.)

84 mm

A'

300 mm

84 mm

$\frac{a}{2}$

a

A

a

Fig. P9.136

y

24 in.

38 in.

z

x

Fig. P9.137

9.137   A 0.1-in.-thick piece of sheet metal is cut and bent into the machine component shown. Knowing that the specific weight of steel is 0.284 lb/in³, determine the moment of inertia of the component with respect to each of the coordinate axes.

9.138 A 3-mm-thick piece of sheet metal is cut and bent into the machine component shown. Knowing that the density of steel is 7850 kg/m³, determine the moment of inertia of the component with respect to each of the coordinate axes.

Fig. *P9.138*

Fig. P9.139

9.139 The cover of an electronic device is formed from sheet aluminum that is 2 mm thick. Determine the mass moment of inertia of the cover with respect to each of the coordinate axes. (The density of aluminum is 2770 kg/m³.)

9.140 A framing anchor is formed of 2-mm-thick galvanized steel. Determine the mass moment of inertia of the anchor with respect to each of the coordinate axes. (The density of galvanized steel is 7530 kg/m³.)

9.141 A 2-mm-thick piece of sheet steel is cut and bent into the machine component shown. Knowing that the density of steel is 7850 kg/m³, determine the moment of inertia of the component with respect to each of the coordinate axes.

Fig. *P9.140*

Fig. P9.141

Fig. P9.142

*****9.142** The piece of roof flashing shown is formed from sheet copper that is 0.032 in. thick. Knowing that the specific weight of copper is 558 lb/ft³, determine the mass moment of inertia of the flashing with respect to each of the coordinate axes.

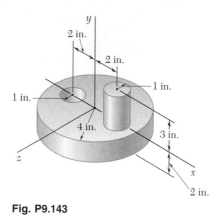

Fig. P9.143

9.143 The machine element shown is fabricated from steel. Determine the mass moment of inertia of the assembly with respect to (a) the x axis, (b) the y axis, (c) the z axis. (The specific weight of steel is 0.284 lb/in³.)

9.144 Determine the mass moment of inertia of the steel machine element shown with respect to the y axis. (The density of steel is 7850 kg/m³.)

Fig. P9.144 and P9.145

9.145 Determine the mass moment of inertia of the steel machine element shown with respect to the z axis. (The density of steel is 7850 kg/m³.)

9.146 An aluminum casting has the shape shown. Knowing that the specific weight of aluminum is 0.100 lb/in³, determine the moment of inertia of the casting with respect to the z axis.

Fig. *P9.146*

Fig. P9.147

9.147 Determine the moment of inertia of the steel machine element shown with respect to (a) the x axis, (b) the y axis, (c) the z axis. (The specific weight of steel is 490 lb/ft³.)

9.148 Aluminum wire with a mass per unit length of 0.049 kg/m is used to form the circle and the straight members of the figure shown. Determine the mass moment of inertia of the assembly with respect to each of the coordinate axes.

Fig. P9.148

9.149   The figure shown is formed of 3-mm-diameter steel wire. Knowing that the density of the steel is 7850 kg/m^3, determine the mass moment of inertia of the wire with respect to each of the coordinate axes.

360 mm

360 mm

360 mm

Fig. P9.149

9.150   A homogeneous wire with a weight per unit length of 0.041 lb/ft is used to form the figure shown. Determine the moment of inertia of the wire with respect to each of the coordinate axes.

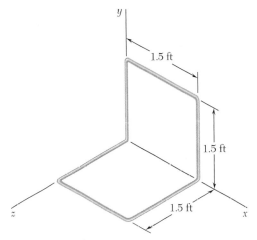

1.5 ft

1.5 ft

1.5 ft

Fig. *P9.150*

*9.16. MOMENT OF INERTIA OF A BODY WITH RESPECT TO AN ARBITRARY AXIS THROUGH *O*. MASS PRODUCTS OF INERTIA

In this section you will see how the moment of inertia of a body can be determined with respect to an arbitrary axis OL through the origin (Fig. 9.29) if its moments of inertia with respect to the three coordinate axes, as well as certain other quantities to be defined below, have already been determined.

The moment of inertia I_{OL} of the body with respect to OL is equal to $\int p^2\, dm$, where p denotes the perpendicular distance from the element of mass dm to the axis OL. If we denote by $\boldsymbol{\lambda}$ the unit vector along OL and by $\mathbf{r}$ the position vector of the element dm, we observe that the perpendicular distance p is equal to $r \sin \theta$, which is the magnitude of the vector product $\boldsymbol{\lambda} \times \mathbf{r}$. We therefore write

$$I_{OL} = \int p^2\, dm = \int |\boldsymbol{\lambda} \times \mathbf{r}|^2\, dm \qquad (9.43)$$

Expressing $|\boldsymbol{\lambda} \times \mathbf{r}|^2$ in terms of the rectangular components of the vector product, we have

$$I_{OL} = \int [(\lambda_x y - \lambda_y x)^2 + (\lambda_y z - \lambda_z y)^2 + (\lambda_z x - \lambda_x z)^2]\, dm$$

where the components λ_x, λ_y, λ_z of the unit vector $\boldsymbol{\lambda}$ represent the direction cosines of the axis OL and the components x, y, z of $\mathbf{r}$ represent the coordinates of the element of mass dm. Expanding the squares and rearranging the terms, we write

$$I_{OL} = \lambda_x^2 \int (y^2 + z^2)\, dm + \lambda_y^2 \int (z^2 + x^2)\, dm + \lambda_z^2 \int (x^2 + y^2)\, dm$$

$$- 2\lambda_x\lambda_y \int xy\, dm - 2\lambda_y\lambda_z \int yz\, dm - 2\lambda_z\lambda_x \int zx\, dm \qquad (9.44)$$

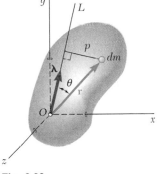

Fig. 9.29

Referring to Eqs. (9.30), we note that the first three integrals in (9.44) represent, respectively, the moments of inertia I_x, I_y, and I_z of the body with respect to the coordinate axes. The last three integrals in (9.44), which involve products of coordinates, are called the *products of inertia* of the body with respect to the x and y axes, the y and z axes, and the z and x axes, respectively. We write

$$I_{xy} = \int xy \, dm \qquad I_{yz} = \int yz \, dm \qquad I_{zx} = \int zx \, dm \qquad (9.45)$$

Rewriting Eq. (9.44) in terms of the integrals defined in Eqs. (9.30) and (9.45), we have

$$I_{OL} = I_x\lambda_x^2 + I_y\lambda_y^2 + I_z\lambda_z^2 - 2I_{xy}\lambda_x\lambda_y - 2I_{yz}\lambda_y\lambda_z - 2I_{zx}\lambda_z\lambda_x \quad (9.46)$$

We note that the definition of the products of inertia of a mass given in Eqs. (9.45) is an extension of the definition of the product of inertia of an area (Sec. 9.8). Mass products of inertia reduce to zero under the same conditions of symmetry as do products of inertia of areas, and the parallel-axis theorem for mass products of inertia is expressed by relations similar to the formula derived for the product of inertia of an area. Substituting the expressions for x, y, and z given in Eqs. (9.31) into Eqs. (9.45), we find that

$$\begin{aligned} I_{xy} &= \bar{I}_{x'y'} + m\bar{x}\bar{y} \\ I_{yz} &= \bar{I}_{y'z'} + m\bar{y}\bar{z} \\ I_{zx} &= \bar{I}_{z'x'} + m\bar{z}\bar{x} \end{aligned} \qquad (9.47)$$

where $\bar{x}$, $\bar{y}$, $\bar{z}$ are the coordinates of the center of gravity G of the body and $\bar{I}_{x'y'}$, $\bar{I}_{y'z'}$, $\bar{I}_{z'x'}$ denote the products of inertia of the body with respect to the centroidal axes x', y', z' (Fig. 9.22).

*9.17. ELLIPSOID OF INERTIA. PRINCIPAL AXES OF INERTIA

Let us assume that the moment of inertia of the body considered in the preceding section has been determined with respect to a large number of axes OL through the fixed point O and that a point Q has been plotted on each axis OL at a distance $OQ = 1/\sqrt{I_{OL}}$ from O. The locus of the points Q thus obtained forms a surface (Fig. 9.30). The equation of that surface can be obtained by substituting $1/(OQ)^2$ for I_{OL} in (9.46) and then multiplying both sides of the equation by $(OQ)^2$. Observing that

$$(OQ)\lambda_x = x \qquad (OQ)\lambda_y = y \qquad (OQ)\lambda_z = z$$

where x, y, z denote the rectangular coordinates of Q, we write

$$I_x x^2 + I_y y^2 + I_z z^2 - 2I_{xy}xy - 2I_{yz}yz - 2I_{zx}zx = 1 \quad (9.48)$$

The equation obtained is the equation of a *quadric surface*. Since the moment of inertia I_{OL} is different from zero for every axis OL, no point Q can be at an infinite distance from O. Thus, the quadric

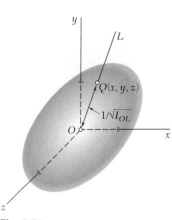

Fig. 9.30

surface obtained is an *ellipsoid*. This ellipsoid, which defines the moment of inertia of the body with respect to any axis through O, is known as the *ellipsoid of inertia* of the body at O.

We observe that if the axes in Fig. 9.30 are rotated, the coefficients of the equation defining the ellipsoid change, since they are equal to the moments and products of inertia of the body with respect to the rotated coordinate axes. However, the *ellipsoid itself remains unaffected*, since its shape depends only upon the distribution of mass in the given body. Suppose that we choose as coordinate axes the principal axes x', y', z' of the ellipsoid of inertia (Fig. 9.31). The equation of the ellipsoid with respect to these coordinate axes is known to be of the form

$$I_{x'}x'^2 + I_{y'}y'^2 + I_{z'}z'^2 = 1 \tag{9.49}$$

which does not contain any products of the coordinates. Comparing Eqs. (9.48) and (9.49), we observe that the products of inertia of the body with respect to the x', y', z' axes must be zero. The x', y', z' axes are known as the *principal axes of inertia* of the body at O, and the coefficients $I_{x'}$, $I_{y'}$, $I_{z'}$ are referred to as the *principal moments of inertia* of the body at O. Note that, given a body of arbitrary shape and a point O, it is always possible to find axes which are the principal axes of inertia of the body at O, that is, axes with respect to which the products of inertia of the body are zero. Indeed, whatever the shape of the body, the moments and products of inertia of the body with respect to x, y, and z axes through O will define an ellipsoid, and this ellipsoid will have principal axes which, by definition, are the principal axes of inertia of the body at O.

If the principal axes of inertia x', y', z' are used as coordinate axes, the expression obtained in Eq. (9.46) for the moment of inertia of a body with respect to an arbitrary axis through O reduces to

$$I_{OL} = I_{x'}\lambda_{x'}^2 + I_{y'}\lambda_{y'}^2 + I_{z'}\lambda_{z'}^2 \tag{9.50}$$

The determination of the principal axes of inertia of a body of arbitrary shape is somewhat involved and will be discussed in the next section. There are many cases, however, where these axes can be spotted immediately. Consider, for instance, the homogeneous cone of elliptical base shown in Fig. 9.32; this cone possesses two mutually perpendicular planes of symmetry OAA' and OBB'. From the definition (9.45), we observe that if the $x'y'$ and $y'z'$ planes are chosen to coincide with the two planes of symmetry, all of the products of inertia are zero. The x', y', and z' axes thus selected are therefore the principal axes of inertia of the cone at O. In the case of the homogeneous regular tetrahedron $OABC$ shown in Fig. 9.33, the line joining the corner O to the center D of the opposite face is a principal axis of inertia at O, and any line through O perpendicular to OD is also a principal axis of inertia at O. This property is apparent if we observe that rotating the tetrahedron through $120°$ about OD leaves its shape and mass distribution unchanged. It follows that the ellipsoid of inertia at O also remains unchanged under this rotation. The ellipsoid, therefore, is a body of revolution whose axis of revolution is OD, and the line OD, as well as any perpendicular line through O, must be a principal axis of the ellipsoid.

Fig. 9.31

Fig. 9.32

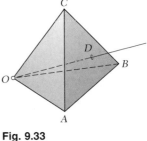

Fig. 9.33

*9.18. DETERMINATION OF THE PRINCIPAL AXES AND PRINCIPAL MOMENTS OF INERTIA OF A BODY OF ARBITRARY SHAPE

The method of analysis described in this section should be used when the body under consideration has no obvious property of symmetry.

Consider the ellipsoid of inertia of the body at a given point O (Fig. 9.34); let $\mathbf{r}$ be the radius vector of a point P on the surface of the ellipsoid and let $\mathbf{n}$ be the unit vector along the normal to that surface at P. We observe that the only points where $\mathbf{r}$ and $\mathbf{n}$ are collinear are the points P_1, P_2, and P_3, where the principal axes intersect the visible portion of the surface of the ellipsoid, and the corresponding points on the other side of the ellipsoid.

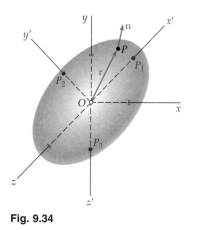

Fig. 9.34

We now recall from calculus that the direction of the normal to a surface of equation $f(x, y, z) = 0$ at a point $P(x, y, z)$ is defined by the gradient ∇f of the function f at that point. To obtain the points where the principal axes intersect the surface of the ellipsoid of inertia, we must therefore write that $\mathbf{r}$ and ∇f are collinear,

$$\nabla f = (2K)\mathbf{r} \tag{9.51}$$

where K is a constant, $\mathbf{r} = x\mathbf{i} + y\mathbf{j} + z\mathbf{k}$, and

$$\nabla f = \frac{\partial f}{\partial x}\mathbf{i} + \frac{\partial f}{\partial y}\mathbf{j} + \frac{\partial f}{\partial x}\mathbf{k}$$

Recalling Eq. (9.48), we note that the function $f(x, y, z)$ corresponding to the ellipsoid of inertia is

$$f(x, y, z) = I_x x^2 + I_y y^2 + I_z z^2 - 2I_{xy}xy - 2I_{yz}yz - 2I_{zx}zx - 1$$

Substituting for $\mathbf{r}$ and ∇f into Eq. (9.51) and equating the coefficients of the unit vectors, we write

$$\begin{aligned}
I_x x &- I_{xy}y - I_{zx}z = Kx \\
-I_{xy}x &+ I_y y - I_{yz}z = Ky \\
-I_{zx}x &- I_{yz}y + I_z z = Kz
\end{aligned} \tag{9.52}$$

Dividing each term by the distance r from O to P, we obtain similar equations involving the direction cosines λ_x, λ_y, and λ_z:

$$
\begin{aligned}
I_x\lambda_x - I_{xy}\lambda_y - I_{zx}\lambda_z &= K\lambda_x \\
-I_{xy}\lambda_x + I_y\lambda_y - I_{yz}\lambda_z &= K\lambda_y \\
-I_{zx}\lambda_x - I_{yz}\lambda_y + I_z\lambda_z &= K\lambda_z
\end{aligned}
\tag{9.53}
$$

Transposing the right-hand members leads to the following homogeneous linear equations:

$$
\begin{aligned}
(I_x - K)\lambda_x - I_{xy}\lambda_y - I_{zx}\lambda_z &= 0 \\
-I_{xy}\lambda_x + (I_y - K)\lambda_y - I_{yz}\lambda_z &= 0 \\
-I_{zx}\lambda_x - I_{yz}\lambda_y + (I_z - K)\lambda_z &= 0
\end{aligned}
\tag{9.54}
$$

For this system of equations to have a solution different from $\lambda_x = \lambda_y = \lambda_z = 0$, its discriminant must be zero:

$$
\begin{vmatrix}
I_x - K & -I_{xy} & -I_{zx} \\
-I_{xy} & I_y - K & -I_{yz} \\
-I_{zx} & -I_{yz} & I_z - K
\end{vmatrix} = 0
\tag{9.55}
$$

Expanding this determinant and changing signs, we write

$$
K^3 - (I_x + I_y + I_z)K^2 + (I_xI_y + I_yI_z + I_zI_x - I_{xy}^2 - I_{yz}^2 - I_{zx}^2)K \\
- (I_xI_yI_z - I_xI_{yz}^2 - I_yI_{zx}^2 - I_zI_{xy}^2 - 2I_{xy}I_{yz}I_{zx}) = 0
\tag{9.56}
$$

This is a cubic equation in K, which yields three real, positive roots K_1, K_2, and K_3.

To obtain the direction cosines of the principal axis corresponding to the root K_1, we substitute K_1 for K in Eqs. (9.54). Since these equations are now linearly dependent, only two of them may be used to determine λ_x, λ_y, and λ_z. An additional equation may be obtained, however, by recalling from Sec. 2.12 that the direction cosines must satisfy the relation

$$
\lambda_x^2 + \lambda_y^2 + \lambda_z^2 = 1
\tag{9.57}
$$

Repeating this procedure with K_2 and K_3, we obtain the direction cosines of the other two principal axes.

We will now show that *the roots K_1, K_2, and K_3 of Eq. (9.56) are the principal moments of inertia of the given body*. Let us substitute for K in Eqs. (9.53) the root K_1, and for λ_x, λ_y, and λ_z the corresponding values $(\lambda_x)_1$, $(\lambda_y)_1$, and $(\lambda_z)_1$ of the direction cosines; the three equations will be satisfied. We now multiply by $(\lambda_x)_1$, $(\lambda_y)_1$, and $(\lambda_z)_1$, respectively, each term in the first, second, and third equation and add the equations obtained in this way. We write

$$
I_x^2(\lambda_x)_1^2 + I_y^2(\lambda_y)_1^2 + I_z^2(\lambda_z)_1^2 - 2I_{xy}(\lambda_x)_1(\lambda_y)_1 \\
- 2I_{yz}(\lambda_y)_1(\lambda_z)_1 - 2I_{zx}(\lambda_z)_1(\lambda_x)_1 = K_1[(\lambda_x)_1^2 + (\lambda_y)_1^2 + (\lambda_z)_1^2]
$$

Recalling Eq. (9.46), we observe that the left-hand member of this equation represents the moment of inertia of the body with respect to the principal axis corresponding to K_1; it is thus the principal moment of inertia corresponding to that root. On the other hand, recalling Eq. (9.57), we note that the right-hand member reduces to K_1. Thus K_1 itself is the principal moment of inertia. We can show in the same fashion that K_2 and K_3 are the other two principal moments of inertia of the body.

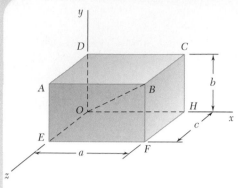

SAMPLE PROBLEM 9.14

Consider a rectangular prism of mass m and sides a, b, c. Determine (a) the moments and products of inertia of the prism with respect to the coordinate axes shown, (b) its moment of inertia with respect to the diagonal OB.

SOLUTION

a. Moments and Products of Inertia with Respect to the Coordinate Axes. *Moments of Inertia.* Introducing the centroidal axes x', y', z', with respect to which the moments of inertia are given in Fig. 9.28, we apply the parallel-axis theorem:

$$I_x = \bar{I}_{x'} + m(\bar{y}^2 + \bar{z}^2) = \tfrac{1}{12}m(b^2 + c^2) + m(\tfrac{1}{4}b^2 + \tfrac{1}{4}c^2)$$

$$I_x = \tfrac{1}{3}m(b^2 + c^2) \quad \blacktriangleleft$$

Similarly, $\qquad I_y = \tfrac{1}{3}m(c^2 + a^2) \qquad I_z = \tfrac{1}{3}m(a^2 + b^2) \quad \blacktriangleleft$

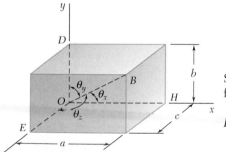

Products of Inertia. Because of symmetry, the products of inertia with respect to the centroidal axes x', y', z' are zero, and these axes are principal axes of inertia. Using the parallel-axis theorem, we have

$$I_{xy} = \bar{I}_{x'y'} + m\bar{x}\bar{y} = 0 + m(\tfrac{1}{2}a)(\tfrac{1}{2}b) \qquad I_{xy} = \tfrac{1}{4}mab \quad \blacktriangleleft$$

Similarly, $\qquad\qquad\qquad\qquad\qquad I_{yz} = \tfrac{1}{4}mbc \qquad I_{zx} = \tfrac{1}{4}mca \quad \blacktriangleleft$

b. Moment of Inertia with Respect to OB. We recall Eq. (9.46):

$$I_{OB} = I_x\lambda_x^2 + I_y\lambda_y^2 + I_z\lambda_z^2 - 2I_{xy}\lambda_x\lambda_y - 2I_{yz}\lambda_y\lambda_z - 2I_{zx}\lambda_z\lambda_x$$

where the direction cosines of OB are

$$\lambda_x = \cos\theta_x = \frac{OH}{OB} = \frac{a}{(a^2 + b^2 + c^2)^{1/2}}$$

$$\lambda_y = \frac{b}{(a^2 + b^2 + c^2)^{1/2}} \qquad \lambda_z = \frac{c}{(a^2 + b^2 + c^2)^{1/2}}$$

Substituting the values obtained for the moments and products of inertia and for the direction cosines into the equation for I_{OB}, we have

$$I_{OB} = \frac{1}{a^2 + b^2 + c^2}[\tfrac{1}{3}m(b^2 + c^2)a^2 + \tfrac{1}{3}m(c^2 + a^2)b^2 + \tfrac{1}{3}m(a^2 + b^2)c^2$$
$$- \tfrac{1}{2}ma^2b^2 - \tfrac{1}{2}mb^2c^2 - \tfrac{1}{2}mc^2a^2]$$

$$I_{OB} = \frac{m}{6}\frac{a^2b^2 + b^2c^2 + c^2a^2}{a^2 + b^2 + c^2} \quad \blacktriangleleft$$

Alternative Solution. The moment of inertia I_{OB} can be obtained directly from the principal moments of inertia $\bar{I}_{x'}$, $\bar{I}_{y'}$, $\bar{I}_{z'}$, since the line OB passes through the centroid O'. Since the x', y', z' axes are principal axes of inertia, we use Eq. (9.50) to write

$$I_{OB} = \bar{I}_{x'}\lambda_x^2 + \bar{I}_{y'}\lambda_y^2 + \bar{I}_{z'}\lambda_z^2$$

$$= \frac{1}{a^2 + b^2 + c^2}\left[\frac{m}{12}(b^2 + c^2)a^2 + \frac{m}{12}(c^2 + a^2)b^2 + \frac{m}{12}(a^2 + b^2)c^2\right]$$

$$I_{OB} = \frac{m}{6}\frac{a^2b^2 + b^2c^2 + c^2a^2}{a^2 + b^2 + c^2} \quad \blacktriangleleft$$

SAMPLE PROBLEM 9.15

If $a = 3c$ and $b = 2c$ for the rectangular prism of Sample Prob. 9.14, determine (a) the principal moments of inertia at the origin O, (b) the principal axes of inertia at O.

SOLUTION

a. Principal Moments of Inertia at the Origin O. Substituting $a = 3c$ and $b = 2c$ into the solution to Sample Prob. 9.14, we have

$$I_x = \tfrac{5}{3}mc^2 \qquad I_y = \tfrac{10}{3}mc^2 \qquad I_z = \tfrac{13}{3}mc^2$$
$$I_{xy} = \tfrac{3}{2}mc^2 \qquad I_{yz} = \tfrac{1}{2}mc^2 \qquad I_{zx} = \tfrac{3}{4}mc^2$$

Substituting the values of the moments and products of inertia into Eq. (9.56) and collecting terms yields

$$K^3 - (\tfrac{28}{3}mc^2)K^2 + (\tfrac{3479}{144}m^2c^4)K - \tfrac{589}{54}m^3c^6 = 0$$

We then solve for the roots of this equation; from the discussion in Sec. 9.18, it follows that these roots are the principal moments of inertia of the body at the origin.

$$K_1 = 0.568867mc^2 \qquad K_2 = 4.20885mc^2 \qquad K_3 = 4.55562mc^2$$
$$K_1 = 0.569mc^2 \qquad K_2 = 4.21mc^2 \qquad K_3 = 4.56mc^2 \qquad \blacktriangleleft$$

b. Principal Axes of Inertia at O. To determine the direction of a principal axis of inertia, we first substitute the corresponding value of K into two of the equations (9.54); the resulting equations together with Eq. (9.57) constitute a system of three equations from which the direction cosines of the corresponding principal axis can be determined. Thus, we have for the first principal moment of inertia K_1:

$$(\tfrac{5}{3} - 0.568867)\,mc^2(\lambda_x)_1 - \tfrac{3}{2}mc^2(\lambda_y)_1 - \tfrac{3}{4}mc^2(\lambda_z)_1 = 0$$
$$-\tfrac{3}{2}mc^2(\lambda_x)_1 + (\tfrac{10}{3} - 0.568867)\,mc^2(\lambda_y)_1 - \tfrac{1}{2}mc^2(\lambda_z)_1 = 0$$
$$(\lambda_x)_1^2 + (\lambda_y)_1^2 + (\lambda_z)_1^2 = 1$$

Solving yields

$$(\lambda_x)_1 = 0.836600 \qquad (\lambda_y)_1 = 0.496001 \qquad (\lambda_z)_1 = 0.232557$$

The angles that the first principal axis of inertia forms with the coordinate axes are then

$$(\theta_x)_1 = 33.2° \qquad (\theta_y)_1 = 60.3° \qquad (\theta_z)_1 = 76.6° \qquad \blacktriangleleft$$

Using the same set of equations successively with K_2 and K_3, we find that the angles associated with the second and third principal moments of inertia at the origin are, respectively,

$$(\theta_x)_2 = 57.8° \qquad (\theta_y)_2 = 146.6° \qquad (\theta_z)_2 = 98.0° \qquad \blacktriangleleft$$

and

$$(\theta_x)_3 = 82.8° \qquad (\theta_y)_3 = 76.1° \qquad (\theta_z)_3 = 164.3° \qquad \blacktriangleleft$$

In this lesson we defined the *mass products of inertia* I_{xy}, I_{yz}, and I_{zx} of a body and showed you how to determine the moments of inertia of that body with respect to an arbitrary axis passing through the origin O. You also learned how to determine at the origin O the *principal axes of inertia* of a body and the corresponding *principal moments of inertia*.

1. **Determining the mass products of inertia of a composite body.** The mass products of inertia of a composite body with respect to the coordinate axes can be expressed as the sums of the products of inertia of its component parts with respect to those axes. For each component part, we can use the parallel-axis theorem and write Eqs. (9.47)

$$I_{xy} = \overline{I}_{x'y'} + m\overline{x}\,\overline{y} \qquad I_{yz} = \overline{I}_{y'z'} + m\overline{y}\,\overline{z} \qquad I_{zx} = \overline{I}_{z'x'} + m\overline{z}\,\overline{x}$$

where the primes denote the centroidal axes of each component part and where $\overline{x}$, $\overline{y}$, and $\overline{z}$ represent the coordinates of its center of gravity. Keep in mind that the mass products of inertia can be positive, negative, or zero, and be sure to take into account the signs of $\overline{x}$, $\overline{y}$, and $\overline{z}$.

    a. From the properties of symmetry of a component part, you can deduce that two or all three of its centroidal mass products of inertia are zero. For instance, you can verify that for a thin plate parallel to the xy plane; a wire lying in a plane parallel to the xy plane; a body with a plane of symmetry parallel to the xy plane; and a body with an axis of symmetry parallel to the z axis, *the products of inertia $\overline{I}_{y'z'}$ and $\overline{I}_{z'x'}$ are zero.*

    For rectangular, circular, or semicircular plates with axes of symmetry parallel to the coordinate axes; straight wires parallel to a coordinate axis; circular and semicircular wires with axes of symmetry parallel to the coordinate axes; and rectangular prisms with axes of symmetry parallel to the coordinate axes, *the products of inertia $\overline{I}_{x'y'}$, $\overline{I}_{y'z'}$, and $\overline{I}_{z'x'}$ are all zero.*

    b. Mass products of inertia which are different from zero can be computed from Eqs. (9.45). Although, in general, a triple integration is required to determine a mass product of inertia, a single integration can be used if the given body can be divided into a series of thin, parallel slabs. The computations are then similar to those discussed in the previous lesson for moments of inertia.

2. Computing the moment of inertia of a body with respect to an arbitrary axis OL. An expression for the moment of inertia I_{OL} was derived in Sec. 9.16 and is given in Eq. (9.46). Before computing I_{OL}, you must first determine the mass moments and products of inertia of the body with respect to the given coordinate axes as well as the direction cosines of the unit vector $\boldsymbol{\lambda}$ along OL.

3. Calculating the principal moments of inertia of a body and determining its principal axes of inertia. You saw in Sec. 9.17 that it is always possible to find an orientation of the coordinate axes for which the mass products of inertia are zero. These axes are referred to as the *principal axes of inertia* and the corresponding moments of inertia are known as the *principal moments of inertia* of the body. In many cases, the principal axes of inertia of a body can be determined from its properties of symmetry. The procedure required to determine the principal moments and principal axes of inertia of a body with no obvious property of symmetry was discussed in Sec. 9.18 and was illustrated in Sample Prob. 9.15. It consists of the following steps.

a. Expand the determinant in Eq. (9.55) and solve the resulting cubic equation. The solution can be obtained by trial and error or, preferably, with an advanced scientific calculator or with the appropriate computer software. The roots K_1, K_2, and K_3 of this equation are the principal moments of inertia of the body.

b. To determine the direction of the principal axis corresponding to K_1, substitute this value for K in two of the equations (9.54) and solve these equations together with Eq. (9.57) for the direction cosines of the principal axis corresponding to K_1.

c. Repeat this procedure with K_2 and K_3 to determine the directions of the other two principal axes. As a check of your computations, you may wish to verify that the scalar product of any two of the unit vectors along the three axes you have obtained is zero and, thus, that these axes are perpendicular to each other.

d. When a principal moment of inertia is approximately equal to a moment of inertia with respect to a coordinate axis, the calculated values of the corresponding direction cosines will be very sensitive to the number of significant figures used in your computations. Thus, for this case we suggest you express your intermediate answers in terms of six or seven significant figures to avoid possible errors.

9.151 Determine the products of inertia I_{xy}, I_{yz}, and I_{zx} of the steel machine element shown. (The specific weight of steel is 490 lb/ft³.)

Fig. P9.151

Fig. P9.152

9.152 Determine the products of inertia I_{xy}, I_{yz}, and I_{zx} of the steel machine element shown. (The specific weight of steel is 0.284 lb/in³.)

9.153 and 9.154 Determine the mass products of inertia I_{xy}, I_{yz}, and I_{zx} of the cast aluminum machine component shown. (The density of aluminum is 2700 kg/m³.)

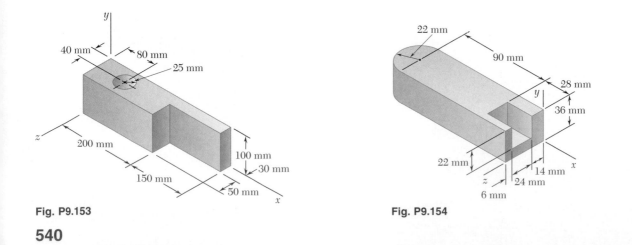

Fig. P9.153

Fig. P9.154

9.155 through 9.157 A section of sheet steel 3 mm thick is cut and bent into the machine component shown. Knowing that the density of the steel is 7860 kg/m³, determine the mass products of inertia I_{xy}, I_{yz}, and I_{zx} of the component.

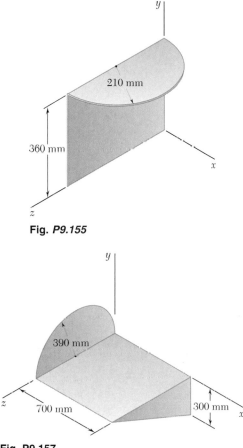

210 mm

360 mm

x

z

Fig. P9.155

36 mm

80 mm

45 mm

45 mm

r

z

Fig. *P9.156*

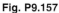

390 mm

z

700 mm

300 mm

x

Fig. P9.157

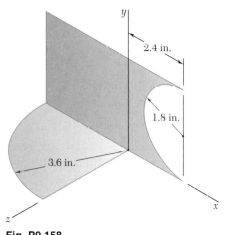

2.4 in.

1.8 in.

3.6 in.

x

z

Fig. P9.158

9.158 A section of sheet steel 0.08 in. thick is cut and bent into the machine component shown. Knowing that the specific weight of steel is 490 lb/ft³, determine the mass products of inertia I_{xy}, I_{yz}, and I_{zx} of the component.

9.159 and *9.160* Brass wire with a weight per unit length w is used to form the figure shown. Determine the products of inertia I_{xy}, I_{yz}, and I_{zx} of the wire figure.

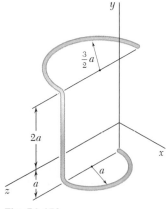

$\frac{3}{2}a$

2a

a

a

x

z

Fig. P9.159

2a

a

a

2a

$\frac{3}{2}a$

z

x

Fig. *P9.160*

9.161 The figure shown is formed of 0.075-in.-diameter aluminum wire. Knowing that the specific weight of aluminum is 0.10 lb/in^3, determine the products of inertia I_{xy}, I_{yz}, and I_{zx} of the wire figure.

Fig. P9.161

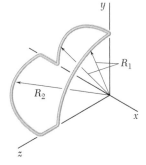

Fig. P9.162

9.162 Thin aluminum wire of uniform diameter is used to form the figure shown. Denoting by m' the mass per unit length of the wire, determine the products of inertia I_{xy}, I_{yz}, and I_{zx} of the wire figure.

9.163 Complete the derivation of Eqs. (9.47), which express the parallel-axis theorem for mass products of inertia.

9.164 For the homogeneous tetrahedron of mass m shown, (a) determine by direct integration the product of inertia I_{zx}, (b) deduce I_{yz} and I_{xy} from the results obtained in part a.

Fig. P9.164

Fig. P9.165

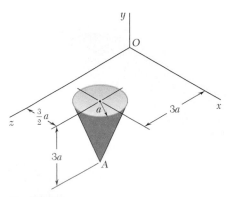

Fig. P9.166

9.165 The homogeneous circular cylinder shown has a mass m. Determine the moment of inertia of the cylinder with respect to the line joining the origin O and point A which is located on the perimeter of the top surface of the cylinder.

9.166 The homogeneous circular cone shown has a mass m. Determine the moment of inertia of the cone with respect to the line joining the origin O and point A.

9.167 Shown is the machine element of Prob. 9.143. Determine its moment of inertia with respect to the line joining the origin O and point A.

Fig. P9.167

9.168 Determine the moment of inertia of the steel machine element of Probs. 9.147 and 9.151 with respect to the axis through the origin which forms equal angles with the x, y, and z axes.

9.169 The thin bent plate shown is of uniform density and weight W. Determine its mass moment of inertia with respect to the line joining the origin O and point A.

Fig. P9.169

Fig. P9.170

9.170 A piece of sheet metal of thickness t and density ρ is cut and bent into the shape shown. Determine its mass moment of inertia with respect to a line joining points A and B.

9.171 Determine the mass moment of inertia of the machine component of Probs. 9.138 and 9.157 with respect to the axis through the origin characterized by the unit vector $\boldsymbol{\lambda} = (-4\mathbf{i} + 8\mathbf{j} + \mathbf{k})/9$.

9.172 through 9.174 For the wire figure of the problem indicated, determine the mass moment of inertia of the figure with respect to the axis through the origin characterized by the unit vector $\boldsymbol{\lambda} = (-3\mathbf{i} - 6\mathbf{j} + 2\mathbf{k})/7$.

    9.172   Prob. 9.150
    9.173   Prob. 9.149
    9.174   Prob. 9.148

9.175 For the rectangular prism shown, determine the values of the ratios b/a and c/a so that the ellipsoid of inertia of the prism is a sphere when computed (a) at point A, (b) at point B.

Fig. P9.175

Fig. P9.177

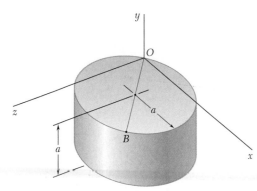

Fig. P9.181

9.176 For the right circular cone of Sample Prob. 9.11, determine the value of the ratio a/h for which the ellipsoid of inertia of the cone is a sphere when computed (a) at the apex of the cone, (b) at the center of the base of the cone.

9.177 For the homogeneous circular cylinder shown, of radius a and length L, determine the value of the ratio a/L for which the ellipsoid of inertia of the cylinder is a sphere when computed (a) at the centroid of the cylinder, (b) at point A.

9.178 Given an arbitrary body and three rectangular axes x, y, and z, prove that the moment of inertia of the body with respect to any one of the three axes cannot be larger than the sum of the moments of inertia of the body with respect to the other two axes. That is, prove that the inequality $I_x \leq I_y + I_z$ and the two similar inequalities are satisfied. Further, prove that $I_y \geq \frac{1}{2}I_x$ if the body is a homogeneous solid of revolution, where x is the axis of revolution and y is a transverse axis.

9.179 Consider a cube of mass m and side a. (a) Show that the ellipsoid of inertia at the center of the cube is a sphere, and use this property to determine the moment of inertia of the cube with respect to one of its diagonals. (b) Show that the ellipsoid of inertia at one of the corners of the cube is an ellipsoid of revolution, and determine the principal moments of inertia of the cube at that point.

9.180 Given a homogeneous body of mass m and of arbitrary shape and three rectangular axes x, y, and z with origin at O, prove that the sum $I_x + I_y + I_z$ of the moments of inertia of the body cannot be smaller than the similar sum computed for a sphere of the same mass and the same material centered at O. Further, using the results of Prob. 9.178, prove that if the body is a solid of revolution, where x is the axis of revolution, its moment of inertia I_y about a transverse axis y cannot be smaller than $3ma^2/10$, where a is the radius of the sphere of the same mass and the same material.

9.181 The homogeneous circular cylinder shown has a mass m, and the diameter OB of its top surface forms 45° angles with the x and z axes. (a) Determine the principal moments of inertia of the cylinder at the origin O. (b) Compute the angles that the principal axes of inertia at O form with the coordinate axes. (c) Sketch the cylinder, and show the orientation of the principal axes of inertia relative to the x, y, and z axes.

9.182 through 9.186 For the component described in the problem indicated, determine (a) the principal moments of inertia at the origin, (b) the principal axes of inertia at the origin. Sketch the body and show the orientation of the principal axes of inertia relative to the x, y, and z axes.

    9.182 Prob. 9.167
    9.183 Probs. 9.147 and 9.151
    9.184 Prob. 9.169
    9.185 Prob. 9.170
    9.186 Probs. 9.150 and 9.172

In the first half of this chapter, we discussed the determination of the resultant **R** of forces $\Delta\mathbf{F}$ distributed over a plane area A when the magnitudes of these forces are proportional to both the areas ΔA of the elements on which they act and the distances y from these elements to a given x axis; we thus had $\Delta F = ky\,\Delta A$. We found that the magnitude of the resultant **R** is proportional to the first moment $Q_x = \int y\,dA$ of the area A, while the moment of **R** about the x axis is proportional to the *second moment*, or *moment of inertia*, $I_x = \int y^2\,dA$ of A with respect to the same axis [Sec. 9.2].

The *rectangular moments of inertia I_x and I_y of an area* [Sec. 9.3] were obtained by evaluating the integrals

Rectangular moments of inertia

$$I_x = \int y^2\,dA \qquad I_y = \int x^2\,dA \tag{9.1}$$

These computations can be reduced to single integrations by choosing dA to be a thin strip parallel to one of the coordinate axes. We also recall that it is possible to compute I_x and I_y from the same elemental strip (Fig. 9.35) using the formula for the moment of inertia of a rectangular area [Sample Prob. 9.3].

$$dI_x = \tfrac{1}{3}y^3\,dx$$
$$dI_y = x^2 y\,dx$$

Fig. 9.35

Fig. 9.36

The *polar moment of inertia of an area A* with respect to the pole O [Sec. 9.4] was defined as

Polar moment of inertia

$$J_O = \int r^2\,dA \tag{9.3}$$

where r is the distance from O to the element of area dA (Fig. 9.36). Observing that $r^2 = x^2 + y^2$, we established the relation

$$J_O = I_x + I_y \tag{9.4}$$

545

The *radius of gyration of an area A* with respect to the x axis [Sec. 9.5] was defined as the distance k_x, where $I_x = k_x^2 A$. With similar definitions for the radii of gyration of A with respect to the y axis and with respect to O, we had

$$k_x = \sqrt{\frac{I_x}{A}} \qquad k_y = \sqrt{\frac{I_y}{A}} \qquad k_O = \sqrt{\frac{J_O}{A}} \qquad (9.5\text{–}9.7)$$

The parallel-axis theorem was presented in Sec. 9.6. It states that the moment of inertia I of an area with respect to any given axis AA' (Fig. 9.37) is equal to the moment of inertia $\bar{I}$ of the area with respect to the centroidal axis BB' that is parallel to AA' *plus* the product of the area A and the square of the distance d between the two axes:

$$I = \bar{I} + Ad^2 \qquad (9.9)$$

This formula can also be used to determine the moment of inertia $\bar{I}$ of an area with respect to a centroidal axis BB' when its moment of inertia I with respect to a parallel axis AA' is known. In this case, however, the product Ad^2 should be *subtracted* from the known moment of inertia I.

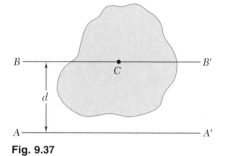

Fig. 9.37

A similar relation holds between the polar moment of inertia J_O of an area about a point O and the polar moment of inertia J_C of the same area about its centroid C. Letting d be the distance between O and C, we have

$$J_O = \bar{J}_C + Ad^2 \qquad (9.11)$$

The parallel-axis theorem can be used very effectively to compute the *moment of inertia of a composite area* with respect to a given axis [Sec. 9.7]. Considering each component area separately, we first compute the moment of inertia of each area with respect to its centroidal axis, using the data provided in Figs. 9.12 and 9.13 whenever possible. The parallel-axis theorem is then applied to determine the moment of inertia of each component area with respect to the desired axis, and the various values obtained are added [Sample Probs. 9.4 and 9.5].

Sections 9.8 through 9.10 were devoted to the transformation of the moments of inertia of an area *under a rotation of the coordinate axes.* First, we defined the *product of inertia of an area A* as

$$I_{xy} = \int xy \, dA \qquad (9.12)$$

and showed that $I_{xy} = 0$ if the area A is symmetrical with respect to either or both of the coordinate axes. We also derived the *parallel-axis theorem for products of inertia.* We had

$$I_{xy} = \bar{I}_{x'y'} + \bar{x}\bar{y}A \qquad (9.13)$$

where $\bar{I}_{x'y'}$ is the product of inertia of the area with respect to the centroidal axes x' and y' which are parallel to the x and y axes, respectively, and $\bar{x}$ and $\bar{y}$ are the coordinates of the centroid of the area [Sec. 9.8].

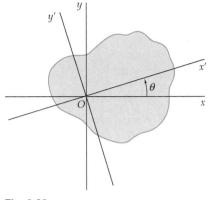

Fig. 9.38

In Sec. 9.9 we determined the moments and product of inertia $I_{x'}$, $I_{y'}$, and $I_{x'y'}$ of an area with respect to x' and y' axes obtained by rotating the original x and y coordinate axes through an angle θ counterclockwise (Fig. 9.38). We expressed $I_{x'}$, $I_{y'}$, and $I_{x'y'}$ in terms of the moments and product of inertia I_x, I_y, and I_{xy} computed with respect to the original x and y axes. We had

Rotation of axes

$$I_{x'} = \frac{I_x + I_y}{2} + \frac{I_x - I_y}{2}\cos 2\theta - I_{xy}\sin 2\theta \quad (9.18)$$

$$I_{y'} = \frac{I_x + I_y}{2} - \frac{I_x - I_y}{2}\cos 2\theta + I_{xy}\sin 2\theta \quad (9.19)$$

$$I_{x'y'} = \frac{I_x - I_y}{2}\sin 2\theta + I_{xy}\cos 2\theta \quad (9.20)$$

The *principal axes of the area about O* were defined as the two axes perpendicular to each other and with respect to which the moments of inertia of the area are maximum and minimum. The corresponding values of θ, denoted by θ_m, were obtained from the formula

Principal axes

$$\tan 2\theta_m = -\frac{2I_{xy}}{I_x - I_y} \quad (9.25)$$

The corresponding maximum and minimum values of I are called the *principal moments of inertia* of the area about O; we had

Principal moments of inertia

$$I_{\text{max,min}} = \frac{I_x + I_y}{2} \pm \sqrt{\left(\frac{I_x - I_y}{2}\right)^2 + I_{xy}^2} \quad (9.27)$$

We also noted that the corresponding value of the product of inertia is zero.

The transformation of the moments and product of inertia of an area under a rotation of axes can be represented graphically by drawing *Mohr's circle* [Sec. 9.10]. Given the moments and product of inertia I_x, I_y, and I_{xy} of the area with respect to the x and y

Mohr's circle

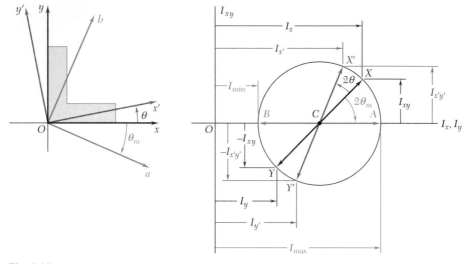

Fig. 9.39

coordinate axes, we plot points X (I_x, I_{xy}) and Y $(I_y, -I_{xy})$ and draw the line joining these two points (Fig. 9.39). This line is a diameter of Mohr's circle and thus defines this circle. As the coordinate axes are rotated through θ, the diameter rotates through *twice that angle,* and the coordinates of X' and Y' yield the new values $I_{x'}$, $I_{y'}$, and $I_{x'y'}$ of the moments and product of inertia of the area. Also, the angle θ_m and the coordinates of points A and B define the principal axes a and b and the principal moments of inertia of the area [Sample Prob. 9.8].

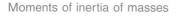

The second half of the chapter was devoted to the determination of *moments of inertia of masses,* which are encountered in dynamics in problems involving the rotation of a rigid body about an axis. The mass moment of inertia of a body with respect to an axis AA' (Fig. 9.40) was defined as

$$I = \int r^2 \, dm \tag{9.28}$$

where r is the distance from AA' to the element of mass [Sec. 9.11]. The *radius of gyration* of the body was defined as

$$k = \sqrt{\frac{I}{m}} \tag{9.29}$$

The moments of inertia of a body with respect to the coordinate axes were expressed as

$$I_x = \int (y^2 + z^2) \, dm$$

$$I_y = \int (z^2 + x^2) \, dm \tag{9.30}$$

$$I_z = \int (x^2 + y^2) \, dm$$

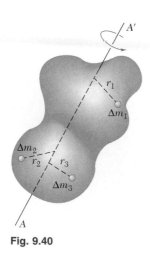

Fig. 9.40

We saw that the *parallel-axis theorem* also applies to mass moments of inertia [Sec. 9.12]. Thus, the moment of inertia I of a body with respect to an arbitrary axis AA' (Fig. 9.41) can be expressed as

$$I = \bar{I} + md^2 \qquad (9.33)$$

Parallel-axis theorem

where $\bar{I}$ is the moment of inertia of the body with respect to the centroidal axis BB' which is parallel to the axis AA', m is the mass of the body, and d is the distance between the two axes.

Fig. 9.41　　　　　　**Fig. 9.42**　　　　　　**Fig. 9.43**

The moments of inertia of *thin plates* can be readily obtained from the moments of inertia of their areas [Sec. 9.13]. We found that for a *rectangular plate* the moments of inertia with respect to the axes shown (Fig. 9.42) are

Moments of inertia of thin plates

$$I_{AA'} = \tfrac{1}{12}ma^2 \qquad I_{BB'} = \tfrac{1}{12}mb^2 \qquad (9.39)$$
$$I_{CC'} = I_{AA'} + I_{BB'} = \tfrac{1}{12}m(a^2 + b^2) \qquad (9.40)$$

while for a *circular plate* (Fig. 9.43) they are

$$I_{AA'} = I_{BB'} = \tfrac{1}{4}mr^2 \qquad (9.41)$$
$$I_{CC'} = I_{AA'} + I_{BB'} = \tfrac{1}{2}mr^2 \qquad (9.42)$$

When a body possesses *two planes of symmetry,* it is usually possible to use a single integration to determine its moment of inertia with respect to a given axis by selecting the element of mass dm to be a thin plate [Sample Probs. 9.10 and 9.11]. On the other hand, when a body consists of *several common geometric shapes,* its moment of inertia with respect to a given axis can be obtained by using the formulas given in Fig. 9.28 together with the parallel-axis theorem [Sample Probs. 9.12 and 9.13].

Composite bodies

In the last portion of the chapter, we learned to determine the moment of inertia of a body *with respect to an arbitrary axis OL* which is drawn through the origin O [Sec. 9.16]. Denoting by λ_x,

Moment of inertia with respect to an arbitrary axis

λ_y, λ_z the components of the unit vector $\boldsymbol{\lambda}$ along OL (Fig. 9.44) and introducing the *products of inertia*

$$I_{xy} = \int xy \, dm \qquad I_{yz} = \int yz \, dm \qquad I_{zx} = \int zx \, dm \quad (9.45)$$

we found that the moment of inertia of the body with respect to OL could be expressed as

$$I_{OL} = I_x\lambda_x^2 + I_y\lambda_y^2 + I_z\lambda_z^2 - 2I_{xy}\lambda_x\lambda_y - 2I_{yz}\lambda_y\lambda_z - 2I_{zx}\lambda_z\lambda_x \quad (9.46)$$

Fig. 9.44

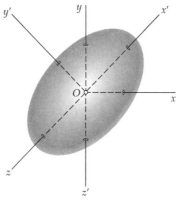

Fig. 9.45

Ellipsoid of inertia

By plotting a point Q along each axis OL at a distance $OQ = 1/\sqrt{I_{OL}}$ from O [Sec. 9.17], we obtained the surface of an ellipsoid, known as the *ellipsoid of inertia* of the body at point O. The principal axes x', y', z' of this ellipsoid (Fig. 9.45) are the *principal axes of inertia* of the body; that is, the products of inertia $I_{x'y'}$, $I_{y'z'}$, $I_{z'x'}$ of the body with respect to these axes are all zero. There are many situations when the principal axes of inertia of a body can be deduced from properties of symmetry of the body. Choosing these axes to be the coordinate axes, we can then express I_{OL} as

Principal axes of inertia
Principal moments of inertia

$$I_{OL} = I_{x'}\lambda_{x'}^2 + I_{y'}\lambda_{y'}^2 + I_{z'}\lambda_{z'}^2 \qquad (9.50)$$

where $I_{x'}$, $I_{y'}$, $I_{z'}$ are the *principal moments of inertia* of the body at O.

When the principal axes of inertia cannot be obtained by observation [Sec. 9.17], it is necessary to solve the cubic equation

$$K^3 - (I_x + I_y + I_z)K^2 + (I_xI_y + I_yI_z + I_zI_x - I_{xy}^2 - I_{yz}^2 - I_{zx}^2)K$$
$$- (I_xI_yI_z - I_xI_{yz}^2 - I_yI_{zx}^2 - I_zI_{xy}^2 - 2I_{xy}I_{yz}I_{zx}) = 0 \qquad (9.56)$$

We found [Sec. 9.18] that the roots K_1, K_2, and K_3 of this equation are the principal moments of inertia of the given body. The direction cosines $(\lambda_x)_1$, $(\lambda_y)_1$, and $(\lambda_z)_1$ of the principal axis corresponding to the principal moment of inertia K_1 are then determined by substituting K_1 into Eqs. (9.54) and solving two of these equations and Eq. (9.57) simultaneously. The same procedure is then repeated using K_2 and K_3 to determine the direction cosines of the other two principal axes [Sample Prob. 9.15].

Review Problems

9.187 Determine by direct integration the moment of inertia of the shaded area with respect to the y axis.

Fig. P9.187

Fig. P9.188

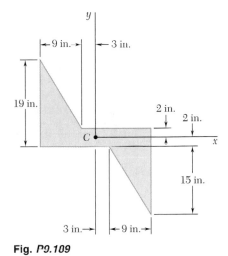

Fig. P9.109

9.188 Determine the moments of inertia $\bar{I}_x$ and $\bar{I}_y$ of the area shown with respect to centroidal axes respectively parallel and perpendicular to side AB.

9.189 Using the parallel-axis theorem, determine the product of inertia of the area shown with respect to the centroidal x and y axes.

9.190 A piece of thin, uniform sheet metal is cut to form the machine component shown. Denoting the mass of the component by m, determine its moment of inertia with respect to (*a*) the x axis, (*b*) the y axis.

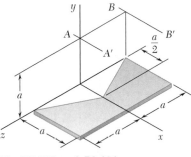

Fig. P9.190 and *P9.191*

9.191 A piece of thin, uniform sheet metal is cut to form the machine component shown. Denoting the mass of the component by m, determine its moment of inertia with respect to (*a*) the axis AA', (*b*) the axis BB', where the AA' and BB' axes are parallel to the x axis and lie in a plane parallel to and at a distance a above the zx plane.

9.192 For the $5 \times 3 \times \frac{1}{2}$-in. angle cross section shown, use Mohr's circle to determine (*a*) the moments of inertia and the product of inertia with respect to new centroidal axes obtained by rotating the x and y axes 45° counterclockwise, (*b*) the orientation of new centroidal axes for which $\bar{I}_{x'} = 2$ in^4 and $\bar{I}_{x'y'} < 0$.

$\frac{1}{2}$ in.

0.75 in.

L5 × 3 × $\frac{1}{2}$

5 in.

1.75 in.

$\frac{1}{2}$ in.

3 in.

Fig. P9.192

5 in. 5 in.

L3 × 3 × $\frac{1}{4}$

W8 × 31

Fig. *P9.193*

9.193 Four $3 \times 3 \times \frac{1}{4}$-in. angles are welded to a rolled W section as shown. Determine the moments of inertia and the radii of gyration of the combined section with respect to its centroidal x and y axes.

9.194 For the 2-kg connecting rod shown, it has been experimentally determined that the mass moments of inertia of the rod with respect to the center-line axes of the bearings AA' and BB' are, respectively, $I_{AA'} = 78$ g · m^2 and $I_{BB'} = 41$ g · m^2. Knowing that $r_a + r_b = 290$ mm, determine (*a*) the location of the centroidal axis GG', (*b*) the radius of gyration with respect to axis GG'.

Fig. P9.194

9.195 Determine the mass moment of inertia of the 0.9-lb machine component shown with respect to the axis AA'.

0.4 in.

A'

2.4 in.

A

1.2 in.

1.6 in.

Fig. *P9.195*

9.196 Determine the moments of inertia and the radii of gyration of the steel machine element shown with respect to the x and y axes. (The density of steel is 7850 kg/m^3.)

y

44

120

70   70   120

44

40

20

20

x

z

Dimensions in mm

Fig. P9.196

y

a

O a a x

a

a

Fig. P9.197

9.197 Determine the moments of inertia of the shaded area shown with respect to the x and y axes.

9.198 A 20-mm-diameter hole is bored in a 32-mm-diameter rod as shown. Determine the depth d of the hole so that the ratio of the moments of inertia of the rod with and without the hole with respect to the axis AA' is 0.96.

d 5.8 mm

20 mm A ———————— A' 32 mm

70 mm

Fig. P9.198

Computer Problems

9.C1 A pattern of 6-mm-diameter holes is drilled in a circular steel plate of diameter d to form the drain cover shown. Knowing that the center-to-center distance between adjacent holes is 14 mm and that the holes lie within a circle of diameter $(d - 30)$ mm, determine the moment of inertia of the cover with respect to the x axis when (*a*) $d = 200$ mm, (*b*) $d = 450$ mm, (*c*) $d = 720$ mm.

Fig. P9.C1

9.C2 Many cross sections can be approximated by a series of rectangles as shown. Using computational software, calculate the moments of inertia and the radii of gyration of cross sections of this type with respect to horizontal and vertical centroidal axes. Apply the above software to the cross section shown in (*a*) Fig. P9.31 and P9.33, (*b*) Fig. P9.32 and P9.34, (*c*) Fig. P9.43, (*d*) Fig. P9.44.

9.C3 The panel shown forms the end of a trough which is filled with water to line AA'. Determine the depth d of the point of application of the resultant of the hydrostatic forces acting on the panel (center of pressure). Knowing that $h = 400$ mm, $b = 200$ mm, and $a = nb$, where n varies from 1 to 8, plot the depth of the center of pressure d as a function of n.

Fig. P9.C2

Fig. P9.C3

554

9.C4 For the area shown, plot I_{xy} as a function of n for values of n from 1 to 10 knowing that $a = 80$ mm and $b = 60$ mm.

Fig. P9.C4

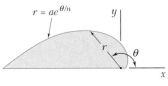

Fig. P9.C5

*9.C5** The area shown is bounded by the logarithmic spiral $r = ae^{\theta/n}$ with $0 \leq \theta \leq \pi$. Using computational software, determine the product of inertia of the area with respect to the x and y axes when $a = 3$ in. and (*a*) $n = 2$, (*b*) $n = 2.5$, (*c*) $n = 4$.

9.C6 An area with known moments and product of inertia I_x, I_y, and I_{xy} can be used to calculate the moments and product of inertia of the area with respect to axes x' and y' obtained by rotating the original axes counterclockwise through an angle θ. Using computational software, compute and plot $I_{x'}$, $I_{y'}$, and $I_{x'y'}$, as functions of θ for $0 \leq \theta \leq 90°$ for the area of Prob. 9.74.

9.C7 A homogeneous wire with a mass per unit length of 0.08 kg/m is used to form the figure shown. Approximating the figure using 10 straight line segments, use computational software to determine the mass moment of inertia I_x of the wire with respect to the x-axis. Using the software, determine I_x when (*a*) $a = 20$ mm, $L = 220$ mm, $h = 80$ mm, (*b*) $a = 40$ mm, $L = 340$ mm, $h = 200$ mm, (*c*) $a = 100$ mm, $L = 500$ mm, $h = 120$ mm.

9.C8 The area shown is revolved about the x-axis to form a homogeneous solid of weight W. Approximating the area using a series of 400 rectangles of the form $bcc'b'$, each of width Δl, use computational software to determine the mass moment of inertia of the solid with respect to the x-axis. Assuming that $W = 4$ lb, $a = 5$ in., and $b = 20$ in., use the software to solve (*a*) Prob. 9.121, (*b*) Prob. 9.123.

Fig. P9.C7

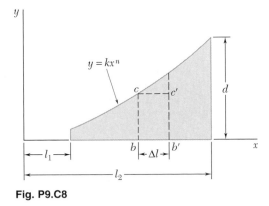

Fig. P9.C8

9.C9 A thin rectangular plate of weight W is welded to a vertical shaft AB with which it forms an angle θ. Knowing that $W = 1$ lb, $a = 10$ in., and $b = 8$ in., plot for values of θ from 0 to 90° the mass moment of inertia of the plate with respect to (*a*) the y axis, (*b*) the z axis.

Fig. P9.C9

Method of Virtual Work

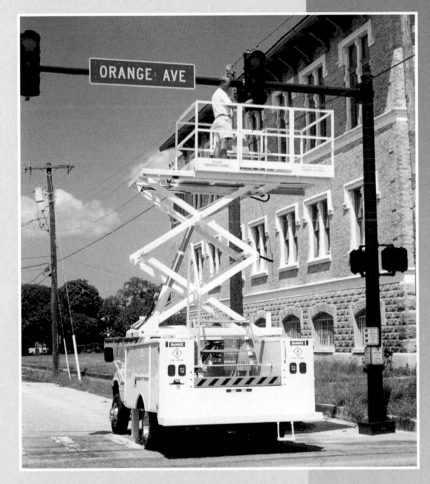

Using the analytical methods discussed in this chapter, it is easily shown that the force exerted by the hydraulic cylinder to raise the scissor lift shown is independent of the worker's location on the platform.

METHOD OF VIRTUAL WORK

10.1 Introduction
10.2 Work of a Force
10.3 Principle of Virtual Work
10.4 Applications of the Principle of Virtual Work
10.5 Real Machines. Mechanical Efficiency
10.6 Work of a Force during a Finite Displacement
10.7 Potential Energy
10.8 Potential Energy and Equilibrium
10.9 Stability of Equilibrium

*10.1. INTRODUCTION

In the preceding chapters, problems involving the equilibrium of rigid bodies were solved by expressing that the external forces acting on the bodies were balanced. The equations of equilibrium $\Sigma F_x = 0$, $\Sigma F_y = 0$, $\Sigma M_A = 0$ were written and solved for the desired unknowns. A different method, which will prove more effective for solving certain types of equilibrium problems, will now be considered. This method is based on the *principle of virtual work* and was first formally used by the Swiss mathematician Jean Bernoulli in the eighteenth century.

As you will see in Sec. 10.3, the principle of virtual work states that if a particle or rigid body, or, more generally, a system of connected rigid bodies, which is in equilibrium under various external forces, is given an arbitrary displacement from that position of equilibrium, the total work done by the external forces during the displacement is zero. This principle is particularly effective when applied to the solution of problems involving the equilibrium of machines or mechanisms consisting of several connected members.

In the second part of the chapter, the method of virtual work will be applied in an alternative form based on the concept of *potential energy*. It will be shown in Sec. 10.8 that if a particle, rigid body, or system of rigid bodies is in equilibrium, then the derivative of its potential energy with respect to a variable defining its position must be zero.

In this chapter, you will also learn to evaluate the mechanical efficiency of a machine (Sec. 10.5) and to determine whether a given position of equilibrium is stable, unstable, or neutral (Sec. 10.9).

*10.2. WORK OF A FORCE

Let us first define the terms *displacement* and *work* as they are used in mechanics. Consider a particle which moves from a point A to a neighboring point A' (Fig. 10.1). If $\mathbf{r}$ denotes the position vector corresponding to point A, the small vector joining A and A' may be denoted by the differential $d\mathbf{r}$; the vector $d\mathbf{r}$ is called the *displacement* of the particle. Now let us assume that a force $\mathbf{F}$ is acting on the particle. The *work of the force* $\mathbf{F}$ *corresponding to the displacement* $d\mathbf{r}$ is defined as the quantity

$$dU = \mathbf{F} \cdot d\mathbf{r} \tag{10.1}$$

obtained by forming the scalar product of the force $\mathbf{F}$ and the displacement $d\mathbf{r}$. Denoting respectively by F and ds the magnitudes of the force and of the displacement, and by α the angle formed by $\mathbf{F}$ and $d\mathbf{r}$, and recalling the definition of the scalar product of two vectors (Sec. 3.9), we write

$$dU = F \, ds \cos \alpha \tag{10.1'}$$

Being a *scalar quantity,* work has a magnitude and a sign, but no direction. We also note that work should be expressed in units

Fig. 10.1

obtained by multiplying units of length by units of force. Thus, if U.S. customary units are used, work should be expressed in ft · lb or in · lb. If SI units are used, work should be expressed in N · m. The unit of work N · m is called a *joule* (J).†

It follows from (10.1′) that the work dU is positive if the angle α is acute and negative if α is obtuse. Three particular cases are of special interest. If the force **F** has the same direction as $d\mathbf{r}$, the work dU reduces to $F\,ds$. If **F** has a direction opposite to that of $d\mathbf{r}$, the work is $dU = -F\,ds$. Finally, if **F** is perpendicular to $d\mathbf{r}$, the work dU is zero.

The work dU of a force **F** during a displacement $d\mathbf{r}$ can also be considered as the product of F and the component $ds \cos \alpha$ of the displacement $d\mathbf{r}$ along **F** (Fig. 10.2a). This view is particularly useful

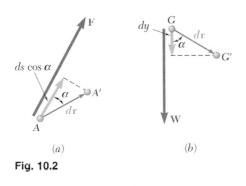

Fig. 10.2

in the computation of the work done by the weight **W** of a body (Fig. 10.2b). The work of **W** is equal to the product of W and the vertical displacement dy of the center of gravity G of the body. If the displacement is downward, the work is positive; if it is upward, the work is negative.

A number of forces frequently encountered in statics *do no work:* forces applied to fixed points ($ds = 0$) or acting in a direction perpendicular to the displacement ($\cos \alpha = 0$). Among these forces are the reaction at a frictionless pin when the body supported rotates about the pin; the reaction at a frictionless surface when the body in contact moves along the surface; the reaction at a roller moving along its track; the weight of a body when its center of gravity moves horizontally; and the friction force acting on a wheel rolling without slipping (since at any instant the point of contact does not move). Examples of forces which *do work* are the weight of a body (except in the case considered above), the friction force acting on a body sliding on a rough surface, and most forces applied to a moving body.

Photo 10.1 The forces exerted by the hydraulic cylinders to position the bucket lift shown can be effectively determined using the method of virtual work since a simple relation exists among the displacements of the points of application of the forces acting on the members of the lift.

---

†The joule is the SI unit of *energy*, whether in mechanical form (work, potential energy, kinetic energy) or in chemical, electrical, or thermal form. We should note that even though N · m = J, the moment of a force must be expressed in N · m, and not in joules, since the moment of a force is not a form of energy.

In certain cases, the sum of the work done by several forces is zero. Consider, for example, two rigid bodies AC and BC connected at C by a *frictionless pin* (Fig. 10.3a). Among the forces acting on AC is the force $\mathbf{F}$ exerted at C by BC. In general, the work of this force

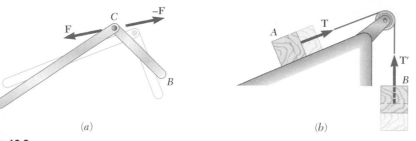

(a)   (b)

Fig. 10.3

will not be zero, but it will be equal in magnitude and opposite in sign to the work of the force $-\mathbf{F}$ exerted by AC on BC, since these forces are equal and opposite and are applied to the same particle. Thus, when the total work done by all the forces acting on AB and BC is considered, the work of the two internal forces at C cancels out. A similar result is obtained if we consider a system consisting of two blocks connected by an *inextensible cord AB* (Fig. 10.3b). The work of the tension force $\mathbf{T}$ at A is equal in magnitude to the work of the tension force $\mathbf{T}'$ at B, since these forces have the same magnitude and the points A and B move through the same distance; but in one case the work is positive, and in the other it is negative. Thus, the work of the internal forces again cancels out.

It can be shown that the total work of the internal forces holding together the particles of a rigid body is zero. Consider two particles A and B of a rigid body and the two equal and opposite forces $\mathbf{F}$ and $-\mathbf{F}$ they exert on each other (Fig. 10.4). While, in general, small dis-

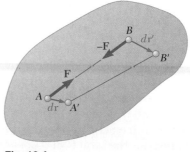

Fig. 10.4

placements $d\mathbf{r}$ and $d\mathbf{r}'$ of the two particles are different, the components of these displacements along AB must be equal; otherwise, the particles would not remain at the same distance from each other, and the body would not be rigid. Therefore, the work of $\mathbf{F}$ is equal in magnitude and opposite in sign to the work of $-\mathbf{F}$, and their sum is zero.

In computing the work of the external forces acting on a rigid body, it is often convenient to determine the work of a couple without considering separately the work of each of the two forces forming the couple. Consider the two forces $\mathbf{F}$ and $-\mathbf{F}$ forming a couple of moment $\mathbf{M}$ and acting on a rigid body (Fig. 10.5). Any small displacement of the rigid body bringing A and B, respectively, into A' and B'' can be divided into two parts, one in which points A and B undergo equal displacements $d\mathbf{r}_1$, the other in which A' remains fixed while B' moves into B'' through a displacement $d\mathbf{r}_2$ of magnitude $ds_2 = r\, d\theta$. In the first part of the motion, the work of $\mathbf{F}$ is equal in magnitude and opposite in sign to the work of $-\mathbf{F}$, and their sum is zero. In the second part of the motion, only force $\mathbf{F}$ does work, and its work is $dU = F\, ds_2 = Fr\, d\theta$. But the product Fr is equal to the magnitude M of the moment of the couple. Thus, the work of a couple of moment $\mathbf{M}$ acting on a rigid body is

$$dU = M\, d\theta \tag{10.2}$$

where $d\theta$ is the small angle expressed in radians through which the body rotates. We again note that work should be expressed in units obtained by multiplying units of force by units of length.

Fig. 10.5

*10.3. PRINCIPLE OF VIRTUAL WORK

Consider a particle acted upon by several forces $\mathbf{F}_1$, $\mathbf{F}_2$, ..., $\mathbf{F}_n$ (Fig. 10.6). We can imagine that the particle undergoes a small displacement from A to A'. This displacement is possible, but it will not necessarily take place. The forces may be balanced and the particle at rest, or the particle may move under the action of the given forces in a direction different from that of AA'. Since the displacement considered does not actually occur, it is called a *virtual displacement* and is denoted by $\delta\mathbf{r}$. The symbol $\delta\mathbf{r}$ represents a differential of the first order; it is used to distinguish the virtual displacement from the displacement $d\mathbf{r}$ which would take place under actual motion. As you will see, virtual displacements can be used to determine whether the conditions of equilibrium of a particle are satisfied.

The work of each of the forces $\mathbf{F}_1$, $\mathbf{F}_2$, ..., $\mathbf{F}_n$ during the virtual displacement $\delta\mathbf{r}$ is called *virtual work*. The virtual work of all the forces acting on the particle of Fig. 10.6 is

$$\delta U = \mathbf{F}_1 \cdot \delta\mathbf{r} + \mathbf{F}_2 \cdot \delta\mathbf{r} + \cdots + \mathbf{F}_n \cdot \delta\mathbf{r}$$
$$= (\mathbf{F}_1 + \mathbf{F}_2 + \cdots + \mathbf{F}_n) \cdot \delta\mathbf{r}$$

or

$$\delta U = \mathbf{R} \cdot \delta\mathbf{r} \tag{10.3}$$

Fig. 10.6

where $\mathbf{R}$ is the resultant of the given forces. Thus, the total virtual work of the forces $\mathbf{F}_1$, $\mathbf{F}_2$, ..., $\mathbf{F}_n$ is equal to the virtual work of their resultant $\mathbf{R}$.

The principle of virtual work for a particle states that *if a particle is in equilibrium, the total virtual work of the forces acting on the particle is zero for any virtual displacement of the particle*. This condition is necessary: if the particle is in equilibrium, the resultant $\mathbf{R}$ of the forces is zero, and it follows from (10.3) that the total virtual work δU is zero. The condition is also sufficient: if the total virtual work δU is zero for any virtual displacement, the scalar product $\mathbf{R} \cdot \delta\mathbf{r}$ is zero for any $\delta\mathbf{r}$, and the resultant $\mathbf{R}$ must be zero.

In the case of a rigid body, the principle of virtual work states that *if a rigid body is in equilibrium, the total virtual work of the external forces acting on the rigid body is zero for any virtual displacement of the body.* The condition is necessary: if the body is in equilibrium, all the particles forming the body are in equilibrium and the total virtual work of the forces acting on all the particles must be zero; but we have seen in the preceding section that the total work of the internal forces is zero; the total work of the external forces must therefore also be zero. The condition can also be proved to be sufficient.

The principle of virtual work can be extended to the case of a *system of connected rigid bodies.* If the system remains connected during the virtual displacement, *only the work of the forces external to the system need be considered,* since the total work of the internal forces at the various connections is zero.

*10.4. APPLICATIONS OF THE PRINCIPLE OF VIRTUAL WORK

The principle of virtual work is particularly effective when applied to the solution of problems involving machines or mechanisms consisting of several connected rigid bodies. Consider, for instance, the toggle vise *ACB* of Fig. 10.7*a*, used to compress a wooden block. We

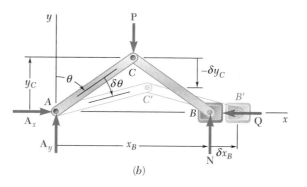

(a)   (b)

Fig. 10.7

wish to determine the force exerted by the vise on the block when a given force **P** is applied at *C* assuming that there is no friction. Denoting by **Q** the reaction of the block on the vise, we draw the free-body diagram of the vise and consider the virtual displacement obtained by giving a positive increment $\delta\theta$ to the angle θ (Fig. 10.7*b*). Choosing a system of coordinate axes with origin at *A*, we note that x_B increases while y_C decreases. This is indicated in the figure, where a positive increment δx_B and a negative increment $-\delta y_C$ are shown. The reactions $\mathbf{A}_x$, $\mathbf{A}_y$, and **N** will do no work during the virtual displacement considered, and we need only compute the work of **P** and **Q**. Since **Q** and δx_B have opposite senses, the virtual work of **Q** is $\delta U_Q = -Q\,\delta x_B$. Since **P** and the increment shown $(-\delta y_C)$ have the same sense, the virtual work of **P** is $\delta U_P = +P(-\delta y_C) = -P\,\delta y_C$. The minus signs obtained could have been predicted by simply noting that the forces **Q** and **P** are directed opposite to the positive *x* and *y* axes, respectively. Expressing the coordinates x_B and y_C in terms of the angle θ and differentiating, we obtain

$$x_B = 2l \sin \theta \qquad y_C = l \cos \theta$$
$$\delta x_B = 2l \cos \theta \, \delta\theta \qquad \delta y_C = -l \sin \theta \, \delta\theta \qquad (10.4)$$

The total virtual work of the forces $\mathbf{Q}$ and $\mathbf{P}$ is thus

$$\delta U = \delta U_Q + \delta U_P = -Q \, \delta x_B - P \, \delta y_C$$
$$= -2Ql \cos \theta \, \delta\theta + Pl \sin \theta \, \delta\theta$$

Setting $\delta U = 0$, we obtain

$$2Ql \cos \theta \, \delta\theta = Pl \sin \theta \, \delta\theta \qquad (10.5)$$
$$Q = \tfrac{1}{2} P \tan \theta \qquad (10.6)$$

The superiority of the method of virtual work over the conventional equilibrium equations in the problem considered here is clear: by using the method of virtual work, we were able to eliminate all unknown reactions, while the equation $\Sigma M_A = 0$ would have eliminated only two of the unknown reactions. This property of the method of virtual work can be used in solving many problems involving machines and mechanisms. *If the virtual displacement considered is consistent with the constraints imposed by the supports and connections, all reactions and internal forces are eliminated and only the work of the loads, applied forces, and friction forces need be considered.*

The method of virtual work can also be used to solve problems involving completely constrained structures, although the virtual displacements considered will never actually take place. Consider, for example, the frame ACB shown in Fig. 10.8a. If point A is kept fixed, while B is given a horizontal virtual displacement (Fig. 10.8b), we need consider only the work of $\mathbf{P}$ and $\mathbf{B}_x$. We can thus determine the

Photo 10.2 The clamping force of the toggle clamp shown can be expressed as a function of the force applied to the handle by first establishing the geometric relations among the members of the clamp and then applying the method of virtual work.

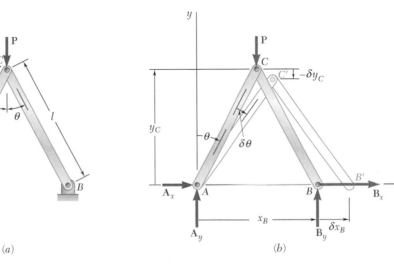

(a)   (b)

Fig. 10.8

reaction component $\mathbf{B}_x$ in the same way as the force $\mathbf{Q}$ of the preceding example (Fig. 10.7b); we have

$$B_x = -\tfrac{1}{2}P \tan \theta$$

Keeping B fixed and giving to A a horizontal virtual displacement, we can similarly determine the reaction component $\mathbf{A}_x$. The components $\mathbf{A}_y$ and $\mathbf{B}_y$ can be determined by rotating the frame ACB as a rigid body about B and A, respectively.

The method of virtual work can also be used to determine the configuration of a system in equilibrium under given forces. For example, the value of the angle θ for which the linkage of Fig. 10.7 is in equilibrium under two given forces $\mathbf{P}$ and $\mathbf{Q}$ can be obtained by solving Eq. (10.6) for $\tan \theta$.

It should be noted, however, that the attractiveness of the method of virtual work depends to a large extent upon the existence of simple geometric relations among the various virtual displacements involved in the solution of a given problem. When no such simple relations exist, it is usually advisable to revert to the conventional method of Chap. 6.

*10.5. REAL MACHINES. MECHANICAL EFFICIENCY

In analyzing the toggle vise in the preceding section, we assumed that no friction forces were involved. Thus, the virtual work consisted only of the work of the applied force $\mathbf{P}$ and of the reaction $\mathbf{Q}$. But the work of the reaction $\mathbf{Q}$ is equal in magnitude and opposite in sign to the work of the force exerted by the vise on the block. Equation (10.5), therefore, expresses that the *output work* $2Ql \cos \theta \, \delta\theta$ is equal to the *input work* $Pl \sin \theta \, \delta\theta$. A machine in which input and output work are equal is said to be an "ideal" machine. In a "real" machine, friction forces will always do some work, and the output work will be smaller than the input work.

Consider, for example, the toggle vise of Fig. 10.7a, and assume now that a friction force $\mathbf{F}$ develops between the sliding block B and the horizontal plane (Fig. 10.9). Using the conventional methods of statics and summing moments about A, we find $N = P/2$. Denoting by μ the coefficient of friction between block B and the horizontal

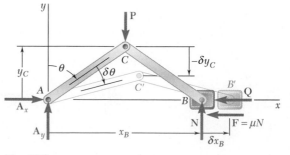

Fig. 10.9

plane, we have $F = \mu N = \mu P/2$. Recalling formulas (10.4), we find that the total virtual work of the forces **Q**, **P**, and **F** during the virtual displacement shown in Fig. 10.9 is

$$\begin{aligned} \delta U &= -Q \, \delta x_B - P \, \delta y_C - F \, \delta x_B \\ &= -2Ql \cos \theta \, \delta\theta + Pl \sin \theta \, \delta\theta - \mu Pl \cos \theta \, \delta\theta \end{aligned}$$

Setting $\delta U = 0$, we obtain

$$2Ql \cos \theta \, \delta\theta = Pl \sin \theta \, \delta\theta - \mu Pl \cos \theta \, \delta\theta \qquad (10.7)$$

which expresses that the output work is equal to the input work minus the work of the friction force. Solving for Q, we have

$$Q = \tfrac{1}{2}P(\tan \theta - \mu) \qquad (10.8)$$

We note that $Q = 0$ when $\tan \theta = \mu$, that is, when θ is equal to the angle of friction ϕ, and that $Q < 0$ when $\theta < \phi$. The toggle vise may thus be used only for values of θ larger than the angle of friction.

The *mechanical efficiency* of a machine is defined as the ratio

$$\eta = \frac{\text{output work}}{\text{input work}} \qquad (10.9)$$

Clearly, the mechanical efficiency of an ideal machine is $\eta = 1$, since input and output work are then equal, while the mechanical efficiency of a real machine will always be less than 1.

In the case of the toggle vise we have just analyzed, we write

$$\eta = \frac{\text{output work}}{\text{input work}} = \frac{2Ql \cos \theta \, \delta\theta}{Pl \sin \theta \, \delta\theta}$$

Substituting from (10.8) for Q, we obtain

$$\eta = \frac{P(\tan \theta - \mu)l \cos \theta \, \delta\theta}{Pl \sin \theta \, \delta\theta} = 1 - \mu \cot \theta \qquad (10.10)$$

We check that in the absence of friction forces, we would have $\mu = 0$ and $\eta = 1$. In the general case, when μ is different from zero, the efficiency η becomes zero for $\mu \cot \theta = 1$, that is, for $\tan \theta = \mu$, or $\theta = \tan^{-1} \mu = \phi$. We note again that the toggle vise can be used only for values of θ larger than the angle of friction ϕ.

SAMPLE PROBLEM 10.1

Using the method of virtual work, determine the magnitude of the couple **M** required to maintain the equilibrium of the mechanism shown.

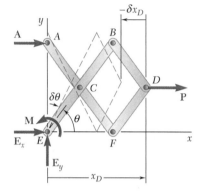

SOLUTION

Choosing a coordinate system with origin at E, we write

$$x_D = 3l \cos \theta \qquad \delta x_D = -3l \sin \theta \, \delta \theta$$

Principle of Virtual Work. Since the reactions **A**, $\mathbf{E}_x$, and $\mathbf{E}_y$ will do no work during the virtual displacement, the total virtual work done by **M** and **P** must be zero. Noting that **P** acts in the positive x direction and **M** acts in the positive θ direction, we write

$\delta U = 0$:
$$+M \, \delta\theta + P \, \delta x_D = 0$$
$$+M \, \delta\theta + P(-3l \sin \theta \, \delta\theta) = 0$$
$$M = 3Pl \sin \theta \quad \blacktriangleleft$$

SAMPLE PROBLEM 10.2

Determine the expressions for θ and for the tension in the spring which correspond to the equilibrium position of the mechanism. The unstretched length of the spring is h, and the constant of the spring is k. Neglect the weight of the mechanism.

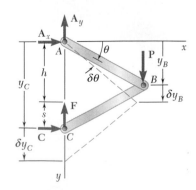

SOLUTION

With the coordinate system shown

$$y_B = l \sin \theta \qquad y_C = 2l \sin \theta$$
$$\delta y_B = l \cos \theta \, \delta\theta \qquad \delta y_C = 2l \cos \theta \, \delta\theta$$

The elongation of the spring is

$$s = y_C - h = 2l \sin \theta - h$$

The magnitude of the force exerted at C by the spring is

$$F = ks = k(2l \sin \theta - h) \tag{1}$$

Principle of Virtual Work. Since the reactions $\mathbf{A}_x$, $\mathbf{A}_y$, and **C** do no work, the total virtual work done by **P** and **F** must be zero.

$\delta U = 0$:
$$P \, \delta y_B - F \, \delta y_C = 0$$
$$P(l \cos \theta \, \delta\theta) - k(2l \sin \theta - h)(2l \cos \theta \, \delta\theta) = 0$$

$$\sin \theta = \frac{P + 2kh}{4kl} \quad \blacktriangleleft$$

Substituting this expression into (1), we obtain

$$F = \tfrac{1}{2}P \quad \blacktriangleleft$$

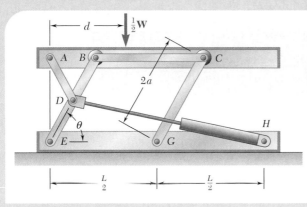

SAMPLE PROBLEM 10.3

A hydraulic-lift table is used to raise a 1000-kg crate. It consists of a platform and of two identical linkages on which hydraulic cylinders exert equal forces. (Only one linkage and one cylinder are shown.) Members EDB and CG are each of length $2a$, and member AD is pinned to the midpoint of EDB. If the crate is placed on the table, so that half of its weight is supported by the system shown, determine the force exerted by each cylinder in raising the crate for $\theta = 60°$, $a = 0.70$ m, and $L = 3.20$ m. This mechanism has been previously considered in Sample Prob. 6.7.

SOLUTION

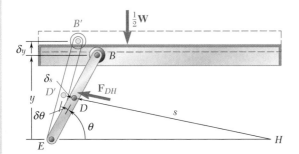

The machine considered consists of the platform and of the linkage, with an input force $\mathbf{F}_{DH}$ exerted by the cylinder and an output force equal and opposite to $\frac{1}{2}\mathbf{W}$.

Principle of Virtual Work. We first observe that the reactions at E and G do no work. Denoting by y the elevation of the platform above the base, and by s the length DH of the cylinder-and-piston assembly, we write

$$\delta U = 0: \qquad -\tfrac{1}{2}W\,\delta y + F_{DH}\,\delta s = 0 \qquad (1)$$

The vertical displacement δy of the platform is expressed in terms of the angular displacement $\delta\theta$ of EDB as follows:

$$y = (EB)\sin\theta = 2a\sin\theta$$
$$\delta y = 2a\cos\theta\,\delta\theta$$

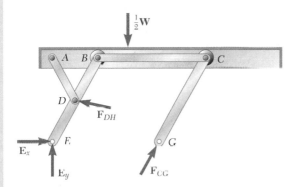

To express δs similarly in terms of $\delta\theta$, we first note that by the law of cosines,

$$s^2 = a^2 + L^2 - 2aL\cos\theta$$

Differentiating,

$$2s\,\delta s = -2aL(-\sin\theta)\,\delta\theta$$
$$\delta s = \frac{aL\sin\theta}{s}\,\delta\theta$$

Substituting for δy and δs into (1), we write

$$(-\tfrac{1}{2}W)2a\cos\theta\,\delta\theta + F_{DH}\frac{aL\sin\theta}{s}\,\delta\theta = 0$$

$$F_{DH} = W\frac{s}{L}\cot\theta$$

With the given numerical data, we have

$$W = mg = (1000\text{ kg})(9.81\text{ m/s}^2) = 9810\text{ N} = 9.81\text{ kN}$$
$$s^2 = a^2 + L^2 - 2aL\cos\theta$$
$$= (0.70)^2 + (3.20)^2 - 2(0.70)(3.20)\cos 60° = 8.49$$
$$s = 2.91\text{ m}$$

$$F_{DH} = W\frac{s}{L}\cot\theta = (9.81\text{ kN})\frac{2.91\text{ m}}{3.20\text{ m}}\cot 60°$$
$$F_{DH} = 5.15\text{ kN} \quad \blacktriangleleft$$

In this lesson you learned to use the *method of virtual work*, which is a different way of solving problems involving the equilibrium of rigid bodies.

The work done by a force during a displacement of its point of application or by a couple during a rotation is found by using Eqs. (10.1) and (10.2), respectively:

$$dU = F \, ds \cos \alpha \qquad\qquad (10.1)$$
$$dU = M \, d\theta \qquad\qquad (10.2)$$

Principle of virtual work. In its more general and more useful form, this principle can be stated as follows: *If a system of connected rigid bodies is in equilibrium, the total virtual work of the external forces applied to the system is zero for any virtual displacement of the system.*

As you apply the principle of virtual work, keep in mind the following:

1. Virtual displacement. A machine or mechanism in equilibrium has no tendency to move. However, *we can cause, or imagine, a small displacement*. Since it does not actually occur, such a displacement is called a *virtual displacement*.

2. Virtual work. The work done by a force or couple during a virtual displacement is called *virtual work*.

3. You need consider only the forces which do work during the virtual displacement. It is important to remember that only the component of a virtual displacement parallel to a given force need be considered when computing the virtual work. This is why, for example, in Sample Prob. 10.2 only the vertical displacement of pin *B* was determined.

4. Forces which do no work during a virtual displacement that is consistent with the constraints imposed on the system include the following:
   a. Reactions at supports
   b. Internal forces at connections
   c. Forces exerted by inextensible cords and cables
None of these forces need be considered when you use the method of virtual work.

5. Be sure to express the various virtual displacements involved in your computations in terms of a *single virtual displacement*. This is done in each of the three preceding sample problems, where the virtual displacements are all expressed in terms of $\delta\theta$. In addition, virtual displacements containing second and higher order terms [for example, $(\delta\theta)^2$ and $(\delta x)^2$] can be ignored when computing the virtual work.

6. Remember that the method of virtual work is effective only in those cases where the geometry of the system makes it relatively easy to relate the displacements involved.

*7. The work of the force **F** exerted by a spring during a virtual displacement δx is* given by $\delta U = F\delta x$. In the next lesson we will determine the work done during a finite displacement of the force exerted by a spring.

Problems

10.1 and 10.2 Determine the vertical force **P** which must be applied at G to maintain the equilibrium of the linkage.

Fig. P10.1 and P10.4

Fig. P10.2 and *P10.3*

10.3 **and 10.4** Determine the couple **M** which must be applied to member *DEFG* to maintain the equilibrium of the linkage.

10.5 An unstretched spring of constant 4 lb/in. is attached to pins at points C and I as shown. The pin at B is attached to member *BDE* and can slide freely along the slot in the fixed plate. Determine the force in the spring and the horizontal displacement of point H when a 20-lb horizontal force directed to the right is applied (*a*) at point C, (*b*) at points G and H.

Fig. P10.5 and *P10.6*

10.6 An unstretched spring of constant 4 lb/in. is attached to pins at points C and I as shown. The pin at B is attached to member *BDE* and can slide freely along the slot in the fixed plate. Determine the force in the spring and the horizontal displacement of point H when a 20-lb horizontal force directed to the right is applied (*a*) at point E, (*b*) at points D and E.

10.7 Knowing that the maximum friction force exerted by the bottle on the cork is 300 N, determine (*a*) the force **P** which must be applied to the corkscrew to open the bottle, (*b*) the maximum force exerted by the base of the corkscrew on the top of the bottle.

Fig. P10.7

Fig. P10.8

10.8 The two-bar linkage shown is supported by a pin and bracket at *B* and a collar at *D* that slides freely on a vertical rod. Determine the force **P** required to maintain the equilibrium of the linkage.

10.9 The mechanism shown is acted upon by the force **P**; derive an expression for the magnitude of the force **Q** required for equilibrium.

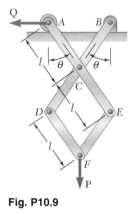

Fig. P10.9

10.10 Knowing that the line of action of the force **Q** passes through point *C*, derive an expression for the magnitude of **Q** required to maintain equilibrium.

10.11 Solve Prob. 10.10 assuming that the force **P** applied at point *A* acts horizontally to the left.

10.12 The slender rod *AB* is attached to a collar *A* and rests on a small wheel at *C*. Neglecting the radius of the wheel and the effect of friction, derive an expression for the magnitude of the force **Q** required to maintain the equilibrium of the rod.

Fig. *P10.10*

Fig. P10.12

10.13 A double scissor lift table is used to raise a 1000-lb machine component. The table consists of a platform and two identical linkages on which hydraulic cylinders exert equal forces. (Only one linkage and one cylinder are shown.) Each member of the linkage is of length 24 in., and pins C and G are at the midpoints of their respective members. The hydraulic cylinder is pinned at A to the base of the table and at F which is 6 in. from E. If the component is placed on the table so that half of its weight is supported by the system shown, determine the force exerted by each cylinder when $\theta = 30°$.

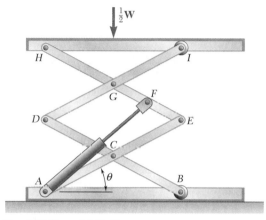

Fig. P10.13 and P10.14

10.14 A double scissor lift table is used to raise a 1000-lb machine component. The table consists of a platform and two identical linkages on which hydraulic cylinders exert equal forces. (Only one linkage and one cylinder are shown.) Each member of the linkage is of length 24 in., and pins C and G are at the midpoints of their respective members. The hydraulic cylinder is pinned at A to the base of the table and at F which is 6 in. from E. If the component is placed on the table so that half of its weight is supported by the system shown, determine the smallest allowable value of θ knowing that the maximum force each cylinder can exert is 8 kips.

10.15 through 10.17 Derive an expression for the magnitude of the couple **M** required to maintain the equilibrium of the linkage shown.

Fig. P10.15   **Fig. P10.16**   **Fig. P10.17**

Fig. P10.18

10.18   The pin at C is attached to member BCD and can slide along a slot cut in the fixed plate shown. Neglecting the effect of friction, derive an expression for the magnitude of the couple **M** required to maintain equilibrium when the force **P** which acts at D is directed (a) as shown, (b) vertically downward, (c) horizontally to the right.

10.19   A 1-kip force **P** is applied as shown to the piston of the engine system. Knowing that $AB = 2.5$ in. and $BC = 10$ in., determine the couple **M** required to maintain the equilibrium of the system when (a) $\theta = 30°$, (b) $\theta = 150°$.

Fig. P10.19 and P10.20

10.20   A couple **M** of magnitude 75 lb · ft is applied as shown to the crank of the engine system. Knowing that $AB = 2.5$ in. and $BC = 10$ in., determine the force **P** required to maintain the equilibrium of the system when (a) $\theta = 60°$, (b) $\theta = 120°$.

10.21   For the linkage shown, determine the force **P** required for equilibrium when $a = 18$ in., $M = 240$ lb · in., and $\theta = 30°$.

10.22   For the linkage shown, determine the couple **M** required for equilibrium when $a = 2$ ft, $P = 30$ lb, and $\theta = 40°$.

Fig. P10.21 and P10.22

10.23   Determine the value of θ corresponding to the equilibrium position of the mechanism of Prob. 10.10 when $P = 60$ lb and $Q = 75$ lb.

10.24   Determine the value of θ corresponding to the equilibrium position of the mechanism of Prob. 10.11 when $P = 20$ lb and $Q = 25$ lb.

10.25   A slender rod of length l is attached to a collar at B and rests on a portion of a circular cylinder of radius r. Neglecting the effect of friction, determine the value of θ corresponding to the equilibrium position of the mechanism when $l = 300$ mm, $r = 90$ mm, $P = 60$ N, and $Q = 120$ N.

Fig. P10.25 and P10.26

10.26   A slender rod of length l is attached to a collar at B and rests on a portion of a circular cylinder of radius r. Neglecting the effect of friction, determine the value of θ corresponding to the equilibrium position of the mechanism when $l = 280$ mm, $r = 100$ mm, $P = 300$ N, and $Q = 600$ N.

10.27 Determine the value of θ corresponding to the equilibrium position of the mechanism of Prob. 10.12 when $l = 600$ mm, $a = 100$ mm, $P = 100$ N, and $Q = 160$ N.

10.28 Determine the value of θ corresponding to the equilibrium position of the mechanism of Prob. 10.13 when $l = 900$ mm, $a = 150$ mm, $P = 75$ N, and $Q = 135$ N.

10.29 Two rods AC and CE are connected by a pin at C and by a spring AE. The constant of the spring is k, and the spring is unstretched when $\theta = 30°$. For the loading shown, derive an equation in P, θ, l, and k that must be satisfied when the system is in equilibrium.

10.30 Two rods AC and CE are connected by a pin at C and by a spring AE. The constant of the spring is 300 N/m, and the spring is unstretched when $\theta = 30°$. Knowing that $l = 200$ mm and neglecting the mass of the rods, determine the value of θ corresponding to equilibrium when $P = 160$ N.

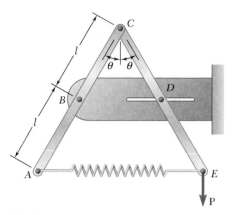

Fig. *P10.29* and P10.30

10.31 Solve Prob. 10.30 assuming that force **P** is moved to C and acts vertically downward.

10.32 For the mechanism shown, block A can move freely in its guide and rests against a spring of constant 15 lb/in. that is undeformed when $\theta = 45°$. For the loading shown, determine the value of θ corresponding to equilibrium.

Fig. P10.32

10.33 A force **P** of magnitude 150 lb is applied to the linkage at B. The constant of the spring is 12.5 lb/in., and the spring is unstretched when AB and BC are horizontal. Neglecting the weight of the linkage and knowing that $l = 15$ in., determine the value of θ corresponding to equilibrium.

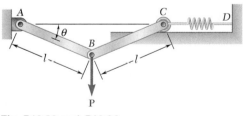

Fig. P10.33 and *P10.34*

10.34 A vertical force **P** is applied to the linkage at B. The constant of the spring is k, and the spring is unstretched when AB and BC are horizontal. Neglecting the weight of the linkage, derive an equation in θ, P, l, and k that must be satisfied when the linkage is in equilibrium.

10.35 and 10.36 Knowing that the constant of spring CD is k and that the spring is unstretched when rod ABC is horizontal, determine the value of θ corresponding to equilibrium for the data indicated.

10.35 $P = 150$ lb, $l = 30$ in., $k = 40$ lb/in.
10.36 $P = 600$ N, $l = 800$ mm, $k = 4$ kN/m.

Fig. *P10.35* and P10.36

10.37 A horizontal force $\mathbf{P}$ of magnitude 160 N is applied to the mechanism at C. The constant of the spring is $k = 1.8$ kN/m, and the spring is unstretched when $\theta = 0$. Neglecting the mass of the mechanism, determine the value of θ corresponding to equilibrium.

Fig. P10.37

Fig. *P10.38*

10.38 A cord is wrapped around drum A which is attached to member AB. Block D can move freely in its guide and is fastened to link CD. Neglecting the weight of AB and knowing that the spring is of constant 4 lb/in. and is undeformed when $\theta = 0$, determine the value of θ corresponding to equilibrium when a downward force $\mathbf{P}$ of magnitude 96 lb is applied to the end of the cord.

10.39 The lever AB is attached to the horizontal shaft BC which passes through a bearing and is welded to a fixed support at C. The torsional spring constant of the shaft BC is K; that is, a couple of magnitude K is required to rotate end B through 1 rad. Knowing that the shaft is untwisted when AB is horizontal, determine the value of θ corresponding to the position of equilibrium when $P = 400$ lb, $l = 10$ in., and $K = 150$ lb · ft/rad.

Fig. P10.39

10.40 Solve Prob. 10.39 assuming that $P = 1.26$ kips, $l = 10$ in., and $K = 150$ lb · ft/rad. Obtain answers in each of the following quadrants: $0 < \theta < 90°$, $270° < \theta < 360°$, and $360° < \theta < 450°$.

10.41 The position of crank BCD is controlled by the hydraulic cylinder AB. For the loading shown, determine the force exerted by the hydraulic cylinder on pin B knowing that $\theta = 60°$.

150 mm

480 N

108 mm

315 mm

120 mm

Fig. P10.41 and P10.42

10.42 The position of crank BCD is controlled by the hydraulic cylinder AB. Determine the angle θ knowing that the hydraulic cylinder exerts a 420-N force on pin B when the crank is in the position shown.

10.43 For the linkage shown, determine the force **P** required for equilibrium when $M = 40$ N · m.

10.44 A cord is wrapped around a drum of radius a that is pinned at A. The constant of the spring is 3 kN/m, and the spring is unstretched when $\theta = 0$. Knowing that $a = 150$ mm and neglecting the mass of the drum, determine the value of θ corresponding to equilibrium when a downward force **P** of magnitude 48 N is applied to the end of the cord.

375 mm

375 mm

35°

35°

P

375 mm

Fig. P10.43

4a

θ

a

A

B

C

P

Fig. P10.44

10.45 The telescoping arm *ABC* is used to provide an elevated platform for construction workers. The workers and the platform together have a mass of 204 kg, and their combined center of gravity is located directly above *C*. For the position when $\theta = 20°$, determine the force exerted on pin *B* by the single hydraulic cylinder *BD*.

Fig. P10.45

10.46 Solve Prob. 10.45 assuming that the workers are lowered to a point near the ground so that $\theta = -20°$.

10.47 A block of weight *W* is pulled up a plane forming an angle α with the horizontal by a force **P** directed along the plane. If μ is the coefficient of friction between the block and the plane, derive an expression for the mechanical efficiency of the system. Show that the mechanical efficiency cannot exceed $\frac{1}{2}$ if the block is to remain in place when the force **P** is removed.

10.48 Denoting by μ_s the coefficient of static friction between collar *C* and the vertical rod, derive an expression for the magnitude of the largest couple **M** for which equilibrium is maintained in the position shown. Explain what happens if $\mu_s \geq \tan \theta$.

Fig. *P10.48* and P10.49

10.49 Knowing that the coefficient of static friction between collar *C* and the vertical rod is 0.40, determine the magnitude of the largest and smallest couple **M** for which equilibrium is maintained in the position shown when $\theta = 35°$, $l = 30$ in., and $P = 1.2$ kips.

10.50 Derive an expression for the mechanical efficiency of the jack discussed in Sec. 8.6. Show that if the jack is to be self-locking, the mechanical efficiency cannot exceed $\frac{1}{2}$.

10.51 Denoting by μ_s the coefficient of static friction between the block attached to rod ACE and the horizontal surface, derive expressions in terms of P, μ_s, and θ for the largest and smallest magnitudes of the force $\mathbf{Q}$ for which equilibrium is maintained.

10.52 Knowing that the coefficient of static friction between the block attached to rod ACE and the horizontal surface is 0.15, determine the magnitudes of the largest and smallest force $\mathbf{Q}$ for which equilibrium is maintained when $\theta = 30°$, $l = 8$ in., and $P = 160$ lb.

10.53 Using the method of virtual work, determine separately the force and the couple representing the reaction at A.

Fig. P10.51 and P10.52

Fig. P10.53 and P10.54

10.54 Using the method of virtual work, determine the reaction at D.

10.55 Referring to Prob. 10.41 and using the value found for the force exerted by the hydraulic cylinder AB, determine the change in the length of AB required to raise the 480-N load 18 mm.

10.56 Referring to Prob. 10.45 and using the value found for the force exerted by the hydraulic cylinder BD, determine the change in the length of BD required to raise the platform attached at C by 50 mm.

10.57 Determine the vertical movement of joint D if the length of member BF is increased by 75 mm. (*Hint:* Apply a vertical load at joint D, and, using the methods of Chap. 6, compute the force exerted by member BF on joints B and F. Then apply the method of virtual work for a virtual displacement resulting in the specified increase in length of member BF. This method should be used only for small changes in the lengths of members.)

Fig. P10.57 and *P10.58*

10.58 Determine the horizontal movement of joint D if the length of member BF is increased by 75 mm. (See the hint for Prob. 10.57.)

*10.6. WORK OF A FORCE DURING A FINITE DISPLACEMENT

Consider a force $\mathbf{F}$ acting on a particle. The work of $\mathbf{F}$ corresponding to an infinitesimal displacement $d\mathbf{r}$ of the particle was defined in Sec. 10.2 as

$$dU = \mathbf{F} \cdot d\mathbf{r} \tag{10.1}$$

The work of $\mathbf{F}$ corresponding to a finite displacement of the particle from A_1 to A_2 (Fig. 10.10a) is denoted by $U_{1 \to 2}$ and is obtained by integrating (10.1) along the curve described by the particle:

$$U_{1 \to 2} = \int_{A_1}^{A_2} \mathbf{F} \cdot d\mathbf{r} \tag{10.11}$$

Using the alternative expression

$$dU = F\, ds \cos \alpha \tag{10.1'}$$

given in Sec. 10.2 for the elementary work dU, we can also express the work $U_{1 \to 2}$ as

$$U_{1 \to 2} = \int_{s_1}^{s_2} (F \cos \alpha)\, ds \tag{10.11'}$$

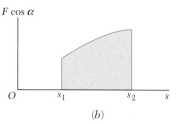

(a)

(b)

Fig. 10.10

where the variable of integration s measures the distance along the path traveled by the particle. The work $U_{1 \to 2}$ is represented by the area under the curve obtained by plotting $F \cos \alpha$ against s (Fig. 10.10b). In the case of a force $\mathbf{F}$ of constant magnitude acting in the direction of motion, formula (10.11') yields $U_{1 \to 2} = F(s_2 - s_1)$.

Recalling from Sec. 10.2 that the work of a couple of moment $\mathbf{M}$ during an infinitesimal rotation $d\theta$ of a rigid body is

$$dU = M\, d\theta \tag{10.2}$$

we express as follows the work of the couple during a finite rotation of the body:

$$U_{1 \to 2} = \int_{\theta_1}^{\theta_2} M \cdot d\theta \tag{10.12}$$

In the case of a constant couple, formula (10.12) yields

$$U_{1 \to 2} = M(\theta_2 - \theta_1)$$

Work of a Weight. It was stated in Sec. 10.2 that the work of the weight **W** of a body during an infinitesimal displacement of the body is equal to the product of W and the vertical displacement of the center of gravity of the body. With the y axis pointing upward, the work of **W** during a finite displacement of the body (Fig. 10.11) is obtained by writing

$$dU = -W \, dy$$

Integrating from A_1 to A_2, we have

$$U_{1 \to 2} = -\int_{y_1}^{y_2} W \, dy = Wy_1 - Wy_2 \tag{10.13}$$

or

$$U_{1 \to 2} = -W(y_2 - y_1) = -W \, \Delta y \tag{10.13'}$$

where Δy is the vertical displacement from A_1 to A_2. The work of the weight **W** is thus equal to *the product of W and the vertical displacement of the center of gravity of the body*. The work is *positive* when $\Delta y < 0$, that is, *when the body moves down*.

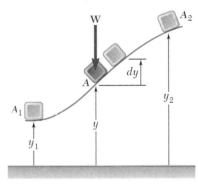

Fig. 10.11

Work of the Force Exerted by a Spring. Consider a body A attached to a fixed point B by a spring; it is assumed that the spring is undeformed when the body is at A_0 (Fig. 10.12a). Experimental evidence shows that the magnitude of the force **F** exerted by the spring on a body A is proportional to the deflection x of the spring measured from the position A_0. We have

$$F = kx \tag{10.14}$$

where k is the *spring constant*, expressed in N/m if SI units are used and expressed in lb/ft or lb/in. if U.S. customary units are used. The work of the force **F** exerted by the spring during a finite displacement of the body from $A_1(x = x_1)$ to $A_2(x = x_2)$ is obtained by writing

$$dU = -F \, dx = -kx \, dx$$

$$U_{1 \to 2} = -\int_{x_1}^{x_2} kx \, dx = \tfrac{1}{2}kx_1^2 - \tfrac{1}{2}kx_2^2 \tag{10.15}$$

Care should be taken to express k and x in consistent units. For example, if U.S. customary units are used, k should be expressed in lb/ft and x expressed in feet, or k in lb/in. and x in inches; in the first case, the work is obtained in ft $\cdot$ lb; in the second case, in in $\cdot$ lb. We note that the work of the force **F** exerted by the spring on the body is *positive* when $x_2 < x_1$, that is, *when the spring is returning to its undeformed position*.

Since Eq. (10.14) is the equation of a straight line of slope k passing through the origin, the work $U_{1 \to 2}$ of **F** during the displacement from A_1 to A_2 can be obtained by evaluating the area of the trapezoid shown in Fig. 10.12b. This is done by computing the values F_1 and F_2 and multiplying the base Δx of the trapezoid by its mean height $\tfrac{1}{2}(F_1 + F_2)$. Since the work of the force **F** exerted by the spring is positive for a negative value of Δx, we write

$$U_{1 \to 2} = -\tfrac{1}{2}(F_1 + F_2) \, \Delta x \tag{10.16}$$

Formula (10.16) is usually more convenient to use than (10.15) and affords fewer chances of confusing the units involved.

(a)

(b)

Fig. 10.12

Fig. 10.11 *(repeated)*

Fig. 10.12a *(repeated)*

*10.7. POTENTIAL ENERGY

Considering again the body of Fig. 10.11, we note from Eq. (10.13) that the work of the weight **W** during a finite displacement is obtained by subtracting the value of the function Wy corresponding to the second position of the body from its value corresponding to the first position. The work of **W** is thus independent of the actual path followed; it depends only upon the initial and final values of the function Wy. This function is called the *potential energy* of the body with respect to the *force of gravity* **W** and is denoted by V_g. We write

$$U_{1 \rightarrow 2} = (V_g)_1 - (V_g)_2 \qquad \text{with } V_g = Wy \qquad (10.17)$$

We note that if $(V_g)_2 > (V_g)_1$, that is, *if the potential energy increases* during the displacement (as in the case considered here), *the work $U_{1 \rightarrow 2}$ is negative.* If, on the other hand, the work of **W** is positive, the potential energy decreases. Therefore, the potential energy V_g of the body provides a measure of *the work which can be done* by its weight **W**. Since only the *change* in potential energy, and not the actual value of V_g, is involved in formula (10.17), an arbitrary constant can be added to the expression obtained for V_g. In other words, the level from which the elevation y is measured can be chosen arbitrarily. Note that potential energy is expressed in the same units as work, that is, in joules (J) if SI units are used† and in ft · lb or in · lb if U.S. customary units are used.

Considering now the body of Fig. 10.12a, we note from Eq. (10.15) that the work of the elastic force **F** is obtained by subtracting the value of the function $\frac{1}{2}kx^2$ corresponding to the second position of the body from its value corresponding to the first position. This function is denoted by V_e and is called the *potential energy* of the body with respect to the *elastic force* **F**. We write

$$U_{1 \rightarrow 2} = (V_e)_1 - (V_e)_2 \qquad \text{with } V_e = \frac{1}{2}kx^2 \qquad (10.18)$$

and observe that during the displacement considered, the work of the force **F** exerted by the spring on the body is negative and the potential energy V_e increases. We should note that the expression obtained for V_e is valid only if the deflection of the spring is measured from its undeformed position.

The concept of potential energy can be used when forces other than gravity forces and elastic forces are involved. It remains valid as long as the elementary work dU of the force considered is an *exact differential.* It is then possible to find a function V, called potential energy, such that

$$dU = -dV \qquad (10.19)$$

Integrating (10.19) over a finite displacement, we obtain the general formula

$$U_{1 \rightarrow 2} = V_1 - V_2 \qquad (10.20)$$

which expresses that *the work of the force is independent of the path followed and is equal to minus the change in potential energy.* A force which satisfies Eq. (10.20) is said to be a *conservative force.*‡

†See footnote, page 559.

‡A detailed discussion of conservative forces is given in Sec. 13.7 of *Dynamics*.

*10.8. POTENTIAL ENERGY AND EQUILIBRIUM

The application of the principle of virtual work is considerably simplified when the potential energy of a system is known. In the case of a virtual displacement, formula (10.19) becomes $\delta U = -\delta V$. Moreover, if the position of the system is defined by a single independent variable θ, we can write $\delta V = (dV/d\theta)\,\delta\theta$. Since $\delta\theta$ must be different from zero, the condition $\delta U = 0$ for the equilibrium of the system becomes

$$\frac{dV}{d\theta} = 0 \qquad (10.21)$$

In terms of potential energy, therefore, the principle of virtual work states that *if a system is in equilibrium, the derivative of its total potential energy is zero.* If the position of the system depends upon several independent variables (the system is then said to possess *several degrees of freedom*), the partial derivatives of V with respect to each of the independent variables must be zero.

Consider, for example, a structure made of two members AC and CB and carrying a load W at C. The structure is supported by a pin at A and a roller at B, and a spring BD connects B to a fixed point D (Fig. 10.13a). The constant of the spring is k, and it is assumed that the natural length of the spring is equal to AD and thus that the spring is undeformed when B coincides with A. Neglecting the friction forces and the weight of the members, we find that the only forces which do work during a displacement of the structure are the weight $\mathbf{W}$ and the force $\mathbf{F}$ exerted by the spring at point B (Fig. 10.13b). The total potential energy of the system will thus be obtained by adding the potential energy V_g corresponding to the gravity force $\mathbf{W}$ and the potential energy V_e corresponding to the elastic force $\mathbf{F}$.

Choosing a coordinate system with origin at A and noting that the deflection of the spring, measured from its undeformed position, is $AB = x_B$, we write

$$V_e = \tfrac{1}{2}kx_B^2 \qquad V_g = Wy_C$$

Expressing the coordinates x_B and y_C in terms of the angle θ, we have

$$x_B = 2l\sin\theta \qquad\qquad y_C = l\cos\theta$$
$$V_e = \tfrac{1}{2}k(2l\sin\theta)^2 \qquad V_g = W(l\cos\theta)$$
$$V = V_e + V_g = 2kl^2\sin^2\theta + Wl\cos\theta \qquad (10.22)$$

The positions of equilibrium of the system are obtained by equating to zero the derivative of the potential energy V. We write

$$\frac{dV}{d\theta} = 4kl^2\sin\theta\cos\theta - Wl\sin\theta = 0$$

or, factoring $l\sin\theta$,

$$\frac{dV}{d\theta} = l\sin\theta(4kl\cos\theta - W) = 0$$

There are therefore two positions of equilibrium, corresponding to the values $\theta = 0$ and $\theta = \cos^{-1}(W/4kl)$, respectively.†

---

†The second position does not exist if $W > 4kl$.

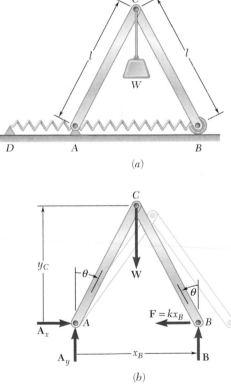

Fig. 10.13

*10.9. STABILITY OF EQUILIBRIUM

Consider the three uniform rods of length $2a$ and weight **W** shown in Fig. 10.14. While each rod is in equilibrium, there is an important difference between the three cases considered. Suppose that each rod is slightly disturbed from its position of equilibrium and then released: rod a will move back toward its original position, rod b will keep moving away from its original position, and rod c will remain in its new position. In case a, the equilibrium of the rod is said to be *stable;* in case b, it is said to be *unstable;* and, in case c, it is said to be *neutral.*

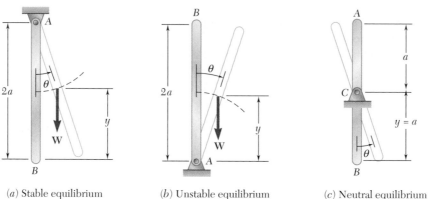

(a) Stable equilibrium   (b) Unstable equilibrium   (c) Neutral equilibrium

Fig. 10.14

Recalling from Sec. 10.7 that the potential energy V_g with respect to gravity is equal to Wy, where y is the elevation of the point of application of **W** measured from an arbitrary level, we observe that the potential energy of rod a is minimum in the position of equilibrium considered, that the potential energy of rod b is maximum, and that the potential energy of rod c is constant. Equilibrium is thus *stable, unstable,* or *neutral* according to whether the potential energy is *minimum, maximum,* or *constant* (Fig. 10.15).

That the result obtained is quite general can be seen as follows: We first observe that a force always tends to do positive work and thus to decrease the potential energy of the system on which it is applied. Therefore, when a system is disturbed from its position of equilibrium, the forces acting on the system will tend to bring it back to its original position if V is minimum (Fig. 10.15a) and to move it farther away if V is maximum (Fig. 10.15b). If V is constant (Fig. 10.15c), the forces will not tend to move the system either way.

Recalling from calculus that a function is minimum or maximum according to whether its second derivative is positive or negative, we can summarize the conditions for the equilibrium of a system with one degree of freedom (that is, a system the position of which is defined by a single independent variable θ) as follows:

$$\frac{dV}{d\theta} = 0 \qquad \frac{d^2V}{d\theta^2} > 0: \text{ stable equilibrium}$$

$$\frac{dV}{d\theta} = 0 \qquad \frac{d^2V}{d\theta^2} < 0: \text{ unstable equilibrium}$$

$$(10.23)$$

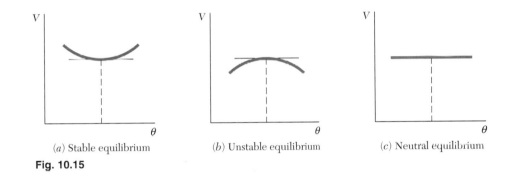

(a) Stable equilibrium   (b) Unstable equilibrium   (c) Neutral equilibrium

Fig. 10.15

If both the first and the second derivatives of V are zero, it is necessary to examine derivatives of a higher order to determine whether the equilibrium is stable, unstable, or neutral. The equilibrium will be neutral if all derivatives are zero, since the potential energy V is then a constant. The equilibrium will be stable if the first derivative found to be different from zero is of even order and positive. In all other cases the equilibrium will be unstable.

If the system considered possesses *several degrees of freedom,* the potential energy V depends upon several variables, and it is thus necessary to apply the theory of functions of several variables to determine whether V is minimum. It can be verified that a system with 2 degrees of freedom will be stable, and the corresponding potential energy $V(\theta_1, \theta_2)$ will be minimum, if the following relations are satisfied simultaneously:

$$\frac{\partial V}{\partial \theta_1} = \frac{\partial V}{\partial \theta_2} = 0$$

$$\left(\frac{\partial^2 V}{\partial \theta_1\, \partial \theta_2}\right)^2 - \frac{\partial^2 V}{\partial \theta_1^2}\frac{\partial^2 V}{\partial \theta_2^2} < 0 \qquad (10.24)$$

$$\frac{\partial^2 V}{\partial \theta_1^2} > 0 \qquad \text{or} \qquad \frac{\partial^2 V}{\partial \theta_2^2} > 0$$

SAMPLE PROBLEM 10.4

A 10-kg block is attached to the rim of a 300-mm-radius disk as shown. Knowing that spring BC is unstretched when $\theta = 0$, determine the position or positions of equilibrium, and state in each case whether the equilibrium is stable, unstable, or neutral.

SOLUTION

Potential Energy. Denoting by s the deflection of the spring from its undeformed position and placing the origin of coordinates at O, we obtain

$$V_e = \tfrac{1}{2}ks^2 \qquad V_g = Wy = mgy$$

Measuring θ in radians, we have

$$s = a\theta \qquad y = b \cos \theta$$

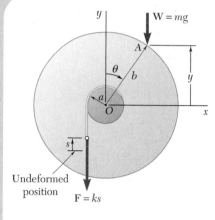

Substituting for s and y in the expressions for V_e and V_g, we write

$$V_e = \tfrac{1}{2}ka^2\theta^2 \qquad V_g = mgb \cos \theta$$
$$V = V_e + V_g = \tfrac{1}{2}ka^2\theta^2 + mgb \cos \theta$$

Positions of Equilibrium. Setting $dV/d\theta = 0$, we write

$$\frac{dV}{d\theta} = ka^2\theta - mgb \sin \theta = 0$$

$$\sin \theta = \frac{ka^2}{mgb}\theta$$

Substituting $a = 0.08$ m, $b = 0.3$ m, $k = 4$ kN/m, and $m = 10$ kg, we obtain

$$\sin \theta = \frac{(4 \text{ kN/m})(0.08 \text{ m})^2}{(10 \text{ kg})(9.81 \text{ m/s}^2)(0.3 \text{ m})}\theta$$

$$\sin \theta = 0.8699 \, \theta$$

where θ is expressed in radians. Solving numerically for θ, we find

$$\theta = 0 \qquad \text{and} \qquad \theta = 0.902 \text{ rad}$$
$$\theta = 0 \qquad \text{and} \qquad \theta = 51.7° \quad \blacktriangleleft$$

Stability of Equilibrium. The second derivative of the potential energy V with respect to θ is

$$\frac{d^2V}{d\theta^2} = ka^2 - mgb \cos \theta$$

$$= (4 \text{ kN/m})(0.08 \text{ m})^2 - (10 \text{ kg})(9.81 \text{ m/s}^2)(0.3 \text{ m}) \cos \theta$$

$$= 25.6 - 29.43 \cos \theta$$

For $\theta = 0$: $\qquad \dfrac{d^2V}{d\theta^2} = 25.6 - 29.43 \cos 0 = -3.83 < 0$

The equilibrium is unstable for $\theta = 0$ ◀

For $\theta = 51.7°$: $\qquad \dfrac{d^2V}{d\theta^2} = 25.6 - 29.43 \cos 51.7° = +7.36 > 0$

The equilibrium is stable for $\theta = 51.7°$ ◀

SOLVING PROBLEMS
ON YOUR OWN

In this lesson we defined the *work of a force during a finite displacement* and the *potential energy* of a rigid body or a system of rigid bodies. You learned to use the concept of potential energy to determine the *equilibrium position* of a rigid body or a system of rigid bodies.

1. The potential energy V of a system is the sum of the potential energies associated with the various forces acting on the system that *do work* as the system moves. In the problems of this lesson you will determine the following:

   a. Potential energy of a weight. This is the potential energy due to *gravity*, $V_g = Wy$, where y is the elevation of the weight W measured from some arbitrary reference level. Note that the potential energy V_g can be used with any vertical force **P** of constant magnitude directed downward; we write $V_g = Py$.

   b. Potential energy of a spring. The potential energy, $V_e = \frac{1}{2}kx^2$, is due to the *elastic* force exerted by a spring, where k is the constant of the spring and x is the deformation of the spring *measured from its unstretched position.*

Reactions at fixed supports, internal forces at connections, forces exerted by inextensible cords and cables, and other forces which do no work do not contribute to the potential energy of the system.

2. Express all distances and angles in terms of a single variable, such as an angle θ, when computing the potential energy V of a system. This is necessary, since the determination of the equilibrium position of the system requires the computation of the derivative $dV/d\theta$.

3. When a system is in equilibrium, the first derivative of its potential energy is zero. Therefore:

   a. To determine a position of equilibrium of a system, once its potential energy V has been expressed in terms of the single variable θ, compute its derivative and solve the equation $dV/d\theta = 0$ for θ.

   b. To determine the force or couple required to maintain a system in a given position of equilibrium, substitute the known value of θ in the equation $dV/d\theta = 0$ and solve this equation for the desired force or couple.

4. Stability of equilibrium. The following rules generally apply:

   a. Stable equilibrium occurs when the potential energy of the system is *minimum*, that is, when $dV/d\theta = 0$ and $d^2V/d\theta^2 > 0$ (Figs. 10.14a and 10.15a).

   b. Unstable equilibrium occurs when the potential energy of the system is *maximum*, that is, when $dV/d\theta = 0$ and $d^2V/d\theta^2 < 0$ (Figs. 10.14b and 10.15b).

   c. Neutral equilibrium occurs when the potential energy of the system is *constant*; $dV/d\theta$, $dV^2/d\theta^2$, and all the successive derivatives of V are then equal to zero (Figs. 10.14c and 10.15c).

See page 583 for a discussion of the case when $dV/d\theta$, $dV^2/d\theta^2$ but *not all* of the successive derivatives of V are equal to zero.

Problems

10.59 Using the method of Sec. 10.8, solve Prob. 10.29.

10.60 Using the method of Sec. 10.8, solve Prob. 10.30.

10.61 Using the method of Sec. 10.8, solve Prob. 10.31.

10.62 Using the method of Sec. 10.8, solve Prob. 10.33.

10.63 Using the method of Sec. 10.8, solve Prob. 10.34.

10.64 Using the method of Sec. 10.8, solve Prob. 10.35.

10.65 Using the method of Sec. 10.8, solve Prob. 10.36.

10.66 Using the method of Sec. 10.8, solve Prob. 10.38.

10.67 Show that the equilibrium is neutral in Prob. 10.1.

10.68 Show that the equilibrium is neutral in Prob. 10.2.

10.69 Two identical uniform rods, each of weight W and length L, are attached to pulleys that are connected by a belt as shown. Assuming that no slipping occurs between the belt and the pulleys, determine the positions of equilibrium of the system and state in each case whether the equilibrium is stable, unstable, or neutral.

Fig. P10.69

Fig. P10.70

10.70 Two uniform rods, each of mass m and length l, are attached to gears as shown. For the range $0 \leq \theta \leq 180°$, determine the positions of equilibrium of the system and state in each case whether the equilibrium is stable, unstable, or neutral.

10.71 Two uniform rods, each of mass m, are attached to gears of equal radii as shown. Determine the positions of equilibrium of the system and state in each case whether the equilibrium is stable, unstable, or neutral.

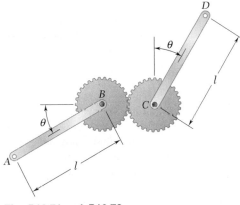

Fig. P10.71 and P10.72

10.72 Two uniform rods, AB and CD, are attached to gears of equal radii as shown. Knowing that $m_{AB} = 3.5$ kg and $m_{CD} = 1.75$ kg, determine the positions of equilibrium of the system and state in each case whether the equilibrium is stable, unstable, or neutral.

10.73 Using the method of Sec. 10.8, solve Prob. 10.39. Determine whether the equilibrium is stable, unstable, or neutral. (*Hint:* The potential energy corresponding to the couple exerted by a torsional spring is $\frac{1}{2}K\theta^2$, where K is the torsional spring constant and θ is the angle of twist.)

10.74 In Prob. 10.40, determine whether each of the positions of equilibrium is stable, unstable, or neutral. (See the hint for Prob. 10.73.)

10.75 Angle θ is equal to 45° after a block of mass m is hung from member AB as shown. Neglecting the mass of AB and knowing that the spring is unstretched when $\theta = 20°$, determine the value of m and state whether the equilibrium is stable, unstable, or neutral.

10.76 A block of mass m is hung from member AB as shown. Neglecting the mass of AB and knowing that the spring is unstretched when $\theta = 20°$, determine the value of θ corresponding to equilibrium when $m = 3$ kg. State whether the equilibrium is stable, unstable, or neutral.

10.77 A slender rod AB, of mass m, is attached to two blocks A and B which can move freely in the guides shown. Knowing that the spring is unstretched when $y = 0$, determine the value of y corresponding to equilibrium when $m = 12$ kg, $l = 750$ mm, and $k = 900$ N/m.

Fig. P10.75 and P10.76

Fig. P10.77

Fig. P10.78

Fig. *P10.79* and P10.80

Fig. P10.81 and P10.82

10.78 The slender rod AB of negligible mass is attached to two 4-kg blocks A and B that can move freely in the guides shown. Knowing that the constant of the springs is 160 N/m and that the unstretched length of each spring is 150 mm, determine the value of x corresponding to equilibrium.

10.79 A slender rod AB, of mass m, is attached to two blocks A and B that can move freely in the guides shown. The constant of the spring is k, and the spring is unstretched when AB is horizontal. Neglecting the weight of the blocks, derive an equation in θ, m, l, and k that must be satisfied when the rod is in equilibrium.

10.80 A slender rod AB, of mass m, is attached to two blocks A and B that can move freely in the guides shown. Knowing that the spring is unstretched when AB is horizontal, determine three values of θ corresponding to equilibrium when $m = 125$ kg, $l = 320$ mm, and $k = 15$ kN/mm. State in each case whether the equilibrium is stable, unstable, or neutral.

10.81 Spring AB of constant 10 lb/in. is attached to two identical drums as shown. Knowing that the spring is unstretched when $\theta = 0$, determine (a) the range of values of the weight W of the block for which a position of equilibrium exists, (b) the range of values of θ for which the equilibrium is stable.

10.82 Spring AB of constant 10 lb/in. is attached to two identical drums as shown. Knowing that the spring is unstretched when $\theta = 0$ and that $W = 40$ lb, determine the values of θ less than 180° corresponding to equilibrium. State in each case whether the equilibrium is stable, unstable, or neutral.

10.83 A slender rod AB of negligible weight is attached to two collars A and B that can move freely along the guide rods shown. Knowing that $\beta = 30°$ and $P = Q = 100$ lb, determine the value of the angle θ corresponding to equilibrium.

Fig. P10.83 and P10.84

10.84 A slender rod AB of negligible weight is attached to two collars A and B that can move freely along the guide rods shown. Knowing that $\beta = 30°$, $P = 40$ lb, and $Q = 10$ lb, determine the value of the angle θ corresponding to equilibrium.

10.85 and *10.86* Collar *A* can slide freely on the semicircular rod shown. Knowing that the constant of the spring is *k* and that the unstretched length of the spring is equal to the radius *r*, determine the value of θ corresponding to equilibrium when $m = 20$ kg, $r = 180$ mm, and $k = 3$ kN/m.

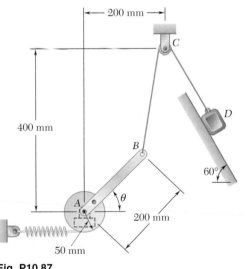

Fig. P10.85

Fig. *P10.86*

10.87 The 12-kg block *D* can slide freely on the inclined surface. Knowing that the constant of the spring is 480 N/m and that the spring is unstretched when $\theta = 0$, determine the value of θ corresponding to equilibrium.

Fig. P10.87

Fig. *P10.88*

10.88 Cable *AB* is attached to two springs and passes through a ring at *C*. Knowing that the springs are unstretched when $y = 0$, determine the distance *y* corresponding to equilibrium.

10.89 Rod *AB* is attached to a hinge at *A* and to two springs, each of constant *k*. If $h = 50$ in., $d = 24$ in., and $W = 160$ lb, determine the range of values of *k* for which the equilibrium of the rod is stable in the position shown. Each spring can act in either tension or compression.

10.90 Rod *AB* is attached to a hinge at *A* and to two springs, each of constant *k*. If $h = 30$ in., $k = 4$ lb/in., and $W = 40$ lb, determine the smallest distance *d* for which the equilibrium of the rod is stable in the position shown. Each spring can act in either tension or compression.

Fig. P10.89 and P10.90

10.91 The uniform plate *ABCD* of negligible mass is attached to four springs of constant *k* and is in equilibrium in the position shown. Knowing that the springs can act in either tension or compression and are undeformed in the given position, determine the range of values of the magnitude *P* of two equal and opposite horizontal forces **P** and −**P** for which the equilibrium position is stable.

Fig. P10.91

10.92 and **10.93** Two bars are attached to a single spring of constant *k* that is unstretched when the bars are vertical. Determine the range of values of *P* for which the equilibrium of the system is stable in the position shown.

Fig. *P10.92* **Fig. *P10.93*** **Fig. P10.94**

10.94 Bar *AC* is attached to a hinge at *A* and to a spring of constant *k* that is undeformed when the bar is vertical. Knowing that the spring can act in either tension or compression, determine the range of values of *P* for which the equilibrium of the system is stable in the position shown.

10.95 The horizontal bar *BEH* is connected to three vertical bars. The collar at *E* can slide freely on bar *DF*. Determine the range of values of *P* for which the equilibrium of the system is stable in the position shown when $a = 300$ mm, $b = 400$ mm, and $Q = 90$ N.

Fig. P10.95 and P10.96

10.96 The horizontal bar *BEH* is connected to three vertical bars. The collar at *E* can slide freely on bar *DF*. Determine the range of values of *Q* for which the equilibrium of the system is stable in the position shown when $a = 480$ mm, $b = 400$ mm, and $P = 600$ N.

***10.97** Bars *AB* and *BC*, each of length *l* and of negligible weight, are attached to two springs, each of constant *k*. The springs are undeformed and the system is in equilibrium when $\theta_1 = \theta_2 = 0$. Determine the range of values of *P* for which the equilibrium position is stable.

Fig. P10.97

***10.98** Solve Prob. 10.97 knowing that $l = 400$ mm and $k = 1.25$ kN/m.

***10.99** Bar *ABC* of length $2a$ and negligible weight is hinged at *C* to a drum of radius *a* as shown. Knowing that the constant of each spring is *k* and that the springs are undeformed when $\theta_1 = \theta_2 = 0$, determine the range of values of *P* for which the equilibrium position $\theta_1 = \theta_2 = 0$ is stable.

***10.100** Solve Prob. 10.99 knowing that $k = 10$ lb/in. and $a = 14$ in.

Fig. P10.99

Work of a force

The first part of this chapter was devoted to the *principle of virtual work* and to its direct application to the solution of equilibrium problems. We first defined the *work of a force* **F** *corresponding to the small displacement* d**r** [Sec. 10.2] as the quantity

$$dU = \mathbf{F} \cdot d\mathbf{r} \tag{10.1}$$

Fig. 10.16

obtained by forming the scalar product of the force **F** and the displacement d**r** (Fig. 10.16). Denoting respectively by F and ds the magnitudes of the force and of the displacement, and by α the angle formed by **F** and d**r**, we wrote

$$dU = F \, ds \cos \alpha \tag{10.1'}$$

The work dU is positive if $\alpha < 90°$, zero if $\alpha = 90°$, and negative if $\alpha > 90°$. We also found that the *work of a couple of moment* **M** acting on a rigid body is

$$dU = M \, d\theta \tag{10.2}$$

where $d\theta$ is the small angle expressed in radians through which the body rotates.

Virtual displacement

Considering a particle located at A and acted upon by several forces $\mathbf{F}_1, \mathbf{F}_2, \ldots, \mathbf{F}_n$ [Sec. 10.3], we imagined that the particle moved to a new position A' (Fig. 10.17). Since this displacement did not actually take place, it was referred to as a *virtual displacement* and denoted by $\delta\mathbf{r}$, while the corresponding work of the forces was called *virtual work* and denoted by δU. We had

$$\delta U = \mathbf{F}_1 \cdot \delta\mathbf{r} + \mathbf{F}_2 \cdot \delta\mathbf{r} + \cdots + \mathbf{F}_n \cdot \delta\mathbf{r}$$

Principle of virtual work

The *principle of virtual work* states that *if a particle is in equilibrium, the total virtual work δU of the forces acting on the particle is zero for any virtual displacement of the particle.*

The principle of virtual work can be extended to the case of rigid bodies and systems of rigid bodies. Since it involves *only forces which do work,* its application provides a useful alternative to the use of the equilibrium equations in the solution of many engineering problems. It is particularly effective in the case of machines and mechanisms consisting of connected rigid bodies, since the work of the reactions at the supports is zero and the work of the internal forces at the pin connections cancels out [Sec. 10.4; Sample Probs. 10.1, 10.2, and 10.3].

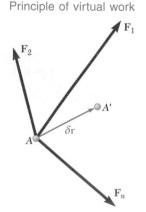

Fig. 10.17

In the case of *real machines* [Sec. 10.5], however, the work of the friction forces should be taken into account, with the result that the *output work will be less than the input work*. Defining the *mechanical efficiency* of a machine as the ratio

$$\eta = \frac{\text{output work}}{\text{input work}} \qquad (10.9)$$

we also noted that for an ideal machine (no friction) $\eta = 1$, while for a real machine $\eta < 1$.

In the second part of the chapter we considered the *work of forces corresponding to finite displacements* of their points of application. The work $U_{1 \rightarrow 2}$ of the force **F** corresponding to a displacement of the particle A from A_1 to A_2 (Fig. 10.18) was obtained by integrating the right-hand member of Eq. (10.1) or (10.1′) along the curve described by the particle [Sec. 10.6]:

$$U_{1 \rightarrow 2} = \int_{A_1}^{A_2} \mathbf{F} \cdot d\mathbf{r} \qquad (10.11)$$

or

$$U_{1 \rightarrow 2} = \int_{s_1}^{s_2} (F \cos \alpha)\, ds \qquad (10.11')$$

Similarly, the work of a couple of moment **M** corresponding to a finite rotation from θ_1 to θ_2 of a rigid body was expressed as

$$U_{1 \rightarrow 2} = \int_{\theta_1}^{\theta_2} M\, d\theta \qquad (10.12)$$

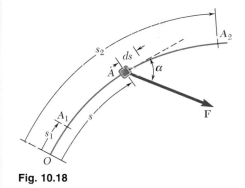

Fig. 10.18

The *work of the weight* **W** *of a body* as its center of gravity moves from the elevation y_1 to y_2 (Fig. 10.19) can be obtained by setting $F = W$ and $\alpha = 180°$ in Eq. (10.11′):

$$U_{1 \rightarrow 2} = -\int_{y_1}^{y_2} W\, dy = Wy_1 - Wy_2 \qquad (10.13)$$

The work of **W** is therefore positive *when the elevation y decreases*.

Fig. 10.19

Fig. 10.20

The *work of the force* **F** *exerted by a spring* on a body A as the spring is stretched from x_1 to x_2 (Fig. 10.20) can be obtained by setting $F = kx$, where k is the constant of the spring, and $\alpha = 180°$ in Eq. (10.11'):

$$U_{1\rightarrow 2} = -\int_{x_1}^{x_2} kx\ dx = \tfrac{1}{2}kx_1^2 - \tfrac{1}{2}kx_2^2 \qquad (10.15)$$

The work of **F** is therefore positive *when the spring is returning to its undeformed position.*

When the work of a force **F** is independent of the path actually followed between A_1 and A_2, the force is said to be a *conservative force,* and its work can be expressed as

Potential energy

$$U_{1\rightarrow 2} = V_1 - V_2 \qquad (10.20)$$

where V is the *potential energy* associated with **F**, and V_1 and V_2 represent the values of V at A_1 and A_2, respectively [Sec. 10.7]. The potential energies associated, respectively, with the *force of gravity* **W** and the *elastic force* **F** exerted by a spring were found to be

$$V_g = Wy \qquad \text{and} \qquad V_e = \tfrac{1}{2}kx^2 \quad (10.17,\ 10.18)$$

Alternative expression for the principle of virtual work

When the position of a mechanical system depends upon a single independent variable θ, the potential energy of the system is a function $V(\theta)$ of that variable, and it follows from Eq. (10.20) that $\delta U = -\delta V = -(dV/d\theta)\,\delta\theta$. The condition $\delta U = 0$ required by the principle of virtual work for the equilibrium of the system can thus be replaced by the condition

$$\frac{dV}{d\theta} = 0 \qquad (10.21)$$

When all the forces involved are conservative, it may be preferable to use Eq. (10.21) rather than to apply the principle of virtual work directly [Sec. 10.8; Sample Prob. 10.4].

Stability of equilibrium

This approach presents another advantage, since it is possible to determine from the sign of the second derivative of V whether the equilibrium of the system is *stable, unstable,* or *neutral* [Sec. 10.9]. If $d^2V/d\theta^2 > 0$, V is *minimum* and the equilibrium is *stable;* if $d^2V/d\theta^2 < 0$, V is *maximum* and the equilibrium is *unstable;* if $d^2V/d\theta^2 = 0$, it is necessary to examine derivatives of a higher order.

Review Problems

10.101 Derive an expression for the magnitude of the force **Q** required to maintain the equilibrium of the mechanism shown.

Fig. P10.101

10.102 The position of boom ABC is controlled by the hydraulic cylinder BD. For the loading shown, determine the force exerted by the hydraulic cylinder on pin B when $\theta = 70°$.

10.103 The position of boom ABC is controlled by the hydraulic cylinder BD. For the loading shown, determine the largest allowable value of the angle θ if the maximum force that the cylinder can exert on pin B is 25 kips.

Fig. P10.102 and *P10.103*

10.104 A vertical bar AD is attached to two springs of constant k and is in equilibrium in the position shown. Determine the range of values of the magnitude P of two equal and opposite vertical forces **P** and $-$**P** for which the equilibrium position is stable if (a) $AB = CD$, (b) $AB = 2CD$.

Fig. P10.104

595

30 mm

90 mm

80 N

C D

B

A

18 N·m 40 mm

E

F 40 mm

G

Fig. P10.105 and *P10.106*

10.105 Determine the vertical force **P** which must be applied at G to maintain the equilibrium of the linkage.

10.106 Determine the couple **M** which must be applied to member *DEFG* to maintain the equilibrium of the linkage.

10.107 Two uniform rods, each of mass m and length l, are attached to drums that are connected by a belt as shown. Assuming that no slipping occurs between the belt and the drums, determine the positions of equilibrium of the system and state in each case whether the equilibrium is stable, unstable, or neutral.

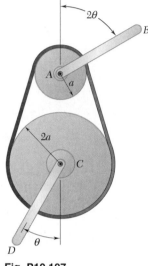

2θ

B

A

a

$2a$

C

D θ

Fig. P10.107

800 N 600 N

1.5 m 1.5 m

A B D E G

C F

1.8 m 2.4 m 1.2 m 1.8 m

Fig. P10.108 and P10.109

10.108 Using the method of virtual work, determine separately the force and the couple representing the reaction at A.

10.109 Using the method of virtual work, determine the reaction at D.

10.110 The slender rod *AB* is attached to a collar A and rests on a small wheel at C. Neglecting the radius of the wheel and the effect of friction, derive an expression for the magnitude of the force **Q** required to maintain the equilibrium of the rod.

P

B

a

C

l

θ

A

Q

Fig. P10.110

10.111 A load **W** of magnitude 100 lb is applied to the mechanism at C. Knowing that the spring is unstretched when $\theta = 15°$, determine the value of θ corresponding to equilibrium and check that the equilibrium is stable.

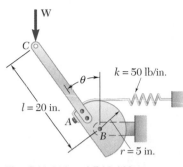

Fig. *P10.111* and P10.112

10.112 A load **W** of magnitude 100 lb is applied to the mechanism at C. Knowing that the spring is unstretched when $\theta = 30°$, determine the value of θ corresponding to equilibrium and check that the equilibrium is stable.

Computer Problems

10.C1 Knowing that $a = 20$ in., $b = 6$ in., $L = 20$ in., and $P = 25$ lb, use computational software to plot the force in member BD as a function of θ for values of θ from 30° to 150°. Determine the range of values of θ for which the absolute value of the force in member BD is less than 100 lb.

10.C2 Collars A and B are connected by the wire AB and can slide freely on the rods shown, where rod DE lies in the xy plane. Knowing that the length of the wire is 500 mm and that the mass of each collar is 2.5 kg, (*a*) use the principle of virtual work to express in terms of the distance x the magnitude of the force **P** required to maintain the system in equilibrium, (*b*) plot P as a function of x for 100 mm $\leq x \leq$ 400 mm.

Fig. P10.C1

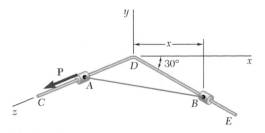

Fig. P10.C2

10.C3 A couple **M** is applied to crank *AB* to maintain the equilibrium of the engine system shown when a force **P** is applied to the piston. (*a*) Knowing that *b* = 48 mm and *l* = 150 mm, use computational software to plot the ratio *M/P* as a function of θ for values of θ from 0 to 180°. (*b*) Determine the value of θ for which the ratio *M/P* is maximum and the corresponding value of *M/P*.

Fig. P10.C3

10.C4 The constant of spring *AB* is *k*, and the spring is unstretched when θ = 0. (*a*) Neglecting the weight of member *BCD*, express the potential energy and its derivative *dV/dθ* in terms of *a*, *k*, *W*, and θ. (*b*) Using computational software and knowing that *W* = 600 N, *a* = 200 mm, and *k* = 15 kN/m, plot the potential energy and *dV/dθ* as functions of θ for values of θ from 0 to 165°. (*c*) Determine the values of θ for which the system is in equilibrium, and state in each case whether the equilibrium is stable, unstable, or neutral.

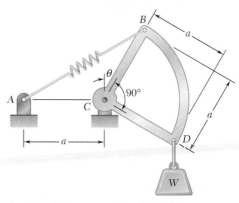

Fig. P10.C4

10.C5 A vertical load **W** is applied to the linkage at *B*. The constant of the spring is *k*, and the spring is unstretched when *AB* and *BC* are horizontal. For *l* = 11 in. and *k* = 12.5 lb/in., plot *W* as a function of θ for values of θ from 0 to 80°.

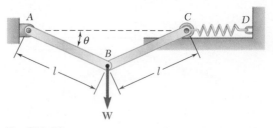

Fig. P10.C5

10.C6 The piston of the boom *ABC* is controlled by the hydraulic cylinder *BD*. Using computational software, plot the magnitude of the force exerted by the hydraulic cylinder on pin *B* as a function of θ for values of θ from 5° to 120°.

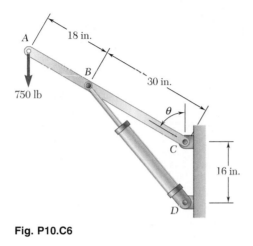

Fig. P10.C6

*10.C7** An 8-kg block is hung from the midpoint *C* of cable *AB* that is attached to two springs as shown. Knowing that the springs are unstretched when *y* = 0, plot the distance *y* corresponding to equilibrium as a function of the spring constant k_2 for 400 N/m ≤ k_2 ≤ 600 N/m.

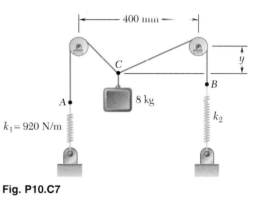

Fig. P10.C7

10.C8 A slender rod *AB* of weight *W* is attached to blocks *A* and *B* which can move freely in the guides shown. The spring constant is *k* and the spring is unstretched when *AB* is horizontal. For *l* = 20 in. and *k* = 1 lb/in., use computational software to determine the three values of the angle θ corresponding to equilibrium for values of *W* from 1 lb to 12 lb in 1 lb increments.

Fig. P10.C8

10.C9 A vertical load **W** is applied to the mechanism at *C*. Neglecting the weight of the mechanism, use computational software to plot θ as a function of *W* for values of *W* from 40 N to 480 N knowing that *k* = 5 kN/m and that the spring is unstretched when θ = 0.

Fig. P10.C9

A P P E N D I X

Fundamentals of Engineering Examination

Engineers are required to be licensed when their work directly affects the public health, safety, and welfare. The intent is to ensure that engineers have met minimum qualifications involving competence, ability, experience, and character. The licensing process involves an initial exam, called the *Fundamentals of Engineering Examination*, professional experience, and a second exam, called the *Principles and Practice of Engineering*. Those who successfully complete these requirements are licensed as a *Professional Engineer*. The exams are developed under the auspices of the *National Council of Examiners for Engineering and Surveying*.

The first exam, the *Fundamentals of Engineering Examination*, can be taken just before or after graduation from a four-year accredited engineering program. The exam stresses subject material in a typical undergraduate engineering program, including statics. The topics included in the exam cover much of the material in this book. The following is a list of the main topic areas, with references to the appropriate sections in this book. Also included are problems that can be solved to review this material.

Concurrent Force Systems (2.2–2.9; 2.12–2.14)
  Problems: 2.33, 2.35, 2.36, 2.38, 2.77, 2.86, 2.93, 2.94, 2.98
Vector Forces (3.4–3.11)
  Problems: 3.16, 3.17, 3.25, 3.31, 3.37, 3.40
Equilibrium in Two Dimensions (2.11; 4.1–4.7)
  Problems: 4.3, 4.11, 4.14, 4.21, 4.34, 4.37, 4.70, 4.81
Equilibrium in Three Dimensions (2.15; 4.8–4.9)
  Problems: 4.102, 4.104, 4.111, 4.114, 4.122, 4.126, 4.138,
    4.141, 4.146
Centroid of an Area (5.2–5.7)
  Problems: 5.6, 5.28, 5.38, 5.51, 5.53, 5.90, 5.96, 5.97, 5.119,
    5.131, 5.133
Analysis of Trusses (6.2–6.7)
  Problems: 6.2, 6.4, 6.33, 6.42, 6.43, 6.57
Equilibrium of Two-Dimensional Frames (6.9–6.11)
  Problems: 6.77, 6.82, 6.84, 6.92, 6.93
Shear and Bending Moment (7.3–7.6)
  Problems: 7.22, 7.25, 7.29, 7.33, 7.43, 7.47, 7.65,
    7.75

Fiction (8.2–8.5; 8.10)
Problems: 8.11, 8.15, 8.21, 8.32, 8.50, 8.52, 8.101, 8.104, 8.105, 8.106
Moments of Inertia (9.2–9.10)
Problems: 9.3, 9.31, 9.32, 9.33, 9.75, 9.78, 9.84, 9.90, 9.103, 9.105

PHOTO CREDITS

CHAPTER 1

Opener (both): © Bettmann/CORBIS; **p. 4:** © NASA; **p. 9:** © Robert Rathe Photography; **p. 10:** Photo copyright 1998, Lockheed Martin.

CHAPTER 2

Opener: © David Parker, 1997/Science Photo Library; **p. 37:** © H. David Seawell/CORBIS; **p. 57:** © D'Arazien/Getty Images.

CHAPTER 3

Opener: © Steve Kaufman/CORBIS; **p. 107:** © Elliot Eisenberg; **p. 112:** © Steve Hix/Getty Images; **p. 124:** © Steve Raymer/CORBIS; **p. 127:** © Elliot Eisenberg; **p. 128:** © Elliot Eisenberg.

CHAPTER 4

Opener: Courtesy Central Artery/Tunnel Project, Boston MA; **p. 159 (both):** © Elliot Eisenberg; **p. 160 (top):** © Lucinda Dowell; **p. 160 (middle and bottom):** Courtesy National Information Services for Earthquake Engineering, University of California, Berkeley; **p. 192 (both):** © Elliot Eisenberg.

CHAPTER 5

Opener: Courtesy Central Artery/Tunnel Project, Boston MA; **p. 220:** © Christies Images/SuperStock; **p. 238:** © Annie Griffiths Belt/CORBIS; **p. 248:** © Ghislain & Marie David de Lossy/Getty Images; **p. 249:** © Bruce Hands/Stock, Boston; **p. 259:** © NASA.

CHAPTER 6

Opener: Natalie Fobes/CORBIS; **p. 286:** Courtesy National Information Services for Earthquake Engineering, University of California, Berkeley; **p. 288:** © Ferdinand Beer; **p. 289:** © Sabina Dowell; **p. 293:** © EyeWire Collection/Getty Images; **p. 331:** Courtesy of Luxo ASA, Oslo, Norway.

CHAPTER 7

Opener (both): © Graeme-Peacock; **p. 355:** © Sabina Dowell; **p. 362:** © Elliot Eisenberg; **p. 383:** © Michael S. Yamashita/CORBIS; **p. 386:** © Brian Yarvin/Photo Researchers; **p. 395:** Steve Kahn/Getty Images.

CHAPTER 8

Opener: Richard Cummins/CORBIS; **p. 416:** © Chuck Savage/ CORBIS; **p. 431:** © Ted Spiegel/CORBIS; **p. 451:** © Adam Woolfitt/ CORBIS; **p. 452:** © Elliot Eisenberg.

CHAPTER 9

Opener: © Steve Dunwell/Getty Images; **p. 484:** © Ed Eckstein/ CORBIS; **p. 513:** Courtesy of Caterpillar Engine Division.

CHAPTER 10

Opener: Courtesy of Stamm Manufacturing, Fort Pierce, FL; **p. 559:** Courtesy of Altec Industries; **p. 563:** Courtesy of DE-STA-CO Industries.

Index

Absolute system of units, 6
Acceleration of gravity, 4
Accuracy, numerical, 13
Action and reaction, 4, 285
Addition
    of couples, 110
    of forces
        concurrent: in a plane, 20, 30
            in space, 49
        nonconcurrent, 122
    of vectors, 18
Angle
    of kinetic friction, 415
    lead, 432
    of repose, 416
    of static friction, 415
Archimedes, 2
Aristotle, 2
Associative property for sums of vectors, 20
Axis
    of symmetry, 224
    of a wrench, 126
Axle friction, 440–441

Ball supports, 192, 193
Ball-and-socket supports, 192, 193
Band brakes, 451
Beams, 361–379
    combined, 361
    loading of, 361
    span of, 362
    supports of, 361–362
    types of, 361
Bearings, 192, 440–441
    collar, 442
    end, 442
    journal, 440–441
    thrust, 442–443
Belt drive, 451–452
Belt friction, 450–452
Bending, 354
Bending moment, 354, 362–374

Bending moment diagram, 364
Bernoulli, Jean, 558
Bollard, 453

Cables, 383–396
    with concentrated loads, 383–384
    with distributed loads, 384–385
    parabolic, 385–386
    reactions at, 160, 192
    span of, 386, 396
Calculators
    accuracy, 13
    use of, 22, 23, 29, 31, 38, 48n.
Catenary, 394–396
Center
    of gravity, 220–221, 259–260
    of pressure, 249, 473–474
    of symmetry, 224
Centimeter, 7
Centroidal axes, principal, 498
Centroids, 220–227, 236–239, 248–249, 259–262
    of areas and lines, 222–227
    of common shapes: of areas, 225
        of lines, 226
        of volumes, 261
    of composite areas and lines, 227
    of composite volumes, 262
    determination of, by integration, 236–238, 262
    of volumes, 259–262
Coefficient
    of kinetic friction, 412–414
    of rolling resistance, 444
    of static friction, 413
Collar bearings, 442
Commutative property
    for scalar products, 93–94
    for sums of vectors, 18
Complete constraints, 164
Components
    of force, 21, 27–28, 45–48
    of moment, 83–84
    of vector product, 79–80

Composite areas
    centroids of, 227
    moments of inertia of, 484–485
Composite bodies
    centroids of, 262
    moments of inertia of, 516
Composition of forces. *See* Addition, of forces
Compound trusses, 304
Compression, 77, 287, 354
Concurrent forces, 20
Conservative force, 580
Constraining forces, 159
Constraints, 563
    complete, 164
    improper, 165, 192
    partial, 164, 192
Coplanar forces, 20
Coplanar vectors, 20
Coulomb friction, 412
Counters, 313
Couple vector, 111
Couples, 107–111
    addition of, 110
    equivalent, 108–110
Cross product. *See* Vector product
Customary units, U.S., 9–13

D'Alembert, Jean, 2
Decimeter, 7
Degrees of freedom, 581, 583
Density, 222*n*., 515
Determinant form
    for mixed triple product, 95–96
    for moment of a force: about an axis, 97–98
        about a point, 84
    for vector product, 80
Determinant reactions, 164
Determinant structures, 318
Determinant trusses, 305
Diagram
    bending-moment, 364
    free-body, 36–67, 159
    shear, 364
Differential elements
    for centroids: of areas, 237
        of volumes, 262
    for moments of inertia: of areas, 475
        of masses, 516

Direction cosines, 47
Direction of a force, 16
Disk clutches, 442
Disk friction, 442
Displacement, 558, 559
    virtual, 561–565
Distributed forces, 220, 473
Distributed loads, 248, 361
Distributive property
    for scalar products, 93–94
    for vector products, 80
Dot product, 98
Dry friction, 412
Dynamics, definition of, 2

Efficiency, 564–565
Ellipsoid of inertia, 532–533
End bearings, 442
Energy, potential, 580–581
Equations of equilibrium
    for a particle, 36–37, 57
    for a rigid body, 158
Equilibrium
    equations (*see* Equations of equilibrium)
    neutral, 582–583
    of a particle: in a plane, 35–36
        in space, 57
    of a rigid body: in a plane, 158–192
        in space, 192
    stability of, 582–583
Equipollent systems of vectors, 124
Equivalent forces, 75–77
Equivalent systems of forces, 123–124
External forces, 74

First moments
    of areas and lines, 223–224
    of volumes, 260
Fixed supports, 160–161, 192, 193
Flexible cords. *See* Cables
Foot, 9
Force, 3
    conservative, 580
    external, 74
    in a member, 289
    on a particle: in a plane, 16–30
        in space, 45–57
    on a rigid body, 74–135

Force systems, 122–133
Force-couple system, 111
Frames, 315–318
Free-body diagram
    of a particle, 36–37
    of a rigid body, 159
Freedom, degrees of, 581
Friction, 412–452
    angles of, 415
    belt, 450–452
    circle of, 442
    coefficient of, 412–414
    Coulomb, 412
    dry, 412
    kinetic, 413
    laws of, 412–414
    static, 412
    wheel, 443–444
Frictionless surfaces, 160, 192, 193

Geometric instability, 165n.
Gram, 6
Gravitation
    constant of, 4
    Newton's law of, 4
Gravitational system of units, 9
Gravity
    acceleration of, 4
    center of, 220–221, 259–260
Guldinus, theorems of, 238–239
Gyration, radius of, 476, 513

Hamilton, Sir William R., 2
Hinges, 160, 192, 193
Hydrostatic forces, 249, 473–474

Ideal machines, 564–565
Improper constraints, 165, 192
Inch, 9
Inclined axes, moments of inertia,
    498, 531, 533
Indeterminate reactions, 164, 192
Indeterminate structures, 318
Indeterminate trusses, 305
Inertia
    ellipsoid of, 532–533

Inertia—*Cont.*
    moments of (*see* Moments of inertia)
    principle axes of: for an area, 498
        for a mass, 532–535
    products of: of areas, 497–498
        of masses, 531–532
        parallel-axis theorem for, 498, 532
Input forces, 331
Input work, 564–565
Internal forces, 75
    in members, 354
    in structures, 285
International system of units, 5–8

Jacks, 431–432
Joints, method of, 285–293
Joule (unit), 559n.
Journal bearings, 440–441

Kilogram, 5
Kilometer, 6
Kilonewton, 6
Kilopound, 9
Kinetic friction, 413
Kip, 9

Lagrange, J. L., 2
Laws
    of friction, 412–414
    Newton's (*see* Newton's law)
Lead angle, 432
Lead of a screw, 432
Line of action, 16, 75–77
Links, 160
Liter, 8
Loading of beams, 361

Machines, 331
    ideal, 564–565
    real, 564–565
Magnitude of a force, 16
Mass, 3
Mass moments of inertia, 512–533
Mass products of inertia, 531–532
Maxwell's diagram, 291

Mechanical efficiency, 564–565
Mechanics
  definition of, 2
  Newtonian, 2
  principles of, 2–5
Megagram, 6
Meter, 5
Metric ton, 6n.
Metric units, 5–7
Mile, 9
Millimeter, 6
Mixed triple product, 95
  determinant form for, 96
Mohr's circle, 506–507
Moment
  bending, 354, 362–374
  of a couple, 107
  first, 223–224, 260
Moment of a force
  about an axis, 97–98
    about a point, 81–82
  second, 473–474
Moments of inertia, 473–535
  of areas, 473–507
    parallel-axis theorem for, 483–484
  of common geometric shapes, 485, 517
  of composite areas, 484–485
  of composite bodies, 516
  determination of, by integration, 474–475, 516
  inclined axes, 498, 531, 533
  of masses, 512–533
    parallel-axis theorem for, 514
  polar, 475
  principal, 498, 532–535
  rectangular, 474
  of thin plates, 515–516
Motion, Newton's laws of. *See* Newton's law
Multiforce members, 315, 354–355

Negative vector, 18
Neutral equilibrium, 582–583
Newton, Sir Isaac, 2–4
Newton (unit), 6
Newtonian mechanics, 2
Newton's law
  of gravitation, 4
  of motion: first, 3, 36
    second, 4
    third, 3, 285

Nonrigid truss, 305
Numerical accuracy, 13

Output forces, 331
Output work, 564–567
Overrigid trusses, 305

Pappus, theorems of, 238–239
Parabolic cable, 385–386
Parallel-axis theorem
  for moments of inertia: of areas, 483–484
    of masses, 514
  for products of inertia: of areas, 498
    of masses, 532
Parallelogram law, 3, 17
Partial constraints, 164, 192
Particles, 3, 16
  equilibrium of: in a plane, 35–36
    in space, 57
  free-body diagram of, 36–37
Pascal (unit), 249n.
Pin-and-bracket supports, 192, 193
Pins, 160–288
Pitch
  of a thread, 432
  of a wrench, 127–128
Plane of symmetry, 260
Point of application of a force, 16, 75
Polar moment of inertia, 475
Pole, 475
Polygon rule, 20
Positive vector, 81
Potential energy, 580–581
Pound force, 9
Pound mass, 11
Pressure, center of, 249, 473–474
Principal axes of inertia
  of areas, 498
  of masses, 532–533
Principal moments of inertia
  of areas, 498
  of masses, 533
Principle
  of transmissibility, 3, 74, 75–77
  of virtual work, 558, 561–564
Principles of mechanics, 2–5
Problems solution, method of, 11–13

Product
  of inertia: of area, 497–498
    of mass, 531–532
  mixed triple, 95–96
  scalar, 93–95
  of a scalar and a vector, 20
  vector, 77–79

Radius of gyration, 476, 513
Reactions at supports and connections, 160, 192, 193
Real machines, 564–565
Rectangular components
  of force, 27–28, 39, 49
  of moment, 83–84
  of vector product, 79–80
Reduction of a system of forces, 122–128
Redundant members, 305
Relativity, theory of, 2
Repose, angle of, 416
Resolution of a force
  into components: in a plane, 21, 27–28
    in space, 45–49
  into a force and a couple, 111–112
Resultant of forces, 16–17, 49, 122–123. *See also*
  Addition, of forces: Addition, of vectors
Revolution
  body of, 238, 516
  surface of, 238
Right-hand rule, 77, 81
Rigid body, 3, 74
  equilibrium of: in a plane, 158–193
    in space, 192
  free-body diagram of, 159
Rigid truss, 288, 305
Rollers, 160, 192, 193
Rolling resistance, 443–444
  coefficient of, 444
Rough surfaces, 160, 192, 193

Sag, 386, 396
Scalar components, 28
Scalar product, 93–95
Scalars, 17
Screws, 431–432
Second, 5, 9
Second moment, 473–474
Sections, method of, 303–304
Self-locking screws, 432
Sense of a force, 16

Shear, 354, 362–374
Shear diagram, 364
SI units, 5–8
Significant figures, 13
Simple trusses, 288, 293
Slug, 9
Space, 2
Space truss, 293
Specific weight, 222, 251, 260
Spring
  force exerted by, 579
  potential energy, 580
Spring constant, 579
Square-threaded screws, 431–432
Stable equilibrium, 582–583
Static friction, 412
  angle of, 415
  coefficient of, 413
Statically determinate reactions, 164
Statically determinate structures, 318
Statically determinate trusses, 305
Statically indeterminate reactions, 164, 192
Statically indeterminate structures, 318
Statically indeterminate trusses, 305
Statics, definition of, 2
Structural shapes, properties of, 486–487
Structures
  analysis of, 285–330
  determinate, 318
  indeterminate, 318
  internal forces in, 285, 354, 355
  two-dimensional, 82, 160
Submerged surfaces, forces on,
  249, 474
Subtraction of vectors, 19
Supports
  ball, 192–193
  ball-and-socket, 192–193
  of beams, 361–362
  reactions at, 160, 161, 192
Surface
  frictionless, 160, 192, 193
  of revolution, 238
  rough, 160, 192, 193
  submerged forces on, 249, 474
Suspension bridges, 385
Symmetry
  axis of, 224
  center of, 224
  plane of, 260

Systems
    of forces, 122–128
    of units, 5–13

Tension, 77, 287, 354
Three-force body, 184
Thrust bearings, 442–443
Time, 2
Toggle vise, analysis of, 562–564
Ton
    metric, 6n.
    U.S., 9
Transfer formula. *See* Parallel-axis theorem
Transmissibility, principle of, 3, 75–76
Triangle rule, 19
Trusses, 286–305
    compound, 304
    determinate, 305
    indeterminate, 305
    overrigid, 305
    rigid, 288, 305
    simple, 288, 293
    space, 293
    typical, 287
Two-dimensional structures, 82, 160
Two-force body, 183–184

Unit vectors, 27–28, 47
Units, 5–13. *See also specific systems of units*
Universal joints, 192, 193, 342
Unstable equilibrium, 582–583
Unstable rigid bodies, 164n.
U.S. customary units, 9–13

V belts, 452
Varignon's theorem, 83
Vector addition, 18–20
Vector components, 30
Vector product, 77–79
    determinant form for, 80
    rectangular components of, 79–80
Vector subtraction, 19
Vectors, 17
    bound, fixed, 17
    coplanar, 20
    couple, 110
    free, 17
    sliding, 18, 76
Virtual displacement, 561–565
Virtual work, 561
    principle of, 558, 561–564

Wedges, 431
Weight, 4
Wheel friction, 443–444
Wheels, 192, 193, 443
Work
    of a couple, 560–561, 578
    of a force, 558–559
    of a force exerted by spring, 579
    of forces on a rigid body, 560–562
    input and output, 564–565
    virtual, 561
    of a weight, 579
Wrench, 127–128

Zero-force member, 292

Answers to Problems

Answers to problems with a number set in straight type are given on this and the following pages. Answers to problems set in italic are not listed. Answers to computer problems are given at www.mhhe.com/beerjohnston7.

CHAPTER 2

2.1	8.40 kN ⬉ 19.0°.
2.2	575 N ⬈ 67.0°.
2.3	37 lb ⬉ 76.0°.
2.5	(a) 107.6 N. (b) 75.0 N.
2.6	(a) 89.2 N. (b) 55.8 N.
2.7	(a) 113.3°. (b) 71.5 lb.
2.8	(a) 22.7°. (b) 69.1 lb.
2.9	(a) 489 N. (b) 738 N.
2.10	(a) 43.5°. (b) 513 N.
2.13	(a) 10.0 lb ↓. (b) 17.32 lb.
2.14	(a) 206 N →. (b) 295 N.
2.15	25.2 lb ⬆ 8.13°.
2.16	8.38 kN ⬉ 18.76°.
2.17	576 N ⬈ 66.4°.
2.18	37.2 lb ⬉ 76.6°.
2.21	(20 kN) 15.32 kN, 12.86 kN; (30 kN) −10.26 kN, 28.2 kN; (42 kN) −39.5 kN, 14.36 kN.
2.22	(40 lb) −30.6 lb, −25.7 lb; (60 lb) 30.0 lb, −52.0 lb; (80 lb) 72.5 lb, 33.8 lb.
2.23	(204 lb) −96.0 lb, 180.0 lb; (212 lb) 112.0 lb, 180.0 lb; (400 lb) −320 lb, −240 lb.
2.25	(a) 176.2 N. (b) 113.3 N.
2.26	(a) 1674 N. (b) 1371 N.
2.27	(a) 339 lb. (b) 218 lb ↓.
2.28	(a) 25.9 lb. (b) 6.70 lb.
2.31	530 N ∡ 38.5°.
2.32	65.2 kN ⬆ 58.2°.
2.33	84.2 lb ⬉ 31.4°.
2.34	327 lb ⬆ 21.5°.
2.35	1007 N ∡ 4.91°.
2.36	1019 N ∡ 26.1°.
2.39	(a) 40.3°. (b) 252 lb.
2.40	(a) 116.0 lb. (b) 60.0 lb.
2.41	(a) 7.29 kN. (b) 9.03 kN.
2.43	(a) 441 lb. (b) 487 lb.
2.44	(a) 5.23 kN. (b) 0.503 kN.
2.45	(a) 169.7 lb. (b) 348 lb.
2.46	(a) 2.81 kN. (b) 2.66 kN.
2.49	F_C = 332 lb; F_D = 368 lb.
2.50	F_B = 672 lb; F_D = 232 lb.
2.53	(a) 750 N. (b) 1190 N.
2.54	0 ≤ W ≤ 609 N.
2.55	(a) 218 kN. (b) 15.11 kN.
2.56	(a) 33.1 kN. (b) 141.7 kN.
2.57	(a) 66.2 N. (b) 208 N.
2.58	(a) 62.8 N. (b) 758 mm.
2.59	(a) 55.0°. (b) 2.41 kN.
2.60	4.14 m.
2.61	(a) 270 lb. (b) 77.5°.

2.64	(a) 90.0°. (b) 68.9 lb.
2.65	40.0 N.
2.66	415 mm.
2.68	(b) 916 N. (d) 687 N.
2.69	149.1 lb ⬆ 32.3° or 274 lb ⬈ 32.3°.
2.70	(a) 24.2°. (b) 146.0 lb.
2.71	(a) 2.30 kN. (b) 3.53 kN.
2.73	(a) +157.0 lb, +100.0 lb, −73.2 lb. (b) 38.3°, 60.0°, 111.5°.
2.75	(a) +2.46 kN, −2.70 kN, +2.07 kN. (b) 54.1°, 130.0°, 60.5°.
2.76	(a) −1.076 kN, −2.55 kN, +2.31 kN. (b) 107.4°, 135.0°, 50.1°.
2.77	(a) 576 N. (b) 67.5°, 30.0°, 108.7°.
2.78	(a) 200 N. (b) 112.5°, 30.0°, 108.7°.
2.79	(a) −30.0 lb, +103.9 lb, +52.0 lb. (b) 104.5°, 30.0°, 64.3°.
2.81	F = 900 lb; θ_x = 27.3°, θ_y = 73.2°, θ_z = 110.8°.
2.82	F = 1300 N; θ_x = 72.1°, θ_y = 157.4°, θ_z = 76.7°.
2.83	(a) 135.0°. (b) F_x = 121.7 N, F_z = 158.5 N; F = 283 N.
2.86	(a) F_y = −359 lb, F_z = −437 lb. (b) θ_x = 70.5°, θ_y = 126.7°.
2.87	+400 N, +2000 N, −500 N.
2.89	+96.0 lb, +144.0 lb, −108.0 lb.
2.90	−75.0 lb, +144.0 lb, +108.0 lb.
2.91	+120.0 N, −140.0 N, −120.0 N.
2.94	701 lb; θ_x = 77.4°, θ_y = 37.4°, θ_z = 55.5°.
2.95	1826 N; θ_x = 48.2°, θ_y = 116.6°, θ_z = 53.4°.
2.96	1826 N; θ_x = 50.6°, θ_y = 117.6°, θ_z = 51.8°.
2.97	379 N; θ_x = 129.3°, θ_y = 40.0°, θ_z = 96.1°.
2.98	(a) 1807 lb. (b) 2.47 kips; θ_x = 90.0°, θ_y = 139.2°, θ_z = 49.2°.
2.100	T_{AB} = 245 lb; T_{AD} = 257 lb.
2.101	279 N.
2.103	239 lb ↑.
2.104	215 lb ↑.
2.105	8.41 kN.
2.107	4.63 kN.
2.108	T_{AB} = 2.63 kN; T_{AC} = 3.82 kN; T_{AD} = 2.43 kN.
2.109	1.603 lb.
2.111	6.66 kN ↑.
2.112	8.81 kN ↑.
2.113	211 lb.
2.115	T_{AD} = T_{CD} = 135.1 N; T_{BD} = 46.9 N.
2.116	T_{AB} = 4.33 kN; T_{AC} = 2.36 kN; T_{AD} = 2.37 kN.
2.117	T_{AB} = 115.9 lb; T_{AC} = 12.78 lb; T_{AD} = 121.9 lb.
2.119	T_{BE} = 0.299 lb; T_{CF} = 1.002 lb; T_{DG} = 1.117 lb.
2.121	T_{AB} = 551 N; T_{AC} = 503 N.
2.122	T_{AB} = 306 N; T_{AC} = 756 N.
2.123	T_{AB} = 46.5 lb; T_{AC} = 34.2 lb; T_{ADE} = 110.8 lb.

2.124 (a) 50.0 lb. (b) 162.0 lb. (c) 468 lb.
2.125 $P = 160.0$ N; $Q = 240$ N.
2.127 (a) 2.27 kN. (b) 1.963 kN.
2.129 (a) 523 lb. (b) 428 lb.
2.131 (a) 454 N. (b) 1202 N.
2.132 (a) 52.0 lb. (b) 45.0 lb.
2.134 $F_C = 6.40$ kN; $F_D = 4.80$ kN.
2.135 $F_B = 15.00$ kN; $F_C = 8.00$ kN.
2.137 68.6 in.
2.139 748 N; $\theta_x = 120.1°$, $\theta_y = 52.5°$, $\theta_z = 128.0°$.
2.140 (a) 368 lb →. (b) 213 lb.

CHAPTER 3

3.1   1.788 N · m ↖.
3.2   14.74 N ⤢ 35.2°.
3.3   50.6° or 59.1°.
3.4   12.21 lb · in. ↖.
3.6   (a) 80.0 N · m ↙. (b) 205 N. (c) 177.8 N ⦨ 20.0°.
3.7   (a) 283 lb · ft ↖. (b) 36.0 lb ⦨ 55.9°.
3.9   27.9 N · m ↖.
3.10   30.8 N · m ↖.
3.11   (a) 7600 lb · in. ↖. (b) 7600 lb · in. ↖.
3.12   250 lb.
3.16   84.0 mm.
3.17   (a) $(-4\mathbf{i} + 7\mathbf{j} + 10\mathbf{k})/\sqrt{165}$.
       (b) $(-2\mathbf{i} + 2\mathbf{j} - 3\mathbf{k})/\sqrt{17}$.
3.19   (a) $-(8$ N · m$)\mathbf{i} - (7$ N · m$)\mathbf{j} - (18$ N · m$)\mathbf{k}$.
       (b) $(17$ N · m$)\mathbf{i} + (4$ N · m$)\mathbf{j} + (31$ N · m$)\mathbf{k}$.
       (c) 0.
3.21 $(5.24$ kN · m$)\mathbf{i} - (3.75$ kN · m$)\mathbf{k}$.
3.23 $(42.2$ lb · in.$)\mathbf{i} + (40.6$ lb · in.$)\mathbf{j} - (50.4$ lb · in.$)\mathbf{k}$.
3.25   (a) $-(1840$ lb · ft$)\mathbf{i} - (368$ lb · ft$)\mathbf{k}$.
       (b) $-(1840$ lb · ft$)\mathbf{i} + (2160$ lb · ft$)\mathbf{j} + (2740$ lb · ft$)\mathbf{k}$.
3.26 $(2.98$ N · m$)\mathbf{i} - (3.16$ N · m$)\mathbf{j} - (2.28$ N · m$)\mathbf{k}$.
3.27   5.49 m.
3.28   4.45 m.
3.30   1.141 m.
3.32   10.96 ft.
3.33   10.96 ft.
3.34   2.08 m.
3.35 $\mathbf{P} \cdot \mathbf{Q} = 17$; $\mathbf{P} \cdot \mathbf{S} = 9$; $\mathbf{Q} \cdot \mathbf{S} = -82$.
3.37   43.6°.
3.39   (a) 134.1°. (b) −230 N.
3.40   (a) 65.0°. (b) 188.3 N.
3.41   (a) 71.1°. (b) 9.73 N.
3.43   (a) 187 in^3. (b) 46 in^3.
3.44   1.700.
3.45 $M_x = -30.7$ N · m; $M_y = 12.96$ N · m;
       $M_z = -2.38$ N · m.
3.46 $M_x = -25.1$ N · m; $M_y = 10.60$ N · m; $M_z = 39.9$ N · m.
3.47   63.0 N.
3.48   49.0 N.
3.49   4.00 ft.
3.51 $P = 40.6$ lb; $\phi = 9.93°$; $\theta = 48.9°$.
3.53   365 lb · in.
3.55   172.9 N · m.
3.56   −187.3 N · m.
3.57 $aP/\sqrt{2}$.
3.58   (b) $a/\sqrt{2}$.
3.59   −18.46 kip · ft.
3.62   4.79 in.

3.63   17.50 in.
3.64   2.75 m.
3.66   295 mm.
3.68   (a) 8.40 N · m ↖. (b) 0.700 m. (c) 31.6°.
3.69   (a) 12.00 lb. (b) 10.29 lb. (c) 6.36 lb.
3.70   (a) 8.62 N · m ↖. (b) 53.1° ⦨. (c) 16.11 N.
3.71   (a) 48.0 N · m ↙. (b) 243 mm.
3.72 $M = 10.00$ N · m; $\theta_x = 90.0°$, $\theta_y = 143.1°$, $\theta_z = 126.9°$.
3.73 $M = 162.3$ N · m; $\theta_x = 67.5°$, $\theta_y = 23.6°$, $\theta_z = 83.2°$.
3.74 $M = 68.8$ lb · in.; $\theta_x = 76.4°$, $\theta_y = 13.56°$, $\theta_z = 90.0°$.
3.76 $M = 350$ N · m; $\theta_x = 33.1°$, $\theta_y = 64.8°$, $\theta_z = 69.9°$.
3.78   (a) 1724 N · m ↙. (b) 958 N · m ↙.
3.79   (a) $\mathbf{F}_B = 20$ lb ←; 100 lb · in. ↖. (b) $\mathbf{F}_C = 25$ lb ↓;
       $\mathbf{F}_D = 25$ lb ↑.
3.81 $\mathbf{F}_B = 100.0$ N ⤡ 30.0°; $\mathbf{F}_C = 140.0$ N ⤡ 30.0°.
3.83 $\mathbf{F}_A = 93.8$ lb ⤧ 60.0°; $\mathbf{F}_C = 156.3$ lb ⤧ 60.0°.
3.84   (a) $\mathbf{F} = (60.0$ lb$)\mathbf{k}$; $\mathbf{M} = (70.0$ lb · ft$)\mathbf{j}$.
       (b) $\mathbf{F} = (60.0$ lb$)\mathbf{k}$; 1.167 ft from A along AB.
       (c) 23.3 lb.
3.85   (a) $\mathbf{F} = 800$ N ↓; $d = 3.00$ m.
       (b) $\mathbf{F} = 800$ N ↓; $d = 0$.
3.86   (a) $\mathbf{F}_A = 110.0$ N ⦨ 20.0°; $\mathbf{M}_A = 3.25$ N · m ↖.
       (b) $\mathbf{F}_E = 110.0$ N ⦨ 20.0°; 31.5 mm below A.
3.88   (a) 10.00 lb; 5.61 in. to the left of B. (b) 77.0° or −13.03°.
3.90 $\mathbf{F} = -(270$ lb$)\mathbf{i}$; $\mathbf{M} = (648$ lb · in.$)\mathbf{j} - (1080$ lb · in.$)\mathbf{k}$.
3.91 $\mathbf{F} = (36.0$ N$)\mathbf{i} - (288$ N$)\mathbf{j} + (144.0$ N$)\mathbf{k}$;
       $\mathbf{M} = (1080$ N · m$)\mathbf{i} - (270$ N · m$)\mathbf{k}$.
3.92 $\mathbf{F} = -(18.00$ N$)\mathbf{i} - (336$ N$)\mathbf{j} + (144.0$ N$)\mathbf{k}$;
       $\mathbf{M} = (1080$ N · m$)\mathbf{i} + (135.0$ N · m$)\mathbf{k}$.
3.93 $\mathbf{F} = (2.00$ lb$)\mathbf{i} + (38.0$ lb$)\mathbf{j} - (24.0$ lb$)\mathbf{k}$;
       $\mathbf{M} = (792$ lb · in.$)\mathbf{i} + (708$ lb · in.$)\mathbf{j} + (1187$ lb · in.$)\mathbf{k}$.
3.95 $\mathbf{F} = (270$ N$)\mathbf{i} - (90.0$ N$)\mathbf{j} + (135.0$ N$)\mathbf{k}$;
       $\mathbf{M} = (88.4$ N · m$)\mathbf{i} + (97.2$ N · m$)\mathbf{j} - (42.0$ N · m$)\mathbf{k}$.
3.96   (a) $-(2.40$ N$)\mathbf{j}$. (b) $x = -16.89$ mm, $z = -24.5$ mm.
3.98   (a) Loading a: $\mathbf{R} = 300$ lb ↓; $\mathbf{M} = 500$ lb · ft ↖.
       Loading b: $\mathbf{R} = 300$ lb ↓; $\mathbf{M} = 450$ lb · ft ↙.
       Loading c: $\mathbf{R} = 300$ lb ↓; $\mathbf{M} = 450$ lb · ft ↖.
       Loading d: $\mathbf{R} = 200$ lb ↑; $\mathbf{M} = 450$ lb · ft ↖.
       Loading e: $\mathbf{R} = 300$ lb ↓; $\mathbf{M} = 100$ lb · ft ↙.
       Loading f: $\mathbf{R} = 300$ lb ↓; $\mathbf{M} = 400$ lb · ft ↖.
       Loading g: $\mathbf{R} = 500$ lb ↓; $\mathbf{M} = 500$ lb · ft ↖.
       Loading h: $\mathbf{R} = 300$ lb ↓; $\mathbf{M} = 450$ lb · ft ↖.
       (b) Loadings c and h.
3.99   Loading f.
3.100 (a) $\mathbf{R} = 300$ lb ↓; 1.500 ft. (b) $\mathbf{R} = 200$ lb ↑; 2.25 ft.
       (c) $\mathbf{R} = 300$ lb ↓; 0.333 ft.
3.101 Force-couple system at E.
3.104 (a) 2.61 m to the right of A. (b) 1.493 m.
3.105 $\mathbf{R} = 362$ N ⤡ 81.9°; $M = 327$ N · m.
3.106 (a) $\mathbf{R} = 45.4$ lb ⤡ 65.4°.
       (b) 4.63 in. to the left of B and 10.09 in. below B.
3.108 (a) $\mathbf{R} = 18.61$ kN ⤢ 55.2°; 1.014 mm to the left of EF.
       (b) 55.9°.
3.109 (a) 22.2°.
       (b) $\mathbf{R} = 18.98$ kN ⤢ 67.8°; 3.28 mm to the left of EF.
3.110 773 N ⤢ 79.0°; 9.54 m to the right of A.
3.111 (a) 1.471 lb ⦨ 55.6°. (b) 19.12 in. to the left of B.
       (c) 1.389 in. above and to the left of A.
3.113 350 N ⤡ 21.4°; 92.6 mm to the left of B and 27.4 mm to the right of F.
3.114 (a) $\mathbf{R} = F$ ⤢ $\tan^{-1}(a^2/2bx)$;
       $\mathbf{M} = 2Fb^2(x - x^3/a^2)/\sqrt{a^4 + 4b^2x^2}$ ↖. (b) 369 mm.

3.115 $\mathbf{R} = -(7\ \text{lb})\mathbf{i} - (8\ \text{lb})\mathbf{j} - (4\ \text{lb})\mathbf{k}$;
$\mathbf{M} = -(12\ \text{lb} \cdot \text{in.})\mathbf{i} + (10.5\ \text{lb} \cdot \text{in.})\mathbf{j} + (7.5\ \text{lb} \cdot \text{in.})\mathbf{k}$.

3.116 $\mathbf{R} = -(840\ \text{N})\mathbf{j} - (679\ \text{N})\mathbf{k}$;
$\mathbf{M} = (4.50\ \text{N} \cdot \text{m})\mathbf{i} + (656\ \text{N} \cdot \text{m})\mathbf{j} - (440\ \text{N} \cdot \text{m})\mathbf{k}$.

3.117 $\mathbf{A} = (8\ \text{N})\mathbf{i} - (160\ \text{N})\mathbf{j} + (10\ \text{N})\mathbf{k}$;
$\mathbf{B} = -(48\ \text{N})\mathbf{i} + (160\ \text{N})\mathbf{j} + (10\ \text{N})\mathbf{k}$.

3.119 (a) 53.1°. (b) $\mathbf{R} = (8\ \text{lb})\mathbf{j} + (1\ \text{lb})\mathbf{k}$;
$\mathbf{M} = -(192\ \text{lb} \cdot \text{in.})\mathbf{i} + (360\ \text{lb} \cdot \text{in.})\mathbf{k}$.

3.120 (a) $\mathbf{R} = (8.66\ \text{lb})\mathbf{j}$;
$\mathbf{M} = (55.0\ \text{lb} \cdot \text{in.})\mathbf{i} + (45.0\ \text{lb} \cdot \text{in.})\mathbf{j} + (77.9\ \text{lb} \cdot \text{in.})\mathbf{k}$.
(b) Clockwise.

3.121 $\mathbf{R} = (13.64\ \text{N})\mathbf{i} - (41.3\ \text{N})\mathbf{j} - (6.36\ \text{N})\mathbf{k}$;
$\mathbf{M} = (10.82\ \text{N} \cdot \text{m})\mathbf{i} - (2.55\ \text{N} \cdot \text{m})\mathbf{j} - (10.03\ \text{N} \cdot \text{m})\mathbf{k}$.

3.122 $\mathbf{R} = -(40\ \text{N})\mathbf{i} + (280\ \text{N})\mathbf{j} - (100\ \text{N})\mathbf{k}$;
$\mathbf{M} = -(135\ \text{N} \cdot \text{m})\mathbf{i} + (205\ \text{N} \cdot \text{m})\mathbf{j} + (741\ \text{N} \cdot \text{m})\mathbf{k}$.

3.123 235 lb; $x = 7.69$ ft, $z = 9.10$ ft.

3.124 $x = 7.03$ ft, $z = 3.55$ ft.

3.125 3400 kN ↓; $x = 30.4$ m, $z = 14.39$ m.

3.126 $a = 19.48$ m; $b = 17.47$ m.

3.129 (a) P; $\theta_x = 90°$, $\theta_y = 90°$, $\theta_z = -180°$. (b) $3a$. (c) Axis of wrench is parallel to the z axis at $x = 0$, $y = a$.

3.131 (a) $-(21\ \text{N})\mathbf{j}$. (b) 0.571 m. (c) Axis of wrench is parallel to the y axis at $x = 0$, $z = 41.7$ mm.

3.133 (a) $-(21\ \text{lb})\mathbf{j} - (20\ \text{lb})\mathbf{k}$. (b) 19.10 in.
(c) $x = 21.1$ in., $z = -7.43$ in.

3.135 (a) $3P(2\mathbf{i} - 20\mathbf{j} - \mathbf{k})/25$. (b) $-0.0988a$.
(c) $x = 2.00a$, $z = -1.990a$.

3.137 $\mathbf{R} = (4\ \text{lb})\mathbf{i} + (6\ \text{lb})\mathbf{j} - (2\ \text{lb})\mathbf{k}$; $p = 1.226$ ft;
$y = -3.68$ ft, $z = -1.031$ ft.

3.138 $\mathbf{A} = (M/b)\mathbf{i} + R(1 + a/b)\mathbf{k}$; $\mathbf{B} = -(M/b)\mathbf{i} - (aR/b)\mathbf{k}$.

3.142 (a) $\mathbf{F} = 360$ N ⟋ 50.0°; $\mathbf{M} = 230$ N · m ↰.
(b) $\mathbf{F}_A = 253$ N ↓; $\mathbf{F}_D = 232$ N ⟋ 5.52°.

3.144 (a) $-(600\ \text{N})\mathbf{k}$; $d = 90$ mm below ED.
(b) $-(600\ \text{N})\mathbf{k}$; $d = 90$ mm above ED.

3.145 (a) 196.2 N · m ↲. (b) 199.0 N ⟍ 59.5°.

3.147 283 lb.

3.149 $\mathbf{R} = (8.40\ \text{lb})\mathbf{i} - (19.20\ \text{lb})\mathbf{j}\ (3.20\ \text{lb})\mathbf{k}$;
$\mathbf{M} = (71.6\ \text{lb} \cdot \text{ft})\mathbf{i} + (56.8\ \text{lb} \cdot \text{ft})\mathbf{j} - (65.2\ \text{lb} \cdot \text{ft})\mathbf{k}$.

3.150 (a) 12.39 N · m ↲. (b) 12.39 N · m ↲. (c) 12.39 N · m ↲.

3.151 185.2 lb ∡ 11.84°; 23.3 in. to the left of the vertical centerline of the motor.

3.152 $\phi = 24.6°$; $d = 34.6$ in.

CHAPTER 4

4.1 (a) 12.13 kN ↑. (b) 16.81 kN ↑.

4.2 (a) 1.182 kN ↓. (b) 2.49 kN ↑.

4.3 (a) 834 lb ↑. (b) 916 lb ↑.

4.4 (a) 751 lb ↑. (b) 874 lb ↑.

4.5 (a) $\mathbf{A} = 20.0$ N ↓; $\mathbf{B} = 150.0$ N ↑.
(b) $\mathbf{A} = 10.0$ N ↓; $\mathbf{B} = 140.0$ N ↑.

4.6 40.0 mm.

4.9 2.32 kg ≤ m_A ≤ 266 kg.

4.10 140.0 N ≤ T_C ≤ 180.0 N.

4.13 6.00 kips ≤ P ≤ 73.5 kips.

4.14 1.00 in. ≤ a ≤ 5.00 in.

4.15 (a) 28.0 N ∡ 60.0°. (b) 28.2 N ⟎ 6.62°.

4.16 (a) 300 N. (b) 584 N ∡ 77.5°.

4.17 (a) $\mathbf{A} = 60.0$ lb ↑; $\mathbf{C} = 170.9$ lb ⟎ 20.6°.
(b) $\mathbf{A} = 97.4$ lb ∡ 60.0°; $\mathbf{C} = 225$ lb ⟎ 22.0°.

4.18 (a) $\mathbf{A} = 40.0$ lb ∡ 30°; $\mathbf{B} = 40.0$ lb ⟍ 30°.
(b) $\mathbf{A} = 115.5$ lb ⟍ 12.55°; $\mathbf{B} = 130.2$ lb ⟍ 30.0°.

4.21 (a) 312 N ↓. (b) 859 N ∡ 21.3°.

4.22 932 N.

4.23 (a) $\mathbf{A} = 47.5$ lb →; $\mathbf{B} = 42.5$ lb ⟍ 28.1°.
(b) $\mathbf{A} = 67.2$ lb ∡ 45.0°; $\mathbf{B} = 46.5$ lb ⟎ 36.3°.

4.24 (a) $\mathbf{A} = 51.5$ lb ∡ 22.8°; $\mathbf{B} = 37.5$ lb ←.
(b) $\mathbf{A} = 50.6$ lb ⟍ 20.2°; $\mathbf{B} = 53.0$ lb ⟍ 45.0°.

4.27 (a) $\mathbf{A} = 37.3$ N ⟍ 58.0°; $\mathbf{E} = 140.5$ N ∡ 60.0°.
(b) $\mathbf{A} = 0$; $\mathbf{E} = 127.3$ N ∡ 45.0°.

4.28 (a) 477 N. (b) 715 N ⟍ 60.5°.

4.29 $T = 128.4$ N; $\mathbf{A} = 145.8$ N ⟍ 81.0°.

4.30 $T = 404$ N; $\mathbf{A} = 304$ N ⟍ 86.9°.

4.31 $T = 20.0$ lb; $\mathbf{C} = 22.4$ lb ∡ 26.6°.

4.33 (a) $1.305P$ ⟍ 40.0°. (b) $1.013P$ ∡ 9.14°.

4.34 (a) 32.5 lb. (b) 56.0 lb ⟎ 2.05°.

4.35 $T = 2P/3$; $\mathbf{C} = 0.577P$ →.

4.37 $T_{BE} = 646$ lb; $T_{CF} = 192.1$ lb; $\mathbf{D} = 750$ lb ←.

4.38 (a) 17.68 N.
(b) $\mathbf{N}_A = 24.1$ N ⟍ 45.0°; $\mathbf{N}_D = 24.1$ N ∡ 45.0°.

4.39 (a) 0°. (b) 12.50 N.
(c) $\mathbf{N}_A = 15.31$ N ⟍ 45.0°; $\mathbf{N}_D = 15.31$ N ∡ 45.0°.

4.41 (a) $\mathbf{N}_B = 2.77$ N ↑; $\mathbf{N}_C = 3.70$ N ⟍ 39.8°.
(b) 5.26 N ∡ 77.3°.

4.42 (a) 6.46 N. (b) $\mathbf{N}_B = 4.47$ N ↑; $\mathbf{N}_C = 5.98$ N ⟍ 39.8°.

4.43 $\mathbf{A} = \mathbf{D} = 0$; $\mathbf{B} = 192.9$ lb ←; $\mathbf{C} = 28.0$ lb →.

4.45 (a) $\mathbf{A} = 20.0$ lb ↑; $\mathbf{M}_A = 30.0$ lb · ft ↰.
(b) $\mathbf{A} = 28.3$ lb ∡ 45.0°; $\mathbf{M}_A = 30.0$ lb · ft ↰.
(c) $\mathbf{A} = 40.0$ lb ↑; $\mathbf{M}_A = 60.0$ lb · ft ↰.

4.46 $\mathbf{C} = 33.9$ N ∡ 45.0°; $\mathbf{M}_C = 0.120$ N · m ↰.

4.47 $\mathbf{C} = 35.8$ N ∡ 26.6°; $\mathbf{M}_C = 0.800$ N · m ↰.

4.48 $T_{\max} = 444$ lb; $T_{\min} = 347$ lb.

4.49 (a) $\mathbf{A} = 14.72$ N ↑; $\mathbf{M}_A = 3.43$ N · m ↰.
(b) $\mathbf{A} = 8.72$ N ↑; $\mathbf{M}_A = 1.034$ N · m ↰.

4.50 2.33 N ≤ F_E ≤ 14.83 N.

4.53 (a) $\sin^{-1}(2M/Wl)$. (b) 22.0°.

4.54 (a) $W/2(1 - \tan \theta)$. (b) 39.8°.

4.55 (a) $\tan^{-1}(2P/mg)$. (b) 63.9°.

4.56 (a) $\sin \theta + \cos \theta = M/Pl$. (b) 17.11°; 72.9°.

4.57 (a) $2 \sin^{-1}[kl/\sqrt{2}(kl - P)]$. (b) 141.1°.

4.59 (a) $\tan \theta - \sin \theta = W/kl$. (b) 58.0°.

4.61 (1) Completely constrained; determinate; equilibrium;
$\mathbf{A} = 116.6$ N ∡ 59.0°, $\mathbf{B} = 60.0$ N ←.
(2) Improperly constrained; indeterminate; no equilibrium.
(3) Partially constrained; determinate; equilibrium;
$\mathbf{A} = \mathbf{C} = 50$ N ↑. (4) Completely constrained;
determinate; equilibrium; $\mathbf{A} = 50$ N ↑, $\mathbf{B} = 78.1$ N
⟍ 39.8°, $\mathbf{C} = 60$ N →. (5) Completely constrained;
indeterminate; equilibrium; $\mathbf{A}_y = 50$ N ↑.
(6) Completely constrained; indeterminate; equilibrium;
$\mathbf{A}_x = 60$ N →, $\mathbf{B}_x = 60$ N ←. (7) Completely constrained;
determinate; equilibrium; $\mathbf{A} = \mathbf{C} = 50$ N ↑.
(8) Improperly constrained; indeterminate; no
equilibrium.

4.63 $P = 28.0$ lb; $\mathbf{A} = 35.0$ lb ⟍ 36.9°.

4.64 $a \geq 2.31$ in.

4.65 (a) 312 N ↓. (b) 859 N ∡ 21.3°.

4.66 932 N.

4.69 $\mathbf{C} = 270$ N ⟍ 56.3°; $\mathbf{D} = 167.7$ N ⟍ 26.6°.

4.70 $\mathbf{A} = 276$ N ⟍ 45.0°; $\mathbf{C} = 379$ N ⟍ 59.0°.

4.71 215 N ⟍ 80.8°.

4.72 189.2 N ∡ 80.9°.

4.73 (a) 183.1 lb. (b) 177.5 lb ⟍ 32.3°.

4.74 (a) 366 lb. (b) 300 lb ⟎ 1.921°.

4.76   77.5 N $\measuredangle$ 30.0°.

4.77   (a) 11.82 kN $\searrow$ 75.0°. (b) 10.68 kN $\swarrow$ 73.4°.

4.79   (a) 45.4 N $\searrow$ 45.0°. (b) 387 N $\searrow$ 85.2°.

4.81 T = 1200 lb; **B** = 1500 lb $\swarrow$ 36.9°.

4.82 T = 544 N; **C** = 865 N $\swarrow$ 33.7°.

4.83   (a) **A** = 40.0 lb $\measuredangle$ 30°; **B** = 40.0 lb $\searrow$ 30°.
      (b) **A** = 115.5 lb $\swarrow$ 12.55°; **B** = 130.2 lb $\searrow$ 30.0°.

4.84   (a) 477 N. (b) 715 N $\searrow$ 60.5°.

4.85   (a) 1.743P $\searrow$ 55.0°. (b) 1.088P $\swarrow$ 23.2°.

4.86   (a) 1.305P $\searrow$ 40.0°. (b) 1.013P $\measuredangle$ 9.14°.

4.87 $\sqrt{(S^2 - L^2)/3}$

4.88   (a) 129.1 mm. (b) 17.10 N. (c) 8.71 N$\leftarrow$.

4.90   (a) 43.0°. (b) **A** = 45.7 N $\leftarrow$; **B** = 108.2 N $\measuredangle$ 65.0°.

4.91   34.5°.

4.94   (a) 18.43°. (b) $mg/5$.

4.95   (a) 23.5°. (b) 136.3 mm.

4.96   **C** = (964 N)**j** − (843 N)**k**; **D** = −(3130 N)**j** + (482 N)**k**.

4.97   **C** = (189.3 N)**j** − (482 N)**k**; **D** = −(4160 N)**j** + (964 N)**k**.

4.99   (a) 34.0 lb. (b) **C** = −(4.20 lb)**j** − (27.2 lb)**k**;
      D = (18.20 lb)**j** − (40.8 lb)**k**.

4.100   (a) 20.6 lb. (b) **C** = −(2.30 lb)**j** − (21.2 lb)**k**;
      D = (9.98 lb)**j** − (35.6 lb)**k**.

4.102 T_A = 92.0 N; T_B = T_C = 76.6 N.

4.103 m = 1.786 kg; x = 200 mm, z = 100 mm.

4.104   (a) $\mathbf{A}_y$ = 2.95 N ↑; $\mathbf{B}_y$ = $\mathbf{C}_y$ = 0.684 N ↑. (b) 54.1°.

4.105 T_A = 5.00 lb; T_C = 3.18 lb; T_D = 21.8 lb.

4.106   (a) 1.600 ft. (b) T_A = 5.00 lb; T_C = 0; T_D = 25.0 lb.

4.107 T_A = 0.211W; T_B = 0.366W; T_C = 0.423W.

4.110   (a) 6.00 lb. (b) 0. (c) 6.00 lb.

4.111   (a) T_A = 24.5 N; T_B = T_C = 36.8 N. (b) 150.0 mm.

4.112 T_{BD} = T_{BE} = 707 N; **C** = −(200 N)**j** + (866 N)**k**.

4.113 T_{BD} = T_{BE} = 5.24 kN; **A** = (5.71 kN)**i** − (2.67 kN)**j**.

4.114 T_{AD} = 486 lb; T_{BE} = 672 lb;
      C = (1008 lb)**i** − (48.0 lb)**j** + (72.0 lb)**k**.

4.116 T_{BD} = 360 lb; T_{BE} = 180.0 lb;
      A = −(90.0 lb)**i** + (540 lb)**j** + (60.0 lb)**k**.

4.118 T_{EG} = 1.758 kN; T_{ICFH} = 2.06 kN;
      D = (4610 N)**i** − (529 N)**j** − (36.4 N)**k**.

4.119 T_{EG} = 1.507 kN; T_{ICFH} = 1.767 kN;
      D = (3950 N)**i** − (373 N)**j** − (31.2 N)**k**.

4.120   (a) 374 N. (b) **E** = (104.9 N)**i** − (152.3 N)**k**;
      F = (115.1 N)**i** + (176.0 N)**j** + (482 N)**k**.

4.122   (a) 97.1 N. (b) **A** = −(23.5 N)**i** + (63.8 N)**j** − (7.85 N)**k**;
      B = (9.81 N)**j** + (66.7 N)**k**.

4.124   (a) 9.19 lb. (b) **A** = (6.93 lb)**j** − (1.106 lb)**k**;
      B = −(1.778 lb)**i** + (1.067 lb)**j** + (5.25 lb)**k**.

4.125   (a) 22.8 lb. (b) **A** = (11.73 lb)**j** − (1.106 lb)**k**;
      B = −(1.244 lb)**i** − (18.13 lb)**j** + (5.25 lb)**k**.

4.126   (a) 100.1 lb. (b) **A** = (33.3 lb)**i** + (109.6 lb)**j** + (41.1 lb)**k**;
      B = (32.9 lb)**j** + (53.9 lb)**k**.

4.127   (a) 180.5 lb. (b) **A** = −(57.0 lb)**i** + (109.6 lb)**j** − (10.96 lb)**k**;
      B = (32.9 lb)**j** + (106.0 lb)**k**.

4.128   (a) 22.4 N. (b) **D** = (6.00 N)**k**;
      $\mathbf{M}_D$ = (0.1320 N · m)**i** + (1.008 N · m)**j** + (0.0896 N · m)**k**.

4.129 T_{CF} = 333 N; T_{DE} = 750 N; **A** = (267 N)**i** + (450 N)**k**;
      $\mathbf{M}_A$ = −(54.0 N · m)**i**.

4.130   (a) 374 N. (b) **E** = (220 N)**i** + (176.0 N)**j** + (330 N)**k**;
      $\mathbf{M}_E$ = −(62.7 N · m)**i** + (14.96 N · m)**k**.

4.131   (a) 9.19 lb. (b) **B** = −(1.778 lb)**i** + (8.00 lb)**j** + (4.15 lb)**k**;
      $\mathbf{M}_B$ = −(33.2 lb · in.)**j** − (208 lb · in.)**k**.

4.132 T_{DI} = 164.8 N; T_{EH} = 933 N; T_{FG} = 187.8 N;
      A = (1094 N)**i** + (98.7 N)**j** − (21.3 N)**k**.

4.133 T_{DI} = 180.6 N; T_{EH} = 960 N; T_{FG} = 100.9 N;
      A = (1067 N)**i** + (110.4 N)**j** − (11.45 N)**k**.

4.136   **B** = (75.0 N)**k**; **C** = +(36.0 N)**j** − (20.0 N)**k**;
      D = −(36.0 N)**j** + (5.00 N)**k**.

4.138   **A** = (120.0 N)**j** − (150.0 N)**k**; **B** = (180.0 N)**i** + (150 N)**k**;
      C = −(180.0 N)**i** + (120.0 N)**j**.

4.140   (a) (3.20 lb)**i** + (4.27 lb)**j**.
      (b) **A** = −(3.20 lb)**i** + (5.73 lb)**j** − (3.00 lb)**k**;
      B = (3.00 lb)**k**.

4.141   (a) 1.575 lb. (b) **A** = −(0.975 lb)**i** + (4.40 lb)**j** − (0.300 lb)**k**;
      $\mathbf{N}_B$ = (0.800 lb)**j** + (0.600 lb)**k**.

4.142   29.1 lb.

4.144   8.73 lb.

4.145   9.05 lb.

4.146   853 N.

4.148   (112.5 N)**j**.

4.149   (a) x = 2.00 m, y = 4.00 m. (b) 131.6 N.

4.151   (a) x = 0, y = 29.6 in., z = 25.0 in. (b) $\mathbf{N}_D$ = (8.98 lb)**i**.

4.152   (a) **D** = 20.0 lb ↓; $\mathbf{M}_D$ = 20.0 lb · ft ↑.
      (b) **D** = 10.00 lb ↓; $\mathbf{M}_D$ = 30.0 lb · ft. $\swarrow$.

4.154   **A** = 244 N →; **D** = 344 N $\searrow$ 22.2°.

4.156   (a) 325 lb ↑. (b) 1175 lb ↑.

4.157   **C** = 7.07 lb $\searrow$ 45.0°; $\mathbf{M}_C$ = 43.0 lb · in. $\swarrow$.

4.158   **C** = 7.07 lb $\searrow$ 45.0°; $\mathbf{M}_C$ = 45.0 lb · in. $\swarrow$.

4.159   (a) 462 N. (b) **C** = −(336 N)**j** + (467 N)**k**;
      D = (505 N)**j** − (66.7 N)**k**.

4.161 T_{BE} = 975 N; T_{CF} = 600 N; T_{DG} = 625 N;
      A = (2100 N)**i** + (175.0 N)**j** − (375 N)**k**.

4.163   (a) 16.00 in. (b) 267 mm.

CHAPTER 5

5.1 $\overline{X}$ = 5.60 in., $\overline{Y}$ = 6.60 in.

5.2 $\overline{X}$ = 96.4 mm, $\overline{Y}$ = 34.7 mm.

5.4 $\overline{X}$ = 1.421 mm, $\overline{Y}$ = 12.42 mm.

5.5 $\overline{X}$ = 49.4 mm, $\overline{Y}$ = 93.8 mm.

5.6 $\overline{X}$ = 0.721 in., $\overline{Y}$ = 3.40 in.

5.7 $\overline{X}$ = $\overline{Y}$ = 31.1 mm.

5.9   1.340.

5.10 $\dfrac{r_1 + r_2}{\pi - 2\alpha} \cos \alpha$.

5.11 $\overline{X}$ = 0, $\overline{Y}$ = 0.632 in.

5.12 $\overline{X}$ = −15.83 mm, $\overline{Y}$ = 3.34 mm.

5.15 $\overline{X}$ = 5.95 in., $\overline{Y}$ = 14.41 in.

5.16 $\overline{X}$ = 3.21 in., $\overline{Y}$ = 3.31 in.

5.17 $(Q_x)_1$ = 25.0 × 10^3 mm^3; $(Q_x)_2$ = −25.0 × 10^3 mm^3.

5.18 $(Q_x)_1$ = 1.393 × 10^6 mm^3; $(Q_x)_2$ = −1.393 × 10^6 mm^3.

5.20   115.3 lb.

5.21 $\overline{X}$ = 4.67 in., $\overline{Y}$ = 6.67 in.

5.23 $\overline{X}$ = 1.441 mm, $\overline{Y}$ = 12.72 mm.

5.25   (a) 1.470 N. (b) 6.13 N $\searrow$ 83.1°.

5.26   63.6°.

5.28   12.00 in.

5.29   (a) 0.472a. (b) 0.387a.

5.31 $\bar{x}$ = $a/3$, $\bar{y}$ = 2$h/3$.

5.32 $\bar{x}$ = 2$a/5$, $\bar{y}$ = 4$h/7$.

5.33 $\bar{x}$ = 16$a/35$, $\bar{y}$ = 16$h/35$.

5.35 $\bar{x}$ = 0, $\bar{y}$ = $\dfrac{2}{3}R \dfrac{3 - \sin^2 \alpha}{2\alpha + \sin 2\alpha} \sin \alpha$.

5.37 $\bar{x}$ = 39$a/50$, $\bar{y}$ = −39$b/175$.

5.38 $\bar{x}$ = $a/4$, $\bar{y}$ = 3$b/10$.

5.39 $\bar{x}$ = 5$L/4$, $\bar{y}$ = 33$a/40$.

5.41	$\bar{x} = 0.546a, \bar{y} = 0.423b.$	5.113	$\overline{X} = \overline{Z} = 0, \overline{Y} = 10.51$ mm above the base.
5.43	$-2r\sqrt{2}/3\pi.$	5.115	$\overline{X} = 274$ mm, $\overline{Y} = -103.6$ mm.
5.44	$2a/5.$	5.116	$\bar{x}_1 = 21a/88; \bar{x}_2 = 27a/40.$
5.45	$\bar{x} = -9.27a, \bar{y} = 3.09a.$	5.117	$\bar{x}_1 = 21h/88; \bar{x}_2 = 27h/40.$
5.46	$\bar{x} = 0.363L, \bar{y} = 0.1653L.$	5.119	$\bar{x} = 5a/16, \bar{y} = \bar{z} = 0.$

5.47 (a) $V = 2.21 \times 10^6$ mm^3; $A = 89.8 \times 10^3$ mm^2.
      (b) $V = 4.36 \times 10^6$ mm^3; $A = 202 \times 10^3$ mm^2.

5.48 (a) $V = 24.0 \times 10^3$ mm^3; $A = 4.50 \times 10^3$ mm^2.
      (b) $V = 26.5 \times 10^3$ mm^3; $A = 5.12 \times 10^3$ mm^2.

5.50 (a) $\pi^2 a^2 b.$ (b) $2\pi^2 a^2 b.$ (c) $2\pi a^2 b/3.$

5.51 $V = 122.0$ in^3; $A = 488$ in^2.

5.53 720 mm^3.

5.54 (a) 3.24×10^3 mm^2. (b) 2.74×10^3 mm^2
      (c) 2.80×10^3 mm^2.

5.56 0.214 lb.

5.57 2.21 kN.

5.59 0.443 lb.

5.60 1013 mm^2.

5.61 (a) $R = 1800$ lb, 10.80 ft to the right of A.
      (b) $\mathbf{A} = 1800$ lb $\uparrow$; $\mathbf{M}_A = 19.44$ kip $\cdot$ ft $\uparrow$.

5.62 (a) $R = 3600$ N, 3.75 m to the right of A.
      (b) $\mathbf{A} = 1350$ N $\uparrow$; $\mathbf{B} = 2250$ N $\uparrow$.

5.63 $\mathbf{A} = 1000$ lb $\uparrow$; $\mathbf{B} = 800$ lb $\uparrow$.

5.65 $\mathbf{A} = 34.0$ N $\uparrow$; $\mathbf{M}_A = 2.92$ N $\cdot$ m $\uparrow$.

5.66 $\mathbf{A} = 270$ N $\uparrow$; $\mathbf{B} = 720$ N $\downarrow$.

5.67 $\mathbf{A} = 3.33$ kN $\uparrow$; $\mathbf{M}_A = 6.33$ kN $\cdot$ m $\uparrow$.

5.69 (a) 1.608 ft. (b) $\mathbf{A} = \mathbf{B} = 152.2$ lb $\uparrow$.

5.71 $\mathbf{B} = 7.04$ kN $\uparrow$; $\mathbf{C} = 15.46$ kN $\uparrow$.

5.72 (a) 3.19 kN/m $\uparrow$. (b) 1.375 kN $\uparrow$.

5.73 $w_A = 6.67$ kN/m; $R_B = 29.0$ kN.

5.75 (a) $\mathbf{H} = 1589$ kN $\rightarrow$, $\mathbf{V} = 5.93$ MN $\uparrow$. (b) 10.48 m to the
      right of A. (c) $\mathbf{R} = 1675$ kN $\nearrow$ $18.43°$.

5.76 (a) $\mathbf{H} = 10.11$ kips $\rightarrow$, $\mathbf{V} = 79.8$ kips $\uparrow$. (b) 15.90 ft to the
      right of A. (c) $\mathbf{R} = 13.53$ kips $\nearrow$ $41.6°$.

5.77 8.32 ft.

5.78 $\mathbf{T} = 9.10$ kips $\leftarrow$; $\mathbf{A} = 21.6$ kips $\leftarrow$.

5.80 4.75%.

5.82 43.5 kN $\nearrow$ $30.0°$.

5.83 $\mathbf{A} = 563$ lb $\nearrow$ $37.1°$; $\mathbf{B} = 1014$ lb $\uparrow$.

5.85 1.110 m.

5.86 1.152 m.

5.87 $\mathbf{A} = 800$ lb $\nwarrow$ $41.8°$; $\mathbf{D} = 11.33$ lb $\rightarrow$.

5.89 (a) 42.1 N $\searrow$ $63.4°$.
      (b) $\mathbf{R}_H = 37.7$ N $\downarrow$; $\mathbf{M}_H = 0.452$ N $\cdot$ m $\downarrow$.

5.90 (a) $-1.118R.$ (b) $0.884.$

5.91 $-(2h^2 - 3b^2)/2(4h - 3b).$

5.92 $-2a(2h - b)/\pi(4h - 3b).$

5.93 (a) $0.548L.$ (b) $2\sqrt{3}.$

5.94 2.44 in.

5.95 0.749 in.

5.96 64.2 mm.

5.97 19.51 mm.

5.100 $\overline{X} = 48.1$ mm, $\overline{Y} = 68.3$ mm, $\overline{Z} = 75.8$ mm.

5.101 $\overline{X} = 20.4$ mm, $\overline{Y} = -4.55$ mm, $\overline{Z} = 29.0$ mm.

5.103 $\overline{X} = 4.54$ in., $\overline{Y} = 6.59$ in., $\overline{Z} = 6.00$ in.

5.104 $\overline{X} = 70.2$ mm, $\overline{Y} = 198.7$ mm, $\overline{Z} = 0.$

5.106 $\overline{X} = 340$ mm, $\overline{Y} = 314$ mm, $\overline{Z} = 283$ mm.

5.107 $\overline{X} = 150.0$ mm, $\overline{Y} = 11.42$ mm, $\overline{Z} = 119.4$ mm.

5.108 $\overline{X} = 5.83$ in., $\overline{Y} = 5.15$ in., $\overline{Z} = 3.94$ in.

5.109 $\overline{X} = 3.06$ ft, $\overline{Y} = 0.539$ ft, $\overline{Z} = 2.99$ ft.

5.110 $\overline{X} = 0.525$ m, $\overline{Y} = 1.241$ m, $\overline{Z} = 0.406$ m.

5.111 $\overline{X} = 0.610$ m, $\overline{Y} = 2.70$ m, $\overline{Z} = 0.$

5.121 $\bar{x} = h, \bar{y} = 0.869a, \bar{z} = 0.$

5.122 $\bar{x} = 1.297a, \bar{y} = \bar{z} = 0.$

5.123 $\bar{x} = \bar{z} = 0, \bar{y} = 0.374b.$

5.125 $\bar{x} = 0, \bar{y} = \bar{z} = -R/2.$

5.127 $V = 9.50$ m^3; $\bar{x} = 3.82$ m.

5.129 $\bar{x} = 0, \bar{y} = 5h/16, \bar{z} = -a/4.$

5.130 $\overline{X} = 7.22$ in., $\overline{Y} = 9.56$ in.

5.132 $\overline{X} = 19.27$ mm, $\overline{Y} = 26.6$ mm.

5.134 0.739 m.

5.135 $V = \pi L^3/6.$

5.137 $\mathbf{B} = 1360$ lb $\uparrow$; $\mathbf{C} = 2360$ lb $\uparrow$.

5.139 (a) $-0.402a.$ (b) $2/5$ and $2/3.$

5.140 $\overline{X} = 0.1452$ m, $\overline{Y} = 0.396$ m, $\overline{Z} = 0.370$ m.

5.141 $\bar{x} = h/6, \bar{y} = \bar{z} = 0.$

CHAPTER 6

6.1 $F_{AB} = 125.0$ N C; $F_{AC} = 260$ N C; $F_{BC} = 100.0$ N T.

6.2 $F_{AB} = 3.40$ kips T; $F_{AC} = 4.00$ kips T; $F_{BC} = 5.00$ kips C.

6.4 $F_{AB} = 1.200$ kN T; $F_{AC} = 3.36$ kN C; $F_{AD} = 3.48$ kN T;
     $F_{BC} = F_{CD} = 2.52$ kN C.

6.5 $F_{AB} = F_{AE} = 6.71$ kN T; $F_{AC} = F_{AD} = 10.00$ kN C;
     $F_{BC} = F_{DE} = 6.00$ kN C; $F_{CD} = 2.00$ kN T.

6.6 $F_{AB} = 4.38$ kips C; $F_{AD} = 8.13$ kips T; $F_{BC} = 3.70$ kips C;
     $F_{BD} = 6.08$ kips C; $F_{CD} = 3.70$ kips T.

6.7 $F_{AB} = F_{DE} = 200$ N C;
     $F_{AC} = F_{BC} = 520$ N T; $F_{BE} = 480$ N C.

6.9 $F_{AB} = F_{FH} = 15.00$ kN C;
     $F_{AC} = F_{CE} = F_{EG} = F_{GH} = 12.00$ kN T; $F_{BC} = F_{FG} = 0$;
     $F_{BD} = F_{DF} = 11.93$ kN C; $F_{BE} = F_{EF} = 0.714$ kN C;
     $F_{DE} = 0.856$ kN T.

6.10 $F_{AB} = F_{FH} = 15.00$ kN C;
      $F_{AC} = F_{CE} = F_{EG} = F_{GH} = 12.00$ kN T;
      $F_{BC} = F_{FG} = 0$; $F_{BD} = F_{DF} = 10.00$ kN C;
      $F_{BE} = F_{EF} = 5.00$ kN C; $F_{DE} = 6.00$ kN T.

6.11 $F_{AB} = F_{FG} = 44.3$ kips C; $F_{AC} = F_{EG} = 40.5$ kips T;
      $F_{BC} = F_{EF} = 11.25$ kips C; $F_{BD} = F_{DF} = 36.9$ kips C;
      $F_{CD} = F_{DE} = 11.25$ kips T; $F_{CE} = 27.0$ kips T.

6.13 $F_{AB} = 3.10$ kips C; $F_{AC} = F_{CE} = 2.93$ kips T;
      $F_{BC} = F_{FG} = 0$;
      $F_{BD} = F_{DF} = 2.02$ kips C; $F_{BE} = 1.000$ kip C;
      $F_{DE} = 1.100$ kips T; $F_{EF} = 0.750$ kip C;
      $F_{EG} = F_{GH} = 2.68$ kips T; $F_{FH} = 2.83$ kips C.

6.15 $F_{AB} = F_{FH} = 11.25$ kN C;
      $F_{AC} = F_{CE} = F_{EG} = F_{GH} = 6.75$ kN T;
      $F_{BC} = F_{FG} = 6.00$ kN T; $F_{BD} = F_{DF} = 9.00$ kN C;
      $F_{BE} = F_{EF} = 3.75$ kN T; $F_{DE} = 0.$

6.16 $F_{AB} = 9.38$ kN C; $F_{AC} = F_{CE} = 5.63$ kN T;
      $F_{BC} = 6.00$ kN T;
      $F_{BD} = F_{DF} = 6.75$ kN C; $F_{BE} = 1.875$ kN C;
      $F_{DE} = F_{FG} = 0$;
      $F_{EF} = 5.63$ kN T; $F_{EG} = F_{GH} = 3.38$ kN T;
      $F_{FH} = 5.63$ kN C.

6.17 $F_{AB} = 5.42$ kips C; $F_{AC} = 6.17$ kips T; $F_{BC} = 1.152$ kips C;
      $F_{BD} = 5.76$ kips C; $F_{CD} = 2.06$ kips T; $F_{CE} = 4.11$ kips T;
      $F_{DE} = 2.30$ kips C.

6.18 $F_{DF} = 6.09$ kips C; $F_{DG} = 2.06$ kips T; $F_{EG} = 4.11$ kips T;
$F_{FG} = 1.152$ kips C; $F_{FH} = 6.43$ kips C;
$F_{GH} = 6.17$ kips T.

6.19 $F_{AB} = 4.95$ kips C; $F_{AC} = F_{CE} = 3.91$ kips T; $F_{BC} = 0$;
$F_{BD} = 3.54$ kips C; $F_{BE} = 1.000$ kip C;
$F_{DE} = 0.500$ kip T; $F_{DF} = 2.52$ kips C;
$F_{DG} = 0.280$ kip C; $F_{EG} = 2.80$ kips T.

6.20 $F_{FG} = 1.750$ kips T; $F_{FH} = 2.52$ kips C;
$F_{GH} = 0.838$ kip T; $F_{GI} = F_{IK} = F_{KL} = 1.677$ kips T;
$F_{HI} = F_{IJ} = F_{JK} = 0$; $F_{HJ} = F_{JL} = 2.12$ kips C.

6.21 $F_{AB} = F_{DF} = 2.86$ kN T; $F_{AC} = F_{EF} = 2.86$ kN C;
$F_{BC} = F_{DE} = 0.750$ kN C; $F_{BD} = 2.76$ kN T;
$F_{BE} = F_{EH} = 0$;
$F_{CE} = 2.76$ kN C; $F_{CH} = F_{EJ} = 1.500$ kN C.

6.24 $F_{AB} = F_{BD} = 41.0$ kN T; $F_{AC} = 40.0$ kN C;
$F_{BC} = 3.00$ kN C;
$F_{CD} = 6.71$ kN T; $F_{CE} = 46.0$ kN C; $F_{DE} = 1.650$ kN C;
$F_{DG} = 47.2$ kN T; $F_{EF} = 50.0$ kN C; $F_{EG} = 4.29$ kN T;
$F_{FG} = 40.0$ kN C; $F_{FH} = 30.0$ kN C.

6.27 Trusses of Probs. 6.13 and 6.25 are simple trusses.

6.28 Trusses of Probs. 6.19, 6.21, and 6.23 are simple trusses.

6.30 *BC, BE, DE, EF, FG, IJ, KN, MN.*

6.32 *BC, LM.*

6.33 *BF, BG, DH, EH, GJ, HJ.*

6.34 *BC, BE, DE, FH, HI, IJ, OQ, QR.*

6.35 (*a*) *BC, FG.* (*b*) *BC, HI, IJ, JK.*

6.36 $F_{AB} = F_{AD} = 2240$ lb C; $F_{AC} = 1755$ lb C;
$F_{BC} = F_{CD} = 422$ lb T; $F_{BD} = 633$ lb T.

6.37 $F_{AB} = F_{AD} = 7.02$ kips T; $F_{AC} = 13.78$ kips C;
$F_{BC} = F_{CD} = 3.31$ kips T; $F_{BD} = 4.77$ kips C.

6.38 $F_{AB} = F_{AD} = 488$ lb T; $F_{AC} = 2080$ lb T;
$F_{BC} = F_{CD} = 1000$ lb C; $F_{BD} = 560$ lb T.

6.39 $F_{AB} = 1342$ N C; $F_{AC} = 0$; $F_{AD} = 1500$ N T;
$F_{AE} = 300$ N T; $F_{BC} = 400$ N T; $F_{BD} = 600$ N C;
$F_{CD} = 566$ N C; $F_{CE} = 200$ N C; $F_{DE} = 100.0$ N C.

6.40 (*b*) $F_{AE} = 504$ N T; $F_{BE} = 746$ N C; $F_{DE} = F_{EF} = 0$;
$F_{EG} = 730$ N T; $F_{EH} = 480$ N C.

6.41 (*b*) $F_{BG} = 696$ N C; $F_{CG} = F_{DG} = F_{FG} = 0$;
$F_{EG} = 730$ N T; $F_{GH} = 550$ N C.

6.42 $F_{CE} = 4.00$ kN T; $F_{DE} = 1.300$ kN T; $F_{DF} = 4.50$ kN C.

6.44 $F_{CE} = 2.87$ kips T; $F_{DE} = 2.28$ kips T; $F_{DF} = 5.13$ kips C.

6.45 $F_{GI} = 5.13$ kips T; $F_{GJ} = 1.235$ kips T; $F_{HJ} = 6.15$ kips C.

6.46 $F_{CF} = 13.00$ kN T; $F_{EF} = 0.559$ kN T; $F_{EG} = 13.50$ kN C.

6.48 $F_{CE} = 14.00$ kips T; $F_{DE} = 1.692$ kips C;
$F_{DF} = 12.73$ kips C.

6.50 $F_{DF} = 13.00$ kN C; $F_{DG} = 4.22$ kN C; $F_{EG} = 16.22$ kN T.

6.51 $F_{GI} = 16.22$ kN T; $F_{HI} = 1.000$ kN T; $F_{HJ} = 17.33$ kN C.

6.52 $F_{BD} = 5.95$ kips C; $F_{CD} = 1.250$ kips T;
$F_{CE} = 4.50$ kips T.

6.53 $F_{EG} = 3.38$ kips T; $F_{FG} = 1.602$ kips T; $F_{FH} = 4.46$ kips C.

6.55 $F_{FH} = 5.10$ kips C; $F_{GI} = 3.89$ kips T; $F_{GJ} = 2.40$ kips T.

6.56 $F_{AB} = 36.4$ kN T; $F_{AG} = 20.0$ kN T; $F_{FG} = 51.6$ kN C.

6.57 $F_{AE} = 77.6$ kN T; $F_{EF} = 51.6$ kN C; $F_{FJ} = 82.0$ kN C.

6.58 $F_{BE} = 0.783$ kN C; $F_{CE} = 3.70$ kN T; $F_{DF} = 3.35$ kN C.

6.59 $F_{GI} = 4.79$ kN T; $F_{HJ} = 5.35$ kN C; $F_{IJ} = 2.97$ kN C.

6.60 $F_{AF} = 2.50$ kips T; $F_{EJ} = 1.500$ kips T.

6.62 $F_{DG} = 3.60$ kN C; $F_{FH} = 3.60$ kN T.

6.63 $F_{GJ} = 0.833$ kN C; $F_{HK} = 3.60$ kN T; $F_{IL} = 3.60$ kN C.

6.64 $F_{BG} = 27.4$ kN T; $F_{DG} = 9.13$ kN T.

6.65 $F_{CF} = 18.25$ kN T; $F_{CH} = 36.5$ kN T.

6.66 $F_{BD} = 3.83$ kips C; $F_{CE} = 3.31$ kips T;
$F_{CD} = 0.606$ kip T.

6.67 $F_{BD} = 2.63$ kips C; $F_{CE} = 2.37$ kips T; $F_{BE} = 0.303$ kip T.

6.68 $F_{CE} = 3.00$ kips T; $F_{DF} = 6.00$ kips C; $F_{DE} = 5.00$ kips T.

6.70 (*a*) Completely constrained, determinate.
(*b*) Partially constrained.
(*c*) Improperly constrained.

6.72 (*a*) Completely constrained, determinate.
(*b*) Completely constrained, indeterminate.
(*c*) Improperly constrained.

6.74 (*a*) Completely constrained, indeterminate.
(*b*) Completely constrained, determinate.
(*c*) Improperly constrained.

6.75 (*a*) Completely constrained, determinate.
(*b*) Partially constrained.
(*c*) Completely constrained, indeterminate.

6.76 (*a*) 112.5 lb ⬈ 36.9°. (*b*) 73.9 lb ⦞ 66.0°.

6.78 (*a*) 29.3 N C. (*b*) **A** = 25.1 N ⦞ 16.69°.

6.79 (*a*) 25.5 N C. (*b*) **A** = 40.6 N ⦞ 53.8°.

6.80 $\mathbf{A}_x = 100.0$ kN ←, $\mathbf{A}_y = 80.0$ kN ↑; $\mathbf{B}_x = 100.0$ kN ←,
$\mathbf{B}_y = 40.0$ kN ↓; $\mathbf{C}_x = 200$ kN →, $\mathbf{C}_y = 40.0$ kN ↓.

6.81 **A** = 100.0 kN ←; $\mathbf{B}_x = 100.0$ kN →, $\mathbf{B}_y = 40.0$ kN ↑;
C = 40.0 kN ↓.

6.82 $\mathbf{A}_x = 4.50$ kips ←, $\mathbf{A}_y = 5.00$ kips ↓; **B** = 2.25 kips →;
$\mathbf{C}_x = 2.25$ kips →, $\mathbf{C}_y = 5.00$ kips ↑.

6.84 (*a*) $\mathbf{A}_x = 96.0$ lb ←, $\mathbf{A}_y = 112.0$ lb ↑; $\mathbf{E}_x = 96.0$ lb →,
$\mathbf{E}_y = 48.0$ lb ↑. (*b*) $\mathbf{A}_x = 96.0$ lb ←, $\mathbf{A}_y = 32.0$ lb ↑;
$\mathbf{E}_x = 96.0$ lb →, $\mathbf{E}_y = 128.0$ lb ↑.

6.85 (*a*) $\mathbf{A}_x = 72.0$ N ←, $\mathbf{A}_y = 48.0$ N ↑; $\mathbf{E}_x = 72.0$ N →,
$\mathbf{E}_y = 72.0$ N ↑. (*b*) $\mathbf{A}_x = 32.0$ N ←, $\mathbf{A}_y = 8.00$ N ↑;
$\mathbf{E}_x = 32.0$ N →, $\mathbf{E}_y = 112.0$ N ↑.

6.86 (*a*) $\mathbf{A}_x = 36.0$ lb ←, $\mathbf{A}_y = 18.00$ lb ↓; $\mathbf{E}_x = 36.0$ lb →,
$\mathbf{E}_y = 18.00$ lb ↑. (*b*) $\mathbf{A}_x = 36.0$ lb ←, $\mathbf{A}_y = 12.00$ lb ↑;
$\mathbf{E}_x = 36.0$ lb →, $\mathbf{E}_y = 12.00$ lb ↓.

6.88 (*a*) **A** = 192.0 N ↓; **B** = 432 N ↑. (*b*) $\mathbf{A}_x = 320$ N →,
$\mathbf{A}_y = 192.0$ N ↓; $\mathbf{B}_x = 320$ N ←, $\mathbf{B}_y = 432$ N ↑.

6.89 (*a*) $\mathbf{B}_x = 128.0$ N →, $\mathbf{B}_y = 40.0$ N ↑; $\mathbf{F}_x = 128.0$ N ←,
$\mathbf{F}_y = 152.0$ N ↑. (*b*) $\mathbf{B}_x = 128.0$ N →, $\mathbf{B}_y = 136.0$ N ↑;
$\mathbf{F}_x = 128.0$ N ←, $\mathbf{F}_y = 56.0$ N ↑. (*c*) $\mathbf{B}_x = 128.0$ N →,
$\mathbf{B}_y = 40.0$ N ↑; $\mathbf{F}_x = 128.0$ N ←, $\mathbf{F}_y = 152.0$ N ↑.

6.92 $\mathbf{A}_x = 45.0$ lb ←, $\mathbf{A}_y = 75.0$ lb ↑;
$\mathbf{E}_x = 45.0$ lb →, $\mathbf{E}_y = 135.0$ lb ↑.

6.93 $\mathbf{B}_x = 175.0$ lb ←, $\mathbf{B}_y = 50.0$ lb ↓;
$\mathbf{E}_x = 175.0$ lb →, $\mathbf{E}_y = 125.0$ lb ↑.

6.94 $\mathbf{A}_x = 1710$ N →, $\mathbf{A}_y = 785$ N ↑;
$\mathbf{E}_x = 1710$ N ←, $\mathbf{E}_y = 1099$ N ↑.

6.95 $\mathbf{A}_x = 1318$ N →, $\mathbf{A}_y = 361$ N ↑;
$\mathbf{E}_x = 1318$ N ←, $\mathbf{E}_y = 1523$ N ↑.

6.96 (*a*) **A** = 31.8 kips ↑; **B** = 31.2 kips ↑.
(*b*) **C** = 8.71 kips ←; **D** = 43.3 kips ⬎ 78.4°.

6.97 (*a*) **A** = 1.778 kips ↑; **B** = 16.22 kips ↑.
(*b*) **C** = 32.7 kips ←; **D** = 35.0 kips ⬎ 20.8°.

6.98 $\mathbf{A}_x = 3.04$ kN ←, $\mathbf{A}_y = 2.70$ kN ↓; $\mathbf{B}_x = 6.08$ kN →,
$\mathbf{B}_y = 1.800$ kN ↑; $\mathbf{E}_x = 3.04$ kN ←, $\mathbf{E}_y = 0.900$ kN ↑.

6.99 $\mathbf{A}_x = 6.5$ kips ←, $\mathbf{A}_y = 2$ kips ↓; $\mathbf{B}_x = 18$ kips →,
$\mathbf{B}_y = 3$ kips ↑; $\mathbf{E}_x = 11.5$ kips ←, $\mathbf{E}_y = 1$ kip ↓.

6.100 $\mathbf{B}_x = 2120$ N →, $\mathbf{B}_y = 1540$ N ↓;
$\mathbf{C}_x = 1195$ N ←, $\mathbf{C}_y = 1540$ N ↑.

6.102 (*a*) $\mathbf{C}_x = 100$ N ←, $\mathbf{C}_y = 100$ N ↑; $\mathbf{D}_x = 100$ N →,
$\mathbf{D}_y = 20$ N ↓. (*b*) $\mathbf{E}_x = 100$ N ←, $\mathbf{E}_y = 180$ N ↑.

6.104 (*a*) 2.02 kN ⦞ 18.18°. (*b*) 714 N T.

6.105 (*a*) 3.17 kN ⦞ 6.52°. (*b*) 2.74 kN T.

6.106 (*a*) $\mathbf{A}_x = 50.0$ kips →, $\mathbf{A}_y = 30.5$ kips ↑.
(*b*) $\mathbf{B}_x = 50.0$ kips ←, $\mathbf{B}_y = 2.50$ kips ↓.

6.108 $\mathbf{A} = 327$ N $\rightarrow$; $\mathbf{B} = 827$ N $\leftarrow$; $\mathbf{D} = 621$ N $\uparrow$;
       $\mathbf{E} = 246$ N $\uparrow$.
6.109 $F_{AF} = P/4$ C; $F_{BG} = F_{DG} = P/\sqrt{2}$ C; $F_{EH} = P/4$ T.
6.111 $F_{AF} = P/\sqrt{2}$ C; $F_{BG} = 0$; $F_{DG} = P\sqrt{2}$ C;
       $F_{EH} = P/\sqrt{2}$ T.
6.112 $F_{AF} = M_0/4a$ C; $F_{BG} = F_{DG} = M_0/a\sqrt{2}$ T;
       $F_{EH} = 3M_0/4a$ C.
6.113 $\mathbf{A} = P/15$ $\uparrow$; $\mathbf{D} = 2P/15$ $\uparrow$; $\mathbf{E} = 8P/15$ $\uparrow$;
       $\mathbf{H} = 4P/15$ $\uparrow$.
6.114 $\mathbf{A} = 3P/13$ $\uparrow$; $\mathbf{D} = P/13$ $\uparrow$; $\mathbf{F} = 9P/13$ $\uparrow$.
6.115   (a) $\mathbf{A} = 2.24P$ ∠ 26.6°; $\mathbf{B} = 2P$ →; frame is rigid.
       (b) Frame is not rigid. (c) $\mathbf{A} = P$ $\uparrow$; $\mathbf{B} = P$ $\downarrow$; $\mathbf{C} = P$ $\uparrow$;
       frame is rigid.
6.116   (a) Frame is not rigid. (b) Reactions can be found for an
       arbitrary value of B_x; frame is rigid.
       (c) $\mathbf{A} = 2.09P$ ⊾ 16.70°; $\mathbf{B} = 2.04P$ ∠ 11.31°; frame is
       rigid.
6.117   (a) $\mathbf{A} = 2.06P$ ∠ 14.04°; $\mathbf{B} = 2.06P$ ⊾ 14.04°; frame is
       rigid. (b) Frame is not rigid. (c) $\mathbf{A} = 1.250P$ ⊾ 36.9°;
       $\mathbf{B} = 1.031P$ ∠ 14.04°; frame is rigid.
6.118   282 N →.
6.120   (a) 179.0 lb $\downarrow$. (b) 135.7 lb ⭦ 61.3°.
6.121   (a) 80.5 lb $\downarrow$. (b) 181.9 lb ⭦ 61.3°.
6.122   455 N ⭦ 16.17°.
6.125   (a) 7.00 kips ←. (b) 17.50 kips ←.
6.127   93.8 lb · in. ↲.
6.128   463 lb · in ↲.
6.129   110.6 N · m ↲.
6.130   14.83°.
6.131   15.22 N · m ↰.
6.132   12.50 N · m ↰.
6.133   3.33 kip · in. ↲.
6.135 $\mathbf{D} = 2.52$ kips ←; $\mathbf{F} = 1.694$ kips ⭦ 16.46°.
6.137 $\mathbf{F} = 3.97$ kN →; $\mathbf{H} = 3.31$ kN ⬈ 20.9°.
6.138 $\mathbf{B} = 1117$ N ⬈ 18.43°; $\mathbf{D} = 1117$ N ⭦ 18.43°.
6.139   1680 N.
6.141   1.408 kips.
6.142   156.6 N.
6.143   50.9 N ⬈ 30.0°.
6.145   60.0 lb.
6.146   133.3 lb.
6.147   (a) 12.15 kN ⭲ 44.7°. (b) 10.63 kN ⭲ 35.7°.
6.148   (a) 14.29 kN ⭲ 0.261°. (b) 14.47 kN ⭲ 9.10°.
6.149   (a) 10.72 kN ∠ 21.8°. (b) 35.4 kN ⭲ 45.0°.
6.151   (a) 30.0 mm. (b) 80.0 N · m ↲.
6.153   (a) 57.7 N · m. (b) $\mathbf{B} = 0$; $\mathbf{D} = -(48.1$ N)$\mathbf{k}$; $\mathbf{E} = (48.1$ N)$\mathbf{k}$.
6.154   (a) 43.3 N · m. (b) $\mathbf{B} = -(50.0$ N)$\mathbf{k}$; $\mathbf{D} = (83.3$ N)$\mathbf{k}$;
       $\mathbf{E} = -(33.3$ N)$\mathbf{k}$.
6.155 $\mathbf{E}_x = 2450$ lb →, $\mathbf{E}_y = 3790$ lb $\uparrow$; $\mathbf{F}_x = 648$ lb →,
       $\mathbf{F}_y = 2890$ lb $\downarrow$; $\mathbf{H}_x = 3100$ lb ←, $\mathbf{H}_y = 900$ lb $\downarrow$.
6.157 $F_{AB} = F_{BD} = F_{DF} = F_{FH} = 32.0$ kips T; $F_{AC} = 34.2$ kips C;
       $F_{BC} = F_{BE} = F_{DE} = F_{DG} = F_{FG} = 0$;
       $F_{CE} = F_{EG} = 34.2$ kips C; $F_{GH} = 48.2$ kips C.
6.159 $F_{BC} = 4.10$ kips T; $F_{BH} = 4.39$ kips T; $F_{GH} = 8.00$ kips C.
6.161 $\mathbf{B}_x = 10.00$ kN →, $\mathbf{B}_y = 13.75$ kN $\uparrow$;
       $\mathbf{D}_x = 22.0$ kN ←, $\mathbf{D}_y = 13.75$ kN $\downarrow$.
6.162 $F_{AB} = 15.90$ kN C; $F_{AC} = 13.50$ kN T; $F_{BC} = 16.80$ kN C;
       $F_{BD} = 13.50$ kN C; $F_{CD} = 15.90$ kN T.
6.164 $F_{DF} = 5.45$ kN C; $F_{DG} = 1.000$ kN T; $F_{EG} = 4.65$ kN T.
6.165 $F_{GI} = 4.65$ kN T; $F_{HI} = 1.800$ kN C; $F_{HJ} = 4.65$ kN C.
6.166   (a) 80.0 lb T. (b) 72.1 lb ⬈ 16.10°.
6.167 $\mathbf{E} = 970$ lb →; $\mathbf{F} = 633$ lb ⬈ 39.2°.

CHAPTER 7

7.1   (On JD) $\mathbf{F} = 0$; $\mathbf{V} = 40.0$ lb $\uparrow$; $\mathbf{M} = 160.0$ lb · in. ↰.
7.2   (On AJ) $\mathbf{F} = 0$; $\mathbf{V} = 60.0$ lb ←; $\mathbf{M} = 60.0$ lb · in. ↰.
7.4   (On AJ) $\mathbf{F} = 129.9$ N ⭨; $\mathbf{V} = 25.0$ N ⬈; $\mathbf{M} = 4.00$ N · m ↲.
7.5   (On AJ) $\mathbf{F} = 624$ N →; $\mathbf{V} = 260$ N $\downarrow$; $\mathbf{M} = 52.0$ N · m ↰.
7.7   (On AJ) $\mathbf{F} = 26.0$ lb ⭦; $\mathbf{V} = 15.00$ lb ⬈; $\mathbf{M} = 234$ lb · in. ↲.
7.9   (On CJ) $\mathbf{F} = 110.4$ N ⭨; $\mathbf{V} = 135.8$ N ⭧; $\mathbf{M} = 50.4$ N · m ↰.
7.10   (a) 140.0 N at C. (b) 156.5 N at B and D.
       (c) 89.6 N · m at C.
7.11   (On AJ) $\mathbf{F} = 48.7$ lb ⭲ 60°; $\mathbf{V} = 64.3$ lb ∠ 30°;
       $\mathbf{M} = 309$ lb · in. ↲.
7.13   (On CJ) $\mathbf{F} = 5.93$ N ⬈; $\mathbf{V} = 412$ N ⭨; $\mathbf{M} = 34.8$ N · m ↲.
7.14   (On KD) $\mathbf{F} = 48.0$ N ⬋; $\mathbf{V} = 83.1$ N ⭨; $\mathbf{M} = 17.86$ N · m ↲.
7.15   (On BJ) $\mathbf{F} = 62.5$ lb ⭨; $\mathbf{V} = 30.0$ lb ⬈; $\mathbf{M} = 90.0$ lb · ft ↰.
7.16   (On AK) $\mathbf{F} = 140.0$ lb ←; $\mathbf{V} = 22.5$ lb $\downarrow$; $\mathbf{M} = 54.0$ lb · ft ↲.
7.19   (On CJ) $\mathbf{F} = 680$ N ⭨; $\mathbf{V} = 87.2$ N ⬋; $\mathbf{M} = 18.97$ N · m ↰.
7.20   (On AK) $\mathbf{F} = 514$ N ⬋; $\mathbf{V} = 87.2$ N ⭦; $\mathbf{M} = 18.97$ N · m ↲.
7.21   (a) (On AJ) $\mathbf{F} = 0$; $\mathbf{V} = P/2$ →; $\mathbf{M} = Pa/2$ ↰.
       (b) (On AJ) $\mathbf{F} = 2P$ $\downarrow$; $\mathbf{V} = 3P/2$ ←; $\mathbf{M} = 3Pa/2$ ↲.
       (c) (On AJ) $\mathbf{F} = 2P/7$ $\downarrow$; $\mathbf{V} = 3P/14$ →; $\mathbf{M} = 3Pa/14$ ↰.
7.22   (a) (On AJ) $\mathbf{F} = P/2$ $\downarrow$; $\mathbf{V} = 0$; $\mathbf{M} = 0$.
       (b) (On AJ) $\mathbf{F} = 11P/14$ $\downarrow$; $\mathbf{V} = 2P/7$ ←; $\mathbf{M} = 2Pa/7$ ↲.
       (c) (On AJ) $\mathbf{F} = 5P/2$ $\downarrow$; $\mathbf{V} = 2P$ ←; $\mathbf{M} = 2Pa$ ↲.
7.23   (On CJ) $0.0666Wr$ ↲.
7.24   (On AJ) $0.1788Wr$ ↲.
7.25   (On AJ) $0.0557Wr$ ↰.
7.27 $0.357Wr$ for $\theta = 49.3°$.
7.29   (b) $wL/4$; $3wL^2/32$.
7.30   (b) $wL/2$; $3wL^2/8$.
7.33   (b) 12.63 kips; 29.8 kip · ft.
7.34   (b) 9.50 kN; 4.80 kN · m.
7.35   (b) 8.00 kips; 33.0 kip · ft.
7.36   (b) 810 N; 162.0 N · m.
7.37   (b) 14.00 kips; 114.0 kip · ft.
7.38   (b) 14.40 kips; 57.6 kip · ft.
7.40   (b) 22.0 kN; 22.0 kN · m.
7.41   (b) 4.50 kN; 13.50 kN · m.
7.43   (b) 540 lb; 203 lb · ft.
7.44   (b) 600 lb; 450 lb · ft.
7.45   (b) 1.833 kN; 799 N · m.
7.47   (a) +800 N; +640 N · m. (b) −400 N; +160 N · m.
7.49   7.50 kN; 7.20 kN · m.
7.50   (b) 768 N; 624 N · m.
7.51   (a) 0.932 ft. (b) 173.7 lb · ft.
7.52   (a) 478 mm. (b) 580 N · m.
7.53   (a) 40.0 kN. (b) 40.0 kN · m.
7.54   (a) 2.13 ft. (b) 4.25 kip · ft.
7.55   (a) 6.93 ft. (b) 402 lb · ft.
7.58   (b) $wL/4$; $3wL^2/32$.
7.59   (b) $wL/2$; $3wL^2/8$.
7.62   (b) 12.63 kips; 29.8 kip · ft.
7.63   (b) 810 N; 162.0 N · m.
7.64   (b) 6.60 kN; 5.82 kN · m.
7.65   (b) 10.50 kips; 30.0 kip · ft.
7.66   (b) 14.00 kips; 114.0 kip · ft.
7.67   (b) 14.40 kips; 57.6 kip · ft.
7.69   (b) 22.0 kN; 22.0 kN · m.
7.71   (b) 7.00 kN; 30.0 kN · m.
7.72   (b) 8.44 kip · ft, 4.25 ft from A.
7.73   (b) 42.2 kip · ft, 8.20 ft from A.
7.74   (a) 35.0 kN · m. (b) 57.1 kN · m.

7.75 (a) 22.5 kN · m, 1.500 m from A.
(b) 30.2 kN · m, 1.350 m from A.

7.78 (b) 25.0 kN · m, 2.50 m from A.

7.79 (b) 13.44 kN · m, 1.833 m from A.

7.80 (a) $V = (w_0/6L)(3x^2 - 6Lx + 2L^2)$;
$M = (w_0/6L)(x^3 - 3Lx^2 + 2L^2x)$. (c) $0.0642w_0L^2$,
at $x = 0.423L$.

7.81 $V = (w_0/L)(-2x^2 + Lx)$;
$M = (w_0/6L)(-4x^3 + 3Lx^2)$. (c) $w_0L^2/6$, at $x = L$.

7.82 (b) $1.359w_0a^2$, at $x = 2.10a$.

7.84 (a) $\mathbf{P} = 850$ lb ↓; $\mathbf{Q} = 1275$ lb ↓. (b) $M_C = -1000$ lb · ft.

7.86 (a) $\mathbf{P} = 6.60$ kN ↓; $\mathbf{Q} = 0.600$ kN ↓.
(b) $V_{max} = 7.40$ kN, at A;
$M_{max} = 7.13$ kN · m, 1.333 m from A.

7.87 (a) $\mathbf{P} = 0.660$ kN ↓; $\mathbf{Q} = 2.28$ kN ↓.
(b) $|V|_{max} = 3.81$ kN, at B;
$M_{max} = 4.23$ kN · m, 3.45 m from A.

7.88 (a) 2.25 ft. (b) $\mathbf{A}_x = 800$ lb ←, $\mathbf{A}_y = 450$ lb ↑.
(c) 1000 lb.

7.90 (a) 3.35 kN ⬐ 17.35°. (b) 3.88 kN ∠ 34.5°.

7.91 (a) 10.67 kN ⬏ 2.15°. (b) 11.26 kN ∠ 18.65°.

7.92 (a) 929 lb ∠ 23.8°. (b) $d_B = 2.22$ ft; $d_D = 3.28$ ft.

7.94 (a) 12.29 kN. (b) 11.00 m.

7.95 (a) 15.62 kN. (b) 8.00 m.

7.96 (a) 113.9 N. (b) 1.626 m.

7.98 32.0 lb.

7.99 34.9 lb.

7.100 $P = 108.0$ kN; $Q = 45.0$ kN.

7.102 (a) 3.69 kN. (b) 60.1 m.

7.104 (a) 50,200 kips. (b) 3575 ft.

7.105 (a) 56,400 kips. (b) 4284 ft.

7.106 (a) 313 mm. (b) 1.433°.

7.108 3.75 ft.

7.109 (a) $\sqrt{3L\Delta/8}$. (b) 3.67 m.

7.110 (a) 58,900 kips. (b) 29.2°.

7.111 (a) 24.0 ft to the left of B. (b) 3450 lb.

7.112 (a) 1.713 m. (b) 6.19 m.

7.113 (a) 2.22 m. (b) 5.36 kN.

7.117 11.00 m.

7.118 8.00 m.

7.120 $y = h[1 - \cos(\pi x/L)]$; $T_0 = w_0L^2/h\pi^2$;
$T_{max} = (w_0L/\pi)\sqrt{(L^2/h^2\pi^2) + 1}$.

7.122 (a) 7.92 ft. (b) 1.164 lb.

7.123 (a) 29.7 ft. (b) 61.5 lb.

7.124 199.5 ft.

7.126 (a) 9.93 N →. (b) 14.83 m.

7.128 (a) 8.83 m. (b) 24.5 N →.

7.130 18.69 ft.

7.131 (a) 4.22 m. (b) 80.3°.

7.132 5.23 m.

7.133 2.68 m and 5.99 m.

7.134 12.14 lb.

7.136 27.3 lb →.

7.137 40.7 lb →.

7.138 (a) $a = 26.8$ m; $b = 20.4$ m. (b) 35.3 m.

7.139 (a) $a = 20.1$ m; $b = 15.29$ m. (b) 26.5 m.

7.140 (a) Point C is 23.9 ft to the left of B. (b) 14.42 lb.

7.141 (a) Point C is 20.7 ft to the left of B. (b) 9.55 lb.

7.143 (a) $1.326T_m/w$. (b) 12.72 km.

7.146 (a) 0.338. (b) $\theta_B = 56.5°$; $T_m = 0.755wL$.

7.147 (b) 11.50 kips; 14.00 kip · ft.

7.149 (a) 2.28 m. (b) $\mathbf{D}_x = 13.67$ kN →, $\mathbf{D}_y = 7.80$ kN ↑.
(c) 15.94 kN.

7.151 (On BJ) $0.289Wr$ ↰.

7.152 (On BJ) $0.417Wr$ ↰.

7.153 (On CJ) $\mathbf{F} = 625$ N ↓; $\mathbf{V} = 120.0$ N ←; $\mathbf{M} = 27.0$ N · m ↰.

7.155 900 N; 675 N · m.

7.156 3040 lb · ft, 4.50 ft from A.

7.157 (a) 48.0 lb. (b) 10.00 ft.

CHAPTER 8

8.1 Block moves; $\mathbf{F} = 193.2$ N ↗.

8.2 Equilibrium; $\mathbf{F} = 246$ N ↗.

8.3 Equilibrium; $\mathbf{F} = 0.677$ lb ↙.

8.5 1.542 lb $\le P \le 12.08$ lb.

8.6 (a) 41.7 lb. (b) 14.04°.

8.7 (a) 50.5°. (b) 66.5°.

8.9 (a) $0 \le \theta \le 54.9°$. (b) $168.7° \le \theta \le 180.0°$.

8.11 (a) 72.0 lb ←. (b) 40.0 lb ←.

8.12 (a) 56.0 lb ←. (b) 40.0 lb ←.

8.13 43.0°.

8.15 (a) 141.3 N. (b) 117.7 N. (c) 50.5 N.

8.17 $Wr\mu_s(1 + \mu_s)/(1 + \mu_s^2)$.

8.18 (a) $0.360Wr$. (b) $0.422Wr$.

8.19 1374 lb · in.

8.21 0.200.

8.22 0.1835.

8.23 (a) 4.62° and 48.2°. (b) $0.526W$ and $0.374W$.

8.25 0.1865.

8.27 0.283.

8.28 (a) 28.1 lb ←. (b) 0.440 in.

8.30 $(\mu_s)_A = 0.1364$; $(\mu_s)_C = 0.1512$.

8.31 $(\mu_s)_A = 0.1875$; $(\mu_s)_C = 0.1977$.

8.32 (a) Plate is in equilibrium.
(b) Plate moves downward.

8.34 (a) $W \le 34.6$ N. (b) 17.82 N $\le W \le 98.2$ N.

8.35 (a) $W \le 4.07$ N and $W \ge 86.4$ N. (b) $W \ge 246$ N.

8.36 -11.71 lb $\le P \le 8.59$ lb.

8.37 -12.31 N $\le P \le 105.3$ N.

8.39 37.9 N.

8.40 24.5 lb · ft ↰.

8.42 0.1757.

8.43 35.8°.

8.45 $1.268W$.

8.46 (a) 4.92 lb. (b) $\mathbf{C}_x = 1.082$ lb ←; $\mathbf{C}_y = 104.7$ lb ↑.

8.47 (a) 10.17 lb. (b) $\mathbf{C}_x = 6.52$ lb →; $\mathbf{C}_y = 105.4$ lb ↑.

8.48 1.957 kN.

8.49 2.13 kN.

8.52 2.46 kips ←.

8.54 (a) 1.647 kN. (b) 1.647 kN.

8.55 (a) 789 N. (b) 1100 N.

8.57 1.509 lb.

8.58 (a) Wedge is forced up and out from between plates.
(b) Wedge will become self-locked at $\theta = 4.20°$.

8.59 305 N.

8.60 (b) 57.7 lb ←.

8.62 (a) 821 N →. (b) Plank moves.

8.63 (a) 818 N ←. (b) Plank does not move.

8.64 0.384.

8.67 14.45 N · m.

8.70 52.2 lb · in.

8.71   36.1 lb · in.

8.72   15.31 N · m.

8.73   4.14 N · m.

8.74   14.48 N · m.

8.75   0.0787.

8.76   90.8 lb.

8.77   83.6 lb.

8.80   (a) 0.344. (b) 120.2 N.

8.81 $T_{AB} = 290$ N; $T_{CD} = 310$ N; $T_{EF} = 331$ N.

8.82 $T_{AB} = 310$ N; $T_{CD} = 290$ N; $T_{EF} = 272$ N.

8.83   (a) 1.177 kips. (b) 1.349°.

8.84   253 N.

8.85   250 N.

8.88   (a) 219 lb. (b) 164.0 lb.

8.89   9.95 in.

8.90   18.39 N.

8.91   15.25 lb · in.

8.95   13.80 N.

8.96   35.0 lb.

8.97   1.200 mm.

8.98   42.7 N.

8.99   (a) 306 lb. (b) 252 lb.

8.101   (a) 0.329. (b) 2.67 turns.

8.102   (a) 0.213. (b) 25.0 lb.

8.103   11.85 lb ≤ W_B ≤ 33.8 lb.

8.104   286 N ≤ P ≤ 4840 N.

8.107   40.1 N · m.

8.108   28.6 N · m.

8.109   (a) 338 lb · in. (b) 168.8 lb.

8.111   (a) 11.98 lb. (b) 1.900.

8.113   (a) 432 N · m. (b) 0.219.

8.114   (a) 646 N. (b) 47.0 N.

8.115   0.361.

8.116   (a) 23.3 lb. (b) 95.1 lb. (c) 68.8 lb.

8.118   (a) 1.847 kg ≤ m_A ≤ 34.7 kg. (b) 2.81 kg ≤ m_A ≤ 22.8 kg.

8.119   (a) 5.55 kg. (b) 8.00 kg.

8.120   (a) 2.28 kg ≤ m_A ≤ 28.1 kg. (b) 3.46 kg ≤ m_A ≤ 18.49 kg.

8.121   (a) 7.79 kg. (b) 5.40 kg.

8.122   1.074 lb.

8.124   (a) 538 N · m ↓. (b) 1.142 kN ↓.

8.125   (a) 538 N · m ↖. (b) 879 N ↓.

8.126   0.350.

8.127   0.266.

8.130   81.7 N · m.

8.131   (a) 638 lb · in. (b) 219 lb.

8.133   48.3° ≤ θ ≤ 78.7°.

8.135   31.0°.

8.137   14.23 kg ≤ m ≤ 175.7 kg.

8.139   665 lb.

8.140   35.1 N · m.

8.141   (b) 2.69 lb.

8.142   143.0 lb ≤ P ≤ 483 lb.

8.143   (a) 11.43 lb. (b) Motion impends at C.

CHAPTER 9

9.1 $5a^3b/33$.

9.2 $3a^3b/10$.

9.3 $b^3h/4$.

9.4 $47a^3b/60$.

9.5 $5ab^3/17$.

9.8 $0.0430ab^3$.

9.9 $4ab^3/15$.

9.10 $ab^3/4$.

9.13 $2.16a^3b$.

9.14 $0.1606a^3b$.

9.15 $ab^3/10$; $b/\sqrt{5}$.

9.17 $a^3b/6$; $a/\sqrt{3}$.

9.18 $15a^3b/91$; $0.524a$.

9.20 $1.523a^3h$; $1.404a$.

9.21 $12a^4$; $1.414a$.

9.22 $4ab(a^2 + 5b^2)/3$; $\sqrt{(a^2 + 5b^2)/3}$.

9.23 $ab(7a^2 + 52b^2)/210$; $\sqrt{(7a^2 + 52b^2)/70}$.

9.26   (b) −10.56%; −0.772%; −0.0488%.

9.27 $31\pi a^4/20$; $1.153a$.

9.28 $bh(12h^2 + b^2)/48$; $\sqrt{(12h^2 + b^2)}/24$.

9.31   2.44 in⁴; 1.093 in.

9.32 7.36×10^6 mm⁴; 32.0 mm.

9.33   0.402 in⁴; 0.444 in.

9.35 $I_x = 60.9a^4$; $I_y = 11.86a^4$.

9.37 $\bar{I} = 20.6$ in⁴; $d_2 = 2.50$ in.

9.38   A = 5.00 in²; $\bar{I} = 12.19$ in⁴.

9.39 $J_B = 608 \times 10^6$ mm⁴; $J_D = 1215 \times 10^6$ mm⁴.

9.40 4.33×10^6 mm⁴.

9.41 $\bar{I}_x = 11.71$ in⁴; $\bar{I}_y = 36.4$ in⁴.

9.43 $\bar{I}_x = 30.6 \times 10^6$ mm⁴; $\bar{I}_y = 12.03 \times 10^6$ mm⁴.

9.44 $\bar{I}_x = 433 \times 10^6$ mm⁴; $\bar{I}_y = 733 \times 10^6$ mm⁴.

9.46   (a) 96.4×10^3 in⁴. (b) 96.2×10^3 in⁴.

9.47   (a) 122.4×10^6 mm⁴. (b) 64.4×10^6 mm⁴.

9.48   (a) 1392×10^6 mm⁴. (b) 729×10^6 mm⁴.

9.49 $\bar{I}_x = 982$ in⁴, $\bar{I}_y = 101.0$ in⁴;
 $\bar{k}_x = 5.66$ in., $\bar{k}_y = 1.814$ in.

9.51 $\bar{I}_x = 126.4 \times 10^6$ mm⁴, $\bar{I}_y = 108.8 \times 10^6$ mm⁴;
 $\bar{k}_x = 94.2$ mm, $\bar{k}_y = 87.5$ mm.

9.52   12.29 mm.

9.53 $\bar{I}_x = 9.34$ in⁴; $\bar{I}_y = 131.2$ in⁴.

9.55   b = 91.2 mm; $\bar{I}_x = 11.34 \times 10^6$ mm⁴,
 $\bar{I}_y = 34.0 \times 10^6$ mm⁴.

9.56   (a) 364 mm.
 (b) $\bar{I}_x = 34.1 \times 10^6$ mm⁴, $\bar{I}_y = 110.2 \times 10^6$ mm⁴.

9.57 $3\pi r/16$.

9.60   1.224r.

9.61 $F_A = F_B = 217$ lb; $F_C = F_D = 222$ lb.

9.62   2.66 m.

9.63   225 mm.

9.64   4.38 in.

9.67 $a^4/2$.

9.68 $a^2b^2/12$.

9.69 $-b^2h^2/8$.

9.70 $a^2b^2(4 + \pi^2)/8\pi^2$.

9.72   501 in⁴.

9.73 138.2×10^6 mm⁴.

9.74 -159.6×10^3 mm⁴.

9.75 1.573×10^6 mm⁴.

9.77 -262×10^6 mm⁴.

9.78   2.81 in⁴.

9.80 $\bar{I}_{x'} = 3.31 \times 10^3$ in⁴; $\bar{I}_{y'} = 2.31 \times 10^3$ in⁴;
 $\bar{I}_{x'y'} = 1.947 \times 10^3$ in⁴.

9.82 $\bar{I}_{x'} = 4.61 \times 10^6$ mm⁴; $\bar{I}_{y'} = 3.82 \times 10^6$ mm⁴;
 $\bar{I}_{x'y'} = -3.83 \times 10^6$ mm⁴.

9.83 $\bar{I}_{x'} = 149.9 \times 10^3$ mm⁴; $\bar{I}_{y'} = 469 \times 10^3$ mm⁴;
 $\bar{I}_{x'y'} = 143.5 \times 10^3$ mm⁴.

9.84 $\bar{I}_{x'} = 5.30$ in^4; $\bar{I}_{y'} = 6.73$ in^4; $\bar{I}_{x'y'} = 4.38$ in^4.

9.86 7.22°; 4820 in^4, 802 in^4.

9.88 12.06°; 8.06×10^6 mm^4, 0.365×10^6 mm^4.

9.89 $-24.0°$; 524×10^3 mm^4, 94.9×10^3 mm^4.

9.90 $-19.64°$; 10.45 in^4, 1.577 in^4.

9.92 $\bar{I}_{x'} = 3.31 \times 10^3$ in^4; $\bar{I}_{y'} = 2.31 \times 10^3$ in^4; $\bar{I}_{x'y'} = 1.947 \times 10^3$ in^4.

9.94 $\bar{I}_{x'} = 4.61 \times 10^6$ mm^4; $\bar{I}_{y'} = 3.82 \times 10^6$ mm^4; $\bar{I}_{x'y'} = -3.83 \times 10^6$ mm^4.

9.95 $\bar{I}_{x'} = 149.9 \times 10^3$ mm^4; $\bar{I}_{y'} = 469 \times 10^3$ mm^4; $\bar{I}_{x'y'} = 143.5 \times 10^3$ mm^4.

9.96 $\bar{I}_{x'} = 5.30$ in^4; $\bar{I}_{y'} = 6.73$ in^4; $\bar{I}_{x'y'} = 4.38$ in^4.

9.97 20.2°; $1.754a^4$, $0.209a^4$.

9.98 7.22°; 4820 in^4, 802 in^4.

9.100 29.7°; 405×10^6 mm^4, 83.9×10^6 mm^4.

9.101 $-24.0°$; 524×10^3 mm^4, 94.9×10^3 mm^4.

9.103 $-20.0°$; 39.6 in^4, 5.37 in^4.

9.104 $-30.1°$; 884×10^6 mm^4, 281×10^6 mm^4.

9.105 (a) -215×10^3 mm^4. (b) $-28.1°$. (c) 568×10^3 mm^4.

9.106 19.64°; 10.45 in^4, 1.577 in^4.

9.107 $-24.0°$; 8.33×10^6 mm^4, 1.509×10^6 mm^4.

9.108 (a) $-7.50°$. (b) 646 in^4, 274 in^4.

9.112 $\pm 159.4 \times 10^3$ mm^4.

9.113 (a) $5mr_2^2/16$. (b) $0.1347mr_2^2$.

9.114 (a) $0.0699mb^2$. (b) $m(a^2 + 0.279b^2)/4$.

9.116 (a) $5ma^2/3$. (b) $11ma^2/18$.

9.117 (a) $ma^2/6$. (b) $m(a^2 + 7b^2)/6$.

9.119 (a) $5ma^2/18$. (b) $3.61ma^2$.

9.120 (a) $0.994ma^2$. (b) $2.33ma^2$.

9.121 $5mb^2/18$.

9.123 (a) $1.085ma^2$. (b) $3.67ma^2$.

9.124 $m(b^2 + h^2)/10$.

9.126 $m(a^2 + b^2)/5$.

9.127 $I_x = I_y = ma^2/4$; $I_z = ma^2/2$.

9.128 (a) $mh^2/6$. (b) $m(a^2 + 4h^2 \sin^2 \theta)/24$. (c) $m(a^2 + 4h^2 \cos^2 \theta)/24$.

9.129 1.964×10^{-6} lb $\cdot$ ft $\cdot$ s^2; 0.345 in.

9.130 1.184×10^{-6} kg $\cdot$ m^2; 8.19 mm.

9.131 $ma^2/2$; $a/\sqrt{2}$.

9.132 2.00 in.

9.133 281×10^{-3} kg $\cdot$ m^2.

9.136 (a) 46.0 mm. (b) 8.54×10^{-3} kg $\cdot$ m^2; 45.4 mm.

9.137 $I_x = 2.14$ lb $\cdot$ ft $\cdot$ s^2; $I_y = 4.83$ lb $\cdot$ ft $\cdot$ s^2; $I_z = 4.03$ lb $\cdot$ ft $\cdot$ s^2.

9.139 $I_x = 612 \times 10^{-6}$ kg $\cdot$ m^2; $I_y = 737 \times 10^{-6}$ kg $\cdot$ m^2; $I_z = 250 \times 10^{-6}$ kg $\cdot$ m^2.

9.141 $I_x = 37.4 \times 10^{-3}$ kg $\cdot$ m^2; $I_y = 90.9 \times 10^{-3}$ kg $\cdot$ m^2; $I_z = 76.7 \times 10^{-3}$ kg $\cdot$ m^2.

9.142 $I_x = 15.40 \times 10^{-3}$ lb $\cdot$ ft $\cdot$ s^2; $I_y = 999 \times 10^{-3}$ lb $\cdot$ ft $\cdot$ s^2; $I_z = 988 \times 10^{-3}$ lb $\cdot$ ft $\cdot$ s^2.

9.143 (a) 34.1×10^{-3} lb $\cdot$ ft $\cdot$ s^2. (b) 50.1×10^{-3} lb $\cdot$ ft $\cdot$ s^2. (c) 34.9×10^{-3} lb $\cdot$ ft $\cdot$ s^2.

9.144 21.2 kg $\cdot$ m^2.

9.145 3.98 kg $\cdot$ m^2.

9.147 (a) 9.88×10^{-3} lb $\cdot$ ft $\cdot$ s^2. (b) 11.53×10^{-3} lb $\cdot$ ft $\cdot$ s^2. (c) 2.19×10^{-3} lb $\cdot$ ft $\cdot$ s^2.

9.148 $I_x = I_z = 6.85 \times 10^{-3}$ kg $\cdot$ m^2; $I_y = 12.63 \times 10^{-3}$ kg $\cdot$ m^2.

9.149 $I_x = 23.2 \times 10^{-3}$ kg $\cdot$ m^2; $I_y = 21.4 \times 10^{-3}$ kg $\cdot$ m^2; $I_z = 17.99 \times 10^{-3}$ kg $\cdot$ m^2.

9.151 $I_{xy} = 0.488 \times 10^{-3}$ lb $\cdot$ ft $\cdot$ s^2; $I_{yz} = 1.184 \times 10^{-3}$ lb $\cdot$ ft $\cdot$ s^2; $I_{zx} = 2.70 \times 10^{-3}$ lb $\cdot$ ft $\cdot$ s^2.

9.152 $I_{xy} = 5.94 \times 10^{-3}$ lb $\cdot$ ft $\cdot$ s^2; $I_{yz} = 3.45 \times 10^{-3}$ lb $\cdot$ ft $\cdot$ s^2; $I_{zx} = 11.19 \times 10^{-3}$ lb $\cdot$ ft $\cdot$ s^2.

9.153 $I_{xy} = 36.2 \times 10^{-3}$ kg $\cdot$ m^2; $I_{yz} = 8.49 \times 10^{-3}$ kg $\cdot$ m^2; $I_{zx} = 20.6 \times 10^{-3}$ kg $\cdot$ m^2.

9.154 $I_{xy} = -691 \times 10^{-6}$ kg $\cdot$ m^2; $I_{yz} = 203 \times 10^{-6}$ kg $\cdot$ m^2; $I_{zx} = -848 \times 10^{-6}$ kg $\cdot$ m^2.

9.157 $I_{xy} = -0.1931$ kg $\cdot$ m^2; $I_{yz} = 0.310$ kg $\cdot$ m^2; $I_{zx} = 2.26$ kg $\cdot$ m^2.

9.158 $I_{xy} = -187.5 \times 10^{-6}$ lb $\cdot$ ft $\cdot$ s^2; $I_{yz} = 0$; $I_{zx} = -102.7 \times 10^{-6}$ lb $\cdot$ ft $\cdot$ s^2.

9.159 $I_{xy} = -11wa^3/g$; $I_{yz} = wa^3(\pi + 6)/2g$; $I_{zx} = -wa^3/4g$.

9.162 $I_{xy} = -m'R_1^3/2$; $I_{yz} = m'R_1^3/2$; $I_{zx} = -m'R_2^3/2$.

9.164 (a) $mac/20$. (b) $I_{xy} = mab/20$; $I_{yz} = mbc/20$.

9.165 $ma^2(10h^2 + 3a^2)/12(h^2 + a^2)$.

9.166 $3.22ma^2$.

9.167 44.3×10^{-3} lb $\cdot$ ft $\cdot$ s^2.

9.168 4.96×10^{-3} lb $\cdot$ ft $\cdot$ s^2.

9.169 $5Wa^2/18g$.

9.170 $5\rho t a^4/12$.

9.171 6.74 kg $\cdot$ m^2.

9.173 25.3×10^{-3} kg $\cdot$ m^2.

9.175 (a) $b/a = 2.00$; $c/a = 2.00$. (b) $b/a = 1.000$; $c/a = 0.500$.

9.176 (a) 2.00. (b) $\sqrt{2/3}$.

9.177 (a) $1/\sqrt{3}$. (b) $\sqrt{7/12}$.

9.179 (a) $ma^2/6$. (b) $I_{x'} = ma^2/6$; $I_{y'} = I_{z'} = 11ma^2/12$.

9.181 (a) $K_1 = 0.363ma^2$; $K_2 = 1.583ma^2$; $K_3 = 1.720ma^2$. (b) $(\theta_x)_1 = (\theta_z)_1 = 49.7°$, $(\theta_y)_1 = 113.7°$; $(\theta_x)_2 = 45.0°$, $(\theta_y)_2 = 90.0°$, $(\theta_z)_2 = 135.0°$; $(\theta_x)_3 = (\theta_z)_3 = 73.5°$, $(\theta_y)_3 = 23.7°$.

9.182 (a) $K_1 = 34.9 \times 10^{-3}$ lb $\cdot$ ft $\cdot$ s^2; $K_2 = 34.0 \times 10^{-3}$ lb $\cdot$ ft $\cdot$ s^2; $K_3 = 50.2 \times 10^{-3}$ lb $\cdot$ ft $\cdot$ s^2. (b) $(\theta_x)_1 = (\theta_y)_1 = 90.0°$, $(\theta_z)_1 = 0$; $(\theta_x)_2 = 3.43°$, $(\theta_y)_2 = 86.6°$, $(\theta_z)_2 = 90.0°$; $(\theta_x)_3 = 93.4°$, $(\theta_y)_3 = 3.41°$, $(\theta_z)_3 = 90.0°$.

9.183 (a) $K_1 = 1.180 \times 10^{-3}$ lb $\cdot$ ft $\cdot$ s^2; $K_2 = 10.72 \times 10^{-3}$ lb $\cdot$ ft $\cdot$ s^2; $K_3 = 11.70 \times 10^{-3}$ lb $\cdot$ ft $\cdot$ s^2. (b) $(\theta_x)_1 = 72.5°$, $(\theta_y)_1 = 83.0°$, $(\theta_z)_1 = 18.89°$; $(\theta_x)_2 = 19.38°$, $(\theta_y)_2 = 83.7°$, $(\theta_z)_2 = 108.3°$; $(\theta_x)_3 = 98.2°$, $(\theta_y)_3 = 9.46°$, $(\theta_z)_3 = 94.7°$.

9.184 (a) $K_1 = 0.1639Wa^2/g$; $K_2 = 1.054Wa^2/g$; $K_3 = 1.115Wa^2/g$. (b) $(\theta_x)_1 = 36.7°$, $(\theta_y)_1 = 71.6°$, $(\theta_z)_1 = 59.5°$; $(\theta_x)_2 = 74.9°$, $(\theta_y)_2 = 54.5°$, $(\theta_z)_2 = 140.5°$; $(\theta_x)_3 = 57.5°$, $(\theta_y)_3 = 138.8°$, $(\theta_z)_3 = 112.4°$.

9.185 (a) $K_1 = 0.203\rho t a^4$; $K_2 = 0.698\rho t a^4$; $K_3 = 0.765\rho t a^4$. (b) $(\theta_x)_1 = 40.2°$, $(\theta_y)_1 = 50.0°$, $(\theta_z)_1 = 86.7°$; $(\theta_x)_2 = 56.2°$, $(\theta_y)_2 = 134.5°$, $(\theta_z)_2 = 63.4°$; $(\theta_x)_3 = 70.8°$, $(\theta_y)_3 = 108.0°$, $(\theta_z)_3 = 153.2°$.

9.186 (a) $K_1 = 10.02 \times 10^{-3}$ lb $\cdot$ ft $\cdot$ s^2; $K_2 = 18.62 \times 10^{-3}$ lb $\cdot$ ft $\cdot$ s^2; $K_3 = 22.9 \times 10^{-3}$ lb $\cdot$ ft $\cdot$ s^2. (b) $(\theta_x)_1 = 35.2°$, $(\theta_y)_1 = (\theta_z)_1 = 65.9°$; $(\theta_x)_2 = 90.0°$, $(\theta_y)_2 = 45.0°$, $(\theta_z)_2 = 135.0°$; $(\theta_x)_3 = 54.7°$, $(\theta_y)_3 = (\theta_z)_3 = 125.3°$.

9.187 $a^3b/6$.

9.188 $\bar{I}_x = 18.13$ in^4; $\bar{I}_y = 4.51$ in^4.

9.190 (a) $7ma^2/18$. (b) $0.819ma^2$.

9.192 (a) $\bar{I}_{x'} = 8.83$ in^4; $\bar{I}_{y'} = 3.20$ in^4; $\bar{I}_{x'y'} = 3.43$ in^4. (b) 57.7° clockwise.

9.194 (a) 176.9 mm. (b) 87.8 mm.

9.196 $I_x = 28.3 \times 10^{-3}$ kg $\cdot$ m^2, $I_y = 183.8 \times 10^{-3}$ kg $\cdot$ m^2; $k_x = 42.9$ mm, $k_y = 109.3$ mm.

9.197 $I_x = 4a^4$; $I_y = 16a^4/3$.

9.198 17.19 mm.

CHAPTER 10

10.1 60 N $\downarrow$.

10.2 40 lb $\downarrow$.

10.4 6.00 N $\cdot$ m $\downarrow$.

10.5 (a) 12.00 lb T; 2.40 in. $\rightarrow$. (b) 28.0 lb T; 5.60 in. $\rightarrow$.

10.7 (a) 75.0 N $\uparrow$. (b) 225 N $\downarrow$.

10.8 500 N $\downarrow$.

10.9 $(3P/2) \tan \theta$.

10.12 $P[(l/a) \cos^3 \theta - 1]$.

10.13 1803 lb.

10.14 5.42°.

10.15 $3Pa/2$.

10.18 (a) $Pl \sin 2\theta$. (b) $3Pl \cos \theta$. (c) $Pl \sin \theta$.

10.19 (a) 126.9 lb $\cdot$ ft $\uparrow$. (b) 81.4 lb $\cdot$ ft $\uparrow$.

10.20 (a) 369 lb $\rightarrow$. (b) 477 lb $\rightarrow$.

10.21 7.70 lb $\searrow$ 30.0°.

10.22 71.5 lb $\cdot$ ft $\uparrow$.

10.25 39.2°.

10.26 32.3°.

10.27 40.8°.

10.28 19.81° and 51.9°.

10.30 25.0°.

10.31 39.7° and 69.0°.

10.32 15.03° and 36.9°.

10.33 40.2°.

10.36 31.0°.

10.37 60.4°.

10.39 61.2°.

10.40 78.7°; 324°; 379°.

10.41 397 N $\nearrow$ 44.7°.

10.42 22.9° and 70.7°.

10.43 31.8 N $\searrow$ 35.0°.

10.44 15.27°.

10.45 9.47 kN $\nwarrow$.

10.46 10.20 kN $\nwarrow$.

10.47 $\eta = 1/(1 + \mu \cot \alpha)$.

10.49 5.00 kip $\cdot$ ft; 1.363 kip $\cdot$ ft.

10.50 $\eta = \tan \theta / \tan (\theta + \phi_s)$.

10.52 150.6 lb; 126.6 lb.

10.53 $\mathbf{A} = 15.00$ lb $\downarrow$; $\mathbf{M}_A = 195.0$ lb $\cdot$ ft $\uparrow$.

10.54 $\mathbf{D} = 315$ lb $\downarrow$.

10.55 21.8 mm (shortened).

10.57 41.7 mm $\downarrow$.

10.60 25.0°.

10.61 39.7° and 69.0°.

10.62 40.2°.

10.65 31.0°.

10.69 $\theta = 14.48°$, unstable; $\theta = 90.0°$, stable; $\theta = 165.5°$, unstable; $\theta = 270°$, stable.

10.70 $\theta = 0$, unstable; $\theta = 137.8°$, stable.

10.73 $\theta = 61.2°$, stable.

10.74 $\theta = 78.7°$, stable; $\theta = 324°$, unstable; $\theta = 379°$, stable.

10.75 1.118 kg, stable.

10.76 109.4°, stable.

10.78 301 mm.

10.80 $\theta = 9.69°$, stable; $\theta = 33.8°$, unstable; $\theta = 90.0°$, stable.

10.81 (a) $0 \le W \le 90.0$ lb. (b) $0 \le \theta \le 45.0°$.

10.82 $\theta = 13.19°$, stable; $\theta = 76.8°$, unstable.

10.83 49.1°.

10.84 70.9°.

10.85 54.9°.

10.87 61.4°.

10.89 $k > 6.94$ lb/in.

10.90 12.25 in.

10.91 $0 \le P < ka$.

10.94 $0 \le P < ka/10$.

10.95 $0 \le P < 320$ N.

10.96 $Q > 432$ N.

10.99 $0 \le P < 0.219ka$.

10.100 $0 \le P < 30.7$ lb.

10.102 17.90 kips $\nearrow$.

10.104 (a) and (b) $P < ka^2/2l$.

10.105 270 N $\uparrow$.

10.107 $\theta = 0$ and $\theta = 180.0°$, unstable; $\theta = 75.5°$ and $\theta = 285°$, stable.

10.108 $\mathbf{A} = 250$ N $\uparrow$; $\mathbf{M}_A = 450$ N $\cdot$ m $\uparrow$.

10.109 1050 N $\uparrow$.

10.110 $(Pl/a) \sin \theta \cos^2 \theta$.

10.112 113.8°.

SI Prefixes

Multiplication Factor	Prefix†	Symbol
$1\ 000\ 000\ 000\ 000 = 10^{12}$	tera	T
$1\ 000\ 000\ 000 = 10^{9}$	giga	G
$1\ 000\ 000 = 10^{6}$	mega	M
$1\ 000 = 10^{3}$	kilo	k
$100 = 10^{2}$	hecto‡	h
$10 = 10^{1}$	deka‡	da
$0.1 = 10^{-1}$	deci‡	d
$0.01 = 10^{-2}$	centi‡	c
$0.001 = 10^{-3}$	milli	m
$0.000\ 001 = 10^{-6}$	micro	μ
$0.\ 000\ 000\ 001 = 10^{-9}$	nano	n
$0.000\ 000\ 000\ 001 = 10^{-12}$	pico	p
$0.000\ 000\ 000\ 000\ 001 = 10^{-15}$	femto	f
$0.000\ 000\ 000\ 000\ 000\ 001 = 10^{-18}$	atto	a

† The first syllable of every prefix is accented so that the prefix will retain its identity. Thus, the preferred pronunciation of kilometer places the accent on the first syllable, not the second.

‡ The use of these prefixes should be avoided, except for the measurement of areas and volumes and for the nontechnical use of centimeter, as for body and clothing measurements.

Principal SI Units Used in Mechanics

Quantity	Unit	Symbol	Formula
Acceleration	Meter per second squared	$\cdots$	m/s^2
Angle	Radian	rad	†
Angular acceleration	Radian per second squared	$\cdots$	rad/s^2
Angular velocity	Radian per second	$\cdots$	rad/s
Area	Square meter	$\cdots$	m^2
Density	Kilogram per cubic meter	$\cdots$	kg/m^3
Energy	Joule	J	$N \cdot m$
Force	Newton	N	$kg \cdot m/s^2$
Frequency	Hertz	Hz	s^{-1}
Impulse	Newton-second	$\cdots$	$kg \cdot m/s$
Length	Meter	m	‡
Mass	Kilogram	kg	‡
Moment of a force	Newton-meter	$\cdots$	$N \cdot m$
Power	Watt	W	J/s
Pressure	Pascal	Pa	N/m^2
Time	Second	s	‡
Velocity	Meter per second	$\cdots$	m/s
Volume, solids	Cubic meter	$\cdots$	m^3
Liquids	Liter	L	$10^{-3}\ m^3$
Work	Joule	J	$N \cdot m$

† Supplementary unit (1 revolution $= 2\pi$ rad $= 360°$).
‡ Base unit.

U.S. Customary Units and Their SI Equivalents

Quantity	U.S. Customary Units	SI Equivalent
Acceleration	ft/s^2	0.3048 m/s^2
	in./s^2	0.0254 m/s^2
Area	ft^2	0.0929 m^2
	in^2	645.2 mm^2
Energy	ft · lb	1.356 J
Force	kip	4.448 kN
	lb	4.448 N
	oz	0.2780 N
Impulse	lb · s	4.448 N · s
Length	ft	0.3048 m
	in.	25.40 mm
	mi	1.609 km
Mass	oz mass	28.35 g
	lb mass	0.4536 kg
	slug	14.59 kg
	ton	907.2 kg
Moment of a force	lb · ft	1.356 N · m
	lb · in.	0.1130 N · m
Moment of inertia		
Of an area	in^4	0.4162 × 10^6 mm^4
Of a mass	lb · ft · s^2	1.356 kg · m^2
Momentum	lb · s	4.448 kg · m/s
Power	ft · lb/s	1.356 W
	hp	745.7 W
Pressure	lb/ft^2	47.88 Pa
	lb/in^2 (psi)	6.895 kPa
Velocity	ft/s	0.3048 m/s
	in./s	0.0254 m/s
	mi/h (mph)	0.4470 m/s
	mi/h (mph)	1.609 km/h
Volume	ft^3	0.02832 m^3
	in^3	16.39 cm^3
Liquids	gal	3.785 L
	qt	0.9464 L
Work	ft · lb	1.356 J

Centroids of Common Shapes of Areas and Lines

Shape		$\bar{x}$	$\bar{y}$	Area
Triangular area			$\dfrac{h}{3}$	$\dfrac{bh}{2}$
Quarter-circular area		$\dfrac{4r}{3\pi}$	$\dfrac{4r}{3\pi}$	$\dfrac{\pi r^2}{4}$
Semicircular area		0	$\dfrac{4r}{3\pi}$	$\dfrac{\pi r^2}{2}$
Semiparabolic area		$\dfrac{3a}{8}$	$\dfrac{3h}{5}$	$\dfrac{2ah}{3}$
Parabolic area		0	$\dfrac{3h}{5}$	$\dfrac{4ah}{3}$
Parabolic spandrel		$\dfrac{3a}{4}$	$\dfrac{3h}{10}$	$\dfrac{ah}{3}$
Circular sector		$\dfrac{2r\sin\alpha}{3\alpha}$	0	αr^2
Quarter-circular arc		$\dfrac{2r}{\pi}$	$\dfrac{2r}{\pi}$	$\dfrac{\pi r}{2}$
Semicircular arc		0	$\dfrac{2r}{\pi}$	πr
Arc of circle		$\dfrac{r\sin\alpha}{\alpha}$	0	$2\alpha r$